Nuclear-Pumped Lasers

Mark Prelas

Nuclear-Pumped Lasers

Springer

Mark Prelas
Electrical & Computer Engineering Department
University of Missouri
Columbia, MO, USA

ISBN 978-3-319-19844-6 ISBN 978-3-319-19845-3 (eBook)
DOI 10.1007/978-3-319-19845-3

Library of Congress Control Number: 2015954350

Springer Cham Heidelberg New York Dordrecht London

Printed on acid-free paper

Springer International Publishing AG Switzerland is part of Springer Science+Business Media (www.springer.com)

Katića and Jure
Born in troubled times and forced into slave labor during World War II, having unimaginable strength, they survived the horror. They were kind and generous and approached life with unyielding compassion, faith, and dignity. They provided love, security, optimism, and a future for their children. Where there was evil they worked towards good; where there was pain and suffering, they provided comfort; where there was hunger they provided sustenance; where there was hate they provided love; where there was injustice and despair, they provided hope; and when they departed they left the world a better place.

Mark Prelas, a grateful son.

Preface

The book *Nuclear-Pump Lasers* evolved from a graduate level course (titled Nuclear-Pumped Lasers and Applications) that I developed and taught during the summer of 1985 as part of the nuclear engineering curriculum at the University of Missouri. There were two motivating factors for developing and teaching this course. The first was that the US Strategic Defense Initiative (SDI) program had been proposed by President Ronald Regan on March 23, 1983, and the Strategic Defense Initiative Organization (SDIO) was set up in 1984 within the US Department of Defense to oversee SDI. The second reason was that two efforts within the SDI program centered on nuclear-pumped laser technologies: the Centaurus and the Falcon programs. The goal of these two programs was to take advantage of the high-energy/high-power capability of nuclear-pumped laser technology to provide a laser platform (space-based or ground-based) with sufficient energy to destroy a ballistic missile during its flight. Nuclear-pumped lasers were viewed as one of the few concepts with sufficient energy/power, scalability, and minimal mass for a space-based platform that met SDI mission requirements.

Nuclear-pumped laser (NPL) technology dates back to 1963. It was always viewed as being a means of beaming large quantities of energy. This technology fits well with SDI objectives, and as a result, Centaurus and Falcon were set up to compete against other SDI technologies such as chemical lasers, free electron lasers, the thermonuclear bomb "pumped" x-ray laser, particle beams, and kinetic energy weapons. NPLs faded from the United States research agenda at the end of the 1980s due to a change in focus within SDO centered on kinetic energy weapons. However, NPLs remained an active research agenda in other countries. Most notably Russia, other nations in the former Soviet Union, and China [1] (which had a cooperative agreement with Russia [2]) remain active in the field. The reason for this broad interest in the technology is that an NPL can scale to high-energy/high-power levels (potentially up to 100 MW continuous-wave (CW) beam power). Military applications have historically driven the NPL research agenda. It should be noted that there are significant humanitarian applications for high-energy/high-power lasers as described in Chap. 6. For example, a high-energy/high-power

CW NPL could be used to deflect a moderately sized asteroid. Other potentially important applications include space propulsion, power transmission, and mining. Despite the promise, nuclear-pumped lasers still have significant problems which need to be solved [3–5], and these issues will be discussed throughout the text.

High-energy/high-power gas lasers by their nature require high pumping power densities (defined as the Watts/cm^3 deposited in the laser medium). That is why powerful gas lasers are typically pulsed. Nuclear-pumping does not generate high-power densities like electrical pumping, but it does generate high-energy densities. This limits the type of high-energy/high-power laser systems that can be driven by nuclear-pumping technologies as discussed in Chap. 4. Other issues involve the design of reactor cores dedicated to the production of laser energy. The nuclear-pumped laser system design is covered in Chap. 5. There are promising approaches that have been developed which could reduce the scale and mass of NPL space-based systems. One such approach is the nuclear-driven flashlamps [6]. A nuclear-driven flashlamp can increase the effective pumping power density through photon focusing while reducing the mass of the system. This capability is applicable to both gaseous- and solid-state candidate laser media.

Due to a renewed interest in nuclear-pumped lasers, the graduate level course was updated and taught during the spring semester of 2014 at the University of Missouri. The course included lectures that covered introductory material, past and present research in the field, logical directions for potential future research, and possible long-term and short-term applications. The field of nuclear-pumped lasers is truly interdisciplinary. It represents the integration of diverse areas such as nuclear engineering, quantum electronics, gaseous electronics, optics, photonics, materials, chemistry, physics, thermodynamics, and astronautics. The graduate students that took the course were nuclear engineers who had some additional experiences in these diverse areas. The students worked on a design project involving an advanced reactor/laser that used a thermal pumping mechanism to drive a carbon dioxide laser. The heat source for the laser was an ultrahigh temperature core reactor concept that was inspired by the nuclear rocket research programs in the United States and Russia. In Chap. 5, there is a section on the design of a thermally pumped carbon dioxide laser driven by an ultrahigh temperature core reactor that emerged from this design study. The diversity of expertise within this group of students was a strategic part of the learning experience. The intended audience for this book includes graduate students (or perhaps senior level undergraduates in an honors program) who have skill sets in the disciplines described above and researchers interested in high-power/high-energy lasers.

This book is organized in six chapters and four Appendices. Chapter 1 focuses on fundamental properties of lasers, nuclear-pumping, and some basic concepts. In Chap. 2, the nuclear reactions that may be used as drivers for nuclear-pumped lasers are discussed. Chapter 3 focuses on the efficient transport of energy from the ionizing radiation to the laser medium and the corresponding processes and mechanisms involved. Once the energy is deposited in the laser medium, the operational characteristics of existing nuclear-pumped lasers are considered, which are discussed in Chap. 4. The integration of the various systems needed to produce the radiation, transport the energy from the radiation source to the laser medium, and cool the system; the gain medium which produces the laser beam; and the Gaussian limited optics necessary to steer the beam from the laser to its intended target is examined in Chap. 5. Finally Chap. 6 covers the long-term applications of high-power/high-energy nuclear-pumped lasers as well as some near-term prospects for the use of technologies developed in nuclear-pumped laser research.

References

1. Yang C, Chen H, Zheng C, Zhao X, Han H (2002) The progress of nuclear pumped laser in CFBR-II reactor. Chin Opt Lett 1:292–293
2. Select-Committee-United-States-House-of-Representatives (1999) U.S. National Security and Military/Commercial Concerns with the People's Republic of China. U.S. Government Printing Office, Washington, DC
3. Miley GH, McArthur DA, Deyoung RJ, Prelas MA (1989) Fission reactor pumped laser: history and prospects. In: Carlson JW, Behrens AD (eds) 50 years of nuclear power. National Academy of Science and NIST, American Nuclear Society, LaGrange Park, IL, pp 333–342
4. Boody FP, Prelas MA, Anderson JH, Nagalingam SJS, Miley GH (1978) Progress in nuclear-pumped lasers. In: Billman K (ed) Radiation energy conversion in space, vol 61. AIAA, New York, pp 379–410
5. Melnikov SP, Sizov AN, Sinyanskii AA, Miley GH (2015) Lasers with nuclear pumping. Springer, New York
6. Prelas MA, Boody FP, Miley GH, Kunze J (1988) Nuclear driven flashlamps. Laser Part Beams 6:25–62

Contents

Chapter 1
Introduction to Nuclear-Pumped Lasers

Abstract Nuclear-Pumped Lasers are lasers that are "pumped" by the products from nuclear reactions. In Chap. 1, the reader is introduced to lasers, the important components of a laser, some basic properties of lasers and how those properties can lead to solutions for problems requiring high-energy/high-power lasers. The development and understanding of these fundamentals are essential for the transitioning of the discussions into an understanding of the principles which are critical to nuclear-pumped laser research.

Keywords Nuclear • Pumped lasers • Laser pumping • Laser kinetics

1.1 Introduction

Nuclear-pumped lasers use the products from nuclear reactions to generate excited states especially the upper laser level. Ionizing radiation refers to different types of radiation that have the capability of creating ion pairs in matter. Ionizing radiation includes ions (such as fission fragments, alpha particles, protons, etc.), beta particles, gamma rays, x-rays, and neutrons. Each type of ionizing radiation source has a characteristic range. Consider, for example, a material in the solid phase. Energetic heavy ions such as fission fragments and alpha particles deposit their energy within the solid on the scale of micrometers. Electrons deposit their energy on the scale of millimeters. Particles which possess high energy and either no rest mass or no net charge, such as gamma rays and neutrons, deposit their energy on the scale of meters. The majority of nuclear-pumped lasers use thermal neutrons to initiate nuclear reactions. These neutrons were supplied by pulsed nuclear reactors. Nuclear-pumped lasers which use nuclear reactors as a thermal neutron source are also referred to as a Reactor-Pumped Laser (or RPL) in the literature [1]. The use of the term nuclear-pumped laser is more general and includes RPLs, thermonuclear bomb pumped lasers, accelerator driven lasers, radioisotope driven lasers or any laser driven by nuclear reactions by any other means.

Ionizing radiation ultimately produces heat and ionization through its interaction with matter. Direct energy conversion systems can either use the heat production directly or the ion pairs generated in the material to produce a current or excited states. If no mechanism puts the ionization to use, the energy content of the ions will

M. Prelas, *Nuclear-Pumped Lasers*, DOI 10.1007/978-3-319-19845-3_1

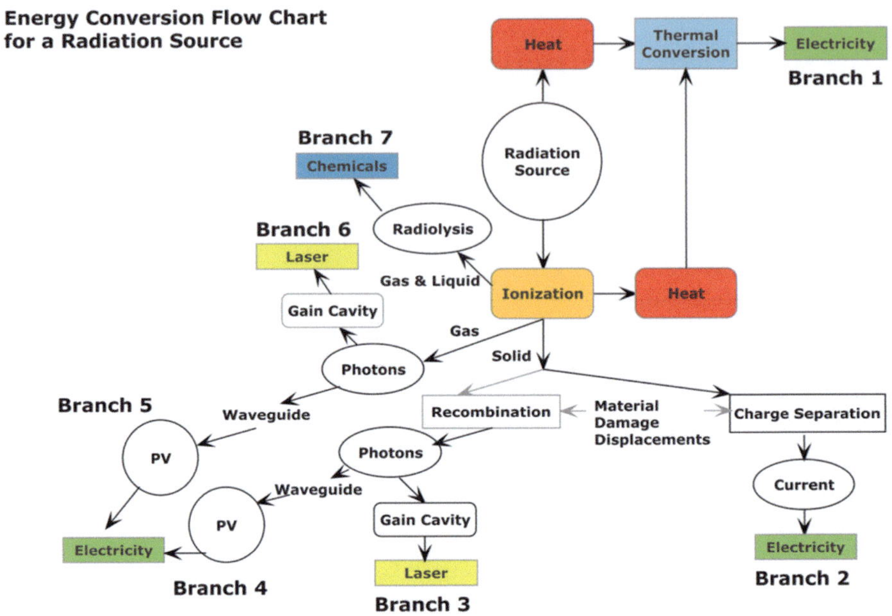

Fig. 1.1 Energy conversion flow chart for radiation sources. Branch 1 uses radiation for heat production. Branch 2 uses the production of charged species in a solid to generate a current flow. Branch 3 uses the production of charged species in a solid to produce laser photons. Branch 4 uses the production of charged species in a solid to produce photons which are used to produce electricity from photovoltaic (PV) cells. Branch 5 uses the production of charged species in a gas to produce photons which then interact with photovoltaic (PV) cells to produce electricity [67]

end up as heat. In typical interactions of ionizing radiation with matter, about 40–50 % of the energy goes into ionization and the remainder is converted to heat. If the mode of energy transduction does not utilize the heat produced by the radiation interactions, but only uses the ion pairs, the maximum theoretical efficiency will be limited to between 40 and 50 %. The efficiency for each conversion mechanism will be eroded further by process inefficiencies specific to the system.

As shown Fig. 1.1, the two basic forms of energy produced by ionizing radiation, heat and ion pairs, can be used in energy conversion through a number of processes illustrated as branches. The radiation source is shown as a circle. In an energy conversion system, the radiation first interacts with matter in a solid, liquid or gaseous phase, and then branches into heat production and ion pair production. The ion pairs can be used with different transducers to produce useful products, such as electricity, laser light, or chemicals [2].

In Branch 1, the ions recombine and eventually produce heat. This heat is then combined with the heat initially produced by the interactions between the radiation source and the material. This heat is used to produce electricity in a variety of ways.

- The most common use is in the steam cycle of a commercial nuclear power plant. However, there are other uses for the heat such as the Radioisotope Thermoelectric Generator or RTG [3] which produces electricity from the Seebeck effect.
- Another potential use for the heat is in thermionic energy conversion [4] in which a hot electrode emits electrons due to the low work function of the electrode material and the electrons are collected over a potential barrier by a cool electrode thus producing a current flow.
- A different approach is thermophotovoltaics [5] in which a hot material emits grey body radiation which is harvested by photovoltaic cells that have an appropriate band-gap for maximizing power production from the grey body radiation.
- The Alkali Metal Thermal to Electric Converter (AMTEC) [6] is based on a thermally regenerative electrochemical cell.
- It is also feasible to use heat to drive a carbon dioxide laser [7].

Branch 2 utilizes only the energy from the ion pairs produced in a solid transducer.

- This is typical of an alphavoltaic cell [8] where the electron-hole pairs are separated by the voltage generated in the depletion region in a p-n junction of a semiconductor material to produce a current flow.
- It is also used by a betavoltaic cell [9] where the electron-hole pairs are separated by the voltage generated in the depletion region in a p-n junction of a semiconductor material to produce a current flow.
- This process is also used in a reciprocating cantilever [10] where charge is collected from the source on a thin cantilever and when sufficient charge is built up the electrostatic force between the tip of the cantilever and the source causes the cantilever to bend and touch thus driving a current in a repetitive process.

Branch 3 utilizes only the energy from the electron-hole pair production in a solid to produce a laser.

- An example is through the interaction of ions with wide direct band-gap materials which use the radiation source to produce the electron-hole pairs which then recombine and form the laser's photons [11].

Branch 4 utilizes the photons from the recombination of electron-hole pairs in a direct band-gap semiconductor to drive a photolytic laser.

- Examples are the Photon Intermediate Direct Energy Conversion (PIDEC) system using a solid-state fluorescence medium to drive a photolytic laser [12].

Branch 5 utilizes the photons from the creation of excited states and ion pairs in a gaseous system to excite a photolytic laser.

- Examples are the PIDEC system used to drive the iodine, xenon fluoride, and Nd^{3+}-doped glass or crystal lasers [12].

Branch 6 utilizes the energy from the excited states and ion pairs to directly drive a gaseous nuclear-pumped laser.

- Examples include the bulk of the gaseous nuclear-pumped lasers directly excited by nuclear reactions [2].

Branch 7 utilizes the energy from the production of excited states and ion pairs for chemical production through radiolysis.

- A variety of radiolytic reactions with water to produce hydrogen and CO_2 to produce CO [2].

The branches which only utilize the energy from ion pair production will inherently be limited to a maximum theoretical efficiency of 40–50 % since the heat produced in the interaction of radiation with matter is wasted. As will be discussed, process inefficiencies for systems represented by each of these branches will determine the actual efficiency.

The word "nuclear" is used because nuclear energy is converted to intermediate states (e.g., ion-pairs or photons) that are used to directly drive the laser. The use of the term "pumped" comes from an analogy between the movement of an electron from one quantum energy state to a higher energy state and pumping water uphill. Imagine that the quantum states of a laser resemble a landscape with hills and valleys. The high energy states of the molecule or atom (like the upper laser level) lie uphill from the ground state (or valley). In contrast, imagine that water is being pumped from a valley floor to the top of a hill and that the water at the top of the hill, due to potential energy, flows back down into the valley. This view of water flow from the valley to the hill and back to the valley is analogous to an electron being moved from the ground state to the upper laser level, then decaying to the lower laser level and then decaying from the lower laser level back into the ground state. This view aptly represents energy flow in the quantum world.

There were several motivations for developing nuclear-pumped lasers. One is that they represent a viable method of directly converting nuclear energy into a useful product (coherent photons). However, the military applications of nuclear-pumped lasers were desired long before the United States Strategic Defense initiative (SDI) program was proposed by President Ronald Regan on March 23, 1983 to develop an Anti Ballistic Missile (ABM) system [13]. Nuclear fuel stores far more energy than chemical fuels and a nuclear reactor capable of directly converting its energy to a coherent beam of photons is worth coveting. The Strategic Defense Initiative Organization (SDIO) was formed in 1984 within the United States Department of Defense to oversee the SDI initiative. Two efforts within the SDI program were centered on nuclear-pump laser technologies, the Centaurus program and the Falcon program. The goal of these two programs was to take advantage of the high-energy/high-power capability of nuclear-pumped laser technology to provide a laser platform, space-based or ground-based, with sufficient energy to destroy a ballistic missile during its flight [14]. A ballistic missile can have two stages (like a Submarine Launched Ballistic Missile or SLBM) or three stages (like an Intercontinental Ballistic Missile or ICBM). It can also have a post

Table 1.1 Burnout times for the various booster components [14]

Booster component	Burnout time (s)	Burnout altitude (km)	Burnout velocity (km/s)
Stage 1	60	25	2.5
Stage 2	120	95	4.5
Stage 3	180	250	6.5
Post boost vehicle	600	$\geq$800	7.1

boost vehicle (PBV) which carries the warhead (or Multiple Independent-targetable Reentry Vehicles or MIRVs). The typical times that a rocket motor can burn, and thus accelerate, for a US ICBM is about 180 s while for a Russian ICBM can be as high as 350 s (see Table 1.1). The dynamics of the ballistic missiles govern how quickly an antiballistic missile system must respond by first identifying the threat, tracking the missile, deploying the antiballistic missile system and how quickly the defensive measure can attack the target.

An ICBM can have a range of 10,000 km and its flight time can vary between 28 and 42 min. The theory behind ballistic missiles defense is that after a missile is launched it can be attacked at four points along its trajectory:

- The Boost Phase: this phase occurs when the missile leaves the launcher. The launcher is typically a silo, a truck, train or a submarine. While the rocket motors are burning during the acceleration phase, the most prominent observable is the high intensity infrared emissions from the plume of the missile. There are other observables such as the large radar cross section of the booster, visible and ultraviolet emission from the plume, solar reflection from the missile and radiation from the shock heated air. Thus with all of the potential signatures, a boost phase interceptor can target missile.
- The Post Boost Phase: after the final stage of the rocket has burned out, the payload is carried in a post-boost phase vehicle (or bus). The bus has sufficient velocity at this point to have a trajectory that is able to reach the target(s). The bus separates from its launch vehicle and then falls in a ballistic trajectory. The bus carries an inertial guidance system, thrusters and a thrust control system. It carries the Reentry Vehicle(s) (RV) and can carry and dispense decoys. For Multiple Independently-targeted Reentry Vehicles (MIRV), thrusters provide small incremental velocity to each warhead to direct them to the target.
- The Midcourse Phase Interception: after the PBV motor has burned out and the missile is beyond the upper atmosphere. During this phase all of the RV's, decoys and bus travel along a nearly ballistic trajectory under the influence of gravity. The midcourse is the longest of the phases (up to 20 min for ICBMs) and ends when the object begin to experience the drag force of the upper atmosphere. The length of time in the midcourse phase gives the defender opportunities to move assets to attack the PBV as well as multiple opportunities to destroy the targets.
- The Reentry Phase Interception: attack the reentry vehicle after it reenters the atmosphere.

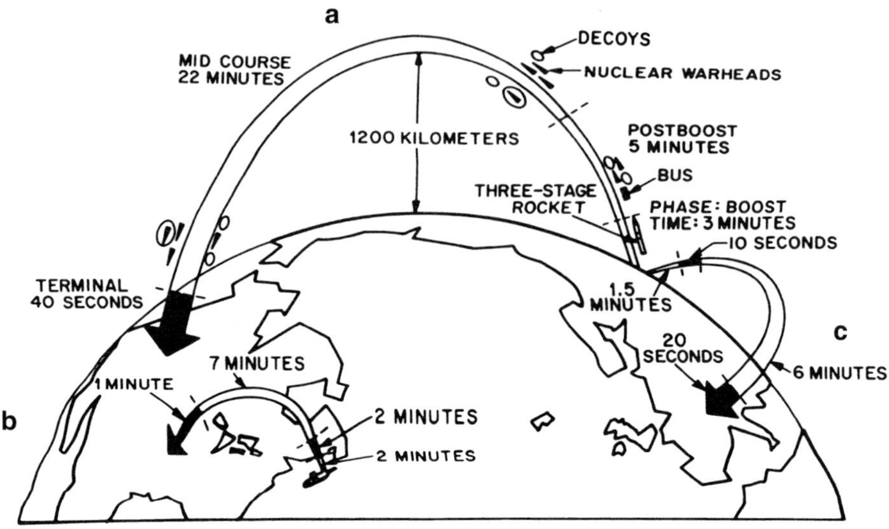

Fig. 1.2 Illustrations of (**a**) the trajectory phases for an Intercontinental Ballistic Missile; (**b**) the trajectory phases of a Submarine Launched Ballistic Missile; (**c**) the trajectory phase for an Intermediate Range Ballistic Missile [14]

There are advantages and disadvantages to attacking the missiles in each of these phases.

Boost Phase Trajectory the burning rocket motor is easy to detect with IR sensors and there are virtually no counter-measures that can be used against the ABM system. Additionally the rocket is full of flammable propellant and thus if hit with an interceptor, the missile will be destroyed. The disadvantage is that the interceptors have to be geographically positioned relatively close to the missile launch site. For short range fast interceptor rockets or theater based laser defense systems, this is not always possible without having to fly over hostile territory. The second disadvantage is that the warning system has to be very fast (2–3 min to discover the launch and then attack the missile) (Fig. 1.2).

Post Boost Phase The observables in the post boost phase are not strong. The radar cross section of the bus is smaller than the booster in the boost phase and the infrared emissions are lower. The bus is a high value target while the reentry vehicles (RVs) remain on the bus. After the RVs are deployed, the bus has no value as a target.

Midcourse Phase Interception The midcourse is the longest of the phases (up to 20 min for ICBMs). It ends when the object begin to experience the drag force of the upper atmosphere. The velocity of the bus is about 25,560 km/h. This long length of time during the midcourse phase gives the defender opportunities to move assets to attack the PBV as well as multiple opportunities to destroy the warheads. The interceptor can be located in space or on the ground (over a wide geographical area). However light weight decoys can be deployed and they will travel along the same trajectory as the PBV. It would be very difficult to distinguish between the RV and decoy. Tracking will require a very powerful radar system as well as space based sensors.

Reentry Phase Interception Atmospheric drag will cause changes in the trajectory, will cause the heat shield to glow, and will destroy any lightweight decoys. This makes the RV easier to track through optical signatures. The RV will begin entering the atmosphere at 25,560 km/h and will begin to slow down. The defender has about 60 s to respond. The interceptors will typically be fast anti-ballistic missile interceptors. The radar system needed to track the missile will be smaller and less sophisticated than what is needed by a mid phase defense system. However, the interceptor system will cover a much smaller geographic area.

There are a number of reasons why lasers are a very desirable candidate for an anti-ballistic missile system. Consider a laser anti-ballistic missile system which has a beam that travels at the speed of light. It can easily hit an RV traveling at a speed of 25,560 km/h if a targeting system could put and keep the beam on the target. In contrast a Terminal High Altitude Area Defense missile (THAAD) traveling at a speed of 10,080 km/h hitting a target traveling 25,560 km/h is far more challenging. This problem is similar to a bicycle traveling at a speed of 20 mph trying to hit a car traveling at 50 mph. The only way this can happen is if the bicycle puts itself in front of the oncoming car's trajectory, and that is easier said than done. As will be discussed, the idea of a laser based ABM system is attractive, but the most significant issue is in achieving a high enough energy on the target (warhead) to render it inoperative. One of the few laser systems that scale to the required energies needed for a laser based ABM system is a nuclear-pumped laser.

1.2 Nuclear-Pumped Lasers

Nuclear-pumped lasers derive their power directly from the radiation emitted in nuclear reactions. They are capable of high-energy/high-power which is an important attribute for SDI. Three major nuclear reactions have been used to power NPL's:

$$^{10}_{5}B + n_{th} \rightarrow {}^{7}_{3}Li + {}^{4}_{2}He + 2.35MeV \tag{1.1}$$

The lithium-7 ion has a kinetic energy of 0.85 MeV and the helium-3 ion has a kinetic energy of 1.5 MeV.

$$^{3}_{2}He + n_{th} \rightarrow {}^{1}_{1}p + {}^{3}_{1}T + 0.76MeV \tag{1.2}$$

The tritium ion has a kinetic energy of 0.19 MeV and the proton has a kinetic energy of 0.57 MeV.

$$ {}^{235}_{92}\mathrm{U}+\mathrm{n_{th}} \rightarrow {}^{236}_{92}\mathrm{U} \rightarrow \mathrm{ff_l} + \mathrm{ff_h} + \nu \cdot \mathrm{n_{fast}} + 210\,\mathrm{MeV} \quad (1.3) $$

In reaction 1.3, the heavy fission fragments and the light fission fragments created from the fission reaction are statistical in nature. There is a distribution of light and heavy fragment masses and energies (Fig. 1.3 and Table 1.2). On average,

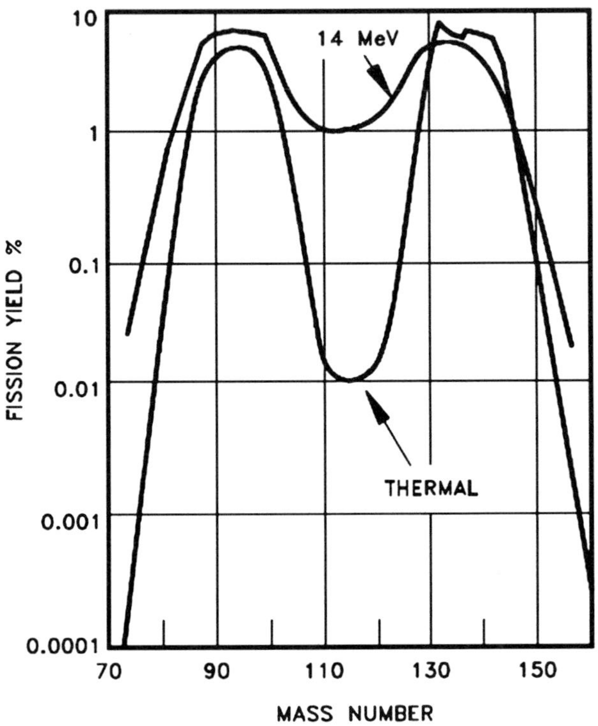

Fig. 1.3 U-235 fission yields for high- and low-energy (thermal) incident neutrons [68]

Table 1.2 Statistical distribution of energy released in the fission of U-235 [68]

Radiation	Energy in MeV
Kinetic energy of fission fragments	167
Fission neutrons	5
Prompt gamma rays	5
Delayed gamma rays – Fission fragments	6
Capture gamma ray energy	10
Beta particles from fission fragments	7
Neutrinos	10
Total energy	210

the heavy fragments carry about 61 MeV and the light fission fragments carry about 100 MeV. There is also a release of fast neutrons in the fission reaction. The exact statistical data shown in the Table shows that of the total energy released in a nuclear fission, which is 210 MeV, about 167 MeV ends up as kinetic energy. The fission neutrons carry off about 5 MeV, prompt gamma rays carry off about

5 MeV, delayed gamma rays from fission fragments carry off about 6 MeV, capture gamma ray energy account for about 10 MeV and beta particles emitted from fission fragments carry off about 7 MeV. The remaining energy comes out as neutrinos which add up to about 10 MeV. However, neutrinos have such a small cross-section for reaction that they are not counted in the energy balance.

The first two nuclear pump lasers that were successfully demonstrated used gamma rays from a thermonuclear weapons explosion to drive the lasers. The lasers were demonstrated by Los Alamos National Laboratory and Lawrence Livermore National Laboratory in 1974. Los Alamos National Laboratory demonstrated a hydrogen fluoride NPL [15] and Lawrence Livermore National Laboratory demonstrated a xenon excimer NPL [16]. The gamma rays given off in a nuclear bomb explosion, which travel at the speed of light, will reach and excite the laser media before the thermal energy completely disintegrates the laser medium and structure. The energy in the laser medium during the short period of time before the laser is destroyed, allows the laser cavity to build up photon intensity through stimulated emission. Part of the photon intensity in the laser cavity is bled off into a fast photon detector which is able to record the time history of the photon intensity build up during the blast.

A NPL in summary is able to convert radiation into atomic or molecular excited states (e.g., the upper laser level) by the interaction of the ionizing radiation with matter and through the buildup of the upper laser level energy state create a laser media which has optical gain. Thus one form of energy is converted into an optical beam which is coherent (meaning the photons have the same frequency, phase and direction). This energy conversion process has a characteristic efficiency which is a significant factor in the suitability of a specific NPL for an application.

Lasers can be "pumped" by a number of types of energy sources. The basic principles of a laser are shown in Fig. 1.4. The laser has an energy source, it has a

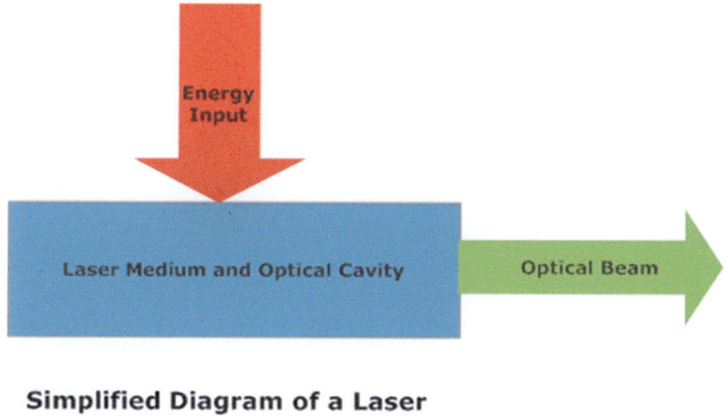

Fig. 1.4 Shown is a simplified diagram of a laser that demonstrates the basic components which make it viable. The first component is an energy source which is shown as an arrow entering the laser medium. The second component is the laser medium and optical cavity which is at the center of the diagram. The third component is the optical beam which exits the optical cavity

laser medium, it has an optical cavity and it has an optical beam. All laser systems are comprised of these basic components albeit each component can vary significantly between different laser systems and drivers.

1.3 Energy Input (Drivers)

The specific driver, or the means by which energy is provided to the laser medium, is very diverse. Energy can be transferred to a laser medium by photons or charged particles. How photons and charged particles are produced is highly variable as will be discussed.

1.3.1 Photon Sources

Photon sources can cover the whole spectrum of electromagnetic radiation which starts from the very high energy (gamma rays) to low energies (Infrared).

1.3.1.1 Ultraviolet to Infrared Photons

Photon sources can be incoherent (flashlamps, x-rays, gamma rays) or may be coherent (lasers). Below some general examples are discussed.

1.3.1.1.1 Flashlamp Driver

The first laser which was discovered is the ruby laser and it was pumped by an incoherent source of photons. This laser was optically pumped with a xenon flashlamp (Fig. 1.5). The photons are produced in the helical xenon flash tube which is wrapped around a ruby rod. An electrical discharge excites the xenon gas in the helical tube and the xenon emits photons which are absorbed by the chromium atoms in the ruby [17]. The ruby laser photon is visible (wavelength of 694.3 nm). The forerunner to the laser was the maser or Microwave Amplification by Stimulated Emission [18, 19] which emitted electromagnetic radiation at microwave frequencies.

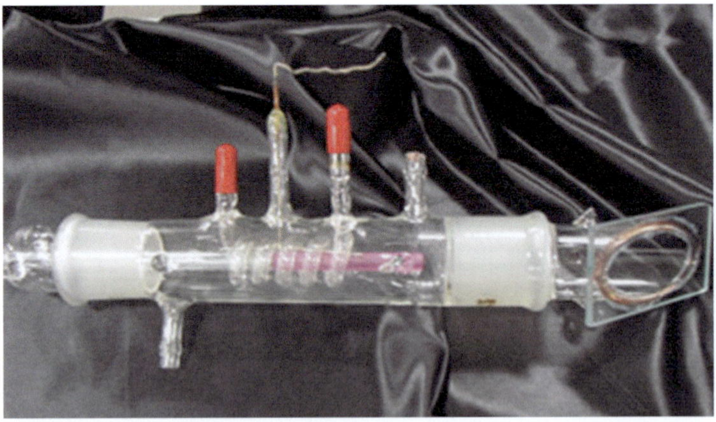

Fig. 1.5 Early ruby laser from the 1960s [69]. The xenon flashlamp is the helical glass tube wrapped around the ruby rod (Picture taken by Stacey Crockett SPIE Newsroom http://spie.org/x39920.xml. Printed and Electronically reproduced by permission of SPIE)

1.3.1.2 Coherent Beam Driver

The use of a coherent beam, or laser, as a driver is a method that is used by the National Ignition Facility for laser fusion research [20]. Figure 1.6 is an illustration of how a coherent beam driver can be used to drive a laser. A GaAlAs diode laser is tunable from about 790–815 nm. The peak absorption peak for a Neodymium doped Yttrium Aluminum Garnet crystal ($Y_3Al_5O_{12}$), or Nd:YAG, occurs at 810 nm which within the tunability range of GaAlAs (Fig. 1.7).

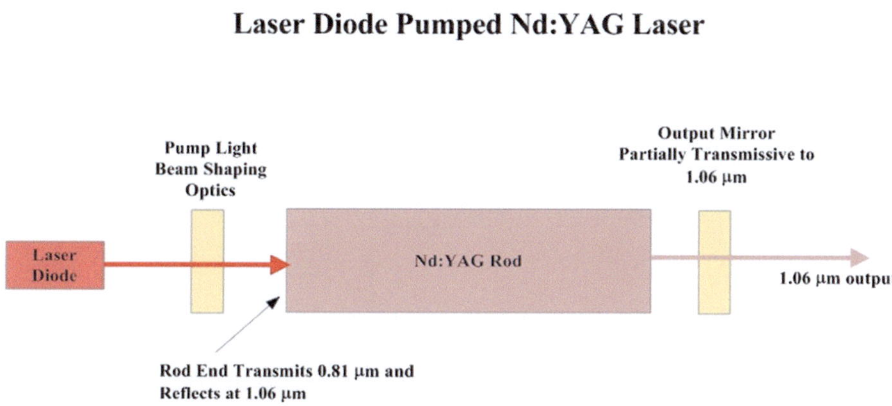

Fig. 1.6 Illustration of a laser being driven by a coherent optical beam [70]. A gallium-aluminum-arsenide (GaAlAs) laser diode is tunable from 790 to 815 nm

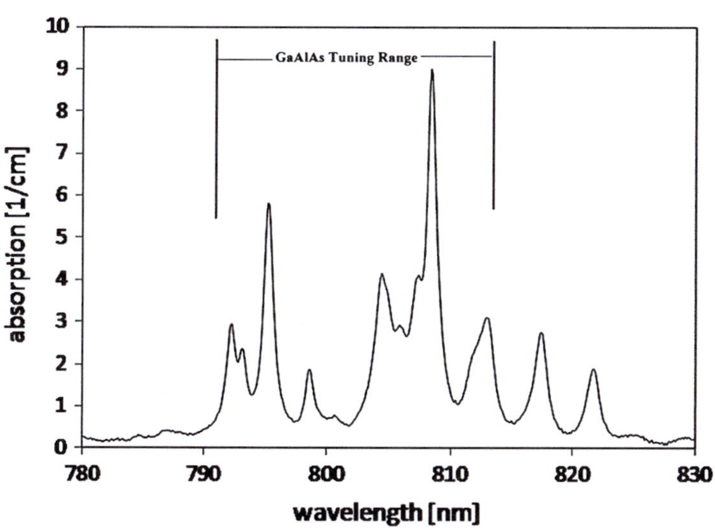

Fig. 1.7 The absorption spectra for a Nd:YAG laser rod is shown along with the tuning range of the GaAlAs diode laser. The peak absorptivity occurs near 810 nm [71]

The Nd^{3+} ion substitutes for Y^{+3} in the cubic YAG crystal and the Nd^{+3} exhibit the energy structure shown in Fig. 1.8. The upper state of the laser is the $^4F_{3/2}$ and the lower laser level state is $^4I_{11/2}$. The lasing transition of the Nd:YAG laser occurs at 1.064 μm as shown in Fig. 1.9.

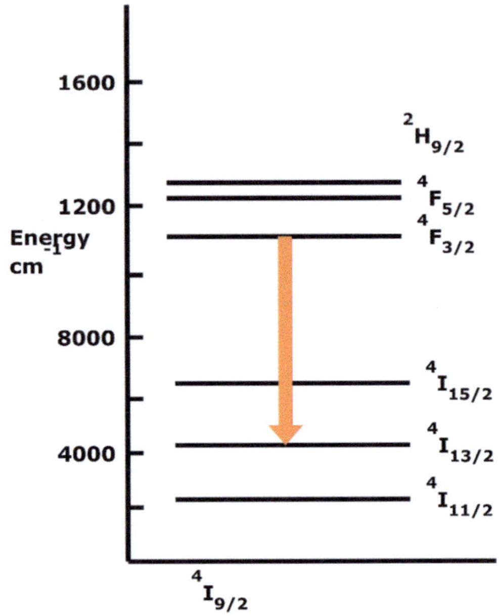

Fig. 1.8 The energy level of the Nd^{3+} in the YAG crystal [71]

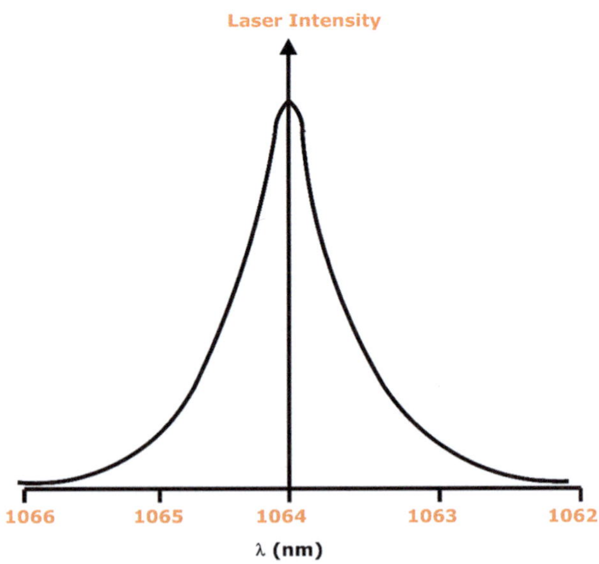

Fig. 1.9 The emission spectrum from a Nd:YAG crystal [71]

1.3.1.3 Gamma Ray Driver

As will be discussed in Chap. 2, gamma rays are a form of ionizing radiation. Thus gamma rays can deposit energy into a laser medium and cause it to lase if the gamma source were strong enough. Edward Teller was a vocal proponent of using thermonuclear weapons for peaceful purposes. President Dwight D. Eisenhower proposed the idea of Atoms for Peace during his address to the 470th Meeting of the United Nations on December 8, 1953 [21]. During August 1955, 84 nations participated in the Atoms for Peace Congress in which peaceful uses of nuclear energy were promoted [22]. Edward Teller promoted a number of concepts in which thermonuclear weapons could be used for peaceful purposes. One such idea was the plowshare program (nuclear explosives for excavation and enhanced gas production) [23].

In the wake of the movement by weapons designers to develop applications for thermonuclear warheads, the Excalibur laser weapon concept was born. Excalibur was an x-ray laser concept that used the gamma rays from a nuclear blast as the laser driver. The laser medium was a very long beryllium rod that would be located some distance from the blast. The beryllium would absorb the energy from the gamma rays produced in the blast. Spontaneous x-ray photons would be produced in the beryllium and would propagate down the length of the rod thus inducing stimulated emission. The x-rays would only have enough time to pass down the length of the rod before the blast destroyed it. The x-rays would be emitted from the end of the rod in a very short pulse at an angle equal to twice the rod width divided by the rod length (see Fig. 1.10). This short-x-ray burst would theoretically have

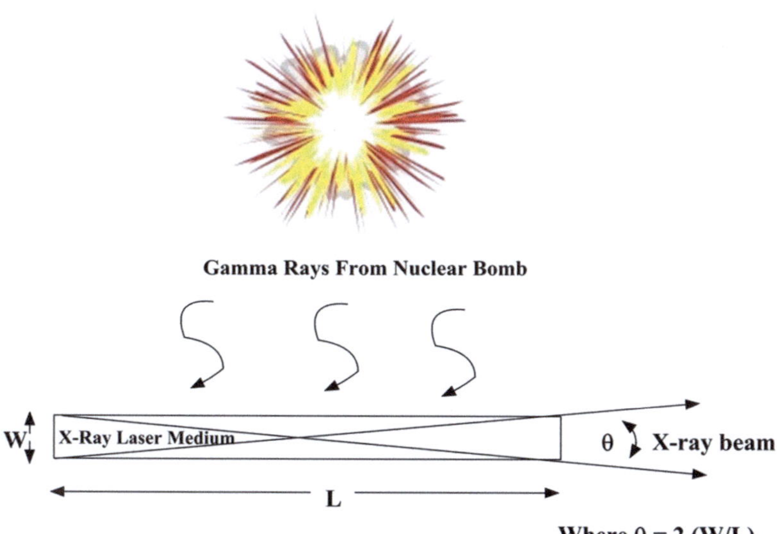

Fig. 1.10 Diagram of the nuclear bomb pumped x-ray laser concept known as Excalibur [72]

sufficient energy to destroy a bus or RV. Edward Teller convinced President Ronald Regan that it was feasible to build an ABM system based on Excalibur to defend against a Soviet ballistic missile attack [24]. President Regan's enthusiasm for the idea motivated his Star Wars speech on March 23, 1983 [25]. There are some obvious issues with Excalibur especially in having an orbiting nuclear warhead.

1.3.2 Electron Swarm Pumping (e.g., D.C., RF, Microwave)

An electron swarm is caused by an electrical discharge, typically in a gas. Electrons in the gas are accelerated in an electric field produced by several different methods: a DC discharge, a pulsed discharge, a radiofrequency discharges (either capacitive coupling or inductive coupling), a microwave resonance cavity, etc. The electron mobility is a constant by which the electron drift velocity is related to the applied electric field ($\vec{E}$) which is responsible for that velocity. In an extremely viscous gas, this constant is low, in a mildly viscous gas this constant is large. The drift velocity ($\vec{u}$) is,

$$\vec{\mathbf{u}} = -\mu_e \vec{\mathbf{E}} \tag{1.4}$$

The electron is accelerated by the field due to the Lorentz force,

$$\vec{F} = m\vec{a} = q\left(\vec{E} + \vec{v} \times \vec{B}\right) \tag{1.5}$$

where $\vec{F}$ is force, m is the electron mass, $\vec{a}$ is acceleration, $\vec{v}$ is electron velocity and $\vec{B}$ is the magnetic field.

The electric field will cause the electron to continue to accelerate until it undergoes a Columbic collision with an electron orbiting an atom or molecule in the gas. These Columbic collisions cause ionization if the drift velocity is high enough. Thus electrons build up in the presence of an electric field and as a result a swarm discharge is created.

1.3.2.1 DC Discharge

A typical set up for a discharge which has an anode and cathode is shown in Fig. 1.11- a diagram of a helium-neon laser which uses a glow discharge.

The current-voltage characteristic of a glow discharge is shown in Fig. 1.12. In this Figure, the Townsend Discharge (or dark discharge) occurs when the ambient electron density is subjected to a voltage which creates a current flow (I_A) up to the point A. Current in this regime increases as voltage increases up to about 1×10^{-6} Amps. Once the breakdown voltage is attained, from point S, the voltage is constant as the

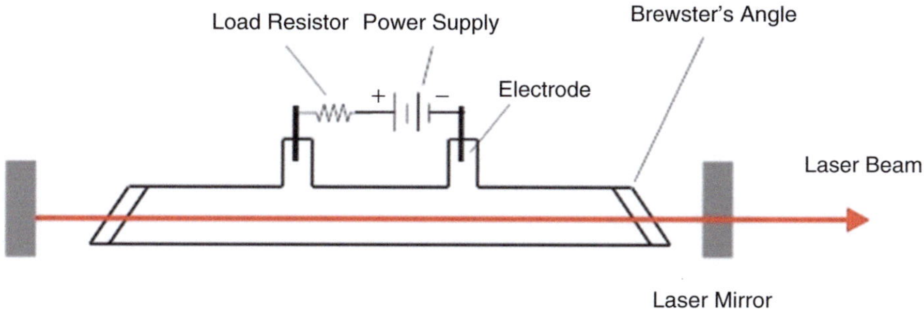

Fig. 1.11 Diagram of a Helium-Neon D.C. discharge "pumped" laser

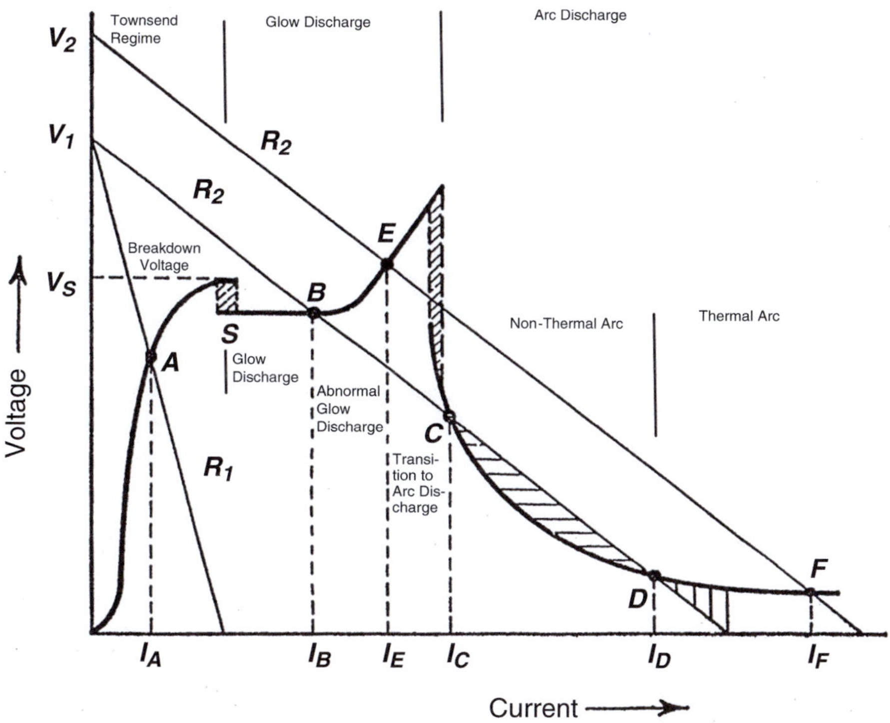

Fig. 1.12 Current voltage characteristics of a glow discharge [73]

current increases, up to point B. This is the glow regime and I_S to I_B is about 1×10^{-3} Amps. The Voltage then begins to rise as current increases from point B to E. This is known as the abnormal glow discharge. The discharge then begins to transition beyond point E to an arc discharge and in the non-thermal arc regime, point C to D, the voltage drops as current flow increases. In the thermal arc regime, point D to F, the voltage drop begins to taper as current increases. Lasers typically operate in the glow discharge regime.

It is possible to form discharges in the transverse direction. This allows the discharge to shift from the high-voltage, low-current regime going from points S to B (cylindrical discharge) to a low voltage, high current regime going from points C to D. This generates a higher power density and can operate at higher pressure (Fig. 1.13).

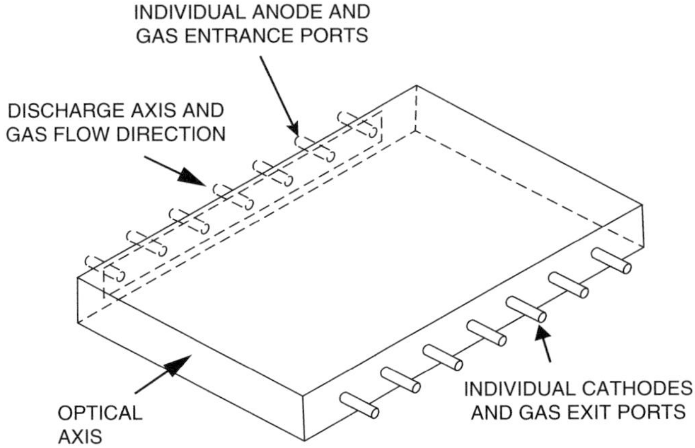

Fig. 1.13 Diagram of a transverse discharge [74]

1.3.2.2 Radiofrequency Discharge

Radiofrequency (RF) discharges can be generated through capacitive coupling or inductive coupling. The advantage of RF discharges is that there are no electrodes in contact with the gas and the gas pressure can be high [26]. Figure 1.14 is an

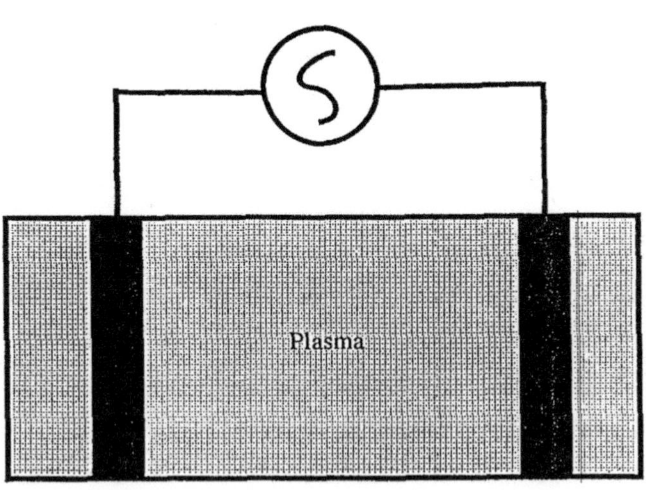

Fig. 1.14 Capacitively coupled RF discharge

illustration of a capacitively coupled RF discharge where the vessel containing the plasma is dielectric and the electrodes are on the outside of the vessel.

An inductively coupled RF discharge, shown in Fig. 1.15, is also electrode less and can create discharges in high pressure gases.

1.3.2.3 Microwave Discharge

A microwave resonance cavity, shown in Fig. 1.16, is electrode less and is able to create a discharge in high pressure gases. A standing electromagnetic wave is created in the resonator cavity which creates very high electric fields which break

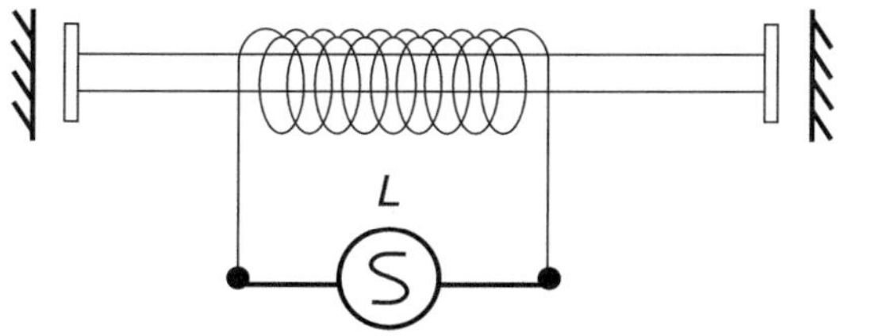

Fig. 1.15 Inductively coupled discharge

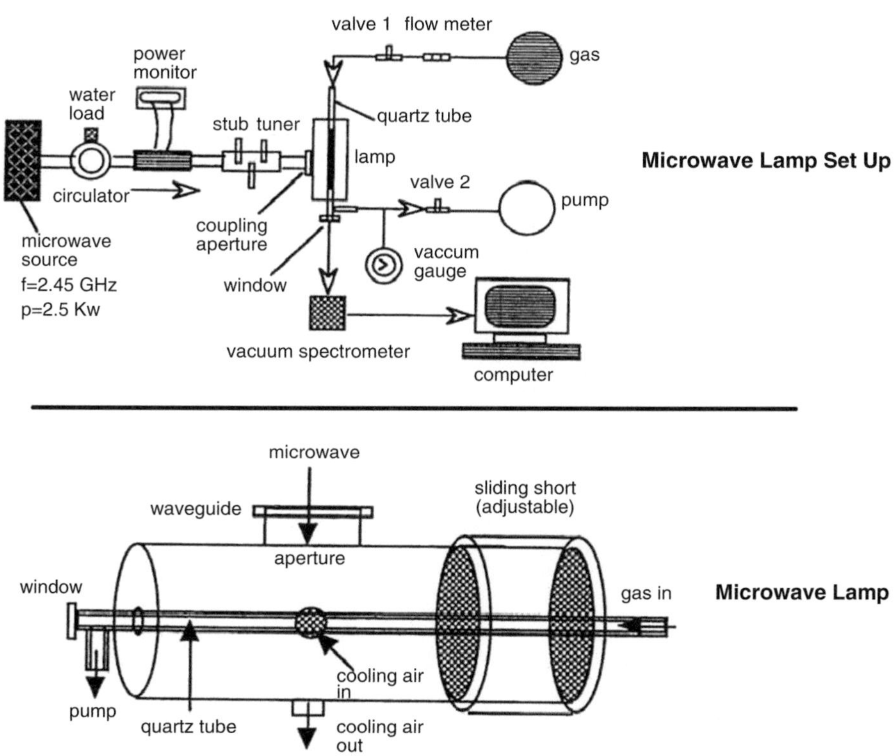

Fig. 1.16 A microwave discharge lamp [27]

down the gas. The set up shown in Fig. 1.16 was used to create various excimer lamps [27].

1.3.2.4 Electron Beam Pumping

An electron beam is created by forcing electron emission by the application of high voltages from a cathode made of metal. The metal has a work function and under high potential electrons will be emitted. The electrons which come from the emitter cathode are then accelerated by a potential and can be focused into a beam by magnetic focusing lenses. The beam can be targeted on a thin membrane like a foil which allows electrons to pass through it with minimal losses (Fig. 1.17) [28]. The laser medium, on the other side of the membrane, is excited by the electron beam. The electron beam is able to generate very high power densities in a laser medium (e.g., Krypton Fluoride Laser) [29, 30].

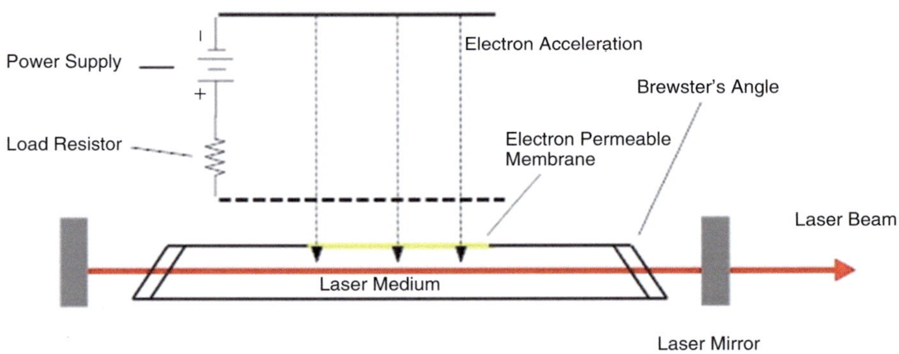

Fig. 1.17 Diagram of an electron beam pumped laser

1.3.3 *Thermal Oven*

It is feasible to use heat to produce a distribution of excited states in a gas. The number of atoms in a given energy state was originally derived by James Clerk Maxwell in 1860 and Ludwig Boltzmann by using statistical mechanics and mathematics to generate a solution based on gas kinetic theory in 1877. The probability that atoms in a gas which occupy a given energy state for a gas in thermodynamic equilibrium is,

$$f(E)dE = \sqrt{\frac{E}{\pi kT}} e^{\left(\frac{-E}{kT}\right)} dE \tag{1.6}$$

where E is energy in eV and k is the Boltzmann constant (8.6173324×10^{-5} eV K^{-1}).

Table 1.3 The fraction of nitrogen molecules (F) with energy greater than 0.117 eV

Temperature (K)	F(0.117 eV)
1000	0.45
1250	0.55
1500	0.62
1750	0.68
2000	0.72

The fraction of atoms or molecules (F(E)) with energy greater than Eo can be calculated from the distribution function.

$$F(Eo) = \frac{\int_{Eo}^{\infty} \sqrt{\frac{E}{\pi kT}} e^{\left(\frac{-E}{kT}\right)} dE}{\int_{0}^{\infty} \sqrt{\frac{E}{\pi kT}} e^{\left(\frac{-E}{kT}\right)} dE} \tag{1.7}$$

The thermally pumped carbon dioxide laser begins by running nitrogen gas through an oven and heating it to a high temperature [7]. The first vibrational level of nitrogen is at 0.117 eV, so by heating the nitrogen to a high temperature, the fraction of nitrogen molecules with energy above 0.117 eV can be calculated from Eq. (1.7) (Table 1.3). Since the gas coming out of the oven will be in thermodynamic equilibrium, the temperature would correspond to both the gas temperature and the vibrational temperature. The fraction of nitrogen molecules with energies greater than 0.117 eV is representative of the nitrogen molecules with energies greater than the first vibrational state of nitrogen. The thermodynamic equilibrium in vibrational temperature comes from the transformation of kinetic energy to vibrational energy through collisions.

The vibrational states in nitrogen are in thermodynamic equilibrium with the gas. In a CO_2 transfer laser a collisional resonance exists between the first several vibrational states of nitrogen ($N_2(v = 1–8)$) and the ground state of carbon dioxide ($CO_2(000)$) state as shown in Fig. 1.18 [31]. The first 8 vibrational levels of nitrogen are capable of populating the upper laser level of carbon dioxide ($CO_2(001)$) [31]. Using the Maxwell Boltzmann distribution, the probability that the nitrogen molecules in the $v = 1$ state is,

$$P(v = 1) = exp[-hc\omega_e(v + ½)/kT] \tag{1.8}$$

The fraction of the nitrogen molecules in vibrational states $v = 1–8$ is given by the sum of the probabilities $P(v = 1–8)$ at the energies, $hc\omega_e\ (v + ½)$ that can occur, divided by the sum of the probabilities for all vibrational states, $v = 0$ to $v = \infty$ (Eq. 1.9) [31]. The frequency, ω_e, from Fig. 1.18 is 2359.1 cm^{-1}, h is Plank's constant ($6.62606957 \times 10^{-34}$ m^2 kg/s), k is Boltzmann's constant ($1.3806488 \times 10^{-23}$ m^2 kg s^{-2} K^{-1}) and T is gas temperature in K.

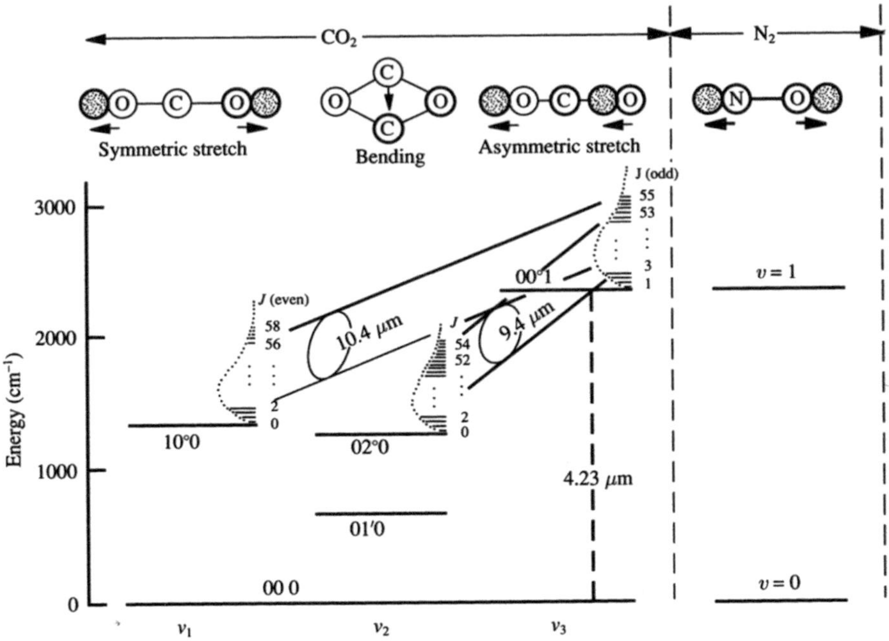

Fig. 1.18 Energy-level diagram of CO_2-N_2 laser (Modified Figure from Verdeyen [75]). There is a resonance between the $N_2(\nu = 1)$ state and the $CO_2(001)$ state (with a separation of 18 cm^{-1}). The minor separation is insignificant and at operational temperatures beyond 300 K, the kinetic energy of the molecule is sufficient to overcome this small differential energy [31, 76]. (Verdeyen, Joseph T., Laser Electronics, 3rd, ©1995. Printed and Electronically reproduced by permission of Pearson Education, Inc., New York, New York.)

$$f = \frac{\sum_{1}^{8} exp[-hc\omega_e(v + ½)/kT]}{\sum_{0}^{\infty} exp[-hc\omega_e(v + ½)/kT]} \tag{1.9}$$

Equation (1.9) can be solved by evaluating the sums in the numerator and denominator. The result of this evaluation is shown in Table 1.4 at various temperatures.

A plot of the nitrogen molecules in vibrational states 1 through 8 divided by the total number of nitrogen states is shown in Fig. 1.19.

The first vibrational level of nitrogen has a resonance with the $CO_2(001)$ state (Fig. 1.18) and by mixing the $CO_2(000)$ ground state with the $N_2(\nu = 1)$ state, a collisional transfer from the vibrational state of nitrogen to the carbon dioxide ground state will create the carbon dioxide upper laser level.

The carbon dioxide laser has been pumped thermally using a set up similar to Fig. 1.20 [7, 32, 33]. In this pumping method, a thermal population of nitrogen gas

Table 1.4 Shown in the table is the ratio of the nitrogen molecules in the 1 through 8 vibrational states divided by the total number of nitrogen molecules in all energy states

T (K)	f Ratio of N_2 in States v = 1 to v = 8 to Total N_2 States
1000	0.0334892
1500	0.103897
2000	0.183
2500	0.257012
3000	0.322293
3500	0.378756
4000	0.427306
4500	0.46899
5000	0.504756
5500	0.535407
6000	0.561613

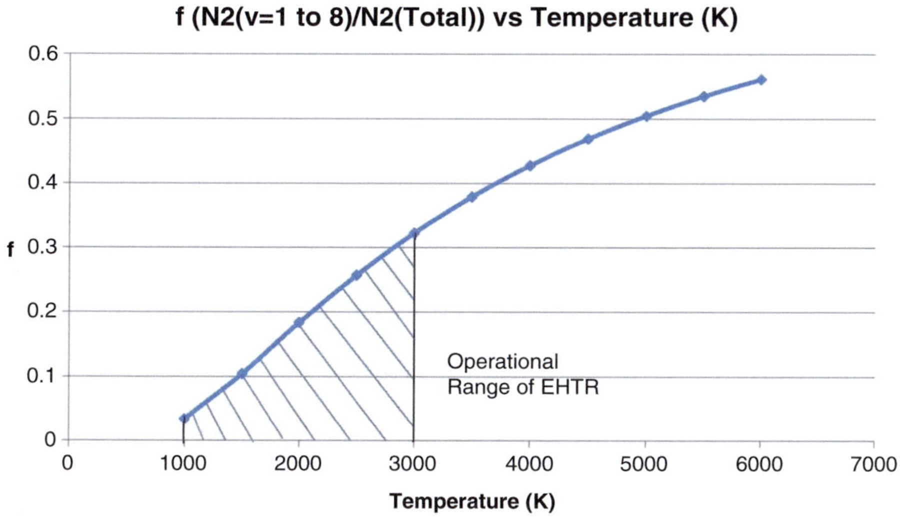

Fig. 1.19 Shown is the fraction of nitrogen in the 1 through 8 vibrational states as a function of temperature. The Potential operational temperature range is shown in the cross hatched area

as well as a thermal distribution of nitrogen vibrational states is created. The high temperature nitrogen gas then flows to a mixing chamber where CO_2 is injected. The nitrogen vibrational states, $v = 1–8$, transfers its energy to CO_2 thus pumping the upper laser level of the carbon dioxide laser.

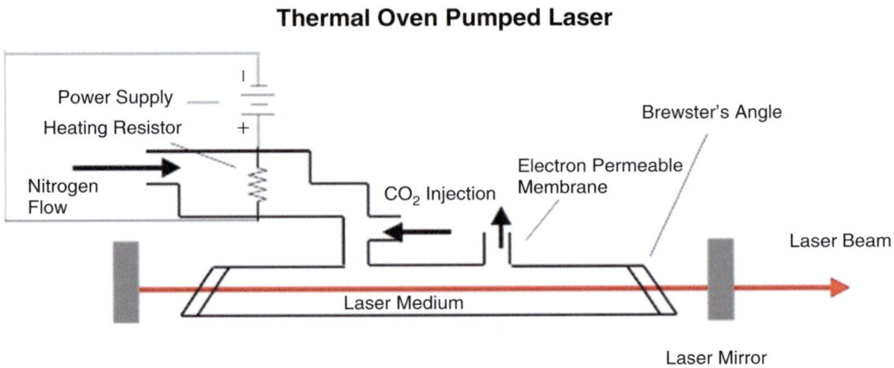

Fig. 1.20 Diagram of a thermally pumped CO_2 laser [7, 32, 33]

1.3.4 Chemical Reaction

Chemical reactions can occur by mixing two streams of individual chemicals which undergo exothermic reactions (Fig. 1.21). The byproduct of these exothermic reactions can be excited molecular states which can lead to laser emission. This method produces energy by using chemical potential of a chemical reaction to cause a burn or flame. The energy from the flame is randomly distributed among the product chemicals. Examples of chemical lasers are deuterium fluoride and the Chemical Oxygen-Iodine Laser (COIL) which have been extensively studied by the US military for directed beam weapons [34]. The deuterium fluoride (DF) laser is based on a reaction which resembles a rocket engine. A fuel that contains hydrogen is mixed with an oxidizer that contains fluorine and the resulting combustion

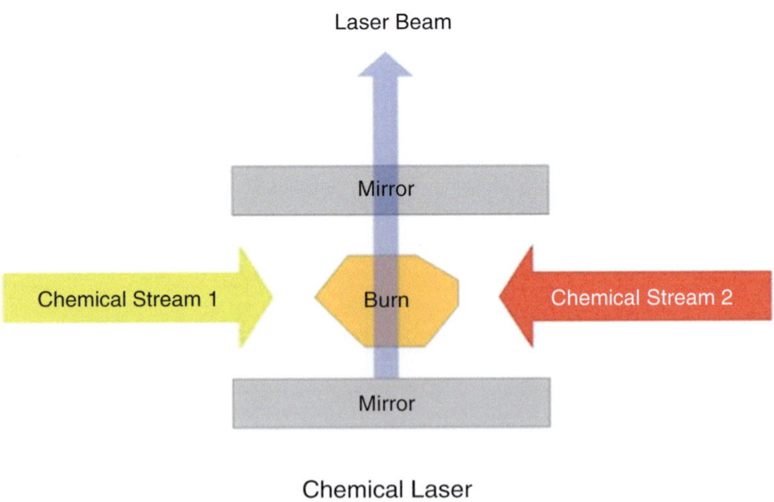

Fig. 1.21 Diagram of a chemical laser

Fig. 1.22 The THEL system with crew [77]. (Tactical High Energy Laser by Dr. Josef Shwartz, Mr. Gerald T. Wilson and Dr. Joel Avidor, *Proc. SPIE* 4632, Laser and Beam Control Technologies, 10 (June 5, 2002). Printed and Electronically reproduced by permission of SPIE.)

produces the upper laser level state. The actual reactions that take place involve a chain of events. The fuel is ethylene (C_2H_4) and the oxidizer is nitrogen trifluoride (NF_3). During combustion fluorine is freed from the molecule. Downstream, deuterium is injected into the exhaust and combines with fluorine to form the DF* (upper laser level) as shown in Eq. 1.10. The DF laser has about 10 lasing lines between 3.6 and 4.0 μm. The DF laser has been used in the Tactical High Energy Laser (THEL) system (Figs. 1.22 and 1.23)

$$D_2 + F \rightarrow DF^* + D \tag{1.10}$$

The COIL is made up of three parts: (1) a singlet oxygen generator; (2) a supersonic nozzle where chemicals are mixed and atomic iodine is produced; (3) the laser cavity. Singlet delta oxygen is produced by mixing H_2O_2, KOH and Cl_2 as shown in Eq. 1.11.

$$Cl_2 + 2KOH + H_2O_2 \rightarrow O_2(1\Delta_g) + 2KCl + 2H_2O \tag{1.11}$$

Atomic iodine is created by first generating atomic fluorine and then mixing the atomic fluorine with HI as shown in reaction 5. The process of generating atomic fluorine is more complex and involves molecular fluorine interacting with NO [35, 36].

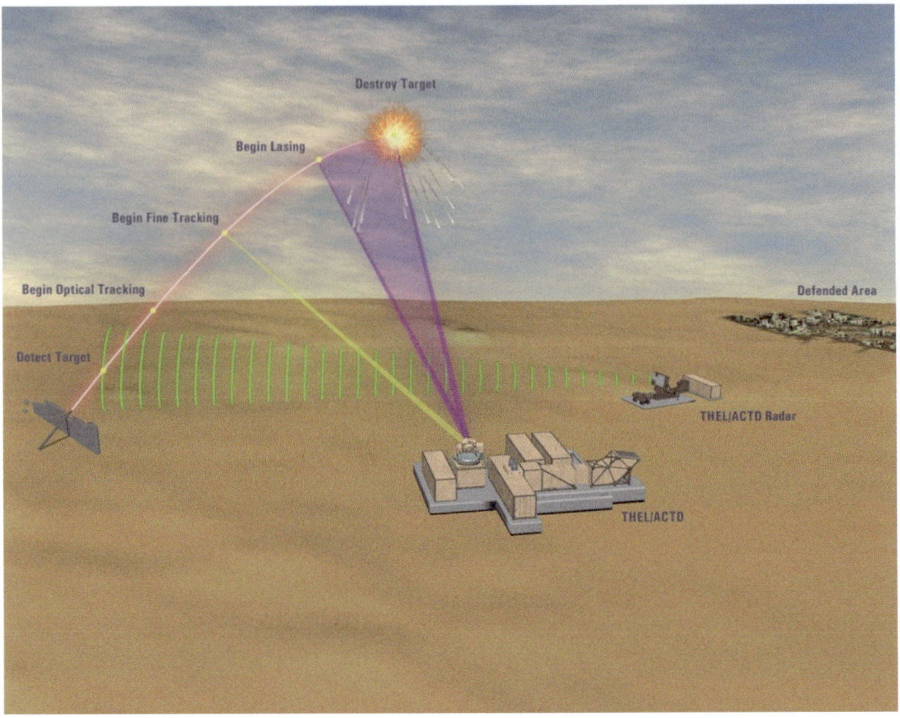

Fig. 1.23 Artist conception of a THEL in an operational environment [77]. (Tactical High Energy Laser by Dr. Josef Shwartz, Mr. Gerald T. Wilson and Dr. Joel Avidor, *Proc. SPIE* 4632, Laser and Beam Control Technologies, 10 (June 5, 2002). Printed and Electronically reproduced by permission of SPIE.)

$$HI + F \rightarrow HF + I \tag{1.12}$$

The oxygen singlet delta state has a very long lifetime (up to 45 min) and thus can be mixed with iodine to create the iodine upper laser level through a transfer reaction (Eq. 1.13).

$$O_2(1\Delta_g) + I \rightarrow O_2(3\Sigma_g) + I^* \tag{1.13}$$

A COIL system was mounted on a Boeing 747 for the Airborne Laser Testbed (ALTB) built by Northrop Grumman [37] (Figs. 1.24 and 1.25). The advantage of an ALLTB is that it is able to fly into different theaters where it is able to target a missile in the boost phase.

1.3.5 Ion Beam Pumped Laser

Heavy ion beams created in an accelerator can be used to drive a laser [38–40]. For example, the pulsed heavy ion beam from the heavy ion synchrotron SIS-18 at the

Fig. 1.24 The Airborne Laser Testbed using a COIL system (Printed and Electronically reproduced by permission of the Missile Defense Agency, US Department of Defense)

Fig. 1.25 Artist's conception of an ALTB system in operation (Printed and Electronically reproduced by permission of the Missile Defense Agency, US Department of Defense)

GSI Helmholtzzentrum fur Schweioneforschung GmbH, Darmstadt, Germany ion accelerator facility (Fig. 1.26) has been used for an ion beam driver. The SIS is capable of pulse widths on the order of 110 ns with peak power densities of around 1 MW/cm^3 [41]. The SIS has been used to pump a KrF excimer laser which has a

Fig. 1.26 Exterior of the heavy ion synchrotron SIS [78]

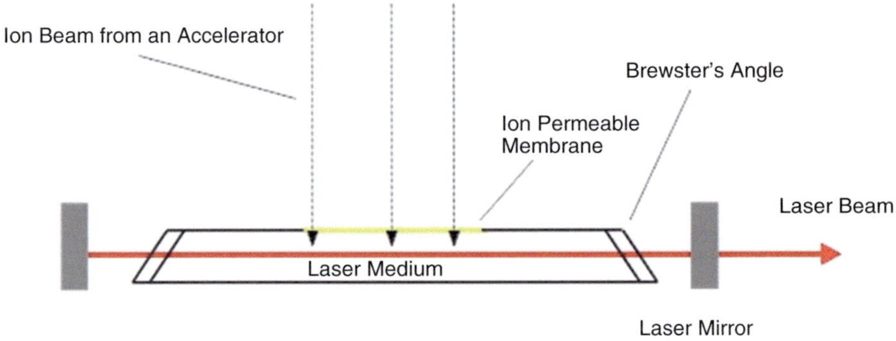

Fig. 1.27 Ion beam driven (accelerator) laser

very high threshold power density. Figure 1.27 is a basic layout of the interface between the accelerator and the laser medium.

1.3.6 Fission Fragment Pumped Laser

Nuclear-Pumped Lasers use the ionizing radiation products from nuclear reactions as the driver for a laser medium. Fission fragments represent one of the most efficient and energetic types of ionizing radiation which can be used a driver. The nuclear-pumping process requires that the nuclear reaction be triggered by some reactant. In the case of fission, that trigger would be a thermal neutron. Uranium-235 has a high cross section for neutrons at thermal energy and when a neutron is captured, the products will be fission fragments [2] as shown in Eq. (1.3). These fission fragments on average carry 165 MeV of the total energy.

Nuclear-pumped lasers have also used other neutron triggered reactions based on boron-10 and helium-3 (Eqs. 1.1 and 1.2).

Two nuclear-pumped lasers were part of the strategic defense initiative program. Centaurus was a cooperative research effort between Lawrence Livermore National Laboratory and Idaho National Engineering Laboratory on a reactor pumped laser system [42]. There is limited published data on the Centaurus program. Falcon was a research effort at Sandia National Laboratory on a reactor pumped laser system [42]. Falcon involved a number of reactor-pumped laser experiments using various mixtures of xenon, argon, neon, and helium. Lasers with wavelengths at 585, 703, 725, 1271, 1733, 1792, 2032, 2630, 2650, and 3370 nm were demonstrated with intrinsic efficiencies as high as 2.5 % [43].

A fission driven nuclear-pumped laser would have to be placed near a thermal neutron source as shown in Fig. 1.28. A thin layer of uranium-235 surrounds the laser medium. When the thermal neutrons are captured by the uranium-235 nuclei, fission fragments are created. These fission fragments stream into the laser medium thus creating ionization and excitation. The neutron source is a nuclear reactor.

The Centaurus and Falcon programs were phased out in the late 1980s despite the advantages of nuclear-pumped lasers in areas such as scaling, energy output and size. SDI strategies changed with a focus on fast interceptors.

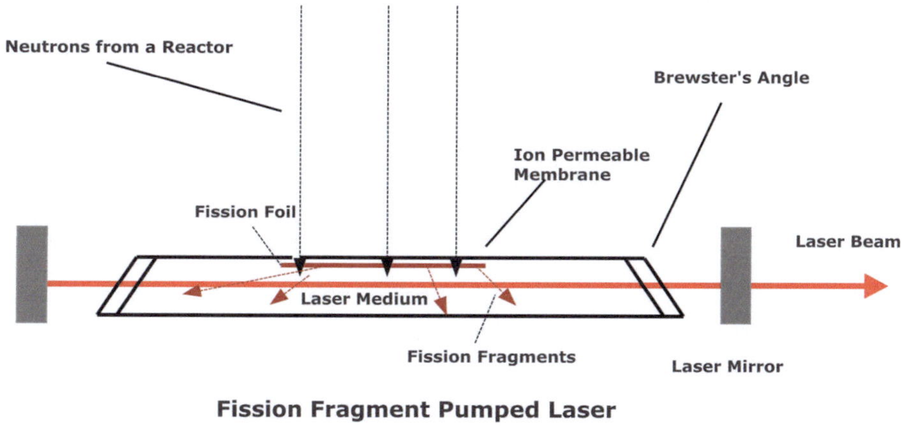

Fig. 1.28 Fission Fragment driven (Nuclear-Pumped) laser

1.4 Gain Medium

The fundamental analytical tool needed to develop a laser was first described by Albert Einstein in 1916 when he derived the fundamental theory of stimulated emission. The process of spontaneous emission, where an atom or molecule in an excited state spontaneously emits a photon isotropically, was known as was the process of absorption where photons passing through a gas will be absorbed. Under the principle that all processes have an inverse process, Einstein reasoned that spontaneous emission is not the reverse process of absorption. Thus another process had to exist. Einstein postulated that a process called stimulated emission is the reverse process of absorption. In Fig. 1.29, an energy level diagram is shown for an atom or molecule where energy level 2 is the upper level and energy level 1 is the lower level.

These 3 processes shown in Fig. 1.29 involve the interactions between photons and atoms or molecules assuming no degeneracies in the energy states:

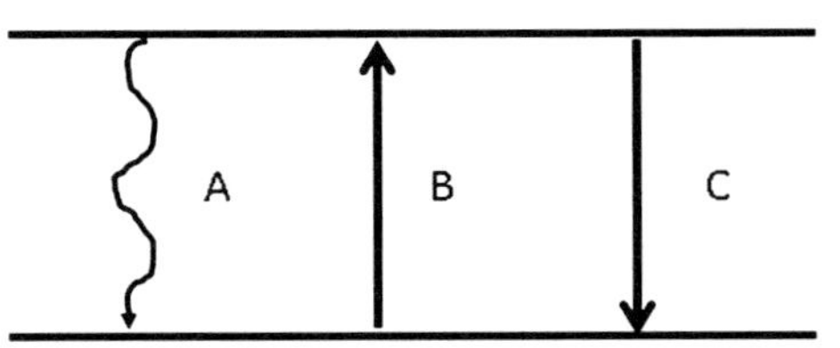

Fig. 1.29 An energy level diagram for two energetic states in an atom. Process A is spontaneous emission, process B is absorption and process C is stimulated emission

(A) Spontaneous Emission (A_{21} (s^{-1}))

$$\text{Rate equation}: \quad -\frac{dN_2}{dt} = -A_{21}N_2 \tag{1.14}$$

where N_2 is the density of the upper state level (atoms or molecules m^{-3}).

(B) Absorption (B_{12} ($J^{-1}m^3s^{-2}$))

$$\text{Rate Equation}: -\frac{dN_1}{dt} = -B_{12}N_1\rho(v)\,\text{or}\,\frac{dN_2}{dt} = B_{12}N_1\rho(v) \tag{1.15}$$

(Note: loss for N_1 and source for N_2)
where $\rho(v)$ is the energy density per unit angular frequency in the gain region ($\rho(v) = \frac{8\pi h v^3}{c^3}$ $\left(\frac{J-s}{m^3}\right)$).

(C) Inverse of the absorption process is Stimulated Emission (B_{21} ($J^{-1}m^3s^{-2}$))

$$\text{Rate Equation}: -\frac{dN_2}{dt} = -B_{21}N_2\rho(v)\,\text{or}\,\frac{dN_1}{dt} = B_{21}N_2\rho(v) \tag{1.16}$$

(Note: source for N_1 and loss for N_2).

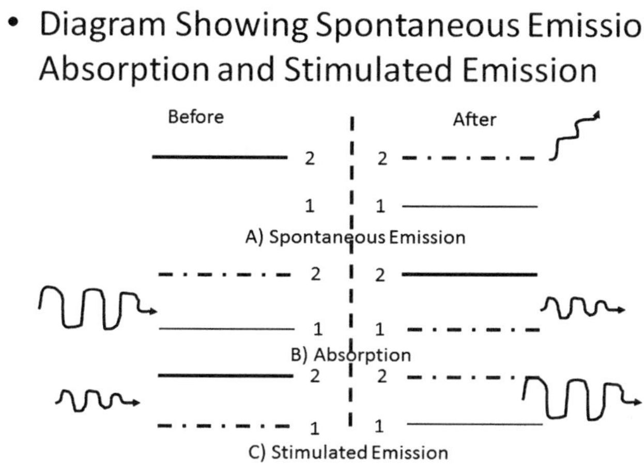

Fig. 1.30 A diagram which illustrates the spontaneous emission, absorption and stimulated emission processes. In the illustration an electron occupies the state if the line is *black*. If the line is *dashed*, the electron does not occupy the state after the interaction. If the line is blank the state is unfilled. In the diagram the *left* side represents the conditions of state 1 and 2 before the interaction and the *right* side after the interaction

The photons which are produced come from spontaneous emission and stimulated emission and the photons which are lost are due to absorption (Fig. 1.30). In spontaneous emission (process A), the photon is emitted from an excited state isotropically. In absorption (process B) a photon which hits the absorbing atom or molecule will cause the atom or molecule to transition from state 1 to state 2 thereby transferring the photon energy completely to the atom or molecule. In stimulated emission (process C), a photon passes an atom or molecule that is in the excited state and will cause that atom or molecule to emit a photon from state 2 to state 1 which will have the same direction, frequency and phase as the incoming photon. The mechanism for absorption and stimulated emission depend on the electromagnetic properties of the photon to interact with an electron orbiting the atom or molecule (which is an electric dipole) causing the electron to oscillate. This behavior is similar to that of an antenna which can emit or absorb electromagnetic waves due the oscillations in the dipole.

The Einstein B coefficient for absorption and stimulated emission are related as shown in Eq. (1.17). If the states have degeneracies, than,

$$g_1 B_{12} = g_2 B_{21} \tag{1.17}$$

where g_1 is the degeneracy of state 1 and g_2 is the degeneracy of state 2.

$$\frac{A_{21}}{B_{21}} = \frac{8\pi h v^3}{c^3} \tag{1.18}$$

where "h" is Plank's constant (6.626×10^{-34} (J-s)), v is frequency (s^{-1}), "c" is the speed of light (3×10^8 m/s).

The lineshape function ($g(\nu)$ (s)), is the spectral width of the emission. The spectral width occurs naturally and can be seen by using the Heisenberg uncertainty principle,

$$\Delta E \Delta t \geq \frac{h}{2\pi} \tag{1.19}$$

where Δt is related to the state lifetime.

$$\Delta E \geq \frac{h}{2\pi\tau} \text{ and } \Delta E = h\Delta\nu \tag{1.20}$$

Thus,

$$\text{h}\Delta\nu \geq \frac{h}{2\pi\tau} \text{ and } \Delta\nu \geq \frac{1}{2\pi\tau} = \nu_1 - \nu_2 \tag{1.21}$$

As shown in Fig. 1.31 Transitions from energy level 1–2 will have a band of energies and a lineshape.

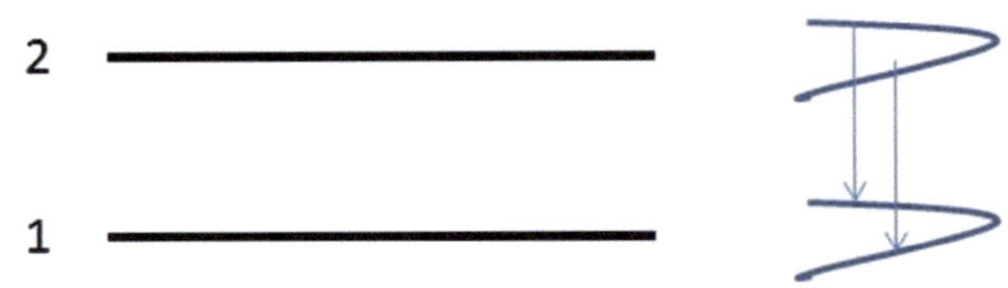

Fig. 1.31 The range of energies of photons given off due to the uncertainty principle in which energies have an uncertainty of ΔE

The natural linewidth, $\Delta\nu_N$ which comes from the energy uncertainty is given in Eq. (1.22),

$$\Delta\nu_N = \frac{1}{\pi\left(\frac{1}{\tau_2} + \frac{1}{\tau_1}\right)} \tag{1.22}$$

The natural spectral lineshape due this uncertainty is shown in Fig. 1.32 and is known as a Lorentzian lineshape (Eq. 1.23). The lineshape has the property that the integral of the lineshape over all possible frequencies is unity (Eq. 1.24).

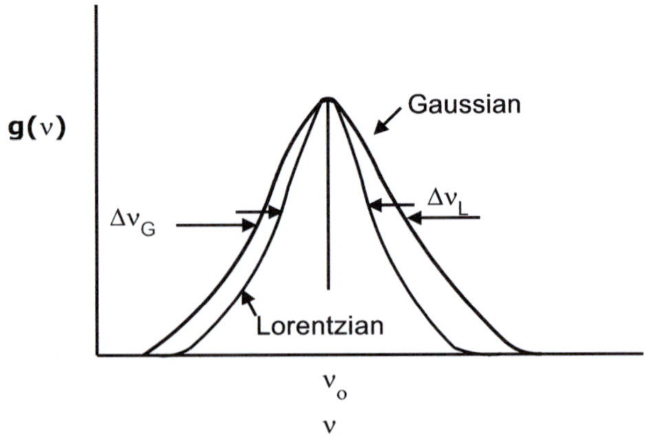

Fig. 1.32 Lorentzian and Gaussian lineshapes

$$g_L(v) = \frac{\Delta v}{2\pi\left[(v - v_0)^2 + \left(\frac{\Delta v}{2}\right)^2\right]} \tag{1.23}$$

$$\int_0^\infty g(v)dv = 1 \tag{1.24}$$

There are other factors that can change the lineshape function. Pressure broadening, which also has a Lorentzian lineshape as shown in Eq. (1.23), is due to the interruption of emission due to collisions which leads to broadening of the transition. The linewidth for pressure broadening is shown in Eq. (1.25).

$$\Delta v_{Pressure} = \frac{NQ}{\pi}\sqrt{\frac{16kT}{\pi M}} \tag{1.25}$$

where N is the density of the atoms (m^{-3}), M is the atomic mass of the colliding atoms of the same mass (kg), Q is the collision cross section (m^2) and k is Boltzmann's constant (1.3806×10^{-23} m^2 kg s^{-2} K^{-1}).

Temperature broadening has a Gaussian lineshape which occurs because the atoms or molecules are in motion and if the atom or molecule has a velocity component in the direction of the emission, the emission can be Doppler shifted or Doppler broadened. The lineshape follows Eq. (1.26). It has a linewidth that is shown in Eq. (1.37). A Gaussian lineshape is broader than a Lorentzian lineshape and the Gaussian linewidth is wider than the Lorentzian linewidth.

$$g_G(v) = \frac{2\left(\ln(2)\right)^{1/2} e^{-\left[4(\ln(2))(v-v_0)^2/(\Delta v)^2\right]}}{\pi \Delta v} \tag{1.26}$$

$$\Delta v_D = 2v_0\sqrt{\frac{2kln(2)}{Mc^2}} \tag{1.27}$$

where c is the speed of light (3×10^8 cm s^{-1}) and k is Boltzmann's constant (1.3806×10^{-23} m^2 kg s^{-2} K^{-1}).

Rate Equation for the Upper Laser Level

The rate of change in the density of energy level 2 for only processes B and C is

$$\frac{dN_2}{dt} = -N_2B_{21}g(v)\rho_v + N_1B_{12}g(v)\rho_v \tag{1.28}$$

The relationships between, $\rho_v, \rho(v), I_v, B_{12}, B_{21}$, and A_{21} are shown in Eqs. (1.29), (1.30), (1.31), (1.32), and (1.33).

$$\rho_v = energy\,density\left(\frac{J}{m^3}\right) = \rho(v)/_{g(v)} \tag{1.29}$$

$$I_v = Intensity\left(\frac{W}{m^2}\right) \tag{1.30}$$

$$\rho_v = \frac{I_v}{c} \tag{1.31}$$

$$B_{12} = B_{21}\,(\text{assuming no degeneracies}) \tag{1.32}$$

$$B_{21} = \left(\frac{c^3}{8\pi h v^3}\right)A_{21} \tag{1.33}$$

The Stimulated Emission cross section can be derived from Eq. (1.28). Equation (1.34) is a rearrangement of Eq. (1.28). Realizing that $\lambda = \frac{c}{v}$, Eq. (1.28) may be rewritten (Eq. 1.35).

$$\frac{dN_2}{dt} = -A_{21}\left(\frac{c^3}{8\pi h v^3}\right)(N_2 - N_1)g(v)\frac{I_v}{c} \tag{1.34}$$

$$\frac{dN_2}{dt} = -A_{21}\frac{\lambda^2}{8\pi}g(v)(N_2 - N_1)\frac{I_v}{hv} \tag{1.35}$$

where λ is wavelength in meters.

Grouping terms in Eq. (1.35),

$$\sigma_{SE} = A_{21}\frac{\lambda^2}{8\pi}g(v) \tag{1.36}$$

σ_{SE}, the stimulated emission coefficient, is thus derived.

The stimulated emission coefficient can be used to find both small signal gain and gain in a laser medium. As shown in Fig. 1.33, gain in a laser can be measured by directing a beam of known intensity and frequency into a gain medium and then measuring the differential intensity of the output beam.

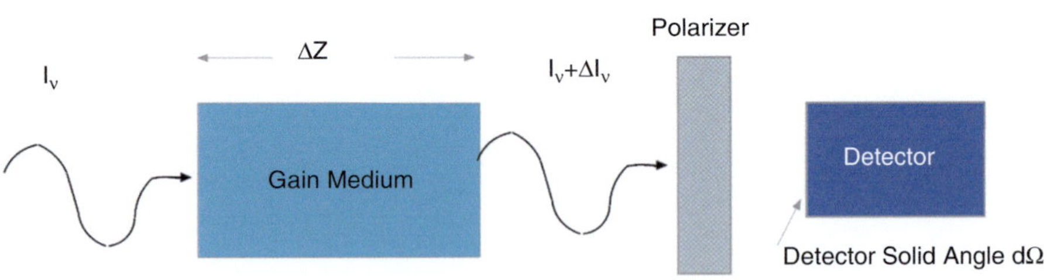

Fig. 1.33 Diagram of a gain experiment

The ΔI_v in the experiment can be found by summing up the source and loss contributions to the differential intensity (Eq. 1.37). The first part of the equation is for the gain in intensity due to stimulated emission. The first term in this equation is energy of the photon (hv), term 2 is emission rate per atom $(B_{21}\frac{I_v}{c})$, term 3 is the lineshape $(g(v))$, term 4 is the polarization probability (taken to be 1), term 5 is the solid angle probability (taken to be 1) and term 6 is number of atoms in the upper laser level (N_2) along the path (Z) of the beam. The second part of the equation is absorption of photons. The first term in this equation is the energy of the photon (hv), term 2 is absorption rate per atom $(B_{12}\frac{I_v}{c})$, term 3 is the lineshape $(g(v))$, term 4 is the polarization probability (taken to be 1), term 5 is the solid angle probability (taken to be 1) and term 6 is number of atoms in the lower laser level (N_1) in the path (Z) of the beam. The final part of the equation is the photon contribution due to spontaneous emission which is made up of term 1 times the energy of the photon (hv), term 2 is the spontaneous emission rate per atom in a frequency range Δv $(A_{21}\Delta v)$, term 3 is the lineshape $(g(v))$, term 4 is the polarization probability (taken to be 1/2), term 5 is the solid angle probability $(\frac{d\Omega}{4\pi})$ and term 6 is the number of atoms in the lower laser level (N_2) in the path (Z) of the beam.

$$\Delta I_v = hv*B_{21}\frac{I_v}{c}*g(v)*1*1*N_2\Delta Z - hv*B_{12}\frac{I_v}{c}*g(v)*1*1*N_1\Delta Z + hv*A_{21}\Delta v*g(v)*\frac{1}{2}*\frac{d\Omega}{4\pi}*N_2\Delta Z \tag{1.37}$$

The small signal gain coefficient can be found by rearranging Eq. (1.37) and taking the limits of the differential elements (Eq. 1.38).

$$\frac{dI_v}{dz} = \left[\frac{hv}{c}(B_{21}N_2 - B_{12}N_1)g(v)\right]I_v + \frac{1}{2}\left[hvA_{21}N_2g(v)\Delta v\frac{d\Omega}{4\pi}\right] \tag{1.38}$$

In Eq. (1.38), the first term is gain and the last term is noise. Assuming that the noise is negligible, Eq. (1.39) can be found.

$$\frac{dI_v}{dz} = \left[A_{21}\frac{\lambda^2}{8\pi}(N_2 - N_1)g(v)\right]I_v \tag{1.39}$$

From Eq. (1.39), call, $\gamma(v) = \left[A_{21}\frac{\lambda^2}{8\pi}(N_2 - N_1)g(v)\right]$, the small signal gain coefficient (m^{-1}).

Rearranging Eq. (1.39) into Eq. (1.40), the gain (G(v)) can be found by solving Eq. (1.40) (Eq. 1.41).

$$\frac{dI_v}{dz} = \gamma(v)I_v \tag{1.40}$$

$$I_v(z) = I_0(z)exp[\gamma(v)z] \tag{1.41}$$

The gain is shown in Eq. (1.42).

$$G(v) = exp[\gamma(v)z] \tag{1.42}$$

A laser oscillator is an optical cavity in which a beam of photons passes through a gain medium multiple times (using mirrors) in order to extract energy from the gain medium by building up beam intensity (Fig. 1.34).

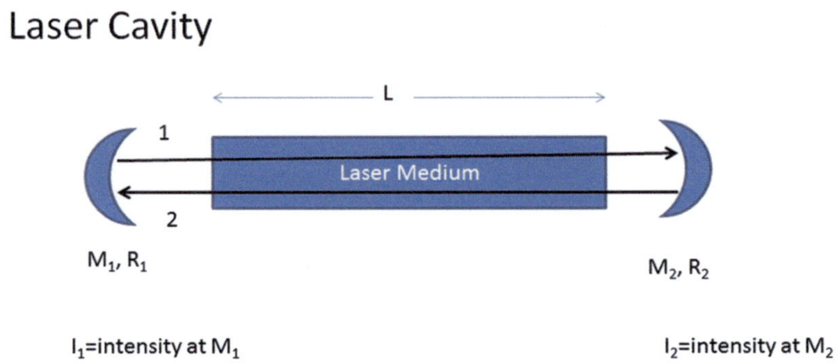

Fig. 1.34 A diagram of a laser oscillator or laser cavity is shown. Two mirrors allow the laser beam to oscillate passing through a gain medium (laser medium) in order to extract energy from the medium. The length of the gain cavity is L, the reflectivity of mirror 1 is R_1 and the reflectivity of mirror 2 is R_2

The minimum criteria for oscillation can be established by looking at the forward traveling beam (beam 1) and the backward traveling beam (beam 2) in Fig. 1.34. The beam bouncing backward from mirror 2 has an intensity of I_2 which is equal to the initial intensity I_v times the gain from its initial pass through the cavity (Eq. 1.43). The beam will move backward through the laser medium (path 2) until it bounces off of mirror 1 at which point its intensity will be I_2 times the gain (Eq. 1.44). By grouping terms, Eq. (1.39) can be simplified (Eq. 1.45).

$$I_2 = I_v e^{[\gamma(v)L]} R_2 \tag{1.43}$$

$$I_1 = \left(I_v e^{[\gamma(v)L]} R_2\right)\left(e^{[\gamma(v)L]} R_1\right) \tag{1.44}$$

$$I_1 = I_v e^{[\gamma(v)2L]} R_2 R_1 \tag{1.45}$$

In order to have a laser system it must produce a beam in which after one complete round trip, $I_1 \geq I_v$. Using this inequality along with Eq. (1.45) for I_1, the relationship in Eq. (1.46) is derived. Dividing Eq. (1.46) by Iv, results in Eq. (1.47). Rearranging Eq. (1.47) isolates the term related to small signal gain to the left side of Eq. (1.48) and the reflectivity coefficients of the mirrors (R_1 and R_2) on the right side of Eq. (1.48). By taking the natural log of both sides of Eq. (1.48) and dividing by 2 L, Eq. (1.49) is derived. This inequality is the minimum small signal gain that the system must have in order for a laser beam to be formed.

$$I_v e^{[\gamma(v)2L]} R_2 R_1 \geq I_v \tag{1.46}$$

$$e^{[\gamma(v)2L]}R_2R_1 \geq 1 \tag{1.47}$$

$$e^{[\gamma(v)2L]} \geq \frac{1}{R_2R_1} \tag{1.48}$$

$$\gamma(v) \geq \frac{ln\left[\frac{1}{R_1R_2}\right]}{2L} = \alpha \tag{1.49}$$

Note that α is the minimum small signal gain for laser action to occur.

1.5 Three and Four Level Lasers

In order for a laser to work, the density of the upper laser level state must be greater than the density of the lower laser level state. This is known as a population inversion. In a three level laser (Fig. 1.34), the lower laser level is the ground state. Thus in order to obtain a population inversion, enough power must be applied to the laser in order to deplete the ground state and populate the upper laser level so that an inversion can be achieved. A system at moderate operating temperatures would favor the ground state in thermodynamic equilibrium (having a Maxwell-Boltzmann distribution). The laser has to be driven to a non-thermodynamic equilibrium condition in order to favor the upper laser level. By necessity, three level lasers require high power densities in order to obtain a population inversion. Thus three level lasers are typically pulsed.

In contrast, a four level laser has a lower laser level which can be depleted by decay processes. Such a laser is in a non-thermodynamic equilibrium and can maintain a population inversion if the mechanism which pumps the upper laser level is more efficient than mechanisms which populate the lower laser level. Additionally, if the lower laser level has a very fast decay time as compared to the upper laser level, it is feasible to achieve an inversion even in steady-state.

1.6 Power Buildup in a Laser

The laser is a dynamic system at the atomic level. The laser medium, the pumping source and the optical cavity are coupled systems which each have unique characteristic time constants. The power source can be steady-state or can be pulsed, thus the variability of the pumping source time constant is significant. The laser medium is dominated by its atomic time constants and collision rates. The laser cavity has a characteristic photon life time constant. The coupling of these three dynamic systems determines the time dependent behavior of the system. A general model for four level lasers can be developed if the following are known,

Pumping rate, S atoms cm^{-3} s^{-1}
State 3 lifetime, τ_3 (s)

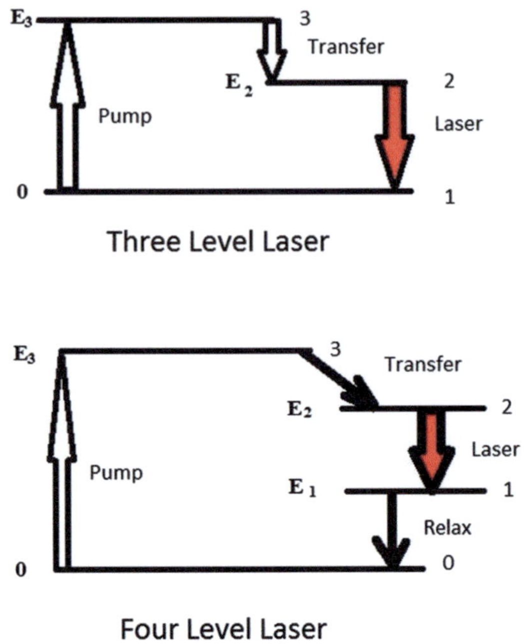

Fig. 1.35 An illustration of energy level diagrams for a three level laser and a four level laser. In the three level laser the lower laser level (state 1) is the ground state. Energy pumps the lower laser level to an intermediate state 3 where that state decays into the upper laser level (state 2). Photons are emitted from state 2 to state 1 to create the laser. In the four level laser the ground state (state 0) is pumped to the intermediate level (state 3) which then decays to the upper laser level (state 2). Photons are emitted from state 2 to state 1 to create the laser. The lower laser level then relaxes through a decay process to the ground state (state 0)

State 2 lifetime, τ_2 (s)
State 1 lifetime, τ_1 (s)
Laser Wavelength, λ (cm)
Linewidth, Δv (s)
Mirror 1 Reflectivity, R_1
Mirror 2 Reflectivity, R_2
Cavity Length, L_c (cm)
Cavity Radius, r_c (cm)

As an example, the time dependent behavior of a laser is examined. The temporal characteristics are dependent on the lifetimes of energy state 1, energy state 2 and energy state 3 in a four level laser system (Fig. 1.35) as well as the parameters of the optical cavity and the time dependent characteristics of the power source. In Fig. 1.36, a low loss optical cavity is shown where the reflectivity of the back mirror is 100 % and the front mirror is 99 %. It is should be understood by the reader that a 100 % reflective mirror is not feasible, but it is used here for the purpose of illustration. Assuming a pump rate for the laser of 0.4 W/cm^3, it is

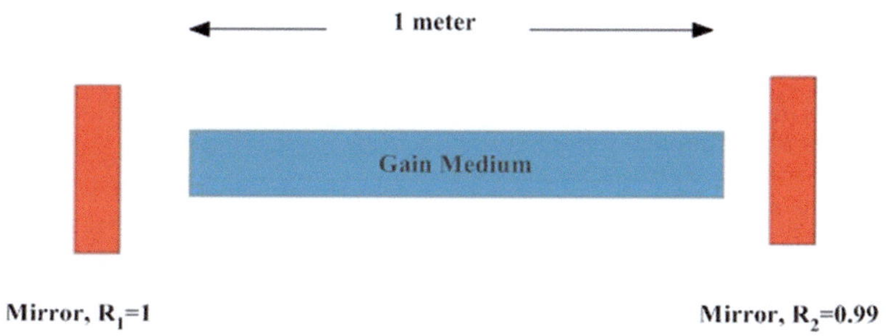

Fig. 1.36 A laser cavity with a gain medium length of 1 m, a back mirror with reflectivity of 100 % and front mirror with reflectivity of 99 %

possible to develop a model for a four level laser which illustrates how the laser intensity would build up.

A laser can be modeled by coupling the laser kinetics equations with optical equations. Starting with a 4 level laser as shown in Fig. 1.34, the model will include the ground state (with a time dependent density of $N_0(t)$ with units atoms or molecules cm^{-3}), energy state 3 (with a time dependent density of $N_3(t)$ with units atoms or molecules cm^{-3}), the upper laser level with energy state 2 (with a time dependent density of $N_2(t)$ with units atoms or molecules cm^{-3}) and the lower laser level with energy state 1 (with a time dependent density of $N_1(t)$ with units atoms or molecules cm^{-3}). At this point it is valuable to recognize that there are two forms of the Einstein B coefficient: one for the density (B_{21}) and one for intensity (B_{21}'). They are related by $B_{21} = \frac{c}{4\pi}B'_{21}$. The net stimulated emission that increases the radiation density ($\rho(\nu)$ J cm^{-3}) per unit time in the domain from hν to h(ν+$\Delta\nu$) is $\frac{h\nu c^3 \Delta N\rho(\nu)g(\nu)d\nu}{8\pi h\nu^3\tau_2} = \frac{h\nu\Delta N c^2 I_\nu \delta(\nu-\nu_0)g(\nu)d\nu}{8\pi h\nu^3\tau_2}$ where $\Delta N = N_2 - N_1$, $c\rho(\nu) = I_\nu\delta(\nu-\nu_0)$ and $\delta(\nu-\nu_0)$ is the Dirac delta function. By integrating the delta function over all frequencies the right hand side becomes $\frac{\Delta N c^2 I_\nu \delta g(\nu_0)}{8\pi\nu_0^3\tau_2}$.

The optical portion of the model, using Eqs. (1.28), (1.29) and (1.31), can be written as the rate of change of the optical intensity, I_ν. The pumping rate (S atoms or molecules created by the energy source cm^{-3} s^{-1}) is directly related to the power density in the laser medium (P Watts cm^{-3}). S and P can have a temporal and spatial dependency which would complicate the model. By assuming that S and P are uniform throughout the volume of the laser, the model is simplified considerably. The first step is to write out the processes involved in the 4 level laser model. The first process is the pumping process in which energy is input into the system and atoms or molecules in the ground state N_0 are pumped into the energy state 3 (Eq. 1.50).

$$N_0 + energy \xrightarrow{S} N_3 \tag{1.50}$$

The next step is that the N_3 state decays into the upper laser level state N_2 (Eq. 1.51).

$$N_3 \xrightarrow{\tau_3} N_2 \tag{1.51}$$

Once the state N_2 is created, it can decay by spontaneous emission (Eq. 1.52) or it can decay by stimulated emission (Eq. 1.53).

$$N_2 \xrightarrow{\tau_2} N_1 \tag{1.52}$$

$$N_2 + photon \xrightarrow{B_{21}} N_1 + 2\,coherent\ photons \tag{1.53}$$

where $B_{21} = \frac{c^3 A_{21}}{8\pi h v_0^3}$, c is the speed of light (3×10^{10} cm s^{-1}), $A_{21} = 1/\tau_2$, v_0 is the laser frequency (s^{-1}), h is plank's constant (6.626×10^{-34} J s), Δv is the linewidth (s^{-1}) and the lineshape for natural broadening is $g(v_0) = \frac{\Delta v}{2\pi\left(\left(\frac{\Delta v}{2}\right)^2 + (v - v_0)^2\right)}$ (s).

The lower laser level, or state N_1, decays by spontaneous emission to the ground state (Eq. 1.54) or can undergo absorption to transition into the upper laser level (Eq. 1.55)

$$N_1 \xrightarrow{\tau_1} N_0 \tag{1.54}$$

$$N_1 + photon \xrightarrow{B_{12}} N_2 \tag{1.55}$$

The intensity I_v builds up in the cavity by stimulated emission (Eq. 1.53), the intensity is reduced by absorption (Eq. 1.55), the intensity is increased by spontaneous emission which is emitted in the correct solid angle ($\Delta\Omega$) in the right phase.

The processes involved in laser kinetics are statistical and thus can be modeled by the Verhulst equation [44]. In this equation, the rate of change in the density of an energy state (N cm^{-3}) is equal to the summation of source rates minus the summation of loss rates (Eq. 1.56).

$$\frac{dN}{dt} = \sum Source\,Rates - \sum Loss\,Rates \tag{1.56}$$

Looking at the N_0 state in the 4 level laser model, a rate equation can be written for the processes in Eqs. (1.50), (1.51), (1.52), (1.53) and (1.54) (Eq. 1.57).

$$\frac{dN_0(t)}{dt} = \frac{N_1(t)}{\tau_1} - S \tag{1.57}$$

The initial condition for the N_0 state is,

$$N_0(t) = 1 \times 10^{15} \tag{1.58}$$

The N_3 state rate equation can be written for the processes in Eqs. (1.50) and (1.51) (Eq. 1.59).

$$\frac{dN_3(t)}{dt} = S - \frac{N_3(t)}{\tau_3} \tag{1.59}$$

The initial condition for the N_3 state is,

$$N_3(t) = 0 \tag{1.60}$$

The N_2 state rate equation can be written for the processes in Eqs. (1.51), (1.52) and (1.53) (Eq. 1.61).

$$\frac{dN_2(t)}{dt} = \frac{N_3(t)}{\tau_3} - B_{21} I_\nu \frac{g(\nu)}{c}(N_2(t) - N_1(t)) - \frac{N_2(t)}{\tau_2} \tag{1.61}$$

where c is the speed of light (3×10^{10} cm s^{-1}).

The initial condition for the N_2 state is,

$$N_2(t) = 0 \tag{1.62}$$

The N_1 state rate equation can be written for the processes in Eqs. (1.52), (1.53) and (1.54) (Eq. 1.63).

$$\frac{dN_1(t)}{dt} = \frac{N_2(t)}{\tau_2} + B_{12} I_\nu \frac{g(\nu)}{c}(N_2(t) - N_1(t)) - \frac{N_1(t)}{\tau_1} \tag{1.63}$$

The initial condition for the N_1 state is,

$$N_1(t) = 0 \tag{1.64}$$

The I_ν rate equation can be written (Eq. 1.65).

$$\frac{dI_\nu(t)}{dt} = \Delta\Omega \frac{N_2(t)}{\tau_2}\frac{c}{K} + B'_{21} I_\nu (N_2(t) - N_1(t)) - \frac{I_\nu(t)}{\tau_c} \tag{1.65}$$

where K is the number of photons per Joule, $B'_{21} = B_{21}\frac{g(\nu_0)}{c}\frac{ch\nu}{4\pi}$, the cavity lifetime is $\tau_c = \frac{2L_c}{cln\left(R_1 R_2 (1-L_i)^2\right)}$, L_c is the cavity length, R_1 is the reflectivity of mirror 1, R_2 is the reflectivity of mirror 2 and L_i are the diffraction losses.

The differential intensity calculated in this equation represents the intensity of the photon beam passing through the differential area of the differential volume. The intensity is thus the beam intensity inside of the laser cavity. If the cross section of interest is at the exit of the gain medium, than it is simply a matter of summing up the differential areas to find the total beam intensity. Note that as the cavity length increases the cavity lifetime increases. Thus the beam intensity will increase as cavity length increases. Thus the power output adjusts as the gain medium volume changes.

The initial condition for the I_ν state is,

$$I_\nu(t) = 0 \tag{1.66}$$

Equations (1.56), (1.57), (1.58), (1.59), (1.60), (1.61), (1.62), (1.63), and (1.64) are known as simultaneous differential equations. The solution can be found numerically with a chemical kinetic code such as Chemkin or using a math program such as Mathematica. The laser kinetics equations are stiff, meaning that the numerical solution is unstable unless the time step is very small. The shortest time constant in a laser kinetics model will typically dictate the time step that is needed to find a stable numerical solution.

Equation (1.57) can be rewritten in terms of finite differences (Eq. 1.62).

$$\frac{dN_0(t)}{\mathrm{d}t} \approx \frac{N_0(t+\Delta t) - N_0(t)}{\Delta t} = \left(\frac{N_1(t)}{\tau_1} - S\right) \tag{1.67}$$

Equation (1.57) can be rewritten as,

$$N_0(t+\Delta t) = \left(\frac{N_1(t)}{\tau_1} - S\right)\Delta t + N_0(t) \tag{1.68}$$

Equations (1.54), (1.56), (1.58) and (1.60) can be rewritten in terms of finite differences (Eqs. 1.69, 1.70, 1.71, 1.72).

$$N_3(t+\Delta t) = \left(S - \frac{N_3(t)}{\tau_3}\right)\Delta t + N_3(t) \tag{1.69}$$

$$N_2(t+\Delta t) = \left(\frac{N_3(t)}{\tau_3} - B_{21}I_\nu\frac{g(\nu)}{c}(N_2(t) - N_1(t)) - \frac{N_2(t)}{\tau_2}\right)\Delta t + N_2(t) \tag{1.70}$$

$$N_1(t+\Delta t) = \left(\frac{N_2(t)}{\tau_2} + B_{12}I_\nu\frac{g(\nu)}{c}(N_2(t) - N_1(t)) - \frac{N_1(t)}{\tau_1}\right)\Delta t + N_1(t) \tag{1.71}$$

$$I_\nu(t+\Delta t) = \left(\Delta\Omega\frac{N_2(t)}{\tau_2}\frac{c}{K} + B'_{21}I_\nu(N_2(t) - N_1(t)) - \frac{I_\nu(t)}{\tau_c}\right)\Delta t + I_\nu(t) \tag{1.72}$$

Equations (1.68), (1.69), (1.70), (1.71), and (1.72) can be solved numerically by a chemical kinetics program like ChemKin or with a general math program like Mathematica (which was used for the following example problems). Using the parameters of a simplified 4 level laser based on atomic carbon where the upper laser level, or state 2, is the $C(3p^1P_1)$ state and the lower laser level is the $C(3s^1P_1^0)$ state [45]. The constants used in generating Fig. 1.37 are,

$S = 8.06 \times 10^{16}$ atoms $cm^{-3}\ s^{-1}$
$\tau_3 = 7.24638 \times 10^{-7}$ s
$\tau_2 = 1.76 \times 10^{-4}$ s
$\tau_1 = 1.00 \times 10^{-7}$ s

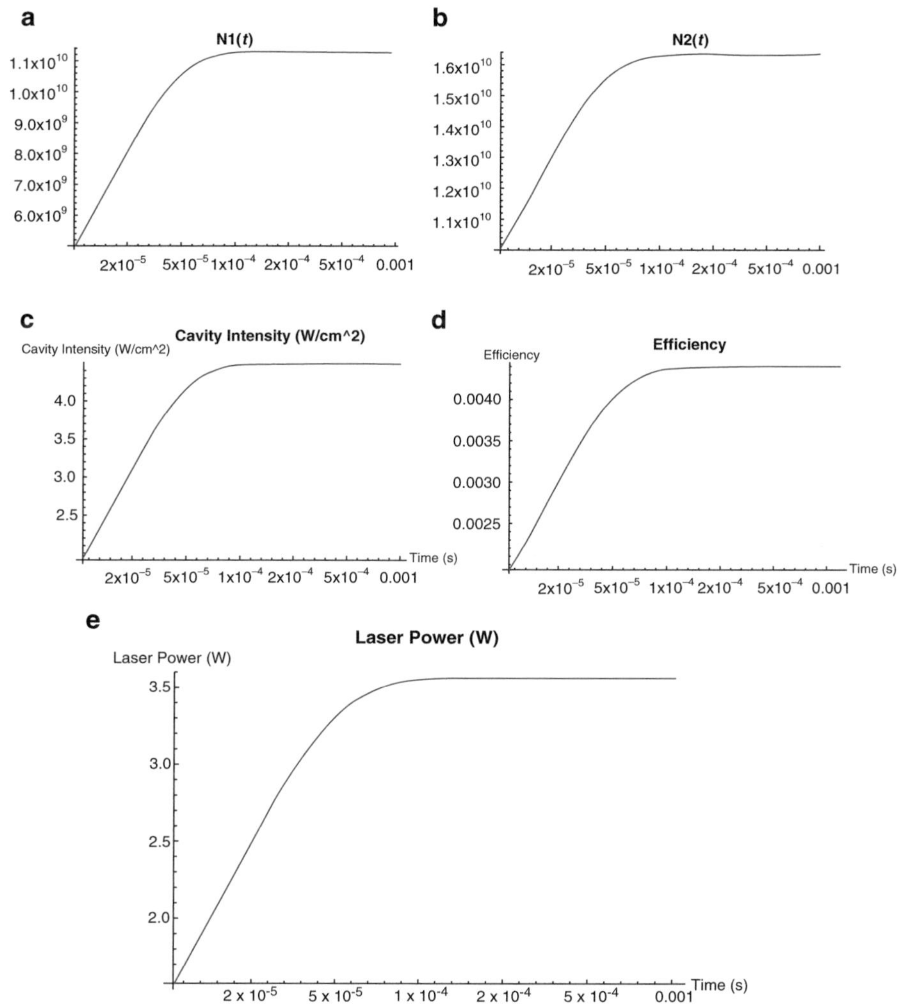

Fig. 1.37 The figure represents results for the atomic carbon laser with a pumping power density of 0.4 W/cm^3. Shown are (**a**) the lower laser level density change with time, (**b**) the upper laser level density with time, (**c**) the cavity intensity as a function of time (**d**) the laser efficiency as a function of time and (**e**) the laser power as a function of time. The cavity intensity saturates at 4.4 W/cm^2. With a 97 % reflective mirror, the laser power output is 3.5 W (with a cavity volume of 2026.8 cm^3). The laser efficiency is 0.44 %

$\tau_c = 1.32 \times 10^{-7}$ s
Laser Wavelength $= \lambda = 1.454$ μm
$g(\nu_0) = 1.06 \times 10^{-7}$ s
$\nu_0 = 2.063 \times 10^{14}\ \text{s}^{-1}$
$B_{21} = 1.05 \times 10^{-16}\ \text{cm}^3\ \text{s}^{-1}\ \text{J}^{-1}$
$\Delta\nu_N = \frac{1}{2\pi}\left\{\frac{1}{\tau_2} + \frac{1}{\tau_1}\right\} = 1.59 \times 10^6\ \text{s}^{-1}$ (Natural Broadening)

$B_{21}' = 0.00121\ cm^3\ s^{-1}$
$B_{21}\ g(\nu_0)/c = 3.71 \times 10^6\ cm^2\ J^{-1}$
$R_1 = 1$
$R_2 = 0.97$
$L_c = 100$ cm
Cavity Radius $= r_c = 2.54$ cm
Cavity Volume $= Lc*\pi(r_c)^2 = V_c = 2025.8\ cm^3$

If the time step is made small enough, then the equations converge and the buildup time of the cavity intensity can be found as well as changes in energy state densities (Fig. 1.37). As can be seen in the figure, it takes time for the laser to reach equilibrium in steady-state pumping conditions. The time constants τ_3, τ_2, τ_1 and τ_c determine the temporal characteristics and the equilibrium condition for the laser. In the case of the atomic carbon laser equilibrium occurs in the 100 μs range.

The atomic carbon laser has a low quantum efficiency (Q.E.) which is defined by Eq. (1.73).

$$Q.E. = \frac{(E_2 - E_1)}{E_3} \tag{1.73}$$

The energy level diagram for the atomic carbon laser is shown in Fig. 1.38. The pumping species is the $He(2^3S)$ state with an energy of 19.80 eV. The 1.454 μm photon has an energy of 0.8527 eV. Thus the quantum efficiency for the laser is, $Q.E. = \frac{0.8527}{19.80} = 0.0431$. A Q.E. of 4.31 % is not very high and it represents the maximum efficiency that the laser can achieve. The calculated intensity is 4.4 W/cm^2 and the laser power transmitted through the mirror is 3.5 W with an efficiency of 0.44 %. These results are consistent with the experimental data for the atomic carbon laser [45].

In contrast, the CO_2 laser (Fig. 1.19) is pumped by a collision between the $N_2(v = 1)$ state with a frequency of 2326 cm^{-1} (energy of 0.2884 eV) and ground state carbon dioxide. The 10.6 μm emission from the CO_2 laser has an energy of 0.1170 eV. Thus the quantum efficiency of the CO_2 laser is, $Q.E. = \frac{0.1170}{0.2884} = 0.406$. The four level laser model (above) was applied to a CO_2 laser using some simplifications. The model was based on a nuclear-pumped laser. This type of laser can scale with volume, thus making possible large scale systems. An NPL can operate steady-state. The model, which was simplified considerably, uses the following parameters:

S = Pump Rate $= 2.5 \times 10^{18}$ atoms $cm^{-3}\ s^{-1}$molecules $cm^{-3}\ s^{-1}$ (a pumping power density of 1 W/cm^3)
$\tau_3 = \frac{1}{1.9 \times 10^4 P_{N_2}(Torr)} s = 1.38 \times 10^{-8}$ s for $P_{N_2} = 3800\,Torr$
$\tau_2 = 1.0$ s
Nitrogen density 5 atm $= 1.345 \times 10^{21}$ molecules/cm^3
$\tau_1 = 2 \times 10^{-4}$ s (due to collisions with He, $\tau_{1spontaneous} = 1.1$ s)
$\tau_c = 1.57 \times 10^{-7}$ s

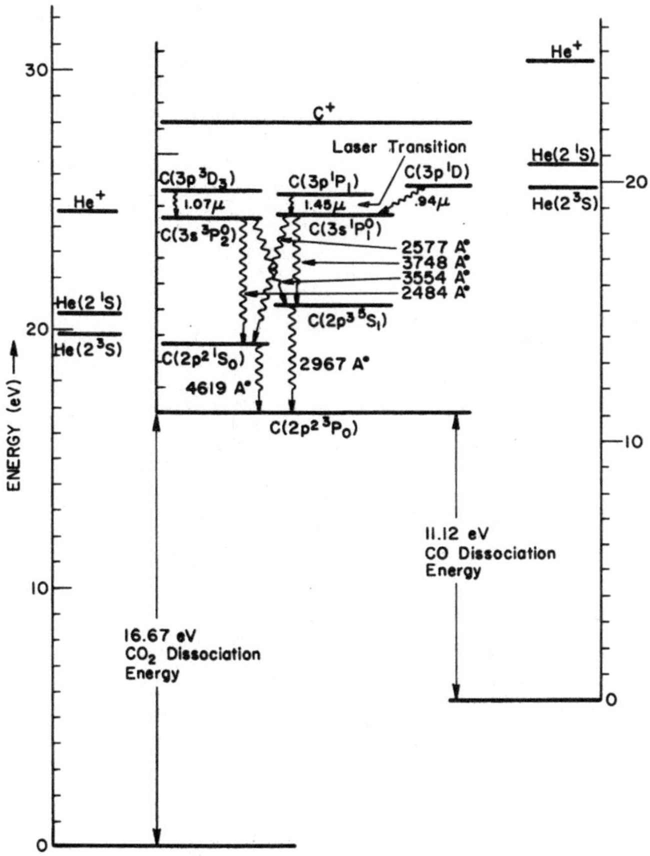

Fig. 1.38 Energy level diagram for the atomic carbon laser

Laser Wavelength = λ = 10.6 μm
$g(\nu_0) = 1.27 \times 10^{-9}$ s
$\nu_0 = 2.83019 \times 10^{13}$ s^{-1}
$B_{21} = 7.15 \times 10^{22}$ cm^3 s^{-1} J^{-1}
$\Delta\nu = 125 \times 10^6$ s^{-1} (Natural Broadening + Doppler + Pressure)
$B_{21}' = 6.83 \times 10^{-6}$ cm^3 s^{-1}
$B_{21}\, g(\nu_0)/c = 3035.4$ cm^2 J^{-1}
$R_1 = 1$
$R_2 = 0.8$
$L_c = 500$ cm
Cavity Radius = r_c = 100 cm
Cavity Volume = $Lc^*\pi(r_c)^2 = V_c = 1.57 \times 10^7$ cm^3

In running the model for the CO_2 laser, the lower laser level density, the upper laser level density and the laser cavity intensity are shown in Fig. 1.39. It should be

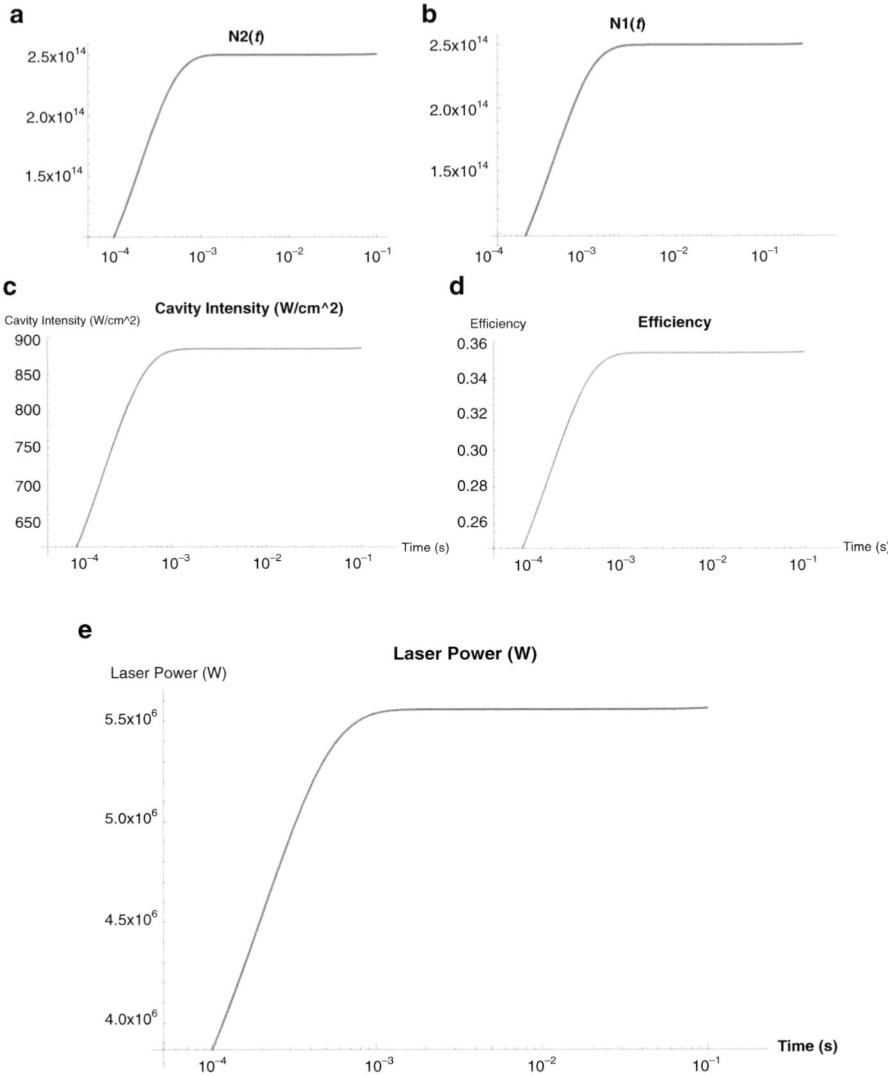

Fig. 1.39 Shown are (**a**) the upper laser level density for the carbon dioxide laser (N_2), (**b**) lower laser level density for the carbon dioxide laser (N_1), (**c**) the cavity intensity from the CO_2 laser model (I_ν), (**d**) the laser efficiency and (**e**) the laser power. Note that at equilibrium $N_2 = N_1$ and that the cavity intensity, I_ν, has saturated at 890 W/cm^2

noted that the upper laser level density and lower laser level density are in equilibrium with one another which is an indication that the laser cavity intensity has saturated. Thus the power leaking out of the laser cavity through the front mirror has extracted the maximum energy possible from the gain medium. Even though the model gives results that match the expected behavior of a typical four level laser, the simplifications used in the model lead to values that are not particularly accurate when compared to more comprehensive models for the CO_2 laser.

The pumping power density for the system is 1 W/cm^3, the laser/reactor core volume is 15.7 m^3, the output power is 1.57 MW and the laser efficiency is 35 %.

A nuclear-pumped CO_2 laser is desirable because the CO_2 laser has some attractive features such as high quantum efficiency, high gain and low pumping power density. The detail in the laser interface with the fissile fuel is challenging and will be discussed in Chaps. 4 and 5.

1.7 Fluorescence Experiments

The search for potential nuclear-pumped laser media began with spectral studies of gas mixtures excited by ionizing radiation from nuclear reactions [46–51]. For example Figs. 1.40, 1.41, and 1.42 shows a fluorescence experimental set up used at the University of Missouri.

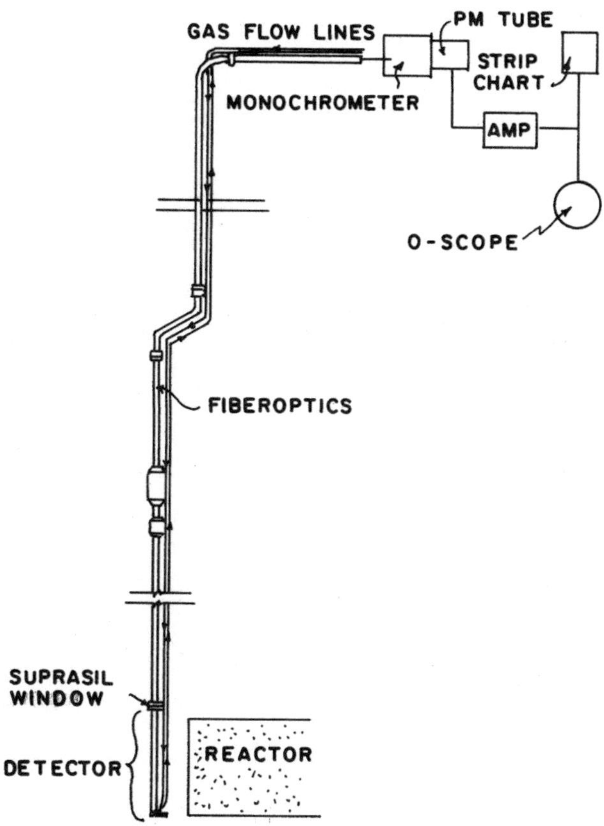

Fig. 1.40 Experimental set up for fluorescence tube using boron-10 driven by the a steady-state reactor (the University of Missouri Research Reactor) with a thermal neutron flux of 1×10^{14} n cm^{-2} s^{-1} [49]

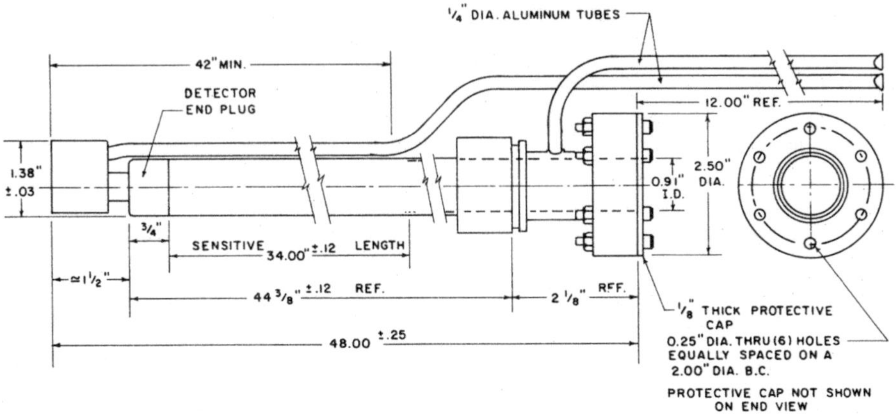

Fig. 1.41 Diagram of the fluorescence tube used in the spectral studies at the University of Missouri [49]

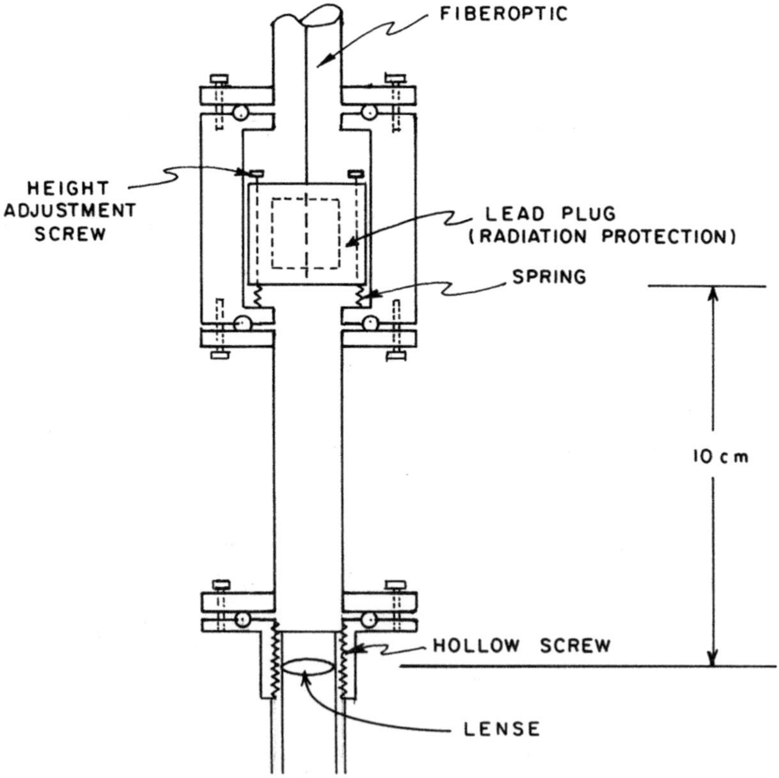

Fig. 1.42 Lens assembly used to focus light from the fluorescence tube into a fused silica (UV) fiberoptic cable [49]

A helium gas flow of 13 SCFH (using research grade gases) was established for 6 min (operating pressure 200 T) in the fluorescence tube. Afterward, the system had been pumped down to its base pressure of 10 mT to purge the reaction cell of impurities. At flow rates ranging from 1 to 30 SCFH, the emission was predominantly from atomic helium (He I), the molecular nitrogen 2P and 1N systems, atomic nitrogen (N I) and atomic oxygen (O I) as seen in Fig. 1.43. The nitrogen and neon impurity lines were due to outgassing from the walls and impurity content in the research grade helium. The molecular nitrogen lines are strong. The transfer of energy from helium to nitrogen or neon impurities is significant (given that the molecular nitrogen nuclear-pumped laser and the neon nuclear-pumped laser operate with impurity levels of nitrogen and neon in a He buffer gas [51–55]) (Fig. 1.43).

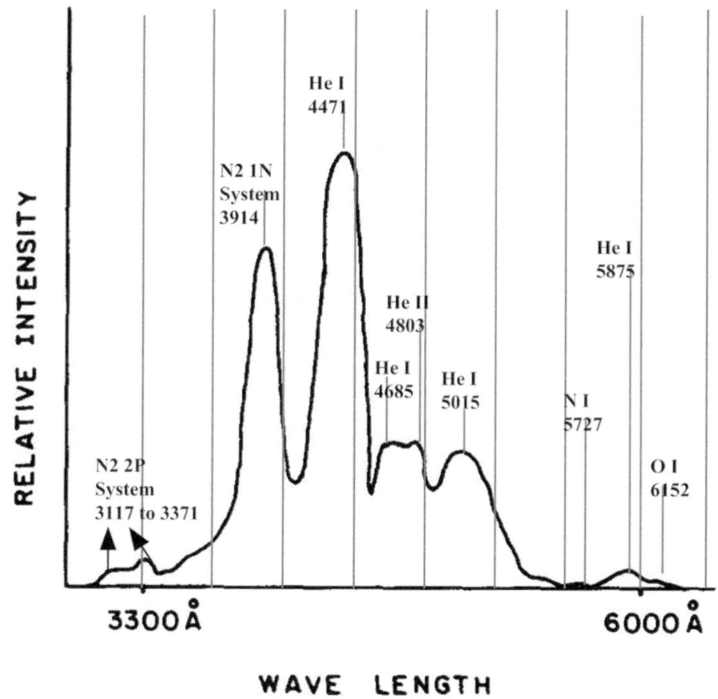

Fig. 1.43 Helium spectra from the boron-10 excited nuclear-pumped plasma using the University of Missouri Research Reactor [49]. The spectra is due to atomic helium, molecular nitrogen and atomic oxygen

When nitrogen was flowed through the fluorescence tube, the predominant fluorescing species was found to be the N_2^+ first negative band (Fig. 1.44). The second positive spectral series of N_2 was observed. About 40 % of the energy deposited in N_2 goes into the formation of N_2^+, of which ~10 % is in the $A\pi$, $B\Sigma$ and $C\Sigma$ states (Table 1.5). A much smaller fraction, on the other hand, goes into the $B\Sigma$ and $C\Sigma$ state of the neutral ~1 %. Hence, one would expect to see the first negative band ($B\Sigma_u^1 \rightarrow X^2\Sigma_g^+$). Since the second negative spectral series ($C\Sigma_u^1 \rightarrow X^2\Sigma_g^+$) lies below the usable range of the fluorescence optical system, even though the lines should exist, they were not seen. The positive transitions, which Dondes, Harteck, and Kunz observed [47] were found in the fluorescence tube experiments.

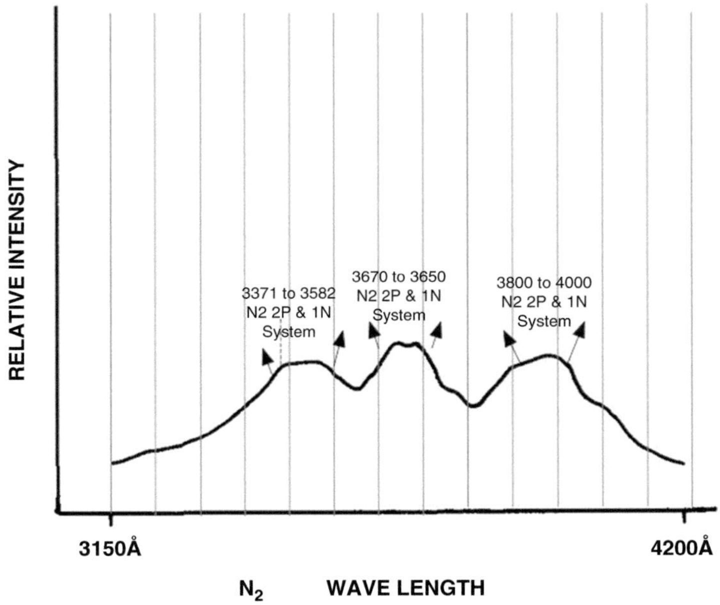

Fig. 1.44 Nitrogen spectra the boron-10 excited nuclear-pumped plasma using the University of Missouri Research Reactor [49]. (2P is the second positive band of N_2^+, 1N is the first negative band of N_2^+)

Table 1.5 Formation energy for various species of nitrogen due to the interaction of nitrogen gas with ionizing radiation

Reaction	Energy used in formation (eV/100 eV deposited in gas)
Ionic processes:	
$N_2 \rightarrow N_2^+(X^2\Sigma_g^+) + e^-$	24.1
$N_2 \rightarrow N_2^+(A^2\ \pi_u) + e^-$	6.1
$N_2 \rightarrow N_2^+(B^2\Sigma_u^+) + e^-$	2.3
$N_2 \rightarrow N_2^+(C^2\Sigma_u^+) + e^-$	2.7
$N_2 \rightarrow N_2^+(D^2\ \pi_g) + e^-$	3.2
$N_2 \rightarrow N(4S) + N^+(^3P) + e^-$	18.2
$N_2 \rightarrow N(4S) + N^+(^1D) + e^-$	
Total for ionization	56.6
Neutral processes:	
$N_2 \rightarrow N(^4S) + N(^4S)$	31.9
$N_2 \rightarrow N(^4S) + N(^1D)$	
$N_2 \rightarrow N(^4S) + N(^2P)$	
$N_2 \rightarrow N(^2D) + N(^2D)$	
$N_2 \rightarrow N_2(A^3\Sigma_u^+)$	0.16
$N_2 \rightarrow N_2(B^3\pi_g)$	0.16
$N_2 \rightarrow N_2(C^3\pi_u)$	0.67
$N_2 \rightarrow N_2(E^3\Sigma_g^+)$	0.24
$N_2 \rightarrow N_2(a^1\pi_g)$	1.56
Total for neutron excitation	34.69

Miley et al. used a boron-10 excitation source to drive mixtures of He:Xe:NF_3 [56, 57]. The mixtures demonstrated strong emission of XeF* at 353 nm. Gain experiments with the XeF* laser were also made using the He:Xe:NF_3 mixtures. The XeF* excimer laser did show gain in experiments conducted in 1977 [13]. The copious amount of high efficiency XeF* excimer fluorescence was a critical observation in this work. This high efficiency fluorescence production from excimer systems (XeF* and XeBr*) was the poignant observation which led Miley et al. to developed the first experiments that used nuclear-driven fluorescers as a flashlamp to drive a photolytic laser in 1977 [12, 58–60].

DeYoung and Weaver made spectral measurements using ^{3}He with various minority gases. The high pressure ^{3}He served as the excitation source for the plasma driven by a fast burst nuclear reactor. The ^{3}He was mixed with 1 % Ar, 1 % Xe, 0.5 % Kr, 1 % Ne, 0.0083 % Cl_2, 1.7 % F_2 and 0.1 % N_2. The nuclear-pumped spectra was basically the same as that of a high-pressure electrically pulsed afterglow plasma with the same mixture [48]. DeYoung and Weaver also examined nuclear-pumped excimer emission from XeF* and KrF* using ^{3}He as the excitation source. The experiments with XeF* were similar to the work of Miley et al. which used a boron-10 excitation source [13, 58, 60]. Both experiments demonstrated XeF* emission at 353.5 nm. The KrF* spectra demonstrated strong emission at 248.4 nm. DeYoung and Weaver also measured spectra from He and Ar plasmas excited by the $^{235}UF_6$(n,ff)FF reaction using a fast burst reactor. The spectra demonstrated quenching of the atomic excited state species by the 10 % UF_6 concentration in the mixture.

A method for absolute spectral calibration of ^{3}He mixed with Xe or Kr was developed by Boody and Prelas [61, 62]. The goal of this spectral calibration work was to develop more efficient spectral emissions from a flashlamp to drive a nuclear-pumped Nd:YAG laser and other potential solid-state laser media [63]. The experimental set up is shown in Fig. 1.45. The first step in the experiment was to determine the Radiation Induced Fluorescence (RIF) and Radiation Induced Absorption (RIA) in the windows, lenses and fiberoptics of the optical system. The RIF in the optical system was examined over a wide range of reactor power ranging from 50 kW (steady-state) to 1500 MW (pulsed). The RIF was linear with reactor power levels. A typical RIF spectra is show in Fig. 1.46. Shown in Fig. 1.47 is the effect of RIF and

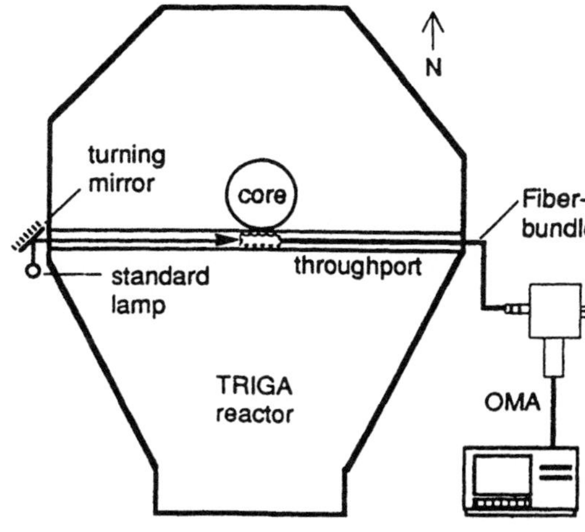

Fig. 1.45 Nuclear driven Xe, Kr and Ar lamp calibrated spectra experiments. The lamp used mixtures of ^{3}He with Xe, Kr and Ar as a minority species [63, 65]

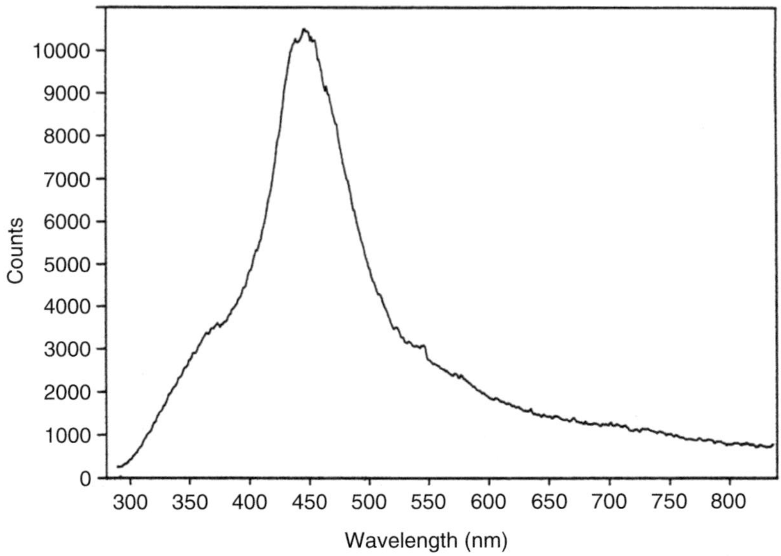

Fig. 1.46 A typical Radiation Induced Fluorescence spectrum from the optics system of the rare gas flashlamp experiment [63, 65]

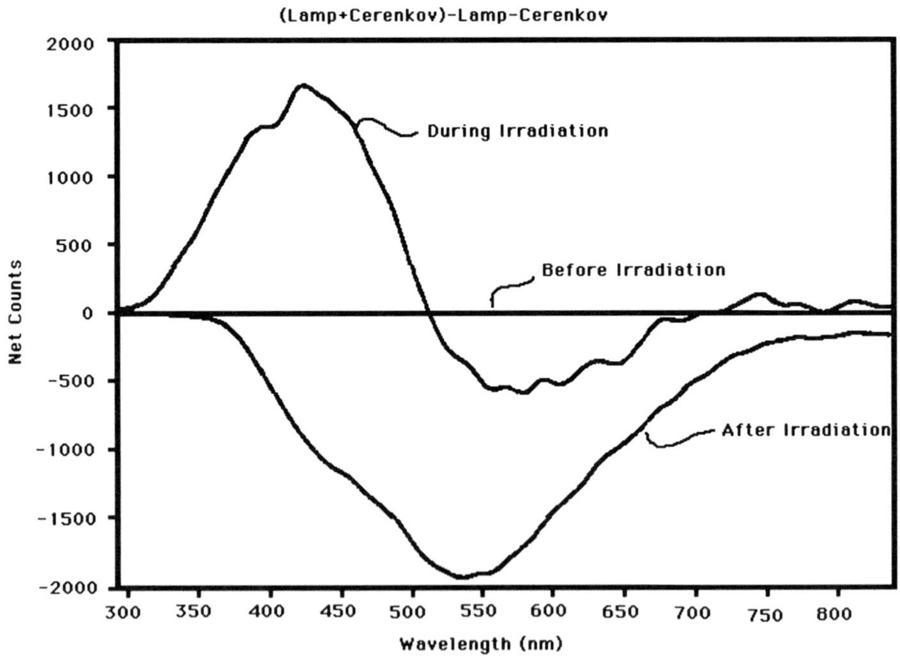

Fig. 1.47 The RIF and RIA calculated by adding the calibrated lamp and Cerenkov then subtracting the calibrated lamp emission and the calibrated lamp and the Cerenkov emission before, during and after irradiation. Before irradiation, the net counts are zero. During irradiation, the RIF is responsible for the net positive counts from 300 to 500 nm and RIA is responsible for net negative counts from 500 to 700 nm. After irradiation, the net negative counts are due to transient absorption centers in the optics caused by radiation damage [63, 65]

RIA before, during and after irradiation. This Figure shows that the effects of radiation on the optics (RIF and RIA) can be predicted for any power level. This predictability is used to find the absolutely calibrated spectra at any reactor power level by subtracting out the effects of RIF and compensating for the RIA.

A demonstration of the method of compensating for RIF and RIA is shown in Figs. 1.48 and 1.49 where the absolutely calibrated spectra for a 3He (760 Torr):Kr (100 Torr) is shown for reactor power levels of 50 kW (7 mW cm^{-3}) and 1600 MW (20 W cm^{-3}).

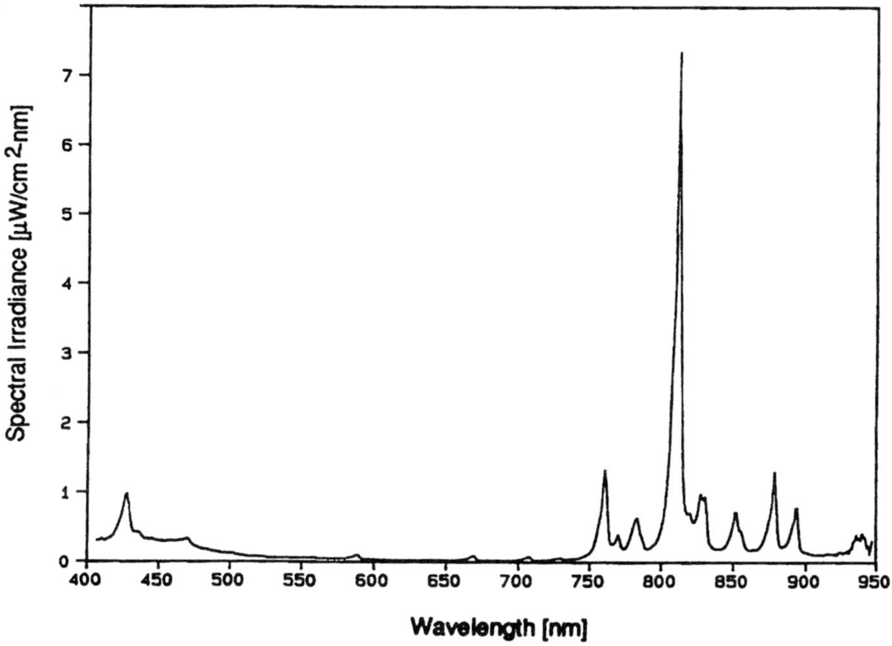

Fig. 1.48 Spectral output of an absolutely calibrated ^{3}He:Kr nuclear-pumped lamp with a power density of 7 mW cm^{-3} (reactor steady-state power level of 50 kW). This spectra is typical of a Kr flashlamp pumped by an electrical discharge [65]

Absolutely calibrated spectra were taken for ^{3}He:Ar (750 Torr:45 Torr) at various power levels. Figure 1.50 is the argon spectra taken for a 1600 MW TRIGA reactor pulse. This spectra is typical of the observed argon spectra taken at different power levels.

Absolutely calibrated spectra were taken for mixtures of ^{3}He:Xe (700 Torr: 22 Torr). A spectra taken with a 1600 MW peak TRIGA pulse is shown in Fig. 1.51. This spectra is typical of xenon at different power levels.

Absolutely calibrated spectra of pure ^{3}He (at 760 Torr) with a power density of 20 W cm^{-3} were taken (Fig. 1.52). The spectra is dominated by the N_2^+ bands. The

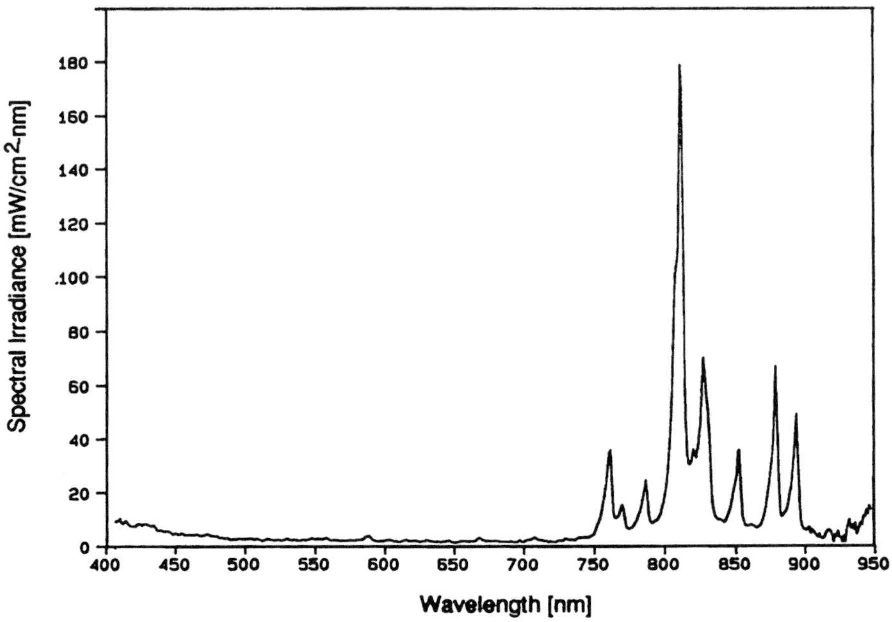

Fig. 1.49 Spectral output of an absolutely calibrated ^{3}He:Kr nuclear-pumped lamp with a power density of 20 W cm^{-3} (reactor pulse peak power of 1600 MW) [65]

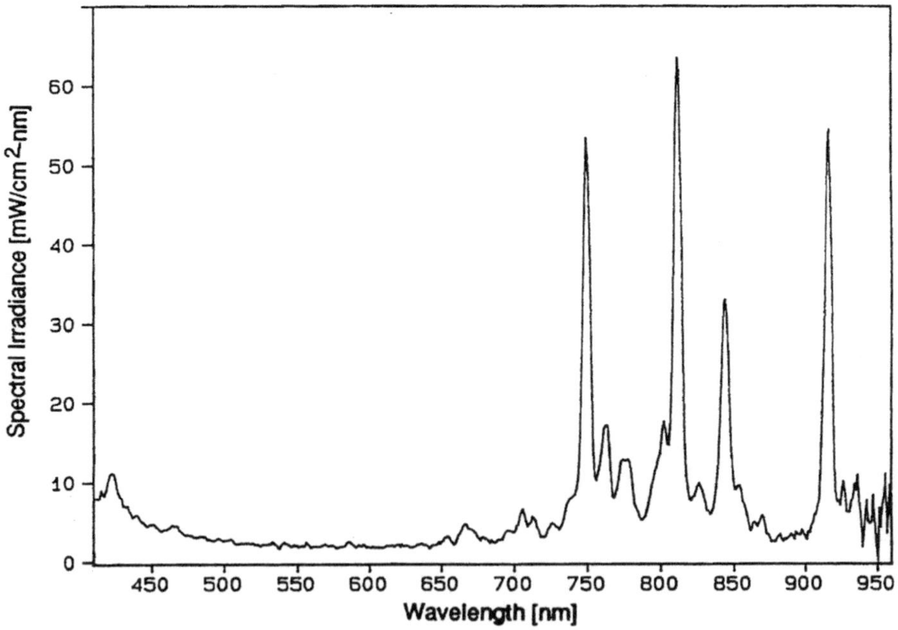

Fig. 1.50 Spectral output of an absolutely calibrated ^{3}He:Ar nuclear-pumped lamp with a power density of 20 W cm^{-3} (reactor pulse peak power of 1600 MW) [65]

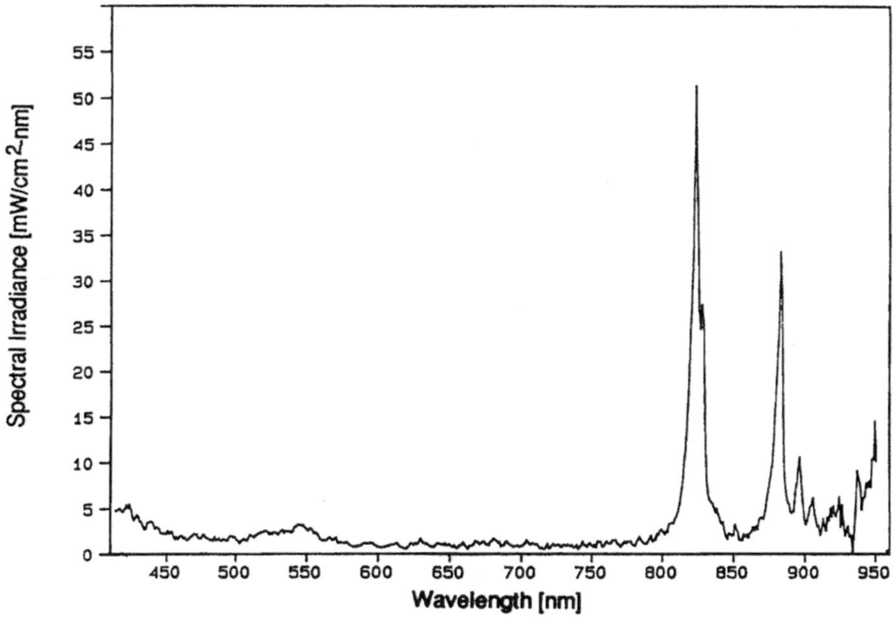

Fig. 1.51 Spectral output of an absolutely calibrated ^{3}He:Xe nuclear-pumped lamp with a power density of 20 W cm^{-3} (reactor pulse peak power of 1600 MW) [65]

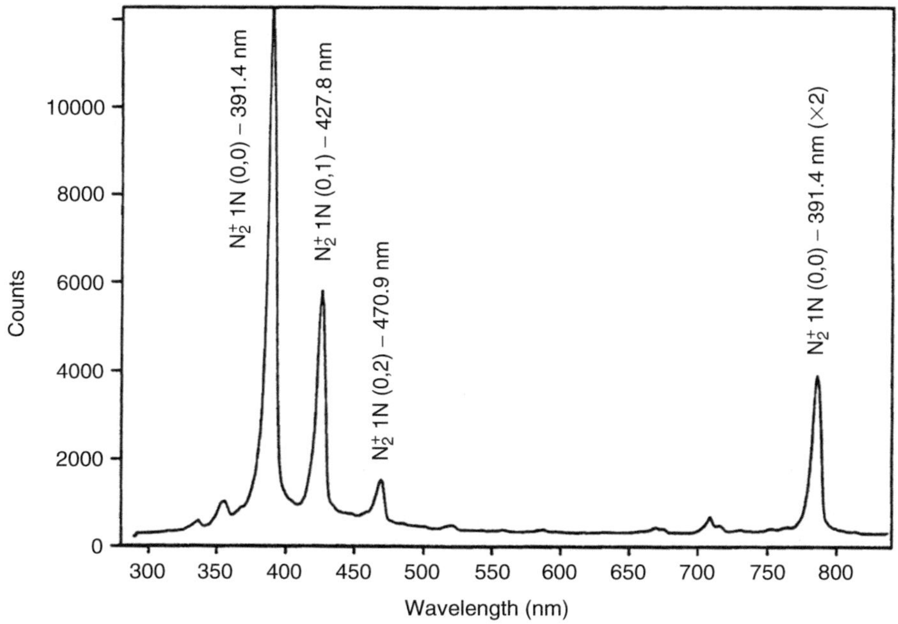

Fig. 1.52 Spectral output of an absolutely calibrated ^{3}He:N_2 nuclear-pumped lamp with a power density of 20 W cm^{-3} (reactor pulse peak power of 1600 MW) [65]

nitrogen is an impurity present in ^{3}He gas. As can be seen, Fig. 1.52 is very similar to Fig. 1.44 (nitrogen excited by boron-10). This result is significant in that spectral output tends to maintain similar structure regardless of reactor power (where spectral intensity changes with reactor power) or pumping source (^{3}He or ^{10}B).

A number of studies on nuclear driven excimer fluorescers were done in the 1980s and 1990s. In 1983, a nuclear driven xenon excimer lamp was tested at the University of Missouri Research Reactor [64]. The experiments demonstrated that the xenon excimer fluorescer at 172 nm was 49.3 % efficient which is very near the theoretical limit of 50 %. Alkali metal excimers have excellent theoretical efficiencies as shown in Table 1.6. In 1990, the sodium excimer was nuclear-pumped. The sodium had to be heated to create a partial pressure [65]. In Fig. 1.53, a heat pipe was designed as an optical cavity for the production of the sodium excimer at 436 nm. The heat pipe was placed in the through port of the University of Illinois TRIGA reactor as shown in Fig. 1.54. The ^{3}He pressure was 815 Torr. The sodium was heated to 651 C which produced a 60 Torr partial pressure of sodium vapor. Strong sodium excimer emission at 436 nm was observed.

The sodium dimer potential curve is shown in Fig. 1.55.

Table 1.6 Properties of Alkali metal excimers [66]

Excimer	Diffuse band (nm)	Main peak (nm)	Efficiency (%)
Li_2*	410–465	458	42
Na_2*	410–455	436	46
K_2*	565–578	575	42
Rb_2*	590–620	601, 603, 606	41
Cs_2*	700–720	707, 713, 718	37

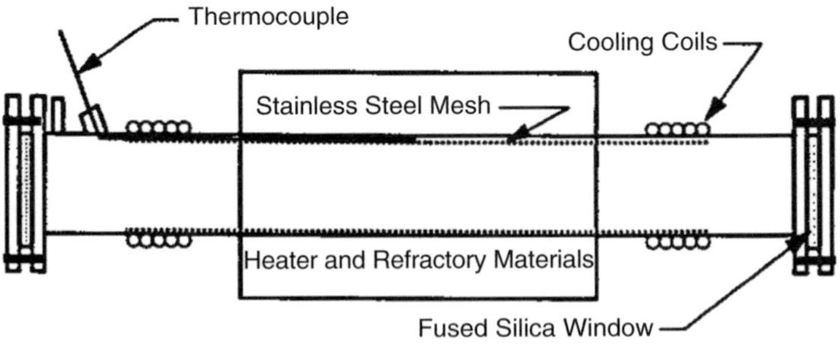

Fig. 1.53 Diagram of a ^{3}He:Na nuclear-pumped lamp used in TRIGA reactor experiment at the University of Illinois [65, 66]

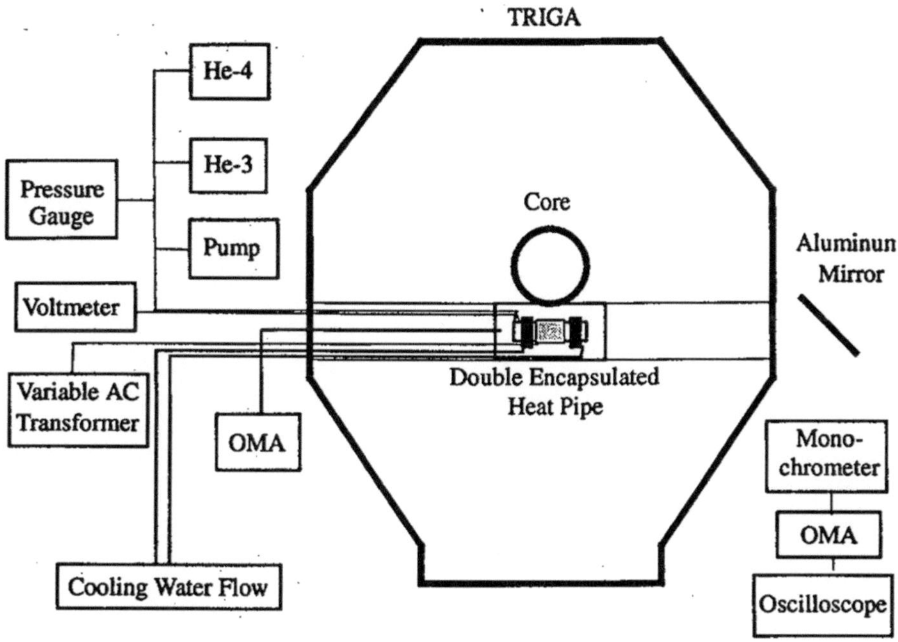

Fig. 1.54 The experimental set up of the alkali excimer laser experiment at the University of Illinois TRIGA reactor [65, 66]

Fig. 1.55 Potential energy diagram of the sodium dimer

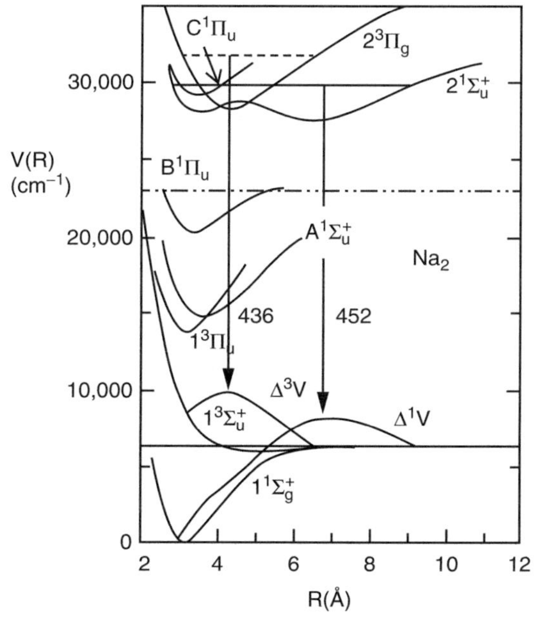

A series of nuclear-pumped sodium lamp experiments were performed on various mixtures of 3He:Na [66]. The characteristic 436 nm emission from the sodium dimer was observed in these experiments (Fig. 1.56).

These experiments demonstrate that the nuclear-excited sodium spectra are very similar to spectra taken with an electrical discharge Fig. 1.57.

In the Former Soviet Union, a number of studies similar to the ones described above were performed using radioisotopes, fission and the ^{3}He(n,p)T reaction [51]. The studies focused on rare gas spectra (He, Ne, Ar, Kr and Xe), nitrogen, carbon dioxide, chlorine, fluorine, cadmium, mercury, cesium, zinc, sodium, and potassium [51].

Fig. 1.56 Nuclear-excited sodium excimer spectrum using 810 Torr of ^{3}He with a sodium temperature of 924 K. The 436 nm line is clearly shown

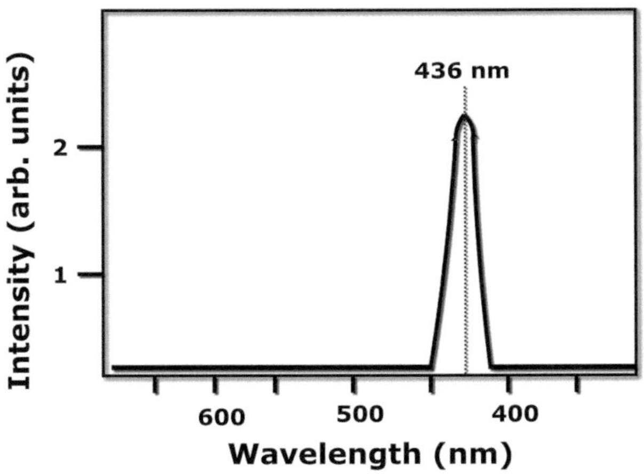

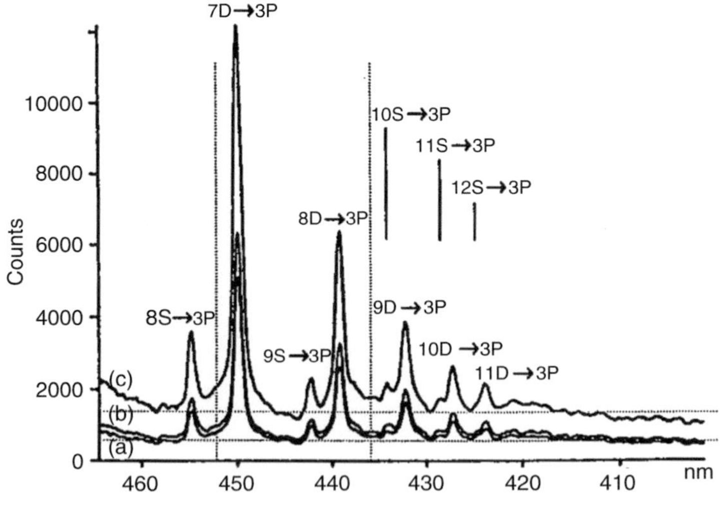

Fig. 1.57 Microwave excited sodium spectra in the ultra violet region [66]

1.8 Summary

A laser system requires a laser medium, a power source and a means of extracting stimulated emission (e.g., oscillator cavity). Ionizing radiation is capable providing the power source. Nuclear-pumped lasers are complex and their feasibility depends upon many factors such as:

- the type of ionizing radiation (ions, electrons, gamma rays and neutrons);
- the range of the ionizing radiation in the laser medium,
- how the ionizing radiation source is interfaced to the laser;
- what triggers the emission of the ionizing radiation source (i.e., spontaneous emission from a radioisotope, neutron capture using a nuclear reactor as the neutron source, or other possible triggers);
- the efficiency of the laser;
- radiation damage to components which can limit the laser lifetime;
- and the optical system which extracts the energy.

In subsequent chapters each of these issues and their impact on the nuclear-pumped laser system will be discussed.

Problems

1. Why do thermal energy conversion systems dominate other types of energy conversion systems such as those described in Fig. 1.1?
2. At what stage of an ICBM launch is the missile most effectively countered by an antiballistic missile system? Why?
3. Why did the Strategic Defense Initiative program pursue nuclear-pumped lasers?
4. What are the main advantages of using a nuclear-driven flashlamp to drive a laser?
5. Use Eqs. (1.6) and (1.7) to generate Table 1.3.
6. Use Eq. (1.9) to generate Table 1.4.
7. If a COIL laser is 10 % efficient, what mass of fuel is needed to produce an integrated beam energy of 50 MJ?
8. If a Deuterium Fluoride laser is 10 % efficient, what mass of fuel is needed to produce an integrated beam energy of 100 MJ?
9. Using the data provided for the simplified 4 level laser model based on atomic carbon, plot the natural broadened line shape. Assume that the operating temperature of the laser is 300 K, plot the Doppler broadened line shape. Assume that the laser gas is made up of 800 Torr of helium and 5 mTorr of carbon dioxide, plot the pressure broadened line shape. Which line shape dominates the atomic carbon laser?
10. Using the data provided for the simplified 4 level laser model based on atomic carbon and Eqs. (1.56), (1.57), (1.58), (1.59), (1.60), (1.61), (1.62), (1.63), and (1.64), reproduce Fig. 1.37.

11. Using the data provided for the simplified 4 level laser model based on the carbon dioxide laser and Eqs. (1.56), (1.57), (1.58), (1.59), (1.60), (1.61), (1.62), (1.63), and (1.64), reproduce Fig. 1.39.
12. Using Fig. 1.49, estimate the integrated spectral emission of the krypton lamp between 800 to 850 nm.
13. Using Fig. 1.50, estimate the integrated spectral emission of the argon lamp between 800 to 850 nm.
14. Using Fig. 1.51, estimate the integrated spectral emission of the xenon lamp between 800 to 850 nm.

References

1. Miley GH, McArthur DA, Deyoung RJ, Prelas MA (1989) Fission reactor pumped laser: history and prospects. In: Carlson JW, Behrens AD (eds) 50 years of nuclear power, National Academy of Science and NIST. American Nuclear Society, Lagrange, IL, pp 333–342
2. Prelas MA, Loyalka SK (1981) A review of the utilization of energetic ions for the production of excited atomic and molecular states and chemical synthesis. Prog Nucl Energy 8:35–52
3. Schmidt GR, Sutliff TJ, Dudzinski LA (2011) Radioisotope power: a key technology for deep space exploration, radioisotopes – applications in physical sciences (Singh N, ed). ISBN: 978-953-307-510-5, InTech. doi:10.5772/22041. Available from: http://www.intechopen.com/books/radioisotopes-applications-in-physical-sciences/radioisotope-power-a-key-technology-for-deep-space-exploration
4. Hatsopoulos GN, Gyftopoulos EP (1973) Thermionic energy conversion, vol. 1: processes and devices. The MIT Press, Cambridge, MA
5. Nelson RE (2003) A brief history of thermophotovoltaic development. Semicond Sci Technol 18:S141–S143
6. Hunt TK, Weber N, Cole T (1981) High efficiency thermoelectric conversion with beta-alumina electrolytes, the sodium heat engine. Solid State Ion 5:263–266
7. Fein ME, Verdeyen JT, Cherrington BE (1969) A thermally pumped CO_2 laser. Appl Phys Lett 14:337–340
8. Department of Energy (2013) In: I. N. Laboratory (ed) Summary of plutonium-238 production alternatives analysis final report. DOE Idaho Falls. Available: http://www5vip.inl.gov/technicalpublications/Documents/5753429.pdf
9. Deus S (2000) Tritium-powered betavoltaic cells based on amorphous silicon. In: Photovoltaic specialists conference, 2000. Conference record of the twenty-eighth IEEE, New York, NY, pp 1246–1249
10. Duggirala R, Li H, Lal A (2008) High efficiency radioisotope energy conversion using reciprocating electromechanical converters with integrated betavoltaics. Appl Phys Lett 92:154104
11. Watermann ML, Prelas MA (2013) Integrated solid-state nuclear pumped laser/reactor design for asteroid redirection. Trans Am Nucl Soc 109:1531
12. Prelas MA, Boody FP, Miley GH, Kunze JF (1988) Nuclear driven flashlamps. Laser Part Beams 6:25–62
13. Boody FP, Prelas MA, Anderson JH, Nagalingam SJS, Miley GH (1978) Progress in nuclear-pumped lasers. In: Billman K, AIAA (eds) Radiation energy conversion in space, vol 61. AIAA, New York, pp 379–410
14. APS_Study_Group_on_Science_and_Technology_of_Directed_Energy_Weapons (1987) Report to the APS of the Study Group on Science and Technology of Directed Energy Weapons. Rev Mod Phys 59:S1–S202

15. Lyons P, Clarke J, Metzger D (1974) Gamma initiated HF laser. Quantum Electron IEEE J 10:736–736
16. Ebert P, Ferderber L, Koehler H, Kuckuck R, Redhead D (1974) Amplified spontaneous emission in xenon pumped by gamma rays. Quantum Electron IEEE J 10:736–736
17. Maiman TH (1960) Stimulated optical radiation in ruby. Nature 187:493–494
18. Gordon JP, Zeiger HJ, Townes CH (1954) Molecular microwave oscillator and new hyperfine structure in the microwave spectrum of NH_3. Phys Rev 95:282–284
19. Gordon JP, Zeiger HJ, Townes CH (1955) The maser—new type of microwave amplifier, frequency standard, and spectrometer. Phys Rev 99:1264–1274
20. Wisoff PJ, Bowers MW, Erbert GV, Browning DF, Jedlovec DR (2004) NIF injection laser system. In: Proceedings of SPIE 5341, Optical engineering at the Lawrence Livermore national laboratory II: the national ignition facility, 28 May 2004. doi:10.1117/12.538466
21. President_Dwight_D._Eisenhower (1953) Atom for peace. Available: http://www.eisenhower.archives.gov/research/online_documents/atoms_for_peace.html. August 15, 2014
22. United_States_Nuclear_Regulatory_Commission (1955) Atoms for peace conference: international conference on the peaceful uses of atomic energy. U.S. Atomic Energy Commission, Geneva, August 1955
23. Kaufman S (2012) Project Plowshare: the peaceful use of nuclear explosives in Cold War America. Cornell University Press, Ithaca, NY
24. Goodchild P (2004) Edward Teller: the real Dr. Strangelove. Harvard University Press, Cambridge, MA
25. Regan PR (1983) Address to the nation on defense and national security. Available: http://www.reagan.utexas.edu/archives/speeches/1983/32383d.htm. August 16, 2014
26. Rakhimova TV, Braginsky OV, Ivanov VV, Kim TK, Kong JT, Kovalev AS et al (2006) Experimental and theoretical study of RF plasma at low and high frequency. IEEE Trans Plasma Sci 34:867–877
27. Peck M, Velez V, Ghosh T, Prelas M (2004) Generation and characterization of krypton and argon excimers from a microwave fluorescence lamp. Trans Am Nucl Soc 90:363–364
28. Ulrich A, Niessl C, Tomizawa H, Wieser J, Murnick DE, Salvermoser M (2000) Low-energy electron-beam-pumped lasers. In: Proceedings of SPIE 4071, International conference on atomic and molecular pulsed lasers III, 2, April 24, 2000. pp 2–8. doi:10.1117/12.383443
29. Harvey EC, Shaw MJ (1991) A simple kinetic model for electron-beam-pumped KrF lasers. Laser Part Beams 9:659–673
30. Sethian JD, Friedman M, Giuliani JL, Lehmberg RH, Obenschain SP, Kepple P et al (2003) Electron beam pumped KrF lasers for fusion energy. Phys Plasmas (1994-present) 10:2142–2146
31. Verdeyen JT (2000) Laser electronics. Prentice Hall, Inc, Upper Saddle River
32. Fein ME, Verdeyen JT, Cherrington BE (1969) ERRATUM: a thermally-pumped CO_2 laser. Appl Phys Lett 15:128
33. Hara H, Nakao S (1978) Enhancement of thermally pumped CO_2 laser power by addition of Ar gas. Jpn J Appl Phys 17:971
34. Hecht J (2011) Understanding lasers: an entry-level guide. Wiley, New York
35. Jirásek V, Čenský M, Špalek O, Kodymová J, Picková I, Jakubec I (2008) Chemical oxygen–iodine laser with atomic iodine generated via fluorine atoms. Chem Phys 345:14–22
36. Truesdell KA, Helms CA, Hager GD (1994) History of chemical oxygen-iodine laser (COIL) development in the USA. In: Proceedings of SPIE 2502, Gas Flow and Chemical Lasers: Tenth International Symposium, 217 (September 23, 1994), pp 217–237. doi:10.1117/12.204917
37. Grumman N (2014) Airborne Laser Testbed (ALTB), August 21, 2014, http://www.mda.mil/news/gallery_altb.html
38. Ulrich A, Busch B, Krötz W, Ribitzki G, Wieser J, Murnick DE (1993) Heavy-ion beam pumping as a model for nuclear-pumped lasers. Laser Part Beams 11:509–519
39. Ulrich A, Wieser J, Brunnhuber A, Krötz W (1994) Heavy ion beam pumped visible laser. Appl Phys Lett 64:1902–1904

40. Ulrich A, Adonin A, Jacoby J, Turtikov V, Fernengel D, Fertman A et al (2006) Excimer laser pumped by an intense, high-energy heavy-ion beam. Phys Rev Lett 97:153901
41. Adonin A (2007) Heavy ion beam pumped KrF* excimer laser. PhD, Physics, Johann Wolfgang Goethe-University, Frankfurt am Main, Germany
42. Repetti TE (1991) Application of reactor-pumped lasers to power beaming (ed), http://www.dtic.mil/dtic/tr/fulltext/u2/a338945.pdf
43. Felty JR, Lipinski RJ, McArthur DA, Pickard PS (1993) DOE reactor-pumped laser program. Department of Energy, Washington, DC
44. Ghosh TK, Prelas MA (2009) Energy resources and systems: volume 1: fundamentals and non-renewable resources. Springer, Dordrecht
45. Prelas M, Miley G (1981) Dynamics of the nuclear and electrically pumped 1.45-μm atomic carbon laser in mixtures of Helium + CO and Helium + CO2. Nucl Technol/Fusion 1(3):402–413
46. Bennett WR Jr (1962) Optical spectra excited in high pressure noble gases by alpha impact. Ann Phys 18:367–420
47. Dondes S, Harteck P, Kunz C (1966) A spectroscopic study of alpha-ray-induced luminescence in gases: part I. Radiat Res 27:174–210
48. DeYoung RJ, Weaver WR (1980) Spectra from nuclear-excited plasmas. J Opt Soc Am 70:500–506
49. Lecours MJ, Prelas MA, Gunn S, Edwards C, Schlapper G (1982) Design, construction, and testing of a nuclear pumping facility at the University of Missouri Research Reactor. Rev Sci Instrum 53:952–959
50. Melnikov SP, Puning VT, Sinyanskii AA (2008) Gas lasers with nuclear pumping: physical processes and experimental techniques. Federal Agency for Education, Moscow Engineering Physics Institute, Moscow, Russia
51. Melnikov SP, Sizov AN, Sinyanskii AA, Miley GH (2015) Lasers with nuclear pumping. Springer, New York
52. Lofthus A, Krupenie PH (1977) The spectrum of molecular nitrogen. J Phys Chem Ref Data 6:113–307
53. NIST (2015) Strong lines in helium. Available: http://physics.nist.gov/PhysRefData/Handbook/Tables/heliumtable2.htm
54. NIST (2015) Strong lines in nitrogen. Available: http://physics.nist.gov/PhysRefData/Handbook/Tables/nitrogentable2.htm
55. NIST (2015) Strong lines in oxygen. Available: http://physics.nist.gov/PhysRefData/Handbook/Tables/oxygentable2.htm
56. Miley GH, Boody FP, Nagalingham SJS, Prelas MA (1978) Production of XeF(B-X) by nuclear-pumping. In: Corcoran VJ (ed) Proceedings of the international conference on laser. Society for Optical & Quantum Electronics STS Press, 1979 – Technology & Engineering, MacLean
57. Boody FP, Miley GH (1979) Data on XeBr* which indicated a 15% efficiency. Nuclear Engineering Department, University of Illinois at Champaign-Urbana (unpublished)
58. Miley GH, Nagalingham SJS, Boody FP, Prelas MA (1978) Production of XeF(B) by nuclear-pumping. In: Proceedings of the international conference on lasers 78, Orlando Florida, Proceedings (A79-51401 23–36). STS Press, McLean, 1979, pp 5–13
59. Prelas M (1979) Nuclear pumping mechanisms in atomic carbon and in excimers. In: Nuclear-pumped lasers, pp 41–43 (SEE N80-13438 04–36), 1979. NASA Langley Research Center, Hampton, VA, pp 41–43
60. Miley GH, Prelas MA (2004) Neutron-pumped excimer flashlamp sources. In: Proceedings of SPIE 5196, Laser-generated and other laboratory X-ray and EUV sources, optics, and applications, January 7, 2004, pp 263–272. doi:10.1117/12.504396
61. Boody FP, Prelas MA (1992) Efficient visible nuclear-driven fluorescer lamps. In: Proceedings of specialist conference on physics of nuclear induced plasmas and problems of nuclear-pumped lasers, USSR, Obninsk, 05/1992, pp 161–165. doi:10.13140/RG.2.1.1599.1525

62. Boody FP, Prelas MA (1992) Absolutely calibrated spectra of nuclear driven rare gases, 400–950 nm. In: Proceedings of specialist conference on physics of nuclear induced plasmas and problems of nuclear-pumped lasers. Institute of Physics and Power Engineering, USSR, Obninsk, pp 149–155. doi:10.13140/RG.2.1.3172.0167
63. Prelas MA (1989) Nuclear-driven solid-state lasers. In: Proceedings of the international conference on lasers 1989, New Orleans, LA. doi:10.13140/RG.2.1.3178.8000
64. Prelas MA (1985) Excimer research using nuclear-pumping facilities, (ed) National Science Foundation: NSF, pp 1–131. doi:10.13140/RG.2.1.3714.7366
65. Prelas MA (1991). In: Department_of_Energy (ed) Remote pumping of solid-state lasers with nuclear driven fluorescers. DOE/ER/13029-T3, pp 1–35
66. Lin L-TS (1994) Microwave and nuclear excitation of Alkali metal vapors. PhD, Nuclear Engineering, University of Missouri, Columbia, MO
67. Prelas MA, Weaver CL, Watermann ML, Lukosi ED, Schott RJ, Wisniewski DA (2014) A review of nuclear batteries. Prog Nucl Energy 75:117–148
68. DOE-HDBK-1019/1-93 (1993) Nuclear physics and reactor theory. Department of Energy, Washington, DC
69. SPIE (2014) SPIE virtual laser exhibit: 1960–1969. Available: http://spie.org/x39920.xml. August 12, 2014
70. Zhou B, Kane TJ, Dixon GJ, Byer RL (1985) Efficient, frequency-stable laser-diode-pumped Nd:YAG laser. Opt Lett 10:62–64
71. Yariv A (1971) Introduction to optical electronics. Holt, Rinehart and Winston, Inc, New York
72. Carter AB (1984) Directed energy missile defense in space–a background paper. NTIS, Alexandria, VA
73. Cobine JD (1958) Gaseous conductors: theory and engineering applications. Dover, New York
74. Freiberg RJ, O'Clark P (1970) CO_2 transverse-discharge lasers. IEEE J Quan Electron QE-6:105–113, February 1970
75. Verdeyen JT (1995) Laser electronics. Prentice Hall, Englewood Cliffs, NJ
76. Yariv A (1976) Introduction to optical electronics. Holt, Rinehart and Winston, New York
77. Shwartz J, Gerald TW, Avidor JM (2002) Tactical high-energy laser. In: Proceedings of SPIE 4632, Laser and beam control technologies, 10 (June 5, 2002). doi:10.1117/12.469758
78. GSI (2014) Images database. Available: https://www.gsi.de/de/presse_medien/mediathek/bilderdatenbank.htm?nr=805#c4441. August 22

Chapter 2
Radiation Interaction with Matter

Abstract The properties of various types of ionizing radiation will be examined in this chapter starting with particles going from the shortest range to the longest range: heavy ions; light ions; beta particles; neutrons; and gamma rays.

Keywords Ionizing radiation • Radiation interactions • Radiation transport

2.1 Fission Fragments

The shortest transport scale lengths are those of ions, and the most massive ions are the fragments produced by fission. The transport scale length of ions decrease with the mass of the ion as will be discussed in subsequent sections. Fission occurs through spontaneous decay of a heavy atom like californium-252, which releases fast neutron energy and fission fragments; the neutron energy spectra and fission yield are shown in Figs. 2.1 and 2.2, respectively. The products of a spontaneous fission event are shown in Eq. 2.1, where ff_l is the light fission fragment, ff_h is the heavy fission fragment, ν is the statistical average number of prompt fission neutrons, n_{fast}, released during fission and are emitted with a high neutron energy distribution [1, 2]. Fission can also be initiated by a neutron capture, whereby a nucleus absorbs an incident neutron, becomes unstable, and breaks apart. An example of fission is the reaction initiated through the interaction of a thermal neutron with a fissile material, such as uranium-235, as shown in Eq. 2.2. Here n_{th} is a thermal neutron having an energy on the order of 25 meV. The reason that thermal neutrons are desired is due to the 1/v (v is velocity) dependence of the neutron capture cross section (Fig. 2.3). Thus the maximum neutron capture cross section occurs at low energies (meaning the operating temperature of the reactor). Some of thermal fission's reaction products are fast neutrons and fission fragments; the neutron energy distribution and bimodal fission yield distribution of U-235 are shown in Figs. 2.4 and 2.5, respectively. The average energy produced by particles released in the fission of U-235, including neutrons, gamma rays, beta particles and neutrinos, is shown in Table 2.1.

$$^{252}_{98}\mathrm{Cf} \rightarrow \mathrm{ff_l} + \mathrm{ff_h} + \nu \cdot \mathrm{n_{fast}} \tag{2.1}$$

M. Prelas, *Nuclear-Pumped Lasers*, DOI 10.1007/978-3-319-19845-3_2

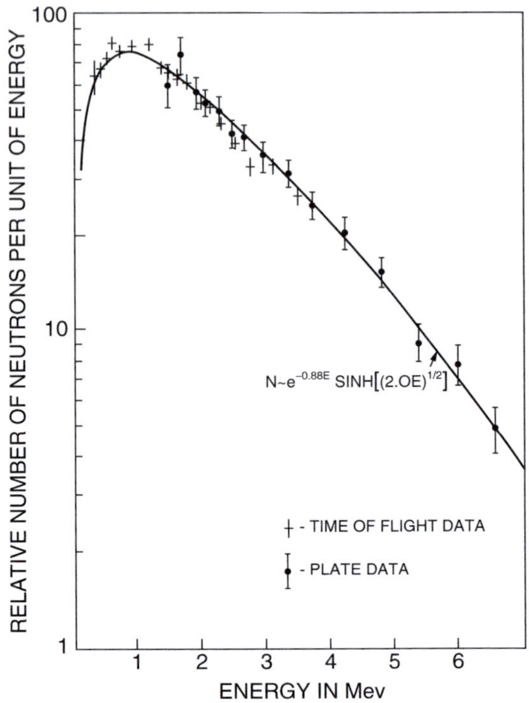

Fig. 2.1 Energy spectrum of neutrons produced by the spontaneous fission of Cf-252 [1]

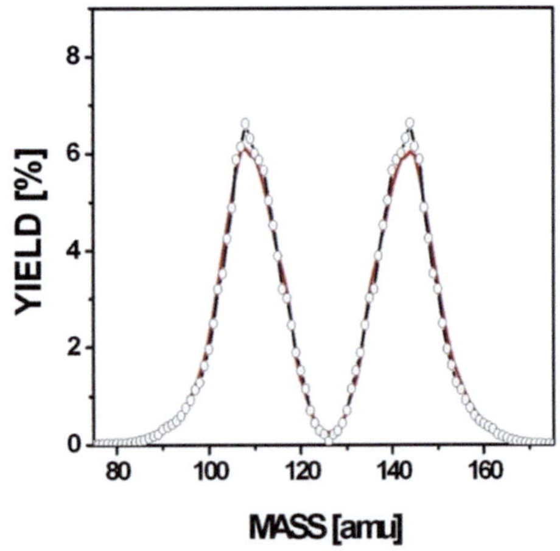

Fig. 2.2 Spontaneous fission yields of Cf-252 [3]

$$^{235}_{92}\text{U} + \text{n}_{\text{th}} \rightarrow {}^{236}_{92}\text{U} \rightarrow \text{ff}_\text{l} + \text{ff}_\text{h} + \nu \cdot \text{n}_{\text{fast}} \quad (2.2)$$

The kinetic energy of each fission fragment depends on the mass of the fragments as shown in Eqs. 2.3 and 2.4, where KE_{ff_l} is the kinetic energy of the light fission fragment, KE_{ff_h} is the kinetic energy of the heavy fission fragment, KE_{ff} is the

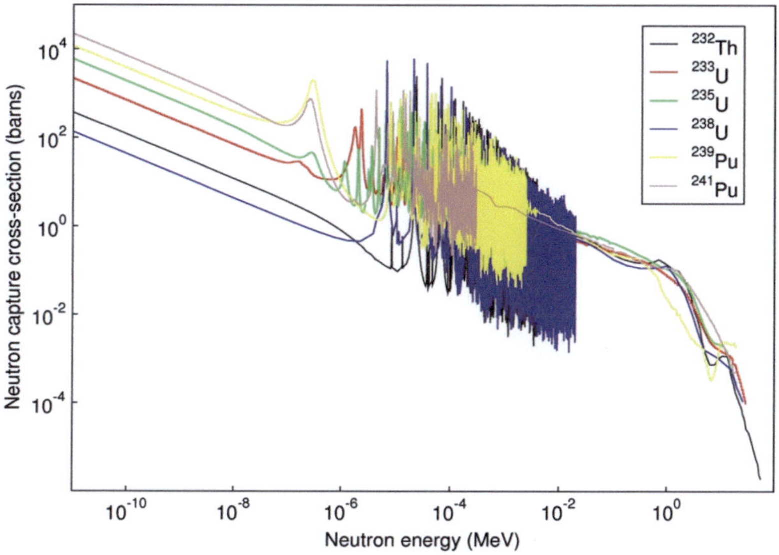

Fig. 2.3 Neutron capture cross sections for some the important actinides (Th-232, U-233, U-235, U-238, Pu-239 and Pu-241) [4]

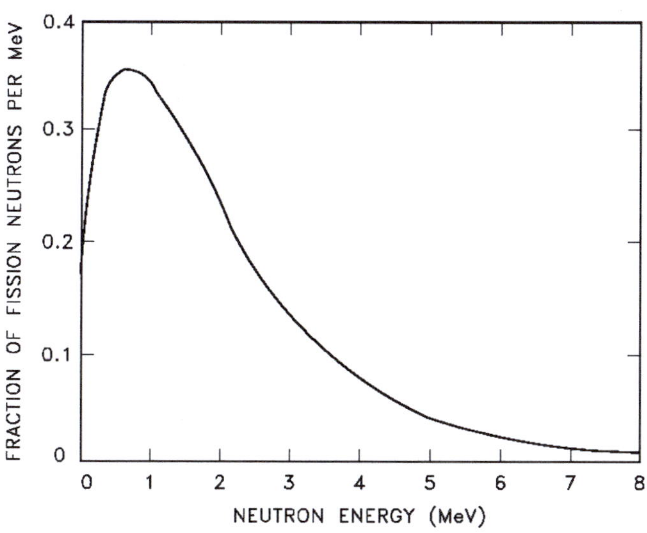

Fig. 2.4 Neutrons energy spectrum produced by the thermal fission of U-235 [5]

total kinetic energy of both fission fragments, m_h is the mass of the heavy fission fragment and m_l is the mass of the light fission fragment. The linear energy transfer of fission fragments and other swift heavy ions can be calculated using the Bethe-Bloch formula.

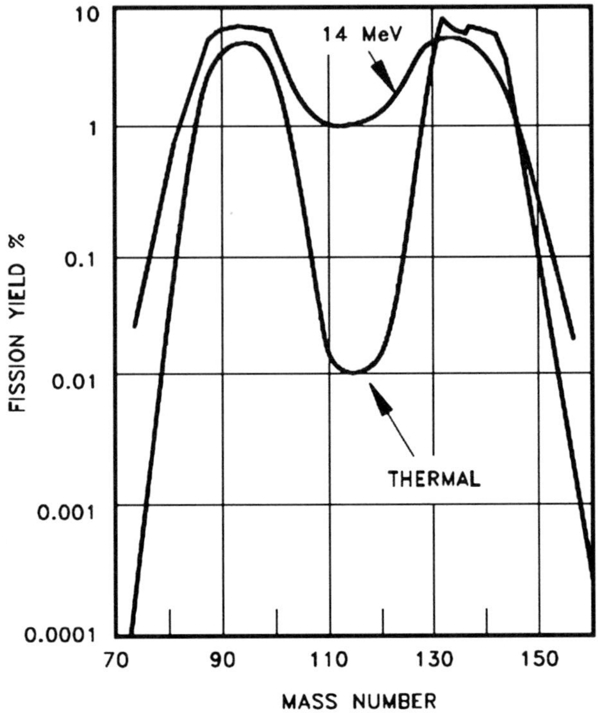

Fig. 2.5 U-235 fission yields for high- and low-energy (thermal) incident neutrons [5]

Table 2.1 Statistical distribution of energy released in the fission of U-235 [5]

Radiation	Energy in MeV
Kinetic energy of fission fragments	167
Fission neutrons	5
Prompt gamma rays	5
Delayed gamma rays from fission fragments	6
Capture gamma ray energy	10
Beta particles from fission fragments	7
Neutrinos	10
Total energy	210

$$KE_{ff_l} = \frac{m_h}{m_h + m_l} KE_{ff} \tag{2.3}$$

$$KE_{ff_h} = \frac{m_l}{m_h + m_l} KE_{ff} \tag{2.4}$$

As an example, consider the specific fission reaction of U-235 shown in Eq. 2.5 that produces La-147 and Br-87. The kinetic energies of the heavy and light fission fragments are calculated in Eqs. 2.6 and 2.7, respectively, and the energy from the fission reaction products are shown in Table 2.2. As stated earlier, the ranges of fission fragments in matter are very short due to their mass and charge; the ranges of the two fission fragments used in this example are pictured in Figs. 2.6 and 2.7. The bromine-87 atom has a range of 6.29 micrometers (μm) in uranium metal and the La-147 atom has a range of 4.22 μm in uranium metal. The spatial energy

Table 2.2 Distribution of energy released during the fission of U-235 which yields the specific fission fragments La-147 and Br-87 [6]

Radiation	Energy in MeV
Kinetic energy of fission fragments	162
Fission neutrons	6
Prompt gamma rays	6
Delayed gamma rays from fission fragments	5
Beta particles from fission fragments	5
Neutrinos	11
Total energy	195

Table 2.1 is an overall statistical yield for fission while this table is specific to a single fission reaction

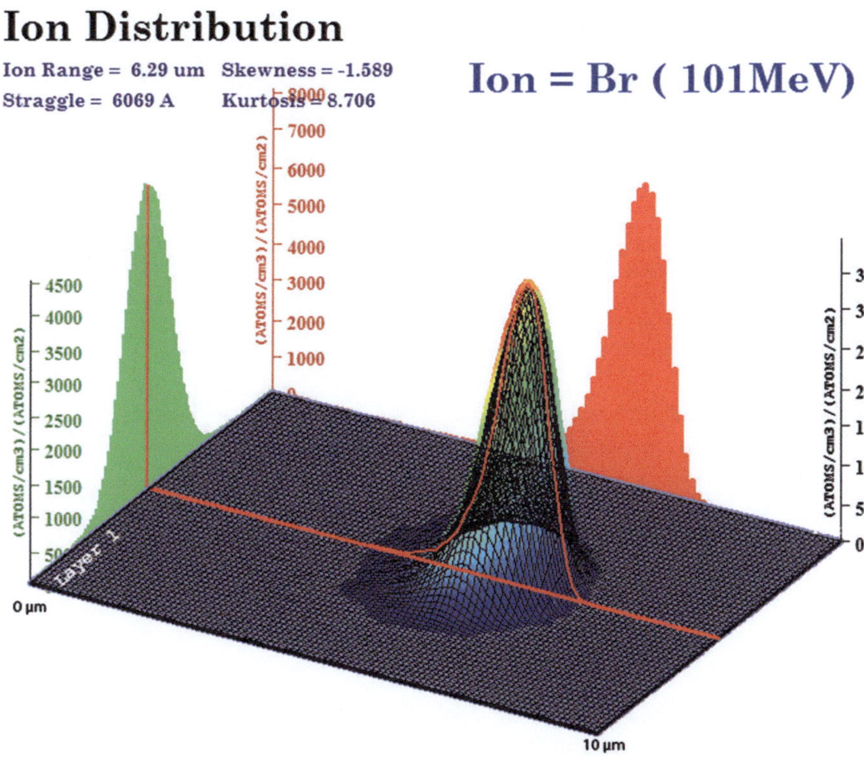

Fig. 2.6 SRIM2011 model showing ion distribution in uranium of 101 MeV bromine-87 ions slowing down [7]. The plot ordinate has units (Atoms/cm^3)/(Atoms/cm^2). By multiplying with ion dose (units of Atoms/cm^2 of bromine-87), the ordinate converts to a density distribution of Br-87 with units of (Atoms/cm^3). The ion source originates from the *left* side so the two dimensional plane indicates the depth and width of the ion distribution

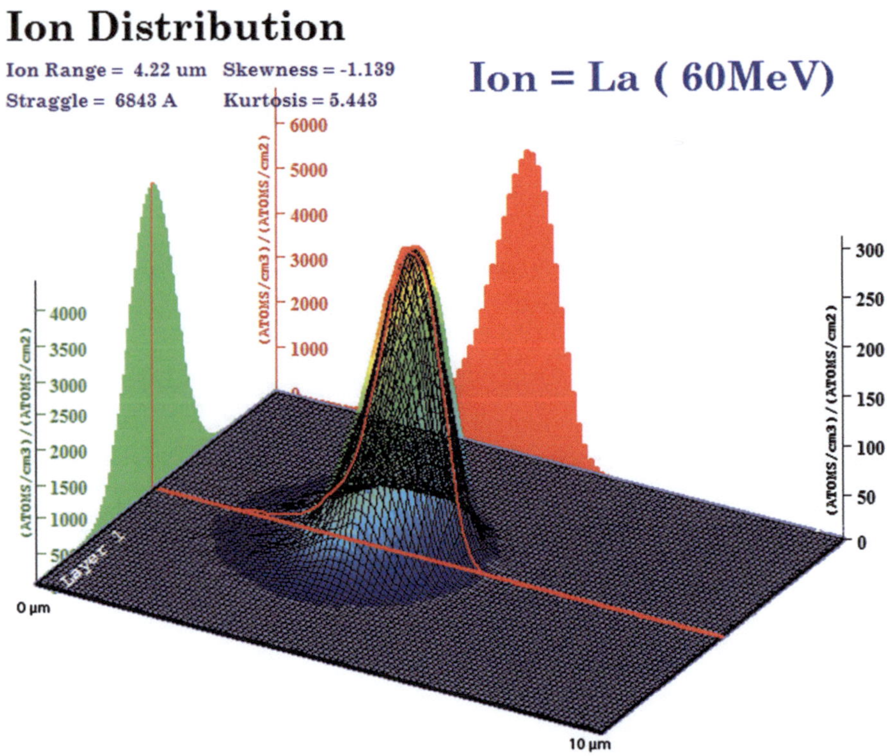

Fig. 2.7 SRIM2011 model showing final ion distribution of 60 MeV lanthanum-147 ions transported through uranium metal [7]. The plot ordinate has units (Atoms/cm^3)/(Atoms/cm^2). By multiplying with ion dose (units of Atoms/cm^2 of La-147), the ordinate converts to a density distribution of La-147 with units of (Atoms/cm^3). The ion source originates from the *left* side so the two dimensional plane indicates the depth and width of the ion distribution

distributions of the light and heavy fission fragments within the material are shown in Figs. 2.8 and 2.9, respectively.

$$^{235}_{92}\text{U} + \text{n}_{\text{th}} \rightarrow {}^{147}_{57}\text{La} + {}^{87}_{35}\text{Br} + 2\text{n}_{\text{fast}} + \text{Q}(195\,\text{MeV}) \tag{2.5}$$

$$\text{KE}_{\text{La}-147} = \frac{87}{147 + 87}162 = 60.23\,\text{MeV} \tag{2.6}$$

$$\text{KE}_{\text{Br}-87} = \frac{147}{147 + 87}162 = 101.77\,\text{MeV} \tag{2.7}$$

The fission reaction shown in Eq. 2.8 is the consequence of the interaction of a fast neutron (energy greater than 1 MeV- see Fig. 2.3) and U-238. Fast fission of U-238, for example, provides a large part of the explosive yield in a thermonuclear weapon. The energy distribution of the fast fission products is similar to that of products in thermal fission.

$$^{238}_{92}\text{U} + \text{n}_{\text{fast}} \rightarrow {}^{239}_{92}\text{U} \rightarrow \text{ff}_{\text{h}} + \text{ff}_{\text{l}} + \nu \cdot \text{n}_{\text{fast}} \tag{2.8}$$

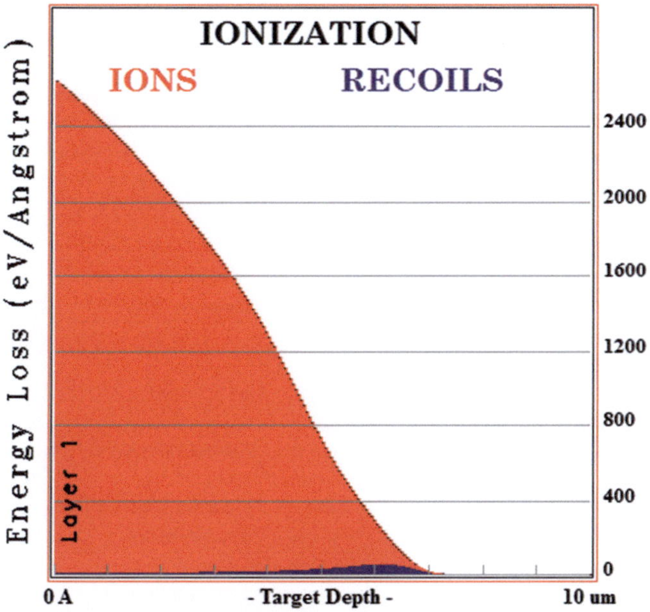

Fig. 2.8 SRIM2011 model showing target ionization in uranium metal by 101 MeV bromine-87 ions [7]. The *left* ordinate is energy loss (eV/Å), the *right* ordinate is the number of recoil atoms

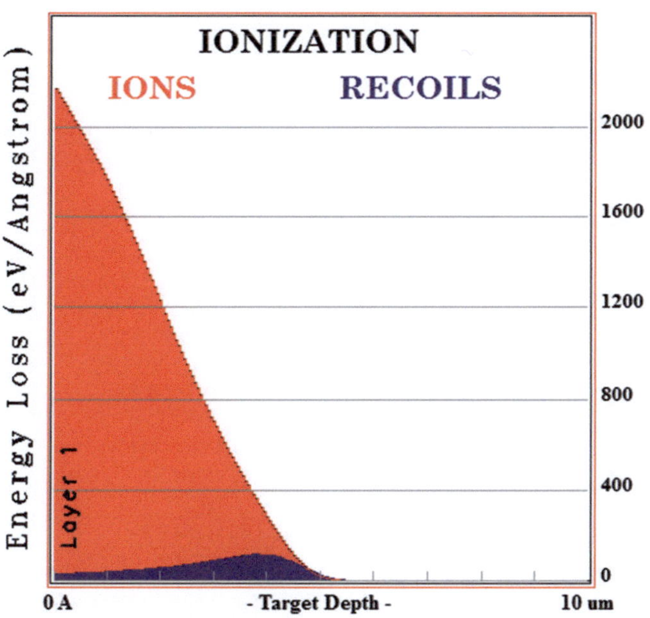

Fig. 2.9 SRIM2011 model of target ionization in uranium metal by 60 MeV lanthanum-147 ion [7] The *left* ordinate is energy loss (eV/Å), the *right* ordinate is the number of recoil atoms

2.2 Alpha Particles

Radioisotopes may be able to drive some types of low threshold power density nuclear-pumped lasers (discussed in Chap. 4). A list of alpha-emitting radioisotopes is shown in Table 2.3. Polonium-210 is used an example in the following text which illustrates the basics of alpha particle interactions (Eq. 2.9).

Table 2.3 Potential α sources for applications in direct energy conversion

Nuclide	Z	N	Decay energy (MeV)	Half life (Years)	Other emissions (MeV, %)		Production reactions
Gd-148	64	84	3.182	74.6	N/A		Sm-147(α, 3n)
							Eu-151(p, 4n)
Po-208	84	124	5.216	2.8979	β^+:	0.3783 (0.00223 %)	Bi-209(d, 3n)
							Bi-209(p, 2n)
Po-210	84	126	5.305	0.379	γ:	0.803 (0.0011 %)	Natural source
Th-228	90	138	5.52	1.9131	α:	5.340 (27.2 %)	Natural source
						5.423 (72.2 %)	
					γ:	0.216 (0.25 %)	
U-232	92	140	5.414	68.9	α:	5.263 (31.55 %)	Pa-232(β)
						5.32 (68.15 %)	Th-232(α, 4n)
					γ:	0.1–0.3 (low %)	
Pu-236	94	142	5.867	2.857	α:	5.721 (30.56 %)	Np-236(β)
						5.768 (69.26 %)	U-235(α, 3n)
Pu-238	94	144	5.593	87.74	α:	5.456 (28.98 %)	Np-238(β)
						5.499 (70.91 %)	Np-237(n, γ)
Am-241	95	146	5.638	432.2	α:	5.442 (13 %)	Pu-241(β)
						5.485 (84.5 %)	
					γ:	0.05954 (35.9 %)	
Cm-243	96	147	6.168	29.1	α:	5.742 (11.5 %)	Multiple-n capture
						5.785 (72.9 %)	U-238, Pu-239
						5.992 (5.7 %)	
						6.058 (4.7 %)	
					γ:	0.2–0.3 (20 %)	
Cm-244	96	148	5.902	18.1	α:	5.762 (23.6 %)	Multiple-n capture
						5.805 (76.4 %)	U-238, Pu-239, Am-243
					γ:	Low percentage	
Bk-248	97	151	5.793	9			Cm-246(α, pn)
Cf-250	98	152	6.128	13.07	α:	6.0304 (84.6 %)	Multiple-n capture
						5.989 (15.1 %)	U-238, Pu-239, Cm-244
					γ:	0.04285 (0.014 %)	
Cf-252	98	154	6.217	2.645	SF:	FF (3.092 %)	Multiple n capture
					α:	6.0758 (15.7 %)	U-238, Pu-239, Cm-244
						6.118 (84.2 %)	
					γ:	0.043–0.155 (0.015 %)	
Es-252	99	153	6.739	1.292	α:	6.5762 (13.6 %)	Bk-249(α, n)
						6.632 (80.2 %)	Cf-252(d, 2n)
					γ:	0.043–0.924 (25 %)	

The criteria used in identifying these isotopes is based on a half-life between 0.379 years and 100 years. Other emissions are shown such as gamma emission (for which additional shielding would be needed)

$$ {}^{210}_{84}Po_{126} \xrightarrow{t_{1/2}=138.376\,d} {}^{206}_{82}Pb_{124} + {}^{4}_{2}He^{2+}(5.305\,MeV) \tag{2.9} $$

Alpha particles are fast moving ions whose interactions with matter are governed by the Bethe-Bloch stopping power equation (Eq. 2.10).

$$ -\frac{dE}{dx} = \frac{4\pi}{m_e c^2}\frac{nz^2}{\beta^2}\left(\frac{e^2}{4\pi\varepsilon_0}\right)^2\left[ln\left(\frac{2m_e c^2\beta^2}{I(1-\beta^2)}\right) - \beta^2\right] \tag{2.10} $$

This equation is for a particle with a speed v, a charge z and an energy E that travels in the x direction in a target material with an electron density of n and an excitation potential of I; c is the speed of light, ε_o is the vacuum permittivity, $\beta = v/c$, e is the charge of an electron and m_e is the rest mass of an electron.

The range of an alpha particle (as shown in Fig. 2.10 is 9.32 μm in uranium) is greater than the range of a fission fragment in uranium metal (4.22 μm for a typical

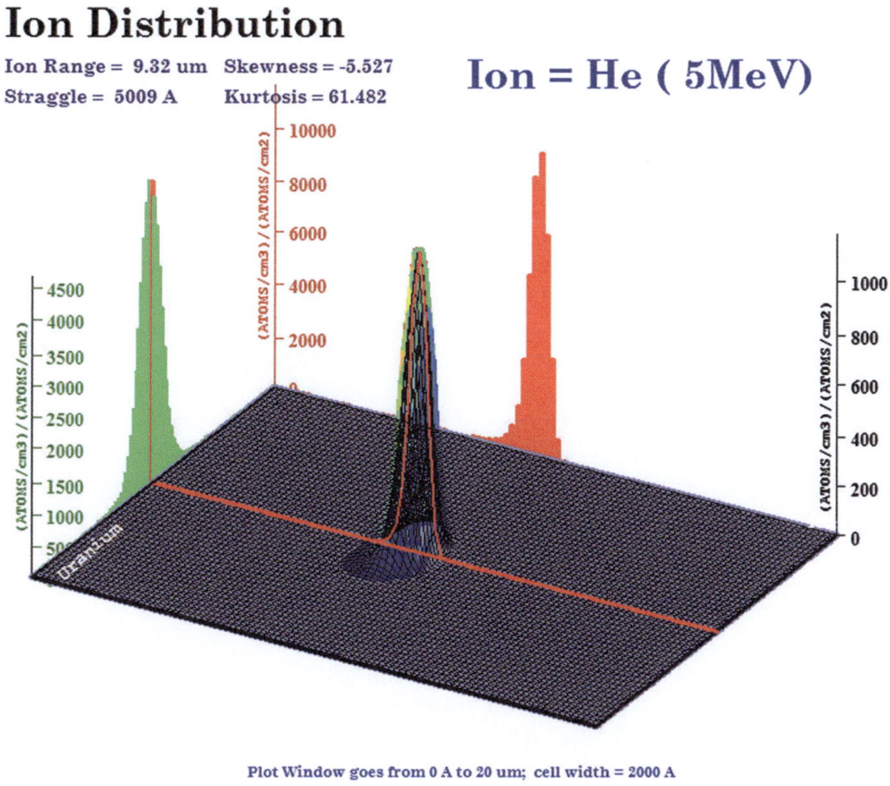

Fig. 2.10 SRIM2011 model showing final ion distribution of 5.3 MeV alpha particles in uranium metal [7]. The plot ordinate has units (Atoms/cm^3)/(Atoms/cm^2). By multiplying with ion dose (units of Atoms/cm^2 of He-4), the ordinate converts to a density distribution of He-4 with units of (Atoms/cm^3). The ion source originates from the *left* side so the two dimensional plane indicates the depth and width of the ion distribution

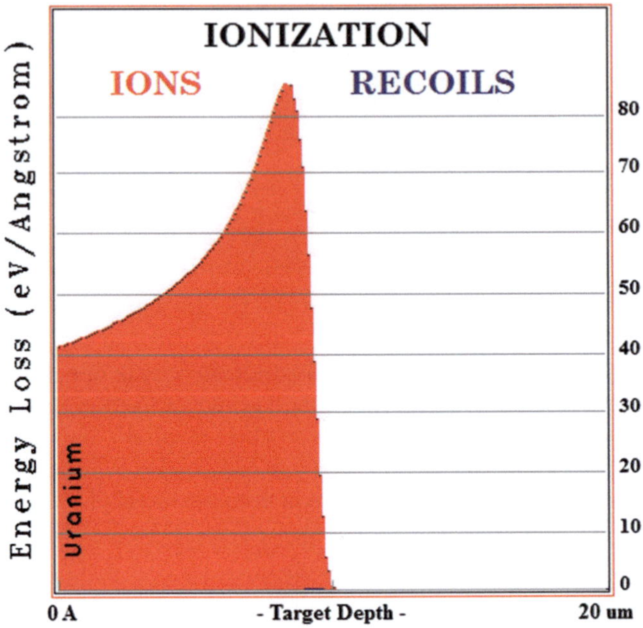

Fig. 2.11 SRIM2011 model showing target ionization produced in uranium metal by 5.3 MeV alpha particles [7] The *left* ordinate is energy loss (eV/Å), the *right* ordinate is the number of recoil atoms

heavy fission fragment and 6.29 μm for a typical light fission fragment) due to its lower charge and mass. The ionization produced by an alpha particle along its path in a solid will follow a classical Bragg curve with a Bragg peak (see Fig. 2.11), whereas a fission fragment has no Bragg peak (see Figs. 2.8 and 2.9), due to the highly changing linear energy transfer of a fission fragments as it picks up electrons during the slowing down process. Further, the range of any charged particle is a function of the electron density of the stopping material, such that less dense materials provide a lower stopping power than higher density materials. For example, the range of 5 MeV alpha particles in air is 40.6 mm (as compared to 9.32 μm in uranium metal). Therefore, it is often instructive to talk about ranges in terms of areal density, which is the linear range divided by the density of the material. The areal density is independent of the electron density of the absorbing materials and is comparable for materials of similar Z.

Alpha particles are also created by thermal neutron capture in B-10 (Eq. 2.11). This reaction has been used as a power source for some nuclear-pumped lasers (as discussed in Chap. 4).

$$^{10}\mathrm{B} + \mathrm{n_{th}} \rightarrow {}^{7}\mathrm{Li} + {}^{3}\mathrm{He} + 2.35\,\mathrm{MeV} \tag{2.11}$$

The lithium-7 ion has a kinetic energy of 0.85 MeV and the helium-3 ion has a kinetic energy of 1.5 MeV.

The three dimensional profile of the ^{3}He ion and ^{7}Li ion distributions in 2 atm of helium gas are shown in Figs. 2.12 and 2.13 respectively. The ^{3}He ion has a range of 2.88 cm in the gas and the ^{7}Li ion has a range of 1.9 cm. In the design of a nuclear-pumped laser, the diameter of the laser tube is determined by the range of the particle in the gas. It also impacts the power density in the laser medium.

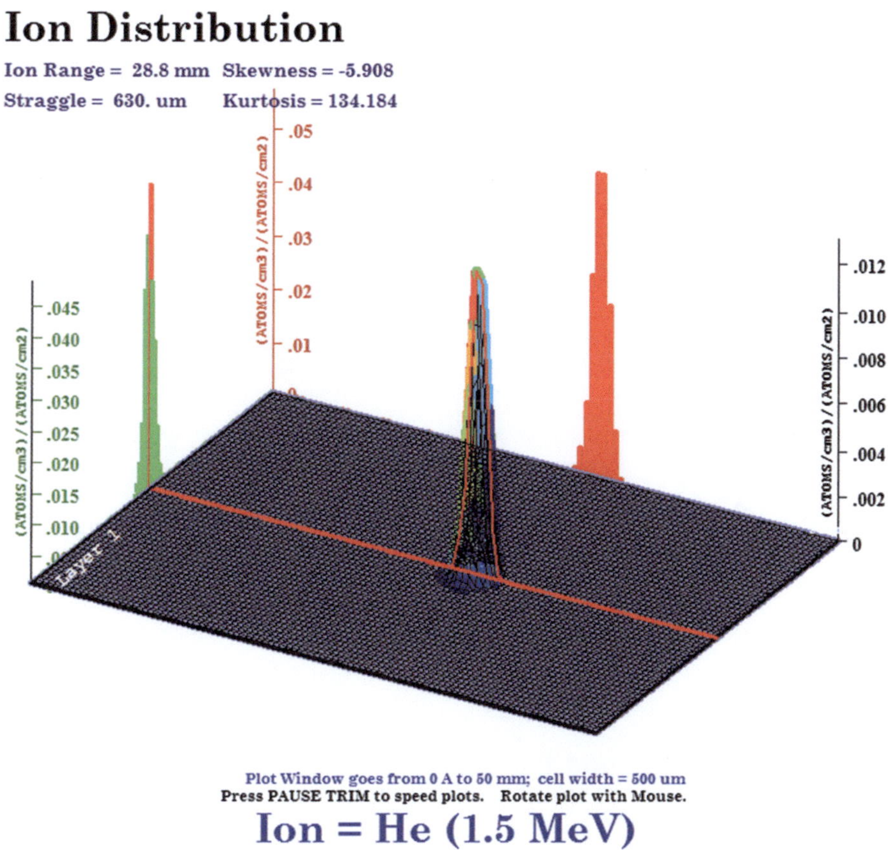

Fig. 2.12 SRIM2011 model showing final ion distribution of 1.5 MeV helium-3 ions transported through two atmospheres of He [7]. The plot ordinate has units (Atoms/cm^3)/(Atoms/cm^2). By multiplying with ion dose (units of Atoms/cm^2 of He-3), the ordinate converts to a density distribution of He-3 with units of (Atoms/cm^3). The ion source originates from the *left* side so the two dimensional plane indicates the depth and width of the ion distribution

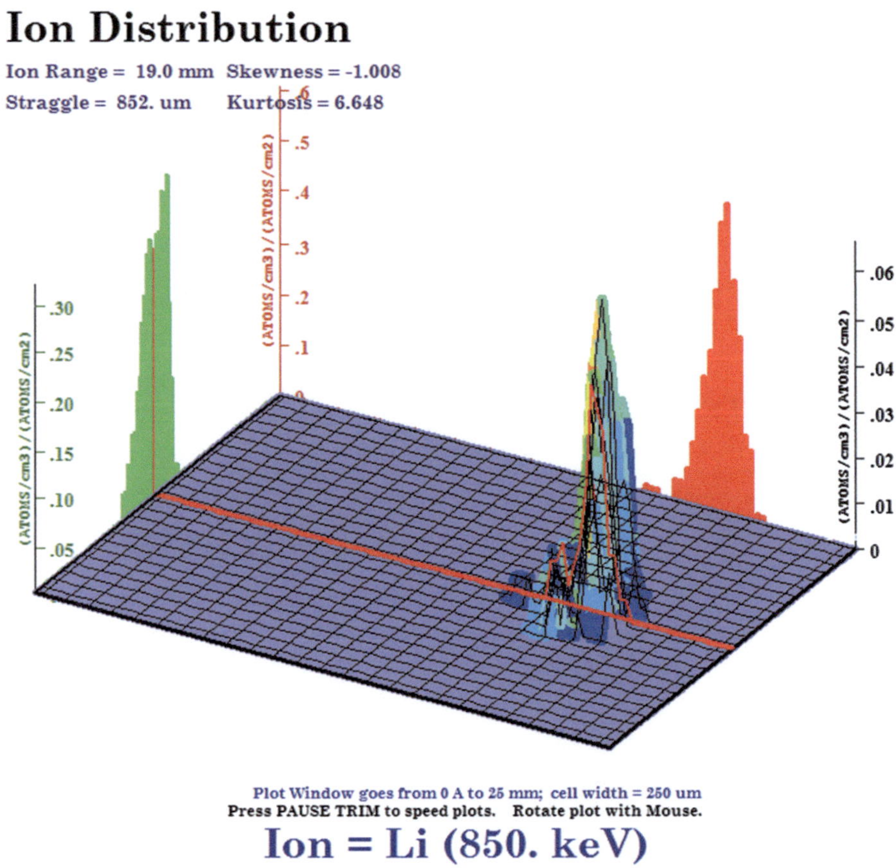

Fig. 2.13 SRIM2011 model showing final ion distribution of 0.85 MeV lithium-7 ions transported through two atmospheres of He [7]. The plot ordinate has units (Atoms/cm^3)/(Atoms/cm^2). By multiplying with ion dose (units of Atoms/cm^2 of Li-7), the ordinate converts to a density distribution of Li-7 with units of (Atoms/cm^3). The ion source originates from the *left* side so the two dimensional plane indicates the depth and width of the ion distribution

2.3 Beta Particles and Positrons

Beta particles have ranges which are substantial in gases and thus would create low power densities in a gaseous medium. On the other hand, the ranges are substantially reduced in solids with high atomic numbers (know as high Z). The power density in a high Z solid may be suitable for nuclear-pumping (possibly semiconductor lasers). The use of beta emitting isotopes produce very low power densities in gases and are not viable for driving gaseous nuclear-pumped lasers but may be feasible for driving high Z solid-state lasers. The physics of beta particle transport is discussed in order to assess the potential of beta emitters as nuclear-pumped laser drivers.

An isotope which produces an electron and antineutrino is a β^- emitter, while an isotope which produces a positron and neutrino is a β^+ emitter. Energetic electrons transfer energy to the electrons orbiting the atoms of the target material via Coulomb scattering and Bremsstrahlung emission as calculated by the modified Bethe formula. Beta-emitting radioisotopes which have useable half lives to possibly drive nuclear-pumped solid-state lasers are shown in Table 2.4.

If an isotope instead emits a positron, the β^+ will encounter some electron in orbit around an atom. The two will then mutually annihilate each other and will produce two energetic gamma rays. These gamma rays then interact with matter using mechanisms which are very different from those of electrons and ions; these interactions are covered in the prior sections.

Compared to ions (which are massive compared to electrons), the path of an electron in matter is complicated. The energetic electron has a mass equal to that of the electrons in orbit of atoms that make up the target material. Thus the electron undergoes significant scattering and follows a random walk-like path as shown in Figs. 2.14 and 2.15. The range *R* of an electron can be estimated using rules of thumb which is routine in the practice of health physics. As an example Eqs. 2.12 and 2.13 can be used to estimate the range of electrons in air. This estimate is useful to conservatively set a safe distance from a radiation source as is done in the practice of health physics but, as discussed below, no rule of thumb is accurate enough and should never be used for the design of a direct energy conversion system like a nuclear-pumped laser.

$$\mathrm{R_{air}(ft)} \approx 12\,\mathrm{ft/MeV} \tag{2.12}$$

$$\mathrm{Aurel\,Density} = R\left(\mathrm{kg/m^2}\right) = \mathrm{R(m)} \cdot \rho\left(\mathrm{kg/m^3}\right) \tag{2.13}$$

An ion pair which is produced from the interaction of beta particles with matter includes an electron which is kicked out of orbit during a Coulomb interaction; this electron is referred to as a secondary electron. Secondary electrons typically have kinetic energies in the range of keV [9]. The secondary electrons further interact with matter to create tertiary electrons through ionization and the tertiary electrons can interact with matter to create quaternary, quinary, or higher-order electrons. Because of their equal mass and charge, electrons can transfer their full energy to a target electron through Coulomb interactions. Nonetheless, the electron energy distribution in an electron beam excited plasma [10] is very similar to the spectra created by light ion bombardment (Fig. 2.16).

For example, the maximum energy for a beta particle produced by the decay of phosphorus-32 (^{32}P) is 1.71 MeV. According to Eq. 2.12 the range of a 1.71 MeV beta particle in air is $1.71 \times 12 = 20.5$ ft ≈ 21 ft. For health physics, this number would be used to determine a very conservative safe distance from a radiation hazard.

Table 2.4 Potential β^- sources for applications in direct energy conversion

Nuclide	Z	N	Decay energy (MeV)	Half life (Years)	β_{max} (MeV)	Other emissions (Units in MeV)		Production method
H-3	1	2	0.019	12.33	0.019	N/A		Li-6(n, α)
Ar-39	18	21	0.565	269	0.565	N/A		Ar-38(n, γ) KCl(n, γ)
Ar-42	18	24	0.6	32.9	0.6	N/A		Ar-40(n, γ) Ar-41(n, γ)
Co-60	27	33	2.824	5.2713	0.318	γ:	1.17 (99 %) 1.33 (0.12 %)	Co-59(n, γ)
Kr-85	36	49	0.67	10.755	0.67 (99.6 %) 0.15 (0.4 %)	γ:	0.514 (0.4 %)	Fission product
Sr-90	38	52	0.546	28.77	0.546	2.281 (Y-90, daughter)		Fission product
Ru-106	44	62	0.039	1.0234	0.039	N/A		Fission product
Cd-113 m	48	65	0.58	14.1	0.58	N/A		Cd-112(n, γ) Cd-113(n, n′)
Sb-125	51	74	0.767	2.73	0.7667	γ:	0.5 (5–20 %)	Sn-124 (n, γ)
Cs-134	55	79	2.058	2.061	0.662 (71 %) 0.089 (28 %)	γ:	0.6–0.8 (97 %)	Cs-133(n, γ)
Cs-137	55	82	1.175	30.1	1.176 (6.5 %) 0.514 (93.5)	γ:	0.6617 (93.5 %)	Fission product
Pm-146	61	85	1.542	5.52	0.795	γ:	0.747 (33 %)	Nd-146(p, n) Nd-148(p, 3n)
Pm-147	61	86	0.225	2.624	0.225	N/A		Nd-146(n, γ)
Sm-151	62	89	0.076	90	0.076	N/A		Fission product
Eu-152	63	89	1.822	13.54	1.818	γ:	0.1–0.3	Eu-151(n, γ)
Eu-154	63	91	1.969	8.592	1.845 (10 %) 0.571 (36.3 %) 0.249 (28.59 %)	γ:	0.123 (38 %), 0.248 (7 %), 0.593 (6 %), 0.724 (21 %), 0.759 (5 %), 0.876 (12 %), 1.0 (31 %), 1.278 (37 %)	Eu-153(n, γ)

(continued)

Table 2.4 (continued)

Nuclide	Z	N	Decay energy (MeV)	Half life (Years)	β_{max} (MeV)	Other emissions (Units in MeV)		Production method
Eu-155	63	92	0.253	4.67	0.147 (47.5 %)	γ:	0.086 (30 %)	Sm-154(n, γ)
							0.105 (21 %)	
					0.166 (25 %)			
					0.192 (8 %)			
					0.253 (17.6 %)			
Tm-171	69	102	0.096	1.92	0.0964 (98 %)	γ:	0.0667 (0.14 %)	Er-170(n, γ)
					0.0297 (2 %)			
Os-194	76	118	0.097	6	0.0143 (0.12 %)	γ:	0.01–0.08	Os-192(n, γ)
								Os-193(n, γ)
					0.0535 (76 %)			
					0.0966 (24 %)			
Tl-204	81	123	0.763	3.78	0.763	N/A		Tl-203(n, γ)
Pb-210	82	128	0.063	22.29	0.0169 (84 %)	γ:	0.046 (4 %)	Natural source
					0.0635 (16 %)			
Ra-228	88	140	0.046	5.75	0.0128 (30 %)	γ:	Low E (low %)	Natural source
					0.0257 (20 %)			
					0.0392 (40 %)			
					0.0396 (10 %)			
Ac-227	89	138	0.044	21.773	0.02 (10 %)	α:	4.953 (47.7 %)	Ra-226(n, γ)
					0.0355 (35 %)		4.940 (39.6 %)	
					0.0448 (54 %)	γ:	0.1 to 0.24 γ	
Pu-241	94	147	0.021	14.35	0.02082	α:	4.853 (12.2 %)	Multiple-n capture
							4.896 (83.2 %)	U-238, Pu-239

The criteria used in identifying these isotopes is based on a half-life between 1 year and 269 years. Other emissions are shown such as gamma emission (for which additional shielding would be needed)

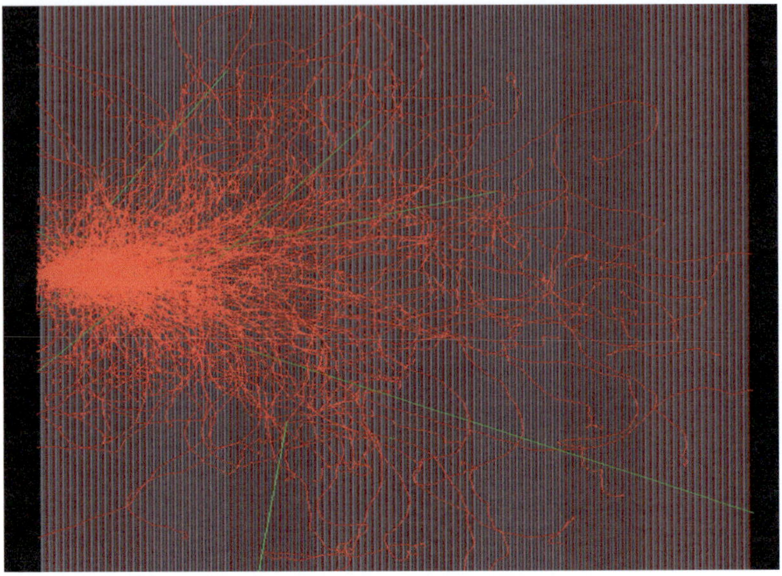

Fig. 2.14 GEANT4 simulation of Sr-90 Beta Decay into SiC of Slab, showing beta particle tracks (*random walk path*) and bremsstrahlung photons (*straight lines*) [8]

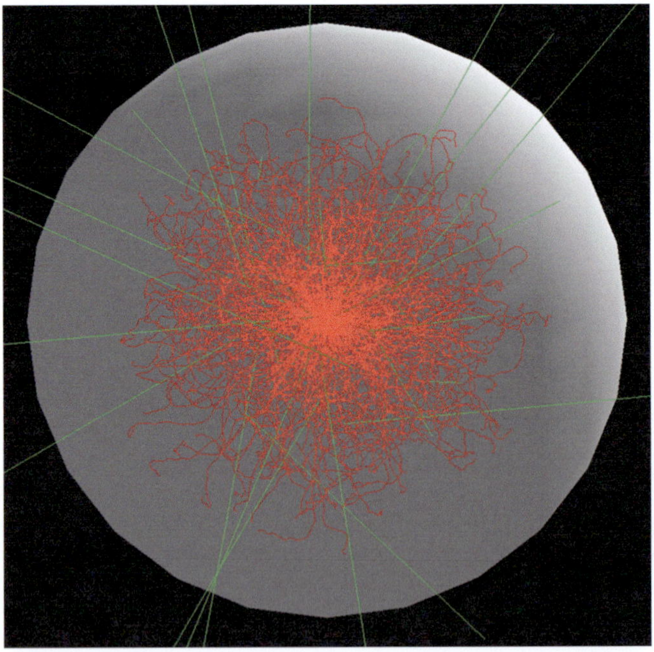

Fig. 2.15 GEANT4 simulation of Sr-90 Beta Decay into SiC of Spherical Model, showing beta particle tracks (*random walk path*) and bremsstrahlung photons (*straight lines*) [8]

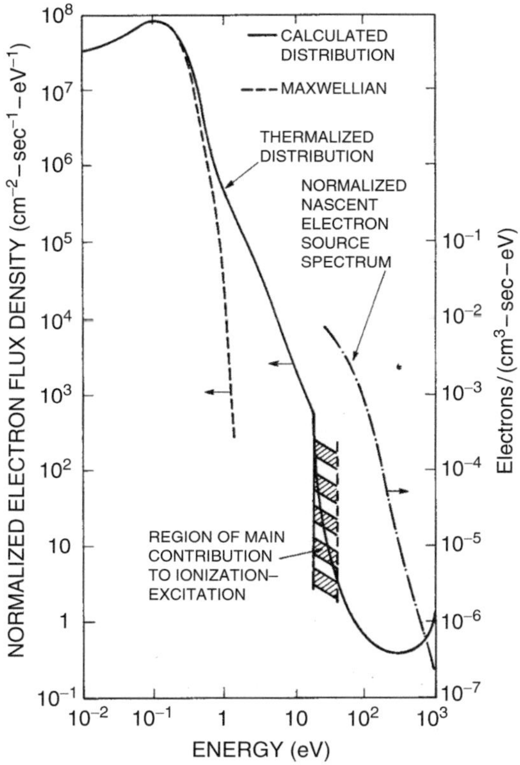

Fig. 2.16 Nascent and asymptotic electron energy distributions for 1-MeV alpha particle bombardment of helium [9]

There are also rules of thumb calculations used by professionals to estimate (over estimate) shielding thicknesses. The concept of areal density is the basis for one such method. Areal density can used to conservatively estimate the range of a beta particle in a target material. The areal density is related to the electron density of a given material which is dependent upon the atomic density and atomic number of that material. Since beta particles interact with the electrons which make up the atom, it also is used to estimate the amount of energy lost by the beta particle as a function of distance (μm^{-1}). Materials with higher electron density have more electrons for the beta particle to interact with and thus will stop the electron in a shorter distance. So is the areal density is calculated for a specific material using Eq. 2.13 where ρ is the mass density of the material in kg/cm^3 and R is the range of the particle in the material in meters. In many texts, the areal density is also referred to as 'the range' and given the symbol R. The reader must sort out how the term 'range' is used in each reference. There are rules of thumb for areal density which are shown in Eqs. 2.14, 2.15, 2.16, 2.17, and 2.18, where E_{max} is the maximum beta energy in MeV, and areal density is given in g/cm^2 (and uses the symbol R).

$$E_{max} \approx 2 \cdot R(g/cm^2) \text{ for } 1 \leq E_{max} \leq 4\,\text{MeV} \quad (2.14)$$

$$E_{max} \approx 1.92 \cdot R^{0.725} \text{ for } R \leq 0.3\,g/cm^2 \quad (2.15)$$

$$R(g/cm^2) \approx 0.407 \cdot (E_{max})^{1.38} \text{ for } E_{max} \leq 0.8\,\text{MeV} \quad (2.16)$$

$$E_{max} \approx 1.85 \cdot R + 0.245 \text{ for } R \geq 0.3\,g/cm^2 \quad (2.17)$$

$$R(g/cm^2) \approx 0.542 \cdot E_{max} - 0.133 \text{ for } E_{max} \geq 0.8\,\text{MeV} \quad (2.18)$$

Models which calculate energy deposition using rules of thumb to find areal density, and therefore particle ranges, will introduce significant errors if used in nuclear-pumped laser power density calculations. The rules of thumb presented above overestimate particle range in order to ensure adequate radiation protection. For example, consider the beta emissions of sulfur-35 ($^{35}_{16}S$), strontium-90 ($^{90}_{38}Sr$), and yttrium-90 ($^{90}_{39}Y$). The maximum energy of the beta particle emitted by ^{35}S is 0.167 MeV, which results in a required areal density of (0.407) $(0.167)^{1.38} \approx 0.034$ (g/cm^2) by Eq. 2.14. For a silicon carbide target with density $\rho = 3.210$ g/cm^3, the estimated range would be R(cm) $\approx R(g/cm^2)/\rho \approx 0.034/3.210 \approx 0.0106$ cm. The maximum energy of a beta particle emitted by ^{90}Sr is 0.546 MeV, which has an estimated areal density of (0.407) $(0.546)^{1.38} \approx 0.177$ (g/cm^2), and the estimated range is R(cm) $\approx R(g/cm^2)/\rho \approx 0.177/3.210 \approx 0.0551$ cm. The 2.28 MeV maximum energy of the beta particle emitted by ^{90}Y is considerably higher than the two previous examples. Eq. 2.18 has an areal density of $2.28/2 \approx 1.14$ (g/cm^2) to stop the particles, which equates a range in silicon carbide of R(cm) $\approx R(g/cm^2)/\rho \approx 1.14/3.21 \approx 0.355$ cm.

When two rules of thumb cover the same energy space, there may be significant difference between them. For example, Eq. 2.18 covers beta energies greater than 0.8 MeV and Eq. 2.14 covers beta energies between 1 and 4 MeV. If Eq. 2.18 is used to estimate the range for a Sr-90 beta, it yields $0.546 \times 2.280 - 0.133 \approx 1.103$ (g/cm^2) and a consequent range in cm of approximately equal to R(cm) $\approx R(g/cm^2)/\rho \approx 1.103/3.210 \approx 0.344$ cm. The difference between the two rules of thumb (Eqs. 2.14 and 2.18) is about 11 μm, which can be significant in the design of an energy conversion device. To reiterate, rules of thumb should never be used to design an energy conversion system.

In addition to the above rules of thumb, there are other simplifications that will misrepresent the energy deposited in a medium and will lead to significant errors in the design of energy conversion devices. One common, but incorrect, assumption is to use the average beta particle energy calculated with the rule of thumb $\beta_{average} = 1/3\ \beta_{max}$ [11]. This simplifies the calculation of spatial energy deposition dependence for an energy conversion device, but it is severely inaccurate. The ionization profile (power density profile) produced by the true spectrum of beta particles emitted from a radioisotope is significantly different from the results of either the simple rules of thumb described above or the average beta energy being 1/3 β_{max}. A second fatal simplification arises when it is assumed that the ionizing particles from a nuclear reaction are not emitted isotropically. The lack of an accurate transport model creates shifts in the energy deposition profile within the energy transducer, so a higher fraction of the incident energy is deposited deeper within the device. This is not what happens [12]. A realistic model for ionizing radiation sources from nuclear reactions is essential to accurately calculate the energy deposition within the energy conversion device. Three example beta-decay reactions with low-, medium- and high-energy are shown in Eqs. 2.19, 2.20, and 2.21.

$$^{35}_{16}S \rightarrow {}^{35}_{17}Cl + \beta^- + \bar{\nu} + 167.47\,\text{keV} \tag{2.19}$$

$$^{90}_{38}Sr \rightarrow {}^{90}_{39}Y + \beta^- + \bar{\nu} + 546\,\text{keV} \tag{2.20}$$

$$^{90}_{39}Y \rightarrow {}^{90}_{40}Zr + \beta^- + \bar{\nu} + 2.28\,\text{MeV} \tag{2.21}$$

Table 2.5 shows pertinent data for the three beta emitters, including half-life, the average beta energy according to the commonly-used rule of thumb (1/3 β_{max}), and the average beta energy calculated exactly from the beta spectrum. As can be seen in Fig. 2.17, the sulfur-35 beta spectrum intensity continuously increases as energy decreases, the beta spectrum intensity of the medium-energy strontium-90 emitter tends to flatten out at low energies, and the high energy yttrium-90 beta spectrum intensity has a distinct maxima and then drops as energy decreases. As can be seen in Table 2.5, the differences between the average energy calculated by the 1/3 β_{max} rule of thumb and the average energy calculated directly from the beta spectrum differs significantly as the maximum energy of the beta particle increases. If the

Table 2.5 Characteristics of common beta-emitting radioisotopes

Isotope	Half-life	Max energy	Average energy: 1/3 βmax rule	Average energy: spectrum	% Difference	Daughter isotope
S-35	87.51 days	167.47 keV	55.8 keV	53.1 keV	+5	Cl-35
Sr-90	28.8 years	546 keV	182 keV	167 keV	+9	Y-90
Y-90	2.67 days	2.28 MeV	760 keV	945 keV	−20	Zr-90

The average energy is calculated using the 1/3 β_{max} rule and using a full spectrum analysis. The differences in the average energy are substantial for high energy beta sources

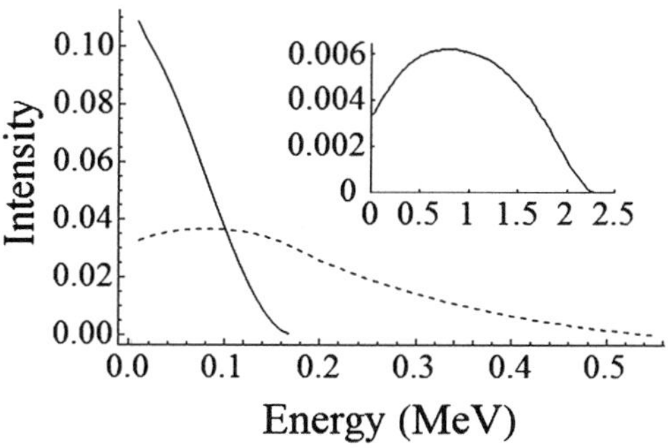

Fig. 2.17 Beta emission energy spectra for S-35 (*solid*), Sr-90 (*dashed*), and Y-90 (*inset*) [8]

average beta energy is calculated using the 1/3 β_{max} rule, the error inherent in using the rule is propagated through the rest of the system calculations. These incorrect average energies will then be used to incorrectly find the particle range and stopping power. Thus to reiterate, the 1/3 β_{max} rule should not be used for design calculations and modeling of nuclear energy conversion devices.

Accurate range calculations are essential when designing a nuclear energy conversion device in order to match the active region of the transducer (L_{trans}) to the optimum position to harvest the energy from the beta particle (λ_{RadTr}). To calculate the range of a beta particle in matter, the full beta energy spectrum should be used in the model (Fig. 2.17). Calculations which use the complete beta spectrum lead to accurate energy deposition profiles. This is clearly shown in the following example. A beta source irradiates silicon carbide semiconductor material. The actual range of beta particles from S-35, Sr-90 and Y-90 decay have been calculated for a beam of beta particles hitting a slab and for a point source in the center of a sphere [8]. These results are significantly different from results that use the average beta energy calculated from the beta spectrum. This further reinforces the statement that unacceptable inaccuracies come from the use of any rules of thumb (Eqs. 2.15, 2.16, 2.17, and 2.18) and the use of averages rather than a full spectrum when designing a nuclear energy conversion device. Results from the calculations using rules of thumb are several orders of magnitude larger than calculations based on the average beta energy (using the average based on the beta spectrum) or the full beta spectrum. It is interesting that there is about a factor of four difference between the range calculated from average beta energy and the range calculated with the full beta spectrum, where the range for the full beta spectrum is greater.

Figures 2.18 and 2.19 represent the energy deposition as a function of distance for a calculation based on the average beta energy and a calculation based on the full beta spectrum, respectively. In the slab geometry, the beta particles were simulated as mono-directional, normal to the stopping material, whereas the point source in the sphere was modeled as isotropic. The results are notably different and again reinforce why designs based on average beta energy have significant errors. Looking at Fig. 2.17, the beta spectrum from each of the isotopes that are represented in this discussion, it is apparent that there is a significant distribution of low energy beta particles which are emitted. By definition, the average beta energy is at the point where the number of beta particles greater than the average energy equal to the number of beta particles less than the average energy. For S-35, the low energy beta population continually increases as the energy approaches zero. For Sr-90, there is a slight peak at 0.08 MeV, but generally the population is flat at lower energies. For Y-90, there is a well-defined maximum in the spectrum at 0.8 MeV. The differences between Figs. 2.18 and 2.19 are not so surprising when the low energy betas in a full spectrum are taken into account.

In the calculations based on average beta energy shown in Fig. 2.18, there are distinct peaks for both the slab and spherical geometries for S-35, Sr-90 and Y-90. For S-35, the peaks occur at 6 μm for the slab geometry and at 10 μm for the

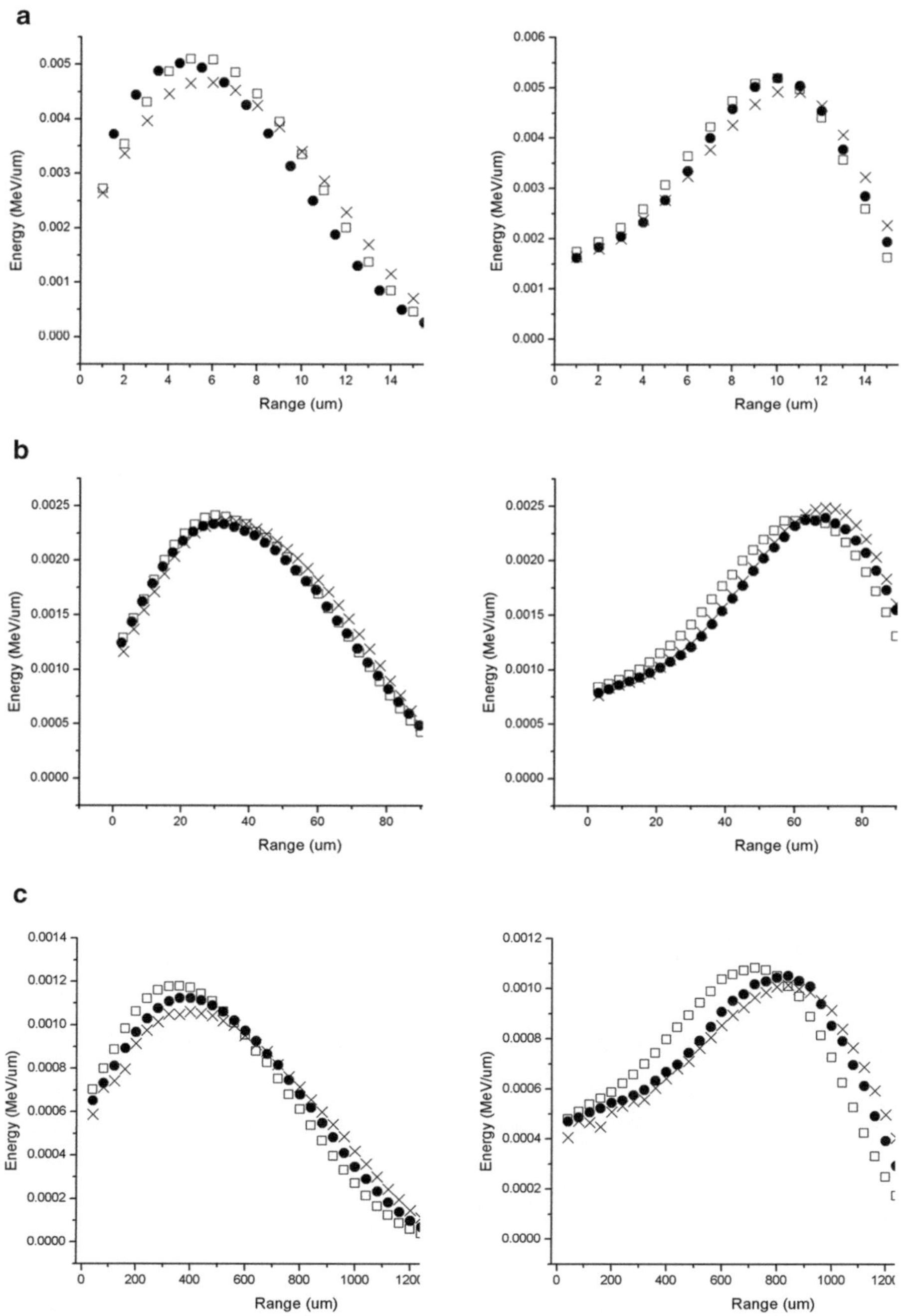

Fig. 2.18 Simulated energy deposition sin silicon carbide based on the average beta energy versus distance in both the monodirectional beta source incident on a slab (*left*) and an isotropic source at the center of a spherical (*right*) geometries using GEANT4, PENELOPE, and MCNPX codes for (**a**) S-35, (**b**) Sr-90, (**c**) Y-90 [8]

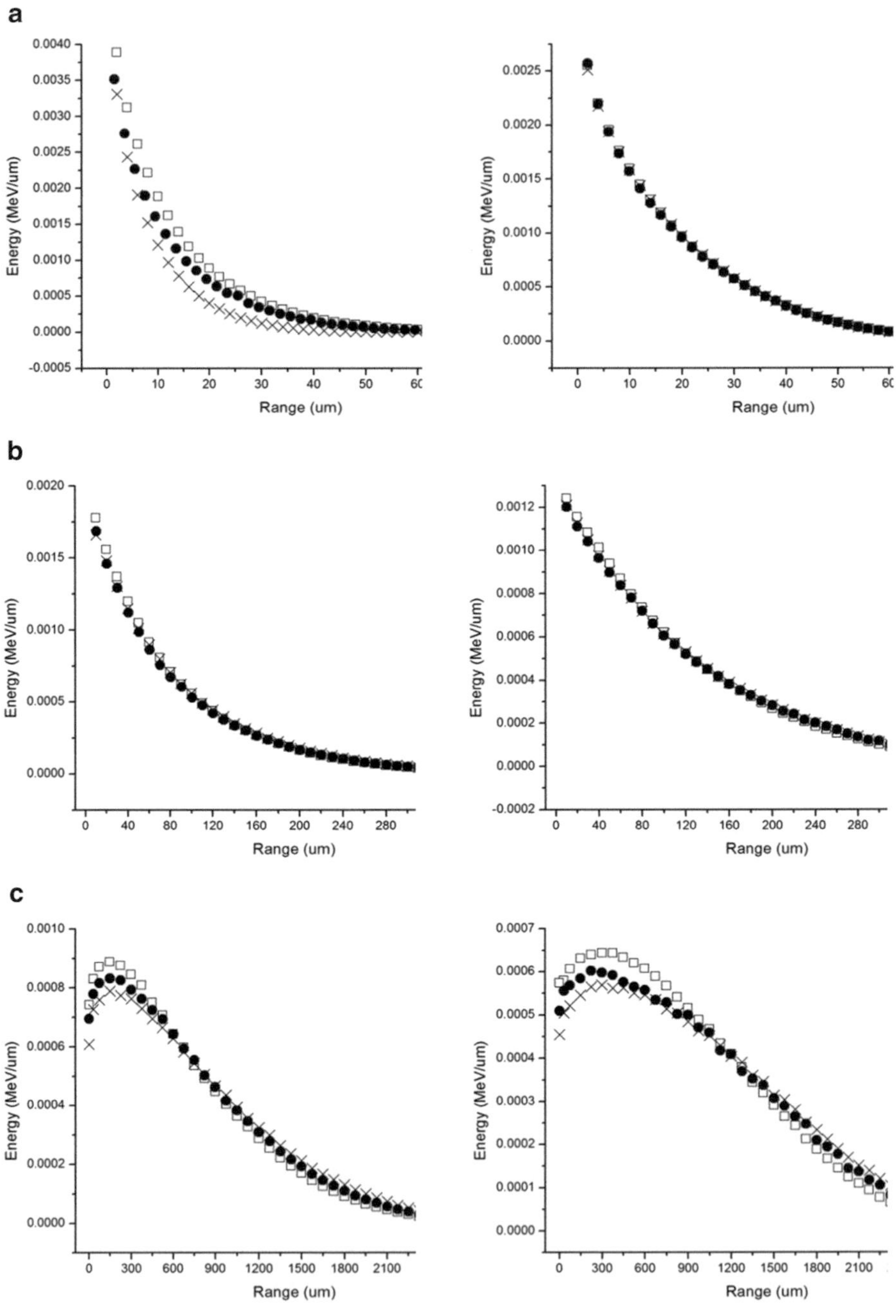

Fig. 2.19 Simulated energy deposition in silicon carbide based on the full beta energy spectrum versus distance for the monodirectional beta source incident on a slab (*left*) and an isotropic source at the center of a spherical (*right*) models using GEANT4, PENELOPE, and MCNPX codes for (**a**) S-35, (**b**) Sr-90, (**c**) Y-90 [8]

spherical geometry. For Sr-90, the peaks occur at 35 μm for the slab geometry and at 65 μm for the spherical geometry. For Y-90, the peaks occur at 400 μm for the slab geometry and at 600 μm for the spherical geometry. In contrast, Fig. 2.19 shows that low energy betas from the spectrum weigh heavily in where the energy is deposited. Low energy betas will have a shorter range in the material than high energy beta particles. So, the energy deposited per unit depth into the stopping material for both S-35 and Sr-90 is highest near the surface of the stopping material and decays exponentially with depth. For the high energy Y-90 beta particles there is a distance where energy deposition peaks. For the slab case it peaks at 150 μm and for the spherical case it peaks at about 300 μm. Another interesting observation is that the maximum amount of energy that is deposited is substantially different (about a factor of two lower) for the full spectrum calculation as opposed to the average energy calculation.

The implications of the above observations are significant. First of all, in using average beta energy to calculate the location of maximum energy deposition, and therefore the optimum location of the transducer, there will be substantial errors in placement of the transducer. For devices such as a nuclear-pumped laser this error is significant. The spatial power deposition in a semiconductor nuclear-pumped laser will define the geometrical position of laser's gain. The challenge is to deposit as much of the energy from the source's beta particles into the semiconductor to produce as high a power density as possible in order to maximize the gain. Realizing that the slab model is idealized in that a mono-directional beta particle beam strikes the cell normal to the surface and that the spherical model is idealized by placing a point source at the center of the sphere where the beta source is equidistant from a shell within the sphere, both models vastly over predict the beta energy being deposited in any given layer. In a realistic device the beta source will be isotropic, thus creating a much greater challenge for maximizing the energy deposited by the beta particles in the volume. In summary, by using average beta energies in design calculations, significant errors are made in locating the optimum position for the gain region. There is also a significant error in calculating the energy transfer rate to the gain region (Table 2.6).

Table 2.6 Range of beta particles in SiC based on the rules of thumb from Eqs. 2.14, 2.15, 2.16, 2.17, and 2.18 compared to results in Figs. 2.18 and 2.19 [15].

	Range in millimeters		
Radioisotope	Rule of thumb	Average beta	Full beta spectrum
S-35	10.6	0.02	0.08
Sr-90	55.1	0.12	0.40
Y-90	344.0	1.6	3.00

The results from the full beta spectrum calculation are exact and this table shows the magnitude of expected errors in calculations which use rules of thumb or average beta energy.

2.4 Neutrons

Neutrons are released during fission (neutron capture induced or spontaneous), fusion, and nuclear reactions between high energy particles and a target material. Neutrons produced in nuclear reactors through the fission reaction, such as Eq. 2.2, are born at energies above 1 MeV (see Fig. 2.4). Neutrons may also be created through spontaneous fission such as in Cf-252 (Fig. 2.20) or released in fusion reactions. Fusion requires that the fusing nuclei in the reaction have sufficient energy in the center of mass frame to overcome the Coulomb barrier. The required energies are generated by fusion reactors (which create high temperature plasmas) or by accelerators. Typical fusion reactions are,

$$\mathrm{D}+\mathrm{D}\rightarrow\begin{cases}\mathrm{T}+\mathrm{p}+4.03\,\mathrm{MeV}\,(50\%)\\ \mathrm{He}^3+\mathrm{n}_{\mathrm{fast}}+3.27\,\mathrm{MeV}\,(50\%)\end{cases} \tag{2.22}$$

$$\mathrm{D}+\mathrm{T}\rightarrow\mathrm{He}^4+\mathrm{n}_{\mathrm{fast}}+17.6\,\mathrm{MeV}$$

The neutron energies from fusion reactions have a very narrow energy spectrum. In the case of the D-D reaction, the neutron energy is 2.45 MeV. In the case of D-T, the neutron energy is 14.1 MeV. Neutrons are not charged and thus will have no electromagnetic interactions with the cloud of electrons in an atom. Neutrons will only collide with a nucleus and will only lose energy through these collisions. Even with solids the volume taken up by nuclei is very small, so the probability of a collision between a neutron and a target nucleus will be low. The amount of energy that a neutron will give up when colliding with a nucleus is based on the principles

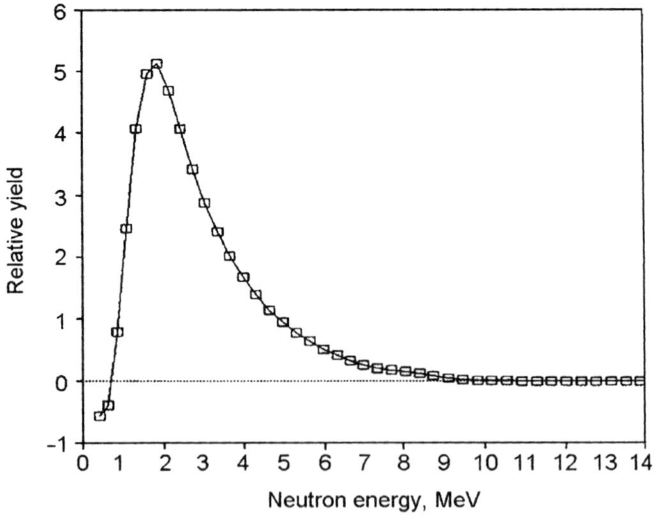

Fig. 2.20 Neutron spectra from the spontaneous fission of Cf-252 [13]

Table 2.7 The half thickness (the thickness where a beam of neutrons will lose half of its intensity) of a plate made of a given material that is exposed to a Po-Be neutron source [14]

Material	Half thickness (cm)
Paraffin	6.6
Water	5.4
12 % Borax in water	5.3
Brass	4.9
Steel (Cold rolled)	4.9
Lead	6.8
Aluminum	7.8

of an elastic collision governed by conservation of momentum and conservation of energy. If the mass of the target nucleus is comparable to the mass of the neutron that collides with it, the neutron will bounce off giving up some of its energy. If the mass of a target nucleus is significantly heavier than the neutron, then the collision does not reduce the energy of the neutron very much. Neutrons will lose more energy if the target atoms are light. The rate of energy loss will be greatest if the target material has a high density of light nuclei. That is why neutron shields are made up of materials with a high density of hydrogen atoms.

The effective range of neutrons in matter is a function of the mass of the target atoms and the density of the material. Neutrons most effectively interact with atoms with low atomic mass like hydrogen, helium, beryllium and carbon. High density, low atomic number materials are most effective for neutrons losing energy by elastic scattering. The range of neutrons in a gas, even at high pressures, is very long and thus the power density that neutrons are able to produce is very low. For example, the range of 2.45 MeV neutrons in 200 atm of helium is on the order of 30 m. Even in solids with a high atomic density of hydrogen, like paraffin (Table 2.7), the range is on the order of a half meter (it takes about six half thicknesses to achieve 99 % energy loss). In terms of nuclear energy conversion devices, neutrons are a very a poor candidate ionizing radiation source for driving a nuclear-pumped laser. Isotopes such as Cf-252 spontaneously fission but the large range of neutrons in matter would require a large volume with a very low power density. With regard to fusion, the only possible reactions on a small scale reported in the literature are those based on "Cold Fusion". This is a controversial area of research which appears to produce heat at levels beyond the capability of chemical reactions. The type(s) of particles that may be created in such hypothetical nuclear reactions have not yet been identified.

2.5 Gamma Rays

Gamma rays are electromagnetic radiation (photons) and possess no charge or rest mass. There are three processes by which gamma rays interact with matter: the photoelectric effect; Compton scattering; and pair production. In the photoelectric effect, the photon, which is incident on the atom, loses all of its energy to an orbital electron. The photon energy is absorbed by the electron and the electron is ejected from the atom leaving behind a positive ion. The probability of the photoelectric effect is highest when the photon energy is low and if the material is made up of high Z atoms. Compton scattering occurs when the photon interacts with an electron orbiting an atom, which ejects the electron and scatters the photon but at a lower energy. This interaction conserves momentum and energy as shown in Eq. 2.23 where E_e is the energy transferred to the electron and E_s is the energy of the photon after the scattering event.

$$hv = E_e + E_s \,(\text{conservation of energy}) \tag{2.23}$$

In addition, momentum is conserved and there is an angular distribution of the reaction's products. At a low energy, most of the initial energy is scattered ($E_s >$ 80 % (hv) when hv <1 keV). When the charge of the atom (Z_e) increases, the probability of interaction also increases. Compton scattering is practically independent of Z because the probability of interaction decreases as the photon energy increases. Pair production occurs when a photon spontaneously transforms into an electron and positron in the presence of an intense electrical field near a nucleus or an electron. The photon energy has to be greater than 1.022 MeV, the rest mass energy of the electron-positron pair. The mass attenuation coefficient for gamma rays is illustrated in Fig. 2.21 [11].

The effective range of gamma rays in matter can be calculated using Eq. 2.24, where I is the intensity or exposure rate, I_0 is the original radiation intensity or exposure rate, μ is the linear attenuation coefficient (cm^{-1}), μ/ρ is the mass attenuation coefficient (cm^2/g), ρ is the absorber density (gm/cm^3), and x is the absorber thickness (cm).

$$I = I_0 e^{-\mu x} = I_o e^{-(\mu/\rho)\rho x} \tag{2.24}$$

If water is chosen as the medium and a gamma ray with 0.6 MeV energy, the mass attenuation coefficient, μ/ρ, is 0.1 cm^2/g (Fig. 2.21a). Assuming that the range can be approximated as the point where the intensity of the beam drops to 1 %, the effective range can be calculated using Eq. 2.24. The range for a 0.6 MeV gamma ray in water is ~46.1 cm. For 100 atm of Ar, the mass is 0.1664 g/cm^3 so the effective range of a 0.6 MeV gamma ray in 100 atm Ar is ~277.4 m. Therefore, although there are many radioisotopes that emit gamma rays, the long range in matter makes their use in nuclear-pumped laser technology unlikely because the systems would be large and would have very low power density. (There is a case however, as discussed in Chap. 4, where a thermonuclear explosion produces an intense enough gamma ray burst to pump a laser.)

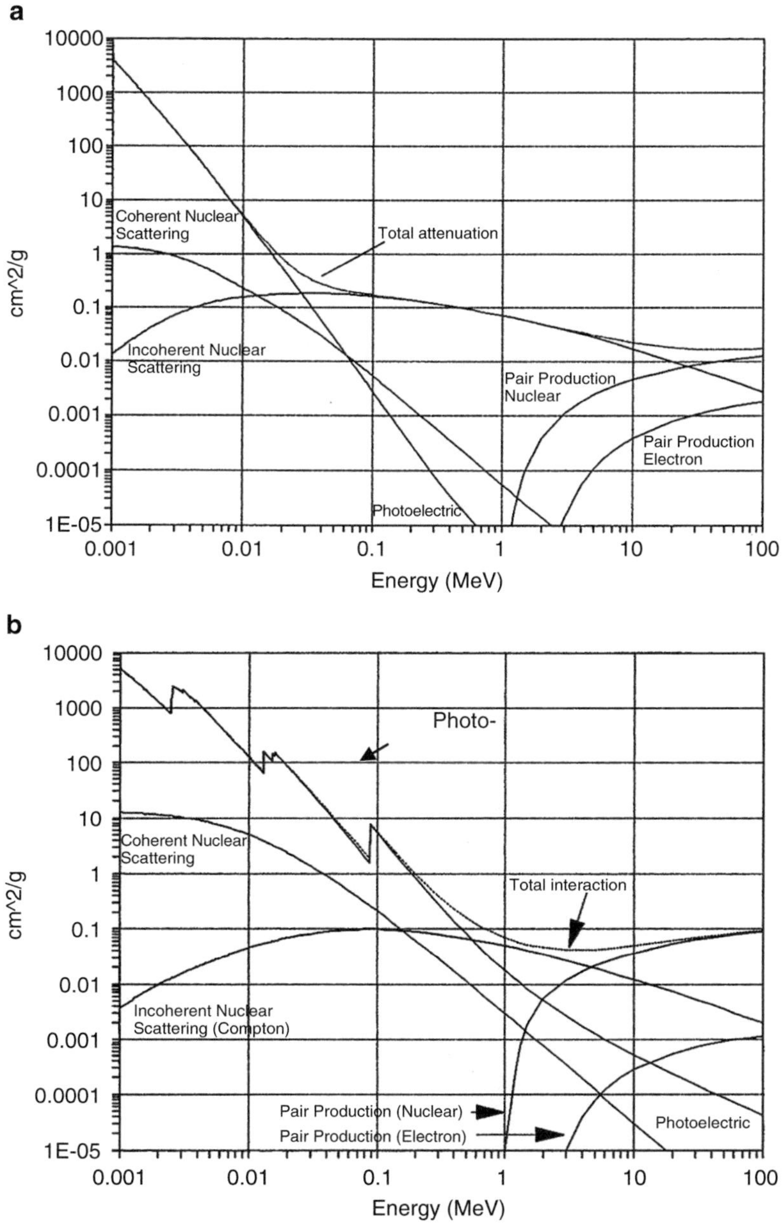

Fig. 2.21 The mass attenuation coefficients for gamma rays of various energies are shown in (**a**) water and (**b**) lead. Note that at low energies the photoelectric effect dominates. At energies near 0.1 MeV, Compton scattering dominates and at very high energies pair production dominates [11]

2.6 Scale Length Matching

Each type of radiation has a characteristic transport scale length in matter (i.e., the range). In designing energy conversion systems based on ionizing radiation, the transport scale length of the radiation must be matched to the scale length of the transducer (e.g., laser media, nuclear battery medium, radiolysis cell, etc.), specifically the region in the transducer where energy deposition contributes to usable energy production. Scale length matching directly impacts the efficiency of the energy conversion system. In larger systems this is not considered; for example, a nuclear power plant stops fission fragments, beta particles, gamma rays and neutrons within the core and reflector to produce heat. RTGs use a solid material to stop the ionizing radiation produced by the alpha decay of Pu-238 and generate heat for a thermoelectric energy conversion system. The alphas completely stop in the solid and the energy released turns into heat.

When an isotropic source is interfaced to a transducer which is geometrically incompatible with the isotropic emitter, then the efficiency of the system will be diminished. For a radiation point source on a planar transducer surface, half of the radiation is emitted in the wrong direction. Furthermore, the angles at which the radiation particle trajectories intersect with the plane varies widely. Thus, the distance that each particle has to travel before intersecting the energy conversion region of the transducer varies widely (Fig. 2.22).

Depending on the type of radiation source, the interactions are complex. If the radiation source is an alpha particle, then the alpha interacts with the electron cloud in the matter through Coulomb interactions. The alpha particle knocks out an energetic electron from its bound state and the energetic electron goes on to create further ionization and excitation (Fig. 2.23). The range of the secondary electron defines how far from the track of the alpha particle that ionization can take place. The range of the electron created in the secondary reaction further enhances ionization beyond the alpha particle track as does the electron from tertiary

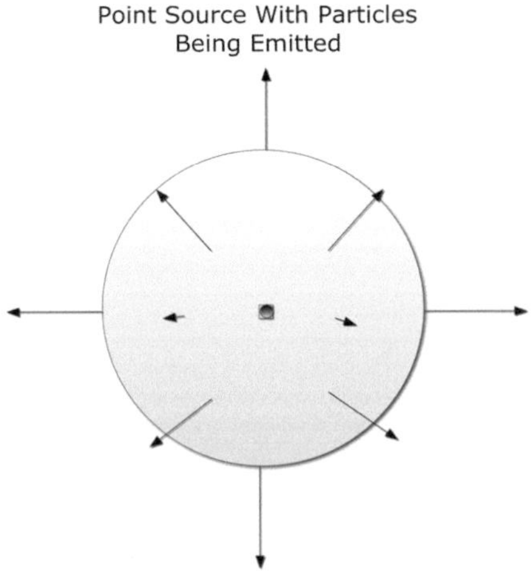

Fig. 2.22 Shown is an illustration of a radiation source. The emission from the source is isotropic [15]

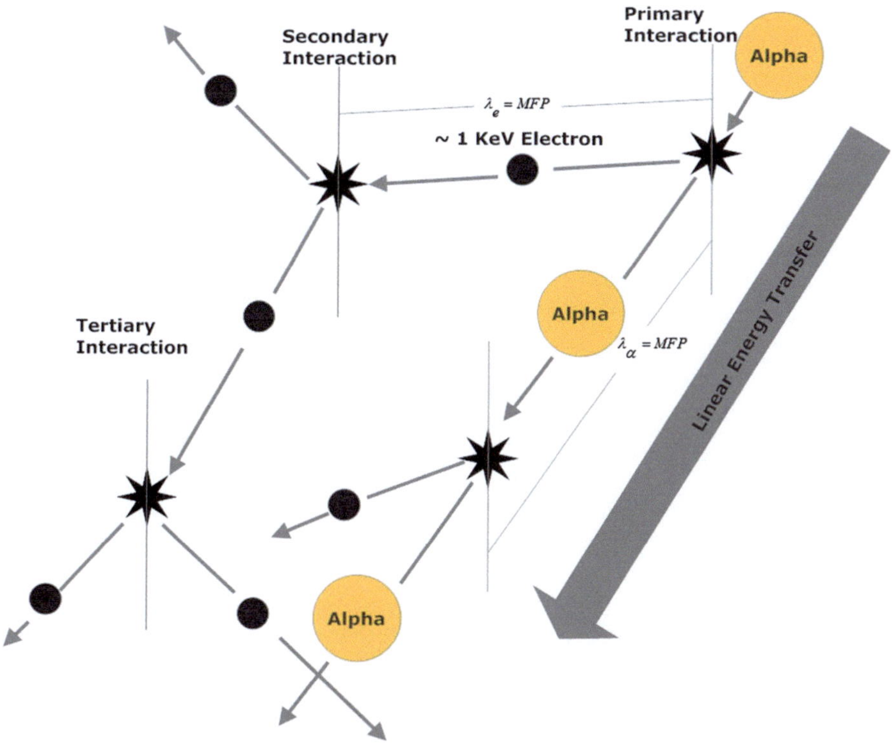

Fig. 2.23 Illustration of an alpha particle interacting with matter. The alpha particle has a linear trajectory through matter and loses energy through Coulomb collisions with electrons in the cloud. These electrons are energetic and can undergo secondary, tertiary and high order interactions as they lose energy [15]

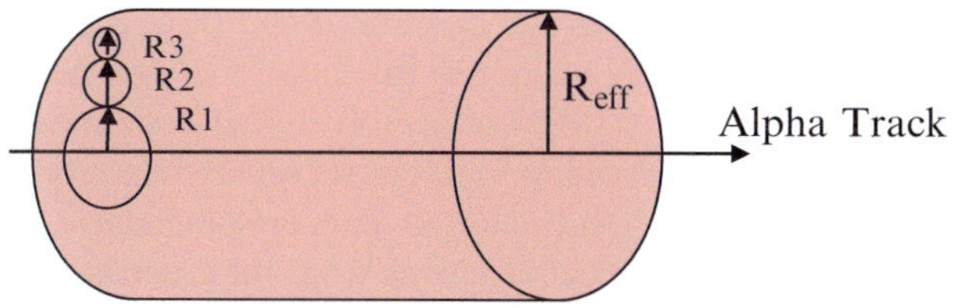

Fig. 2.24 Energy deposition along and radially from the track of an alpha particle is illustrated. The energy of the primary electron is higher than that of the secondary which is higher than that of the tertiary, etc., thus R1 > R2 > R3 > R4 etc. Ionization takes place in a cylinder surrounding the alpha particle path with an effective radius of R_{eff} [15]

electrons and higher order reactions (Fig. 2.24). Ionization occurs beyond the line of trajectory of the alpha particle as illustrated in Fig. 2.24. So, there will be a zone around the path of the alpha particle in which ion pairs are formed. In high density materials such as a solid or a high pressure gas, the effective radius (R_{eff}) of this zone will be small.

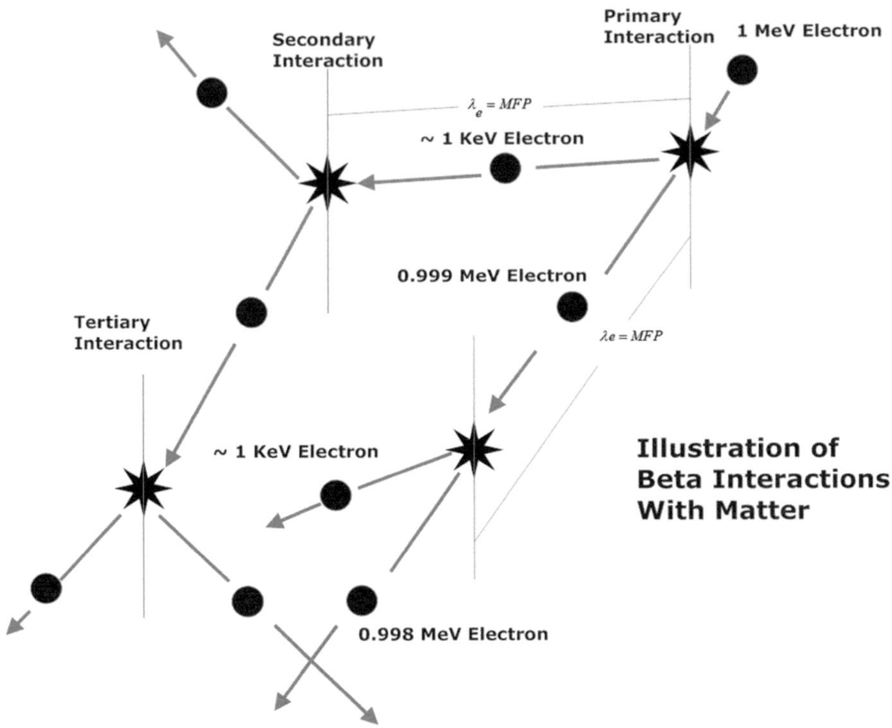

Fig. 2.25 Illustration of a beta particle interacting with matter. The beta particle primarily loses energy through Coulombic collisions with electrons in the cloud. These electrons are energetic and can undergo secondary, tertiary and high order interactions as they lose energy [15]

Energetic electrons also interact by Coulomb interactions with the electron cloud of the matter that it passes through. Thus electrons will undergo primary interactions that kick out an electron from the electron cloud (Fig. 2.25). The incoming electron can lose part or all of its energy to the target electron because they have equal mass. The primary electron will in turn travel about one mean free path and then interact with the electron cloud to kick out secondary electrons and so forth.

Gamma rays will interact with the material via the photoelectric effect, Compton scattering, or pair production to eject electrons from the electron cloud (Fig. 2.26) and this electron in turn will create secondary electrons which then create tertiary reactions and so forth.

Fast neutrons which create ionization through recoil reactions, also create primary, secondary, tertiary and higher order electrons (Fig. 2.27). The interaction commonalities between the various types of ionizing radiation leads to an interesting phenomena, the W values (the energy, measured in electron volts, required to produce one ion pair) for each type of ionizing radiation with a specific form of matter is similar but the interaction scale lengths differ.

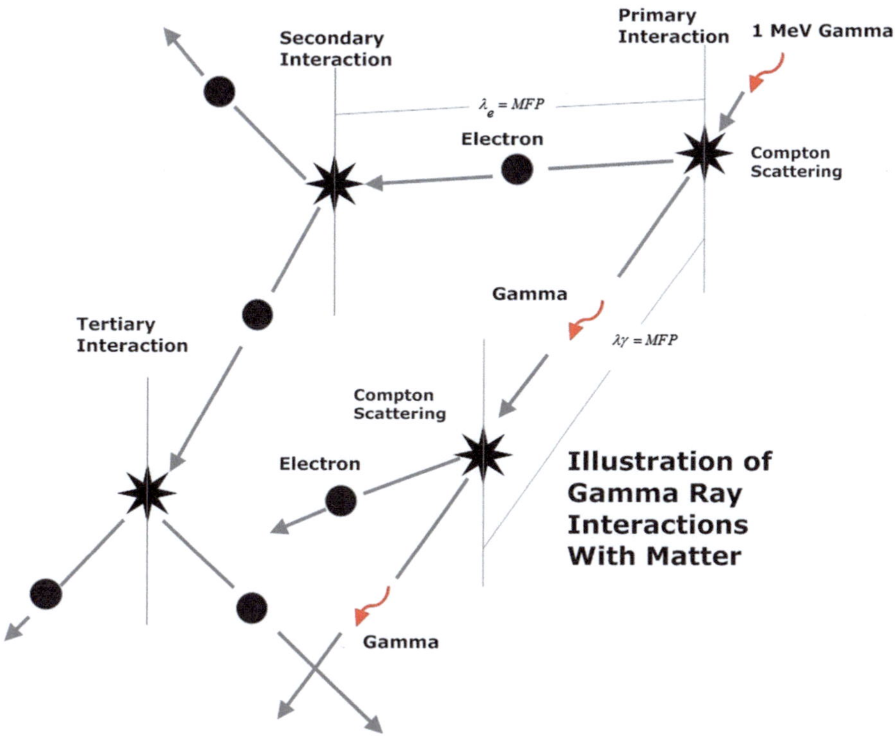

Fig. 2.26 Illustration of a gamma ray interacting with matter. The gamma ray primarily loses energy through the Lorentz force with electrons in the cloud. These electrons are energetic and can undergo secondary, tertiary and higher order interactions as they lose energy [15]

The scale length (λ_{Radtr}) varies with the form of the ionizing radiation, the material that it interacts with, and the phase of the matter (Table 2.8). From Table 2.8, it is clear that the gain region of a semiconductor transducer has a complicated matchup with the transport scale lengths of alpha, beta, gamma or neutron transport in solids.

2.7 Energy Deposition

The formation of secondary, tertiary, quaternary, etc. electrons from the primary event drives the process of ion pair production in radiation interaction with matter [9]. These higher order interactions makes it extremely difficult to model electron-hole pair or ion pair production using available transport codes like MCNP, GEANT4, etc. These transport models are capable of producing results on the energy deposited in a material as the photons, electrons or ions pass through the material. With the knowledge of energy deposition, it is possible to rely on the use

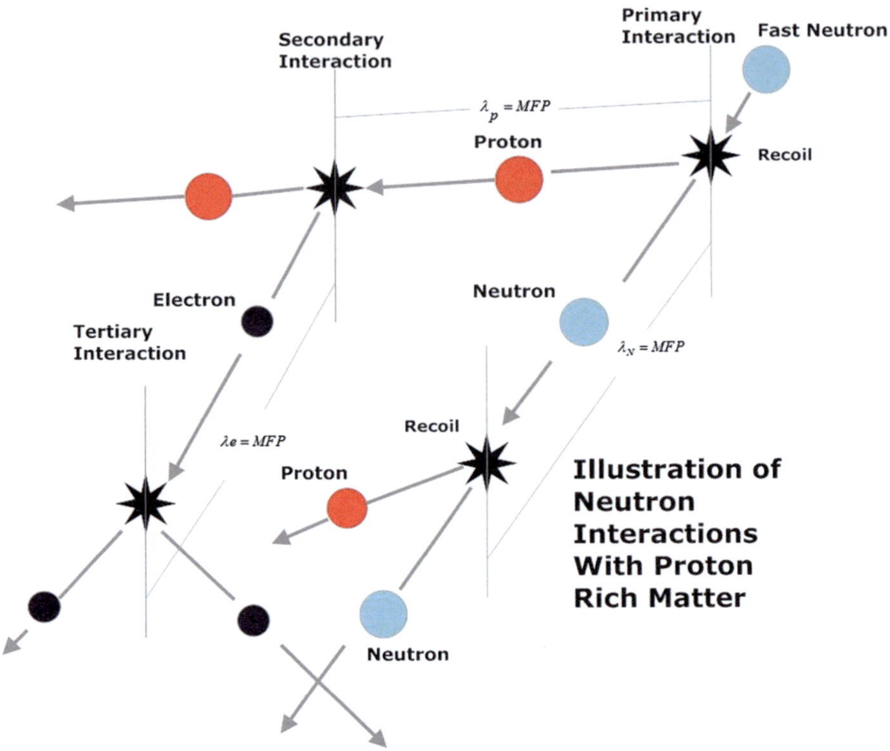

Fig. 2.27 Illustration of fast neutron interactions with a proton rich form of matter. The neutron undergoes an elastic collision with a proton in the target material. The proton recoils and then interacts with the electron cloud, much like the reaction between an alpha particle and matter (Fig. 2.20). The scattered neutron proceeds onward and may have enough energy remaining to undergo another elastic scattering collision with a proton in the material and so forth. On average, a fast neutron will undergo 19 collisions in water before it is thermalized

Table 2.8 Scale length of the radiation source (λ_{Radtr}) in various forms of matter

Radiation	Material	λ_{RadTr} (cm)	Gain region SiC laser (cm)
Alpha (5.4 MeV)	1 Atm. Air	4.6	
	SiC	2.0×10^{-3}	20×10^{-4}
Beta (2.2 MeV)	1 Atm. Air	834	
	SiC	3.0×10^{-1}	20×10^{-4}
Gamma (0.6 MeV)	100 Atm. Ar	2.77×10^{5}	
	SiC	~21.52	20×10^{-4}
Neutron (2.45 MeV)	100 Atm. Helium	3×10^{4}	
	Paraffin	~50	20×10^{-4}

of the W value to find the number of electron-hole pairs or ion pairs produced. W values, which is defined as the average energy required to form an electron-hole pair (or an ion pair), are experimentally measured and have been reliably used in designing calibrated nuclear detectors for decades. The W value has been measured

Table 2.9 The average energy required to produce ion pairs in various gases [17, 19]

Gas	Energy per ion pair, W (eV)	First ionization potential (eV)	Fraction of energy used in ionization (I/W)
H_2	36.3	15.6	0.43
He (pure)	43	24.5	0.58
N_2	36.5	15.5	0.42
O_2	32.5	12.5	0.38
Air	35.0		
Ne (pure)	36.8	21.5	0.58
Ar	26.4	15.7	0.59
Kr	24.1	13.9	0.58
Xe	21.9	12.1	0.55
CH_4	30	14.5	0.48
C_2H_4	29	10.5	0.36
CO	34	14.3	0.42
CO_2	34		
CS_2	26	10.4	0.40
NH_3	39	10.8	0.28

for specific gases, gas mixtures and semiconductor materials (Tables 2.9 and 2.10). All forms of ionizing radiation (gamma rays, neutrons, betas, ions, etc.) have similar W values for any given material that they interact with (i.e., the W value for gammas, betas [16], ions [17] and neutrons interacting with helium, for example, will be similar). The first ionization potential for each material is given in Table 2.9 as well as the ratio of the ionization potential to W-value. Another important point to be made is that the W-value is independent of the distance that the ionizing radiation travels before losing energy. The specific ionization curve for a 5.4 MeV alpha particle in air, for example, is shown in Fig. 2.28.

In addition to forming ion pairs, the alpha particle can lose energy without creating ion pairs, where an electron can receive enough energy by Coulomb interactions to jump to a higher energy state but not sufficiently high to cause ionization. The W-value represents the average energy required to create an ion pair and part of the inefficiency in producing ions is in the production of non-ionized energy states. Non ionizing energy losses go into the creation of excited states and can be accounted for. A rare gas for example has a first electronic state that is a metastable state. This state is typically illustrated with the symbol "*" (e.g. He*). A W*-value is the average energy required to produce a metastable state such as a single He* state in helium gas. For He the W*-value is about 90 eV/He* [18]. In a gaseous system, some of this energy will go into the production of photons through spontaneous emission. Some of the energy ends up as kinetic energy which raises the temperature of the gas. If the photons created in the process are absorbed by the walls of containment or by the gas, then eventually heat is produced.

Table 2.10 Properties for some common semiconductor materials which are useful for direct nuclear energy conversion [20]

Material	Minimum band-gap (E_g) [eV]	Electron drift mobility (μ) [cm^2/V-s]	Fano factor (F)	Density (ρ) [g/cm^3]	Atomic mass [g/mol]	Molar density [moles/cm^3]	Displacement energy (E_d) [eV]	Mean ionization energy (W) [eV]	E_g/W
Silicon	1.12	1450	0.115	2.329	28.1	0.0829	~19	3.63	0.308
Germanium	0.68	3900	0.13	5.323	72.6	0.0733	30	2.96	0.23
Gallium arsenide	1.42	8500	0.1	5.317	144.6	0.0368	10	4.13	0.344
Silicon carbide	2.9	400	0.09	3.22	40.1	0.0803	28	6.88	0.421
Gallium nitride	3.39	1000	—	6.15	83.7	0.0735	24	8.9	0.381
Diamond	5.48	1800	0.08	3.515	12	0.293	43	12.4	0.442

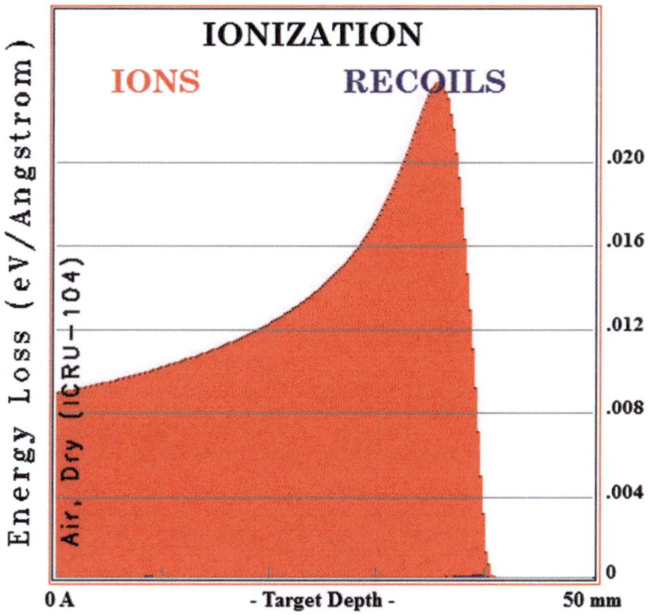

Fig. 2.28 SRIM2011 model showing ionization produced in dry air by 5.4 MeV alpha particle [7]

In the case of radiation interactions with a solid, electron-hole pairs are created as well as heat. The fraction of energy that goes into electron-hole formation depends on the W value and the band-gap energy of the material. In Table 2.10 some common semiconductor materials are shown along with their relevant properties. As above, the mean ionization energy required to form one electron-hole pair in a solid is the W-value. The ratio of the band-gap energy (E_g) to the W value is the effective efficiency for producing electron-hole pairs through the interaction of radiation with matter. As can be seen in the last column of Table 2.10, the electron-hole pair production efficiency has considerable variation from one material to another. Diamond has the highest at 0.442. Thus when ionizing radiation interacts with diamond, 44.2 % of the energy goes into electron-hole pair production. 55.8 % essentially goes into heat production. If nothing is done to use the electron-hole pairs that are being produced, they will recombine and the energy eventually is transformed into heat by a series of processes.

2.8 Spatial Energy Distribution

In a three dimensional object being exposed to ionizing radiation, the spatial energy deposition distribution is complicated. The problem was described by Chung and Prelas for heavy charged particles irradiating gases in a cylindrical configuration [21]. The energy lost from heavy charged particle interaction with the target gas

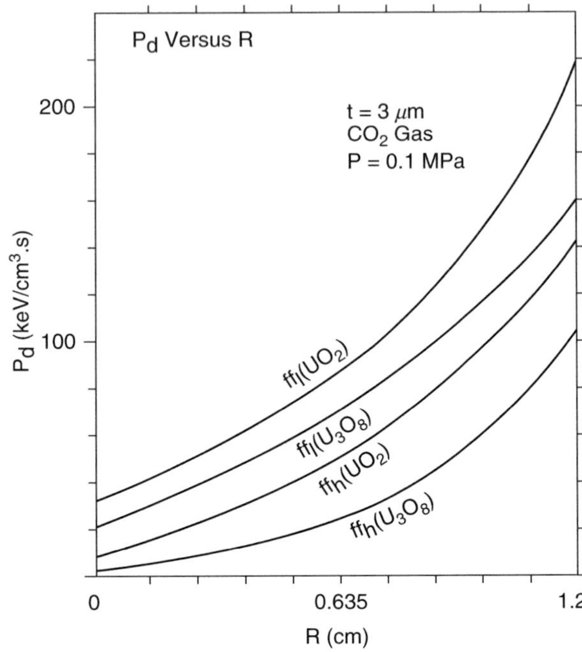

Fig. 2.29 Power deposition (keV/cm^3-s) in a one inch diameter cylinder (with the *center* being R=0) filled with 0.1 MPa of CO_2 gas and the inner radius lined with a 3 μm coating of UO_2 or U_3O_8. The energy deposition contribution from the light fission fragment (ff_l) and the heavy fission fragment (ff_h) are plotted [21]

was calculated as an energy current $\vec{J}(E)$ since primary ionization, which contributes to the energy lost by the heavy charged particle, also deposits energy around the interaction point by secondary and higher-order interactions which occur near the primary interaction point (Fig. 2.24). Chung and Prelas approximated the spatial energy deposition by Eq. 2.25, where $P_d(z)$ is the power density and ∇_r is the gradient in polar coordinates.

$$P_d(z) = \nabla_r \vec{J}(E) \tag{2.25}$$

In using this approximation, they assumed that the effective radius (R_{eff}), over which the energy was deposited through primary, secondary, tertiary and higher order reactions, was small. This assumption is accurate for gas pressures above one atmosphere (Fig. 2.29) and is a very good approximation for solids. Some Monte Carlo programs, like MCNPX and GEANT4, follow primary and higher order interactions of particles with a target material and provide data for energy deposited at specific locations in the geometry. These programs do not model the motion of the ions and electrons in the media (due to diffusion and drift velocities in an electric field). Thus it is difficult to accurately obtain a spatial and temporal distribution of ions and electrons. However, if the density of the material is high enough, such as gas pressures above 1 atm or in solids, then the power density generated by Monte Carlo based codes, such as MCNPX or GEANT4, will be sufficiently accurate for most applications. It is feasible but complicated to couple ion and electron motion in a media with Monte Carlo transport codes by developing customized modules.

2.9 Summary

The interaction of ionizing radiation with matter is a critical component in the understanding of nuclear-pumped laser systems. The range of the ionizing radiation will directly impact the power density deposited in the laser medium which in turn is related to the population inversion of the laser as well as the gain. Heavy and light ions have the best prospect of achieving the required power densities to operate a nuclear-pumped laser. Beta particles have longer ranges and will either require a very intense source or a dense high Z material to stop the particles. Neutrons and gamma rays are too penetrating and are not suitable for driving nuclear-pumped lasers (with the exception of a thermonuclear explosion).

Problems

1. Find 5 other pairs of fragments which can come out of ^{235}U fission besides the example pair of La-147 and Br-87 shown in Eq. 2.5. What are the average kinetic energies for each of the fragments in the five pairs that you identified?
2. Download a copy of SRIM. It is a free program and can be downloaded by anyone at http://www.srim.org/. Using the fission fragment pairs that you found in problem 1, solve a problem in which the a particle beam made up of each fragment enters a slab of made of gold metal normal to the slab surface. Duplicate the plots of Figs. 2.6 and 2.7 for each fragment.
3. What is the range of a 0.85 MeV ^{7}Li ion and a 1.5 MeV ^{3}He ion in 760 Torr of Helium (note that these are ions released in ^{10}B(n$_{th}$, ^{3}He) ^{7}Li reaction shown in Eq. 2.11).
4. How accurate are the Rule of Thumb formulas for beta particle ranges in Eqs. 2.12, 2.13, 2.14, 2.15, 2.16, 2.17, and 2.18?
5. How accurate are power density deposition calculations in a material using the average beta energy emitted from a radioisotope calculated by the 1/3 β_{max} rule?
6. Download a copy of GEANT4 at http://geant4.cern.ch/. Install it on your computer. You will need a Unix environment or a Unix emulator for windows. Do example B1 http://geant4.web.cern.ch/geant4/UserDocumentation/UsersGuides/ForApplicationDeveloper/html/ch10s02.html#sect.ExmpBasic.Summary.
7. What is the effective range of a 1 MeV gamma ray in lead?
8. How can a source which emits ions isotropically be efficiently interfaced to a volume of gas?
9. How is energy transported from ions to a gas?
10. How is energy transported from ions to a liquid?
11. How is energy transported from ions to a solid?
12. There is an isotropic source that emits 100×10^9 ions per second with an ion energy of 5 MeV. A thin layer of the radioisotope is coated on a diamond wafer. How much energy is deposited in the diamond wafer? How many ion pairs are produced in the diamond wafer?

References

1. Smith A, Fields P, Roberts J (1957) Spontaneous fission neutron spectrum of Cf^{252}. Phys Rev 108:411–413
2. Nicodemus DB, Staub HH (1953) Fission neutron spectrum of U^{235}. Phys Rev 89:1288–1290, 15 March 1953
3. Zeynalov S, Hambsch F-J, Obertstedt S (2011) Neutron emission in fission of ^{252}Cf(SF). J Kor Phys Soc 59:1396
4. ThorEA_Wiki (2014) Thermal, epithermal and fast neutron spectra. Available: http://thorea.wikia.com/wiki/Thermal,_Epithermal_and_Fast_Neutron_Spectra, 26 Sept 2014
5. DOE-HDBK-1019/1-93 (1993) Nuclear physics and reactor theory. Department of Energy, Washington DC. Available: http://energy.gov/ehss/downloads/doe-hdbk-10191-93
6. Turner JE, Kelsey CA (1995) Atoms, radiation, and radiation protection. Wiley, New York
7. Ziegler JF, Ziegler MD, Biersack JP (2010) SRIM–The stopping and range of ions in matter (2010). Nucl Inst Methods Phys Res B: Beam Interac Mat Atom 268:1818–1823
8. Oh K, Prelas MA, Rothenberger JB, Lukosi ED, Jeong J, Montenegro DE et al (2012) Theoretical maximum efficiencies of optimized slab and spherical betavoltaic systems utilizing sulfur-35, strontium-90, and yttrium-90. Nucl Technol 179:9
9. Guyot J, Miley G, Verdeyen J (1972) Application of a two-region heavy charged particle model to Noble-gas plasmas induced by nuclear radiations. Nucl Sci Eng 48:373–386
10. Giuliani JL, Petrov GM, Dasgupta A (2002) Electron energy deposition in an electron-beam pumped KrF amplifier: impact of the gas composition. J Appl Phys 92:1200–1206
11. Bernard S, Slaback Jr Lester A, Kent BB (1998) Handbook of health physics and radiological health. Williams & Wilkins, Baltimore
12. Oh K (2011) Modeling and maximum theoretical efficiencies of linearly graded alphavoltaic and betavoltaltaic cells. M.Sc. thesis, Nuclear Science & Engineering Institute, University of Missouri, Columbia
13. Musienko Y, Reucroft S, Ruuska D, Swain J (2000) Studies of neutron irradiation of avalanche photodiodes using ^{252}Cf. Nucl Instrum Methods Phys Res, Sect A 447:437–458
14. Johnson TE, Birky BK (2012) Health physics and radiological health. Lippincott Williams & Wilkins, Philadelphia
15. Prelas MA, Weaver CL, Watermann ML, Lukosi ED, Schott RJ, Wisniewski DA (2014) A review of nuclear batteries. Prog Nucl Energy 75:117–148, August 2014
16. Jesse WP (1958) Absolute energy to produce an ion pair in various ases by beta particles from ^{35}S. Phys Rev 109:2002–2004, March 15 1958
17. Friedlander G (1981) Nuclear and radiochemistry. Wiley, New York
18. Prelas MA, Boody FP, Miley GH, Kunze JF (1988) Nuclear driven flashlamps. Laser Part Beams 6:25–62
19. Friedländer G, Kennedy JW (1955) Nuclear and radiochemistry. Wiley, New York
20. Wrbanek JD, Wrbanek SY, Fralick GC, Chen LY (2007) Micro-fabricated solid-state radiation detectors for active personal dosimetry. NASA/TM 214674. Available: http://www.grc.nasa.gov/WWW/sensors/PhySen/docs/TM-2007-214674.pdf
21. Chung AK, Perelas MA (1984) The transport of heavy charged particles in a cylindrical nuclear-pumped plasma. Nucl Sci Eng 86(3):267–274

Chapter 3
Direct Energy Conversion Principles as Applied to Lasers

Abstract Energetic ions from nuclear reactions have applications in research and industry. These uses include the intricate work of high energy physicists trying to probe the nucleus itself, radioisotope production, radiopharmaceuticals, ion implantation, ion thrusters, particle beam weapons, the initiation of thermonuclear fusion or direct energy conversion. In direct energy conversion the product can take one of several forms: electricity, chemicals (such as hydrogen for energy storage), or lasers [1]. The most exciting application centers about a spectra of ions (between ~1 and ~100 MeV) that are used as drivers for nuclear-pumped lasers.

Keywords Nuclear direct energy conversion • Nuclear-Pumped lasers • Radiation

3.1 Introduction

The nuclear generated ion source may be one of several. Specifically, the ions may be generated by converting a thermal or fast neutron flux, through nuclear reactions, into a source of energetic ions. Typically the nuclear reactions that have been used, as discussed in prior chapters, are:

Neutron driven reactions:

$$^{235}\mathrm{U} + \mathrm{n_{th}}\text{-} \rightarrow \nu \mathrm{n} + \text{fission fragments} + 200\ \mathrm{MeV} \tag{3.1}$$

$$^{10}\mathrm{B} + \mathrm{n_{th}}\text{-} \rightarrow {}^{4}\mathrm{He} + {}^{7}\mathrm{Li} + 2.35\ \mathrm{MeV} \tag{3.2}$$

$$^{3}\mathrm{He}^{3} + \mathrm{n_{th}}\text{-} \rightarrow \mathrm{p} + \mathrm{T} + 0.76\ \mathrm{MeV}. \tag{3.3}$$

Radioactive Decay

A typical alpha emitter,

$$^{210}\mathrm{Po} \rightarrow {}^{206}\mathrm{Pb} + \alpha + 5.307\ \mathrm{MeV} \tag{3.4}$$

A typical beta emitter,

$$^{90}\mathrm{Sr} \rightarrow {}^{90}\mathrm{Y} + \beta + \bar{\nu} + 0.546\ \mathrm{MeV} \rightarrow {}^{90}\mathrm{Zr} + \beta + \bar{\nu} + 2.28\ \mathrm{MeV} \tag{3.5}$$

M. Prelas, *Nuclear-Pumped Lasers*, DOI 10.1007/978-3-319-19845-3_3

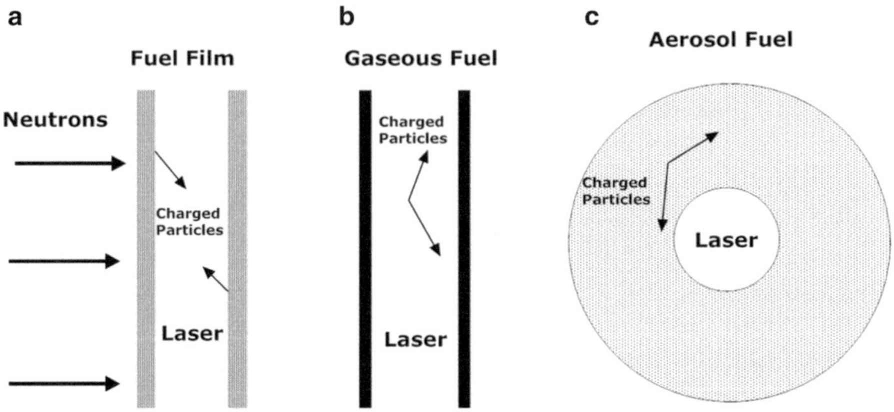

Fig. 3.1 Surface and volume type sources. (**a**) Surface source. (**b**) Volume source. (**c**) Aerosol source (From Prelas et al. [6])

Ultimately, direct energy conversion schemes may be used with ions generated by thermonuclear fusion reactions,

$$\mathrm{D} + \mathrm{T} \rightarrow \mathrm{n} + {}^{4}\mathrm{He} + 17.6\ \mathrm{MeV} \tag{3.6}$$

$$D + D \rightarrow \begin{cases} n + {}^{3}_{2}He + 3.27\ MeV\ (50\%) \\ p + T + 4.04\ MeV\ (50\%) \end{cases} \tag{3.7}$$

$$\mathrm{D} + {}^{3}\mathrm{He} \rightarrow \mathrm{p} + {}^{4}\mathrm{He} + 18.36\ \mathrm{MeV}. \tag{3.8}$$

It has also been suggested that the neutrons generated in a D-T or D-D burn may be utilized in conjunction with fissile materials to generate ions in a hybrid-like configuration [2] (analogous to the fission-fusion hybrid reactor [3]).

Neutrons can also undergo elastic scattering which results in recoil nuclei. Recoil nuclei are ionic and slow down just like any other ionic radiation. Thus fast neutrons are ionizing radiation and the elastic scattering process can be used for energy conversion. The scattering media would work best if it is a light material with a high density.

One can categorize the type of ion source in three ways, a surface, volume, or embedded source. The bounded region where the source interfaces with the reactant species is called a reaction cell. This interface for a surface source consists of a ^{10}B, ^{235}U or other fuel coating on the reaction cell's walls as illustrated by the fission-type surface source in Fig. 3.1. In the fission-type surface configuration, the ions may be emitted in any direction, hence half of the ions are born with a velocity vector in the wrong direction (if the source is coated on a substrate) while the other half lose some of their energy while being transported through the coating (see Fig. 3.1). Alternatively, the transducer or reactant can be located on both sides of the source, but this requires that the source be coated on a very thin film (much less than the range of the ion in the material).

The fission-type surface source coated on a structural material (e.g., a metal substrate) is approximately 20 % efficient in transferring the ion energy generated in the source to a gaseous medium [4].

Ions generated from thermonuclear fusion are in essence surface type sources since the transducer must be located outside of the plasma. Fusion generated ions are more difficult to interface to a direct energy conversion transducer. Suggestions for direct energy conversion using a fusion reactor include the Post generator along with other assorted devices [5]. It may be feasible to achieve efficiencies of 80 % or better in transferring the ion energy to a gaseous medium using a Post generator.

The volume type source is characteristic of the gaseous medium consisting of both the gaseous reactants and target nuclei (e.g. ^{3}He (g), $^{10}BF_3$ (g), uranofullerenes and $^{235}UF_6$ (g)), as shown in Fig. 3.1b. This type of source is essentially 100 % efficient in transporting the product ion energy to the transducer (i.e., the medium which converts energetic ions to a useful energy form). Additionally, there is a fundamental difference in utilizing the specific sources; The surface-type source is more flexible in that almost any gas mixture may be used as a transducer while the volume source requires a significant amount of target nuclei (e.g., ^{3}He (g), $^{10}BF_3$ (g), uranofullerenes and $^{235}UF_6$ (g)), in addition to the transducer medium, in the cell.

Volume sources can produce a much greater energy efficiency, however UF_6 vapor produces strong quenching of the chemical kinetics of most laser and chemical synthesis systems [1, 7]. It also has limited use for fluorescence production applications because of its very high absorption at wavelengths shorter than 420 nm [6] (see Chap. 5).

The multiphase (or aerosol) type source is an embedded source (Fig. 3.1c). For the embedded source, fuel is embedded in the transducer. The fuel has to be geometrically thinner than the range of the ion in the source materials. In this case the fuel, for example, can be small spherical particles suspended in the transducer (an aerosol), it can be thin films stretched across the transducer media or any other type of thin geometrical shape embedded in the transducer that will allow ions to escape the fuel material and deposit the bulk of the ion energy in the transducer. Uranium atoms can be used as a dopant in a glass based fiber laser. It is also possible to trap uranium atoms in fullerene cages and these structures can be mixed in the transducer medium.

For illustrative purposes, this discussion will focus on an embedded aerosol. The use of an aerosol fuel to produce secondary photons for energy transport to a transducer outside of the high radiation environment was conceived in 1981 [8–10] and is a means for scaling up nuclear energy conversion systems while maintaining reasonable efficiencies. Theoretical estimates suggest that 50–80 % of the energy generated in the micropellet (aerosol particle) will be available to provide energy to the photon producing medium [4]. Another major advantage of the aerosol concept, in addition to efficiency, is that the suspended micropellets will not affect the chemical kinetics of the transducers being "pumped." Finally, the aerosol concept can be used with materials which allow high operating temperatures. This last advantage implies that other energy producing cycles can be added to make the system even more efficient (as will be discussed, system efficiencies of approximately 70 % are achievable).

One of the main problems in driving a laser is the maximum achievable power density. Controlled nuclear reactions produce large amounts of energy, but are not capable of producing short (one microsecond or less) pulse widths that are controllable. Nuclear weapons produce shorter pulse widths. But a nuclear weapon is not a desirable pump source. The shortest pulse width produced in a controlled nuclear reactor is about 200 ms (for a fast burst reactor).

Equations can be derived for the power density generated in a surface source configuration, a volume source configuration and an embedded aerosol configuration [1]. The reaction rate is dependent upon the neutron flux ($\Phi(E,\widetilde{r})$) [11],

$$\Phi\left(E,\vec{r}\right) = \int_{\Omega} d\Omega \; n\left(\vec{r},\vec{\Omega},E\right)v(E) \tag{3.9}$$

where $\widetilde{\Omega}$ is the direction of that the neutron is moving, $n(\widetilde{r},\widetilde{\Omega},E)$ is the spatial, directional and energy dependent density of neutrons cm^{-3} and $v(E)$ is the neutron speed. In Eq. 3.9, the directional component is integrated thus giving the neutron flux.

The power density is dependent on the nuclear reaction rate as a function of neutron energy ($RR(E,\widetilde{r})$) which is the product of the atomic density of the target material like uranium ($N(\widetilde{r})$ cm^{-3}), the interaction cross section between the neutrons and target material as a function of energy ($\sigma(E)$ cm^2) and the neutron flux ($\phi(E,\vec{r})$ neutrons $cm^{-2}\,s^{-2}$). The reaction rate is equivalent to the rate at which charged particles are produced per unit volume. In these formulations centimeters will be used for the dimensions.

$$RR\left(E,\vec{r}\right) = N\left(\vec{r}\right)\sigma(E)\Phi\left(E,\vec{r}\right) \tag{3.10}$$

where $N\left(\vec{r}\right)$ is the spatially dependent density of the target material in the nuclear reaction.

Nuclear reaction cross sections have an inverse dependence on the velocity of the neutron (1/v) so the cross sections are very large at low energy. This is why nuclear reactors typically thermalize the neutrons and have a characteristic thermal neutron flux. Thus the dominant energy range is thermal. In conceptualizing the design of a thermal reactor/laser, it is appropriate to use a model based on one group theory [11]. In one group theory, the cross section is dominated by reactions near thermal energy and the one group cross section, σ. Since the dimensions of the foil are very small, the spatial variation of flux within the foil will be small thus one can approximate the flux to be constant in the foil. Also, since the dominant nuclear interaction will be thermal neutron capture, the thermal flux, Φ_{th}, is used to find the fission reaction rate (RR). It is assumed that the core is large and that there are no spatial gradients in the thermal flux. With these simplifying assumptions, the reaction rate becomes,

$$RR = N\sigma\Phi_{th} \tag{3.11}$$

where N is the fuel density in the foil.

3.2 Surface Source

Using the reaction rate equation, 3.11, the power density deposited in the transducer can be calculated for a surface source by using the following process.

- The first step is to multiply the reaction rate by the energy of the charged particles produced in the reaction (Q MeV) to give the charged particle power density (in MeV cm^{-3} s^{-1}) within the foil.
- The second step is to consider the ions that stream out of the foil into the laser medium. Some of the power from the ions is absorbed in the foil. The fraction of power from the charged particles which streams into the laser medium is called the transport efficiency (η_{tr}). Thus the equation is multiplied by the transport efficiency.
- In the third step the charged particle power density produced in the foil is multiplied by the volume of the foil (V_f). The equation up to this point gives the amount of power produced by the foil that is absorbed in the laser medium (in MeV s^{-1}).
- The fourth step assumes that the scale of the laser medium is larger than the range of the charged particle in that material so that all of the power streaming out of the foil is absorbed by the laser medium. By dividing the amount of power which is absorbed by the laser medium by the volume of the laser medium (V_L), the average power density absorbed by the laser medium is found (in MeV cm^{-3} s^{-1}).

  – If the geometry is Cartesian the foil volume is the foil area times the foil thickness (AΔt). The volume of the laser medium will be the area times the laser medium thickness ($A \times r$). Thus if A is large compared to Δt, the losses at the boundaries of the volume will be small and will not be considered. So the ratio of the foil volume to the laser volume is Δt/r.

- In the fifth step, the power density must be converted from MeV cm^{-3} s^{-1} to W cm^{-3}. The conversion factor is 1.6×10^{-13} J/MeV. In a typical nuclear-pumped laser which has a Cartesian geometry, there will be a foil on the top and the bottom of the laser cell. Since 2 foils contribute to the power density of the laser medium, the equation is multiplied by 2.

  Thus the power density for a surface source (P_s) is,

$$P_s = 3.2 \times 10^{-13} N\ \sigma\phi_{th} Q\ \eta_{tr} \Delta t/r\ \left(W/cm^3\right) \tag{3.12}$$

where N = fuel density cm^{-3}, σ = one group fission cross section cm^2, ϕ_{th} = one group flux $cm^{-2}\ s^{-1}$, Q = Mev/reaction, η_{tr} = charged particle transport efficiency, Δt = film thickness cm, r = laser radius cm,

3.3 Volume Source

The power density generated with a gaseous volume source assuming one group theory and no spatial gradients in the flux begins with the reaction rate (Eq. 3.11). If the target nuclei are mixed with the laser medium (e.g., Helium-3 gas mixed with the laser medium), the derivation of the power density is simplified.

- The first step is to multiply the reaction rate by the energy of the charged particles produced in the reaction (Q MeV) to give the charged particle power density (in MeV $cm^{-3}\ s^{-1}$) in the laser medium.
- In the second step, assume that the transport of charged particles to the laser medium is 100 %,
- The final step is to convert the power density from MeV $cm^{-3}\ s^{-1}$ to W cm^{-3}. The conversion factor is 1.6×10^{-13} J/MeV.

The power density for a volume source (P_V) is,

$$P_v = 1.6 \times 10^{-13} N\ \sigma\phi_{th} Q\left(W/cm^3\right) \tag{3.13}$$

3.4 Multiphase Laser Systems

An example of a multiphase system is one in which an aerosol fuel source is embedded in a gaseous laser medium. This strategy can increase power density by generating photons in a large volume (the fluorescer cell) and concentrating them on a small volume (the laser medium). In a photolytically driven laser media powered by fluorescer cell with an embedded source (e.g., aerosol, $^{235}UF_6$, uranofullerenes) the analysis is more complicated than that of a gaseous source [12]. The following model is developed for an aerosol fuel embedded in a fluorescer cell coupled to a photolytic laser:

- The first step is to multiply the reaction rate by the energy of the charged particles produced in the reaction (Q MeV) to give the charged particle power density (in MeV $cm^{-3}\ s^{-1}$) within the aerosol particles.
- The second step is to consider that the ions stream out of the aerosol particles into the fluorescer medium. Some of the power from the ions is absorbed by the aerosol particle. The fraction of power from the aerosol source which streams into the fluorescer medium is the transport efficiency (η_{tr}). Thus the equation is multiplied by the transport efficiency. It is assumed that the range of the ions in the fluorescer medium is less than the average distance between aerosol particles so that there are no additional losses to consider.

- The third step is to find the power density deposited in the fluorescer cell using the fuel dilution factor. The dilution factor can be calculated by dividing the average density of the fuel in the fluorescer (ρ_p) divided by the density of the fuel in an aerosol particle-or the solid-state density of the fuel (ρ_f). The equation up to this point gives the average power deposited in the fluorescer cell by the fission reactions (in MeV cm^{-3} s^{-1}).
- In the fourth step the power density stored in the photons created in the fluorescer cell is found by multiplying the average power density in the medium by the fluorescence efficiency (η_f-defined as fluorescence power generated/ power deposited).
- In the fifth step, the fluorescence is coupled to the laser medium by a waveguide that has a coupling efficiency (η_c). The equation is then multiplied by the coupling efficiency which gives the fraction of the power density in the fluorescence cell that reaches the surface of the laser.
- The sixth step is to assume that all of the fluorescence which is focused on the laser medium is absorbed by the laser medium. Thus the equation is multiplied by the focusing factor which is defined as the volume of the fluorescer cell divided by the volume of the laser medium (V_f/V_o). This gives the average power density absorbed by the laser medium (MeV cm^{-3} s^{-1}).
- In the seventh step, the power density in the laser medium must be converted from MeV cm^{-3} s^{-1} to W cm^{-3}. The conversion factor is 1.6×10^{-13} J/MeV.

Thus the power density absorbed by the laser for a multiphase Photon Intermediate Direct Energy Conversion (PIDEC) system (P_A) is,

$$P_A = 1.6 \times 10^{-13} N\ \sigma\phi_{th} Q\ \eta_{tr}\eta_f\eta_c\left(\rho_p/\rho_f\right)(V_f/V_o)\ \left(\mathrm{W/cm^3}\right) \qquad (3.14)$$

where η_c = coupling efficiency, η_f = fluorescence efficiency, ρ_p = average fuel density in fluorescer gm/cm^3, ρ_f = fuel density gm/cm^{-3}, V_f = volume of fluorescer cm^3, V_o = volume of laser cm^3

3.5 Contrasting the Surface Source, Volume Source and Multiphase Source

In an example which contrasts a surface source to the other types of sources, begin with a He-Ar nuclear-pumped laser (see Chap. 4). In developing the surface source example, consider a reactor design which uses rectangular plates made with a thin foil of uranium metal coated on a thin membrane as the fuel. Between the fuel plates there is a gas flow channel which is the laser medium. The gas which flows through the channels is a mixture of helium and argon. The helium-argon gas mixture is used in a well-known nuclear-pumped laser (NPL) that lases on the 3d[1/2]- 4p[3/2] argon transition at 1.79 μm. This laser requires a pumping power density of about

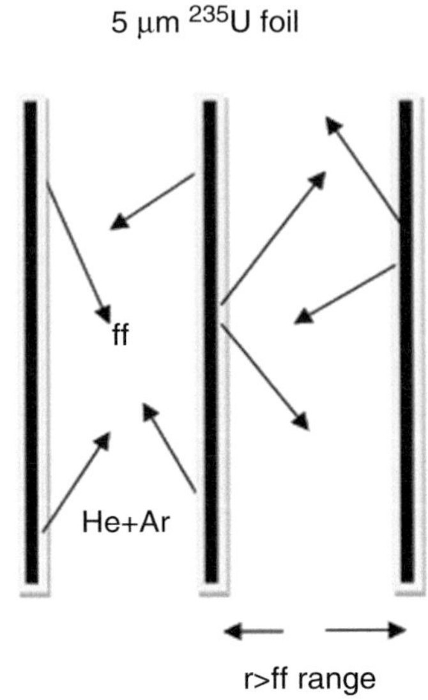

Fig. 3.2 A uranium layer coated on a thin membrane is used as the surface source to drive a He-Ar laser. The foil thickness is 5 μm and the channel width, r, is 0.01 cm

700 W/cm^3 [13]. The pumping power density for a 5 μm thick uranium metal foil in the arrangement shown in Fig. 3.2 will be calculated from first principles [4]. The efficiency for a slab foil can be found using Fig. 3.3. The slab data from this Figure is for a thin film coated on a thick plate (or a "half-cell"). If the foil is a thin film coated on a membrane in which fragments can escape from both sides then it is a "full-cell". The uranium metal foil's midpoint is 2.5 μm. As can be seen in Fig. 3.3, the transport efficiency for a 2.5 μm thick "half-cell" is 25 %. The "full-cell" efficiency is double that of the "half-cell" efficiency thus it is 50 % for this example.

To solve this problem, the following data is needed:

- atomic density of uranium metal is needed (4.82×10^{22} atoms/cm^3),
- the fission cross section (577 barns for a neutron energy of 0.0253 eV-see Appendix 4),
- the fission fragment energy (Q = 168 MeV),
- thermal neutron flux from a TRIGA reactor pulse (2×10^{15} n cm^{-2} s^{-1}),
- laser (or channel) width (r ~ 2 cm-approximate range of fission fragments in gas) and
- transport efficiency for a 5 μm uranium metal foil of 50 %.

So using this data in Eq. 3.12, the pumping power density can be found in Eq. 3.15.

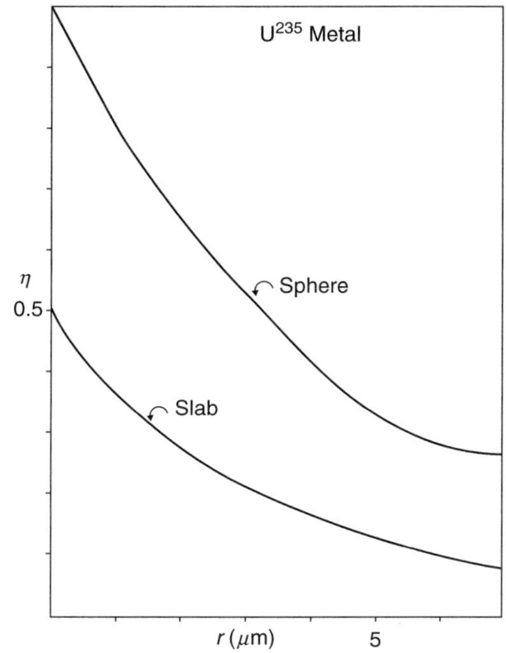

Fig. 3.3 The efficiency of a U-235 metal foil (made with highly enriched uranium) on a thick substrate (or a "half-cell") is shown along with a spherical uranium particle as a function of fuel thickness or radius

$$P_s = 3 \times 10^{-13}(4.82 \times 10^{22})(577 \times 10^{-24})(2 \times 10^{15})(168)(0.5)(5 \times 10^{-4})/1$$
$$= 350.4(\mathrm{W\ cm^{-3}}) \quad (3.15)$$

Thus the pumping power density for this design using a neutron flux of 2×10^{15} n cm^{-2} s^{-1} is 350.4 W cm^{-3}. This power density does not exceed the argon laser's threshold pumping power density of 700 W cm^{-3}. So, highly enriched uranium foils can't be used to pump the He-Ar laser using a TRIGA reactor.

In contrast to the surface source problem a volume source problem can also be done. Assume that a He-Ar laser is used and that the laser gas mixture contains ^{3}He which is mixed with argon and is exposed to a thermal neutron flux of 2×10^{15} n cm^{-2} s^{-1} (assuming no flux depression in the volume). The cross section for the ^{3}He(n, T) p reaction is 5000 barns (for a neutron energy of 0.0253 eV). Q for the reaction is 0.76 MeV. The atomic density of 10 atm of ^{3}He is 24.4×10^{19} atoms/cm^3. Using Eq. 3.13, it is shown that the power density of a volume source does not exceed the minimum required pumping power density for the argon laser (296.7 W $cm^{-3} <$ 700 W cm^{-3}) as seen in Eq. 3.16.

Table 3.1 Power density in a laser volume (W/cm^3) with a flux of 5×10^{16} cm^{-2} s^{-1} and ^{235}U fuel

Source type	Power density in laser (W/cm^{-3})
Surface (^{235}U foil)	8,760
Volume (3 atm ^{3}He)	7,500
Aerosol with black ^{235}U pellets with photon concentration[a]	56,060
Aerosol[a] with reflective ^{235}U pellets with photon concentration[a]	448,500

[a]This value assumes the use of a photon concentration factor (V_f/V_o) of 100 ([12, 14]) and a coupling efficiency for the reflective aerosol fueled flashlamp of 40 % or a coupling efficiency for the black aerosol of 5 %

$$P_v = 1.6 \times 10^{-13}\left(24.4 \times 10^{19}\right)\left(5000 \times 10^{-24}\right)\left(2 \times 10^{15}\right)(0.76)$$
$$= 296.7 \ \left(\mathrm{W/cm^3}\right) \quad (3.16)$$

To demonstrate an embedded source, assume that a photolytic laser is pumped by a nuclear-pumped fluorescence source fueled with a reflective aerosol. This problem is aimed at finding the potential power density deposited in a laser by a nuclear driven flashlamp fueled with an embedded aerosol in the fluorescer medium. The problem assumes that the embedded aerosol will interact with the fluorescence medium and produce photons. The photons are then concentrated on a laser medium by reflective surfaces (even though better concentrator designs are discussed in Chap. 4, this calculation uses the concentrator design in reference [12]). It is assumed that the aerosol fuel is highly enriched uranium metal, the reactor produces a thermal neutron flux of 2×10^{15} n cm^{-2} s^{-1}, the photon transport efficiency is 40 %, the fluorescence efficiency is 50 %, the coupling efficiency is 60 %, the (ρ_p/ρ_f) ratio is 1×10^{-3}, and the concentration factor (V_f/V_o) is 100. Using Eq. 3.14, an achievable power density of 17,940 W cm^{-3} is calculated from Eq. 3.17 which is sufficient to drive many photolytic lasers.

$$\mathrm{P_A} = 1.6 \times 10^{-13}\left(4.82 \times 10^{22}\right)\left(577 \times 10^{-24}\right)\left(2 \times 10^{15}\right)(168)(0.4)(0.5)(0.6)(0.001)(100)$$
$$= 17{,}940 \ \left(\mathrm{W\ cm^{-3}}\right) \quad (3.17)$$

The conventional surface and volume type sources are limited to power densities which are adequate for chemical generation and electricity production but are marginal for driving a short wavelength high-energy, high-efficiency laser (Table 3.1). The embedded aerosol source was conceived of as a way of boosting power densities for pumping high-power/high-efficiency lasers.

Most of the interesting high power/efficient atomic or molecular lasers require large power densities. One of the advantages of using the embedded aerosol source is that the reflective waveguides which transport the photons to a photolytic laser media can be bent around a radiation shield to protect the laser from radiation damage. Thus solid-state laser media, which are susceptible to radiation damage, can use this method. One other benefit of photolytic pumping is that the fluorescence source can be properly tuned to the absorption band(s) of the laser. With a tuned fluorescence source, the efficiency of this type of system increases dramatically. The concept is flexible in that there are other possible ways to embed the fuel source in the fluorescer medium besides aerosols. It is also feasible to use thin foils which are anchored to the fluorescing chamber walls, for example. Other potential fuel shapes with scale lengths less than the range of the ions produced in the fuel source can also be attached to the chamber walls or embedded as an alternative. To keep the active region of the reactor core optically thin, the wall and fuel surfaces need to be coated with reflective material.

3.6 Energy Conversion

In terms of direct energy conversion, the particular method of utilizing the ion source [surface, volume, aerosol or other embedded fuel shapes, $^{235}UF_6$ and uranofullerenes] is important since it affects the scaling of the integrated reactor/laser with respect to size, pressure, mixtures, etc. Hence, one should pay attention, in the following discussion, to the type of coupling that is being described.

Historically, three paths for energetic ion energy conversion have been pursued (see Fig. 1.1 in Chap. 1). The first path is that of particulate radiolysis in Chemical Synthesis Systems (CSSs) in which ionizing radiation induces the formation of chemicals from base reactants. Chemicals can be used to store energy. For example hydrogen is a well-known chemical used for energy storage that is potentially competitive with gasoline [15]. Other potential energy storage chemicals are ammonia, hydrazine and carbon monoxide [7]. Although fundamentally the same concept, the second path, Nuclear-pumped Lasers (NPLs), deals with the production of excited atomic or molecular states for the express purpose of inducing stimulated light emission. This particular path has progressed in two basic directions:

1. Direct ion pumping
2. Indirect ion pumping
   (a) Ion pumping a fluorescer to photolytically drive a laser [1]
   (b) Ion pumping metastable states or long lived states to utilize in a transfer laser [1]

The third path is Direct Electricity Production (DEP). Many of the same considerations in NPLs are also considerations for DEP. In this case the energy produced is in the form of an electrical potential which drives a current flow and provides the energy to drive electrical devices.

3.6.1 Physical Considerations for the Production of Excited Atomic and Molecular States (Lasers)

The use of ionizing radiation to produce excited atomic and molecular states encompasses the fields of radiolysis (the synthesis of chemicals from ionizing radiation), nuclear batteries and nuclear-pumped lasers [1, 16]. This section will focus on nuclear-pumped lasers for illustrative purposes. In the vernacular of laser physics, the term pumping refers to the rate at which the upper state of the laser level is created by the continuous flow of energy into the laser medium. This is analogous to pumping water up-hill into a reservoir and storing energy in the reservoir due to gravitational potential energy. When needed the water can be released to flow through a turbine to create electrical power. Like the water analogy, the potential energy stored in the upper laser level is released through the production of photons using stimulated emission. Stimulated emission requires an optical resonance cavity made up of mirrors which causes the photons to reflect back and forth through the excited laser media thus inducing more stimulated emission.

Nuclear-pumped lasers are driven by ionizing radiation such as the fragments from fission reactions. These types of lasers have received international attention over the years because of their potential to become high energy/high power laser systems. Primarily, nuclear pumping bypasses the need for a secondary energy conversion cycle (e.g., production of electricity with a generator). Nuclear-pumped lasers scale to much higher energies than electrically driven lasers, they use nuclear fuels which store an enormous amount of energy thus allowing the system to operate for years without refueling. Nuclear-pumped lasers can produce a steady-state laser beam capable of beaming power over long distances. The overall mass of the nuclear-pumped laser system will be small when compared to other technologies with similar operational characteristics [1, 5, 6, 17–20]. Nuclear-pumped lasers were envisioned for many applications in which high energy steady-state lasers could be used such as a photon propulsion system for space travel, an antiballistic missile system, a mining device and an asteroid deflector (see Chap. 6).

Nuclear-pumping technology has limitations which can be addressed in the system design. One limitation stems from the physics of typical high-power, high-efficiency laser systems. Most of the desirable lasers require high pumping power density to drive the laser. For example a desirable feature of a laser is to have a short wavelength ($\lambda < 1$ μm). High efficiency short wavelength lasers driven by electrical discharges typically require very large threshold power densities ($P_{th} > 100$ kW/cm^3) to produce the stimulated emission that creates the laser beam. To achieve power densities of this magnitude, electrical driven lasers must have pulse widths on the order of microseconds. The pulsing frequency of electrical driven lasers is usually very slow because the excess heat generated in pulsing can't be transferred out of the system quickly. In contrast nuclear pumping sources do not have narrow pulse widths (except for a nuclear bomb). Even though it is possible to develop pulsed nuclear fission reactors, a nuclear fission reactor is most suitable for steady-state operation. Nuclear pumping can achieve very high energy densities (up to ~50 J/cm^3) but even with the fastest controlled pulsed nuclear fission technology, the energy can only be delivered in a pulse width of about 200×10^{-6} s. This means that the power density from direct nuclear pumping technology is limited (to about ~50 kW/cm^3 or less). Also, due to laser physics principles, the most attractive lasers are generally inefficient (a very good gas laser efficiency is $\eta_L \sim 1$ %). Thus there will be a great deal of waste heat to deal with. As will be discussed in Chap. 6, these limitations can be accommodated in the nuclear-pumped laser's system design.

The theory of direct nuclear pumping dates back to 1963. There is a long history of attempts to drive a laser with radiation from nuclear reactions [1, 6].

One of the first series of NPL experiments were conducted in 1963 and 1964 with Northrop's pulsed TRIGA reactor [21]. These experiments used mixtures of 95 % Ne and 5 % O_2 driven by thin uranium foils (i.e., a surface source). The experiment was based on theoretical predictions that neon metastable states created by excitation from fission fragments slowing down in the gas could transfer energy to molecular oxygen and pump the 8446 nm line. However, the experiment showed emissions from multiple lines in oxygen that were believed to be lasing at 800, 840, 890 and 1000 nm and multiple lines in Ne that were believed to be lasing at 1150, 1520 and 2100 nm. The researchers also observed what they believed were two strong coherent laser beam spots on a photographic color film. The lines were estimated to lase at 340 nm or 440 nm. The 440 nm line was believed to be from a Ne ion transition.

After the 1963/1964 nuclear-pumped laser experiments which were funded by internal Northrop Research & Development dollars, the US Air Force decided to fund further work (Air Force contract AF33(615)-5000) [22–24]. Northrop studied pure noble gases (Ne, Ar, Kr, Xe) and instead of using fission fragments to pump static gas mixtures (like the 1963/1964 experiments), the new experiment was set up to circulate the gases through a fission irradiation tube. There were issues with the flowing system. It was believed that lasing was observed on a Xe line. Due to the financial demands of the Vietnam War, the Air Force cut funding for this program in 1968.

The first published report of a successful demonstration of a direct nuclear-pumped laser occurred in 1974 when both Lawrence Livermore National Laboratory and Los Alamos National Laboratory reported gamma-ray driven lasers pumped by a thermonuclear explosion [25, 26]. Nuclear bomb-driven lasers are only good for one high-energy/power laser burst and are not controllable. The desire to have lasers which can be controlled and operate continuously motivated the effort to develop lasers coupled to fission power reactors. The first published report of a successful and reproducible laser driven by a nuclear reactor (meaning a controlled fission process) was reported by Sandia National Laboratory in 1975 [27]. The CO molecular laser utilized a fast burst reactor as a neutron source and a thin film of UO_2 surrounding the laser gas for the production of fission fragments to drive the laser. This laser is still one of the most efficient nuclear-pumped lasers discovered to date. A problem with the molecular CO laser is that it must be cooled to 77 K in order to achieve its maximum efficiency.

Many other nuclear-pumped lasers have since been discovered (see Chap. 4). These include lasers pumped with ^{3}He gas (used as the ion source for driving the laser through the $^{3}He(n, p)T$ reaction) mixed in with the laser medium, lasers pumped by a thin film of ^{10}B (used as the ion source for driving the laser through the $^{10}B(n, \alpha)^{7}Li$ reaction) surrounding the laser medium and lasers pumped by partial pressures of $^{235}UF_6$. Nuclear-pumped lasers with efficiencies of approximately 1–2.5 % at room temperature have been reported (e.g., atomic Xe @ 1.79 μm and molecular CO_2 @ 10.6. μm [1, 28, 29]).

Nuclear-pumped lasers are envisioned for applications which require photon absorption of very high energy levels (see Chap. 6). A high energy laser should be >1 MJ, it should be about one percent efficient or greater, and it should be able to operate efficiently at the temperature of the working fluid in a fission reactor (which can be potentially high) if it is a direct nuclear-pumped laser or it can operate at modest temperatures if it is an indirect nuclear-pumped laser (e.g., PIDEC). NPLs compare favorably with other high energy laser candidates. The features unique to NPLs which generate these favorable comparisons are: (1) NPLs are driven by the energetic ions from nuclear reactions (e.g., fissile fuel interactions with neutrons), thus an NPL is directly coupled to the core of an operating nuclear reactor; (2) the nuclear pumping scheme scales with volume (limited only by neutron transport) while common electrical pumping schemes scale with surface area; (3) NPLs use a direct energy conversion process and will not be weighed down by a secondary energy conversion system; (4) nuclear-pumped lasers can operate for years with a single fuel loading; (5) nuclear-pumped lasers can operate steady-state producing a high energy laser beam; and (6) unlike electrically-driven laser systems which have engineering difficulties in achieving 1 MJ energies, a self-critical reactor/laser can scale to energies well beyond 1 MJ (see Chap. 5).

To understand the subtleties of making comparisons between NPLs and other types of lasers, it is important to further explore nuances in the various types of reactor/laser interfaces. Since there are two media which need to be considered (the fuel and the transducer), it is possible to look at the state of matter for each medium to categorize the potential types of interfaces (see Table 3.2). Of the 16 possible interfaces between the nuclear fuel and the laser media, only cases 1, 2, 3, 6, and 11 from Table 3.2 can be designed with present day technology. A summary of the state-of-the-art in direct interfaces is discussed in Chap. 5.

3.7 Nuclear Driven Flashlamps

The concept of a Nuclear Driven Flashlamp (NDF) was developed to increase the power density and efficiency of power reactor/laser interfaces and to open up new possibilities for nuclear-pumped lasers. The NDF is based on generating a photon source using the products of nuclear reactions for excitation with the capability of efficiently producing narrow band fluorescence (meaning almost a monochromatic light source). For example, the $6^{1}S_0 \rightarrow 6^{1}P_1$ (254 nm) transition in mercury is very

Table 3.2 Potential types of reactor/laser interfaces

Case	State of fuel	State of laser
1	Solid	Solid
2	Solid	Liquid
3	Solid	Gas
4	Solid	Plasma
5	Liquid	Solid
6	Liquid	Liquid
7	Liquid	Gas
8	Liquid	Plasma
9	Gas	Solid
10	Gas	Liquid
11	Gas	Gas
12	Gas	Plasma
13	Plasma	Solid
14	Plasma	Liquid
15	Plasma	Gas
16	Plasma	Plasma

efficient (~40 %). It emits line radiation at 254 nm that is slightly Doppler broadened (not perfectly monochromatic but close). The problem with line radiation is that as the size of the lamp scales up or the power density scales up, there will be self-absorption of the 254 nm photon (meaning that the density of the 6^1P_1 lower energy state is sufficiently high or the path length is sufficiently long that before the photon exits the medium it has a high probability of being absorbed). All line radiation is subject to self-absorption. The self-absorption mechanism is based on having a high enough density of the lower energy state so that a 254 nm photon released at one point has a high probability of being absorbed along its pathway before exiting the excitation region (Fig. 3.4). A discharge region where the photons escape with a minimal probability of being absorbed is called an optically thin medium. If the probability of absorption is high it is called optically thick. Lamps based on line radiation do not scale well with power density or size [6] and become optically thick rather quickly as the system size is scaled.

In order to produce lamps that are optically thin and able scale with power density and size, fluorescence sources that have an unbound lower energy state are needed. Sources such as these will remain optically thin regardless of power density or size. One such source type exists and is called excited dimmers (excimers). The nature of the excimer is that its upper energy state is bound and the lower energy state is unbound. Table 3.3 shows the fluorescence efficiency for some excimer fluorescers.

Excimer flashlamps can be very efficient and can be used for photolytic processes such as driving a photolytic laser, chemical production by photochemistry, or for electricity production using advanced photovoltaic cells [1, 6, 7]. With regard to the nuclear-pumped laser application, photolytic lasers have a lower power

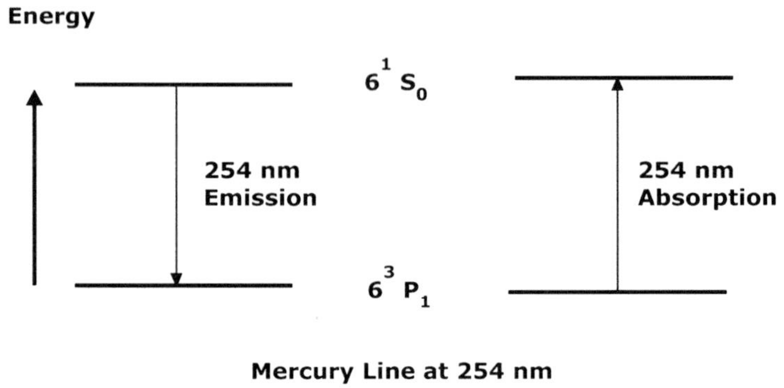

Fig. 3.4 Emission and self-absorption for the mercury $6\,^1S_0 \rightarrow 6\,^3P_1$ (254 nm) transition

Table 3.3 Theoretical and experimental efficiencies for rare-gas and rare-gas halide excimers [6]

Excimer	λ (nm)	η theoretical	η experimental	References
Ar_2	126	50·5	na	[1]
ArF	193	33	na	[1]
Ar_2F	284	22·4	na	[1]
Kr_2	147	47·5	46.2	[30]
KrF	249	34·0	na	[1]
Kr_2F	415	16·8	na	[1]
NeF	108	35	na	
Xe_2	172	47·7	68[a]	[31]
			45[b]	[30]
			39 ± 10	[32]
			43[c], 61[a]	[33]
			45[d], 46[e]	[33]
XeF	346	23·7	12.1	[34]
XeBr	282	29·1	15	[35]
XeI	252	37·2	na	[1]

[a]Fission fragments and gaseous phase reactant
[b]Electron beam and gaseous phase reactant
[c]Gamma rays and gaseous phase reactant
[d]Fission fragments and liquid phase reactant
[e]Gamma rays and liquid phase reactant
na indicates that no data could be found

density threshold than discharged pumped lasers because the photolytic laser takes advantage of an energy resonance in the laser medium with the photon. In establishing the capabilities of the nuclear driven flashlamp, if the photons are not concentrated, then the effective power density of the flashlamp is limited to approximately 5 kW/cm^3 (for a neutron flux of about 2×10^{15} n cm^{-2} s^{-1} [36]). If photon concentration is used then the effective power density from a nuclear driven flashlamp can increase by an order of magnitude or more. Photolytic lasers which are power density dependent, can then be chosen and designed with appropriate photon concentration levels to assure that the minimal power density threshold is met. A method of concentrating the photons generated by a nuclear driven flashlamp was first devised in 1979 [19, 37] as a means to increase the absorbed power density in a laser medium. With high power density a wide class of high-energy, high-efficiency nuclear-pumped lasers became feasible. Examples of photolytic lasers where photon concentration methods can be used are the photolytically driven excimer lasers [12]. Without photon concentration to boost the effective power density, these lasers could not be driven by a NDF. These concepts were expanded in new directions for direct energy conversion of nuclear energy including chemical production from photochemical reactions and electricity production from advanced photovoltaic cells (made with wide band-gap materials) [1, 38].

3.8 Neutron Sources

Neutron sources are used to drive nuclear reactions and are capable of producing the highest power densities in direct energy conversion applications. Neutron sources are classified in three categories:

- Small: less than 1×10^9 n s^{-1} such as radioisotopes or a sealed tube neutron generator,
- Moderately-sized: between 1×10^9 n s^{-1} and 1×10^{12} n s^{-1} such as a plasma focus, light ion accelerator, and high energy photoneutron or photofission source, and
- Large: greater than 1×10^{12} n s^{-1} such as a nuclear fission reactor, a spallation neutron source or a nuclear fusion reactor.

3.8.1 Accelerator Based Neutron Sources

Neutrons are produced by high energy interactions between ions and target nuclei (e.g., Spallation Neutron Source or SNS [39]). In an SNS a negatively charged hydrogen ion is produced and injected into a linear particle accelerator (LINAC). The LINAC accelerates the ion to 90 % of the speed of light. The ions pass through a foil which strips off the two electrons and transforms the H^- ion to a H^+ ion. The protons are injected into a storage ring where the protons are trapped through motion around the ring. After enough protons accumulate in bunches and proton bunches are released in a microsecond proton pulse every second. The proton pulses strikes the nuclei of mercury atoms where 20–30 neutrons are spalled. Figure 3.5

Fig. 3.5 View of the spallation neutron source at Oak Ridge National Laboratory [42]

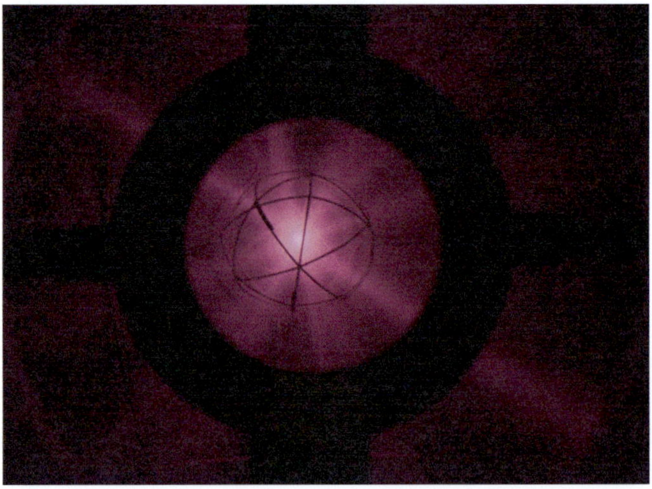

Fig. 3.6 The University of Missouri glow discharge spherical IEC in star mode [43]

shows the scale of a SNS. Other types of accelerators can produce neutrons and some are commercially available such as the Cockcroft-Walton neutron generator and the Inertial Electrostatic Confinement Neutron Source (Fig. 3.6) [40, 41]. The cost per neutron makes the SNS, as well as other accelerator based neutron sources, range from expensive to very expensive.

Fig. 3.7 A Californium 252 radioisotope neutron source [44]

3.8.2 Radioisotope Based Neutron Sources

Radioisotope neutron sources which spontaneously fission such as Cf-252 (Fig. 3.7), radioisotopes that interact with alpha emitters such as PuBe, AmBe and AmLi and photoneutron sources such as Be plus a gamma source are also available. Radioisotopes are considered to be small neutron sources since they typically produce about a million neutrons per second. Radioisotopes constantly emit radiation and thus are hazards which need shielding.

3.8.3 Fission Reactor Based Neutron Sources

The least expensive neutron source, based on the cost per neutron, is a fission reactor. Neutrons are produced as a byproduct of the fission reaction [45]. Due to thermodynamic considerations, steady-state reactors are typically limited to producing a thermal neutron flux in the range of 0.5×10^{12}–1×10^{15} n cm^{-2} s^{-1}. A new generation of steady-state reactors with higher neutron flux output possibly can be designed using fuels that can operate at high temperature and/or that have exceptional heat transport properties. Such reactors could take advantage of the very thin fuel elements designed as part of the embedded source concept for nuclear pumping which allows the bulk of the fission energy to escape to the transducer medium which is typically a gas. A reactor using gases to absorb the energy from

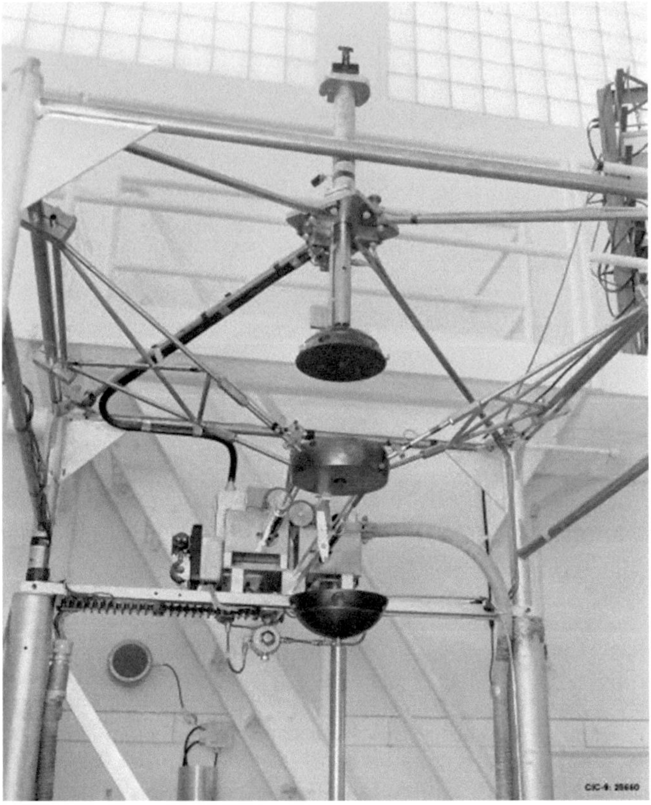

Fig. 3.8 A photo of the Godiva fast burst reactor at Los Alamos National Laboratory is shown. The device uses an unreflected 54 kg sphere of 93.7 % enriched uranium [46]

fission and by using optical emission as a means to rid the gas of a good fraction of its energy content would have beneficial cooling properties.

3.8.3.1 Fast Burst Reactor

A fast burst reactor is made up of two half spheres of highly enriched uranium that are separated (known as a Godiva type reactor (Fig. 3.8)). The name Godiva comes from the fact that the spheres are bare. Each of the spherical halves is subcritical, but when put together, they become critical. The half spears are driven together by a force and then driven apart by thermal forces created by the reactor as it undergoes a rapid power excursion. Godiva type reactors are capable of producing about a 200 μs Full Width Half Maximum (FWHM) pulse with a peak thermal neutron flux (in an optimum polyethylene sleeve) of about 1×10^{17} neutrons cm^{-2} s^{-1} and a peak fast flux of about 1×10^{19} neutrons cm^{-2} s^{-1}. Fast burst reactors are available in the USA at White Sands and in Russia at Chelyabinsk (Table 3.4).

Table 3.4 List of operational fast burst reactors taken from the IAEA database

Country	Facility name	Type	Thermal power (kW)	Status
Russian Federation, Chelyabinsk	FBR-L, Fast burst-laser	Fast burst	800,000	Operational
United States of America, White Sands	Fast burst (FBR)	Fast burst	250,000,000	Operational

3.8.3.2 Transient Reactor Test Facility

The Transient Reactor Test (TREAT) facility was built in the 1950s on the Idaho National Laboratory reactor test site. It was designed for safety experiments on reactor fuels. TREAT uses the original fuel which forms a 19 by 19 array of elements with a cross section of 10 cm^2 at 122 cm long. All of these fuel elements contain fuel blocks which are made from a small amount of highly enriched uranium oxide particles which are finely dispersed in a graphite matrix. The blocks must have at least 1.7 g cm^{-3} of carbon with a boron concentration less than 2 ppm. The carbon to uranium ratio is on the order of 1000. The fuel assemblies can be removed to load various experiments. TREAT is capable of generating many types of transient shaped pulses with a variety of power levels. In order to provide this wide array of pulse shapes TREAT has an Automatic Reactor Control System (ARCS) which can use an open and closed loop computer control that can create transients which are (a) initiated by a step insertion of reactivity and terminated when a predetermined reactor period, power, energy release or time interval has been achieved or (b) follow a predetermined time and power or period profile and terminated when a predetermined experimental parameter, reactor parameter or time interval has been reached. TREAT can have power bursts with Full Width Half Maximum (FWHM) pulses ranging from 66 ms to several seconds (Fig. 3.9).

3.8.3.3 Training, Research, Isotope Production General Atomic (TRIGA) Reactor

During the First Geneva Conference on the Peaceful Uses of Atomic Energy in 1955, General Atomics made a presentation on the Training, Research, Isotope Production General Atomic (TRIGA type reactor). Since that first presentation 60 TRIGA reactors have been constructed. Three TRIGA models were produced,

- Mark-I-an underground pool without beam tubes
- Mark-II-an above ground tank with several beam tubes
- Mark III-an above ground oval tank with a moveable core

A TRIGA reactor usually has 57 standard fuel elements (fueled with low enriched uranium – 20 % enrichment). The fuel/moderator material is 8 % by weight of LEU, 91 % zirconium by weight and 1 % hydrogen by weight. Each

Fig. 3.9 Transient reactor test facility (TREAT) [47].

fuel element is 3.75 cm in diameter and 72.24 cm long. The cladding is made of 0.76 mm of aluminum or 51 mm of steel. The active core diameter is 49.5 cm and core height 35.56 cm. There are 2.3 kg of uranium 235 in the core. The negative temperature coefficient of the TRIGA core is one of its most important features because, as will be described, it allows the core to be pulsed. The negative temperature coefficient is about 9.7×10^{-5} Δk/°C for an aluminum clad fuel element with graphite reflector and 12.1×10^{-5} Δk/°C for stainless steel clad and a water reflector.

The core can be configured for high power operation (greater than 100 kW) by adding more fuel elements and additional graphite reflector elements placed in the outer rings of the core. TRIGA reactors have excess core reactivity which allows the core to be pulsed by using a gas pressure surge to force the central control rod upwards to increase core reactivity up to $4. This leads to a rapid raise in power until the negative temperature coefficient causes the power to drop leading to a pulse with approximate peak core powers of up to 4000 MW (Figs. 3.10 and 3.11).

3.8.3.4 High Flux Steady-State Reactor

There are a number of high flux steady-state reactors in the world. The highest flux for this class of reactor is from the Advanced Test Reactor (ATR). The ATR is a highly flexible high flux nuclear reactor (Fig. 3.12) that is made up of fuel elements which are 94 % U-235. The fuel contains 1.2 wt% (weight percent) U-234, 0.7 wt%

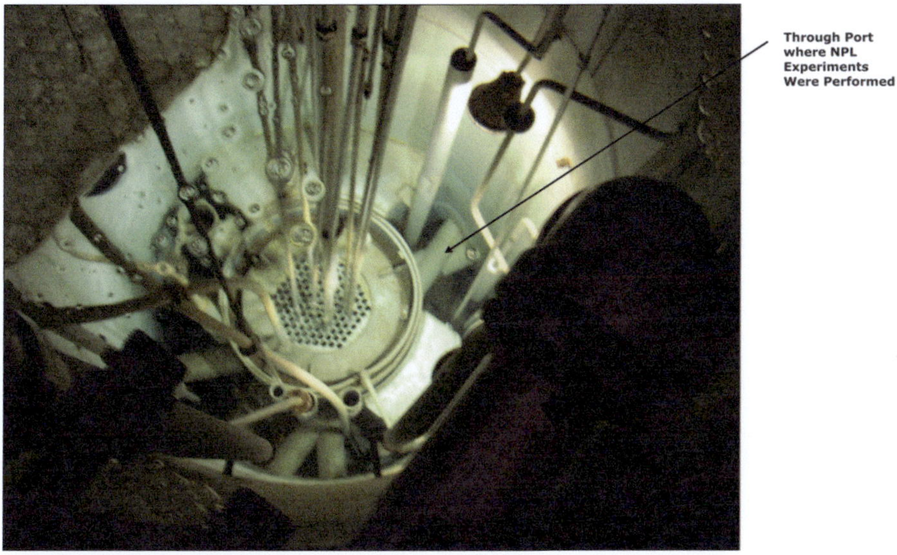

Fig. 3.10 University of Illinois TRIGA reactor core which shows the position of the through port where NPL experiments were placed [48].

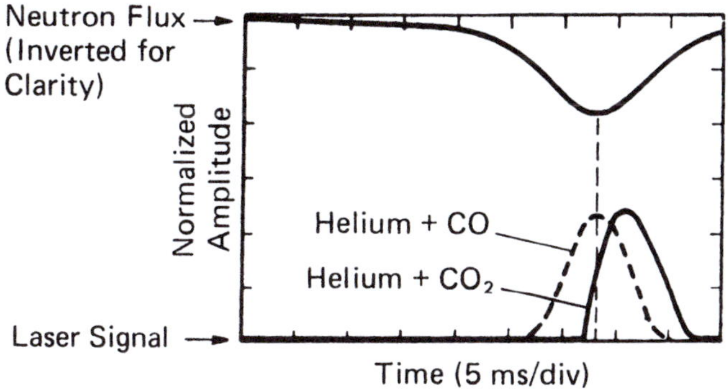

Fig. 3.11 Shown is a typical pulse width for a TRIGA reactor that is used to drive an atomic carbon laser using mixtures of helium + CO_2 and helium plus CO [49]. The full width at half maximum of the pulse is about 10 ms

U-236 and 5–7 wt% U-238. The fuel element is made up of uranium aluminide (UAlx). The total weight of the fuel element is not greater than 25 lb. Four different types of ATR Mark VII fuel elements have been used (7F, 7NB, 7NBH and YA). Each type of fuel element is virtually identical only varying in the content of the fuel matrix. In 7F all 19 fuel plates consist of enriched uranium in an aluminum matrix. The eight outer plates (plates 1–4 and 16–19) contain boron as a burnable poison. Fuel type 7NB contains no burnable poison thus all 19 fuel plates consist only of enriched uranium in an aluminum matrix. The 7NBH fuel is very similar to

Fig. 3.12 The advanced test reactor Cherenkov radiation emission during operation [50]

the 7NB fuel element with the exception that there are one or two borated plates in the elements. The YA plate is identical to the 7F plate with the exception that plate 19 is only an aluminum alloy plate with no uranium or boron. The ATR fuel plates are formed by rolling them into a curved shape (Fig. 3.13). Fuel plate 1 has the smallest radius of curvature while plate 19 has the largest radius of curvature. The uranium aluminide is 0.02 in. thick for all 19 plates. The side plates are made of an aluminum alloy 6061-T6 or 6061-T651 and are 0.19 in. thick. The maximum channel thickness between fuel plates is 0.087 in.

The ATR fuel elements are loaded into four lobe configuration shown in Fig. 3.14. There are 77 irradiation positions which are 48 in. in length and 0.5–5 in. in diameter.

3.8.3.5 New Generation Pulsed Reactors

It may also be possible to build a new generation of pulsed reactors as well with very narrow neutron pulse widths (discussed in Chap. 5). Such reactors would have to have a very high neutron multiplication value (k value—See Chap. 9 of Volume 1 of Energy Resources and Systems [52]). As discussed, fast burst reactors are designed to have very high k values thus having fast pulse rise times. The fast burst reactor is essentially two halves of a fuel assembly that is brought close together rapidly with a spring loaded system creating a critical assembly which generates

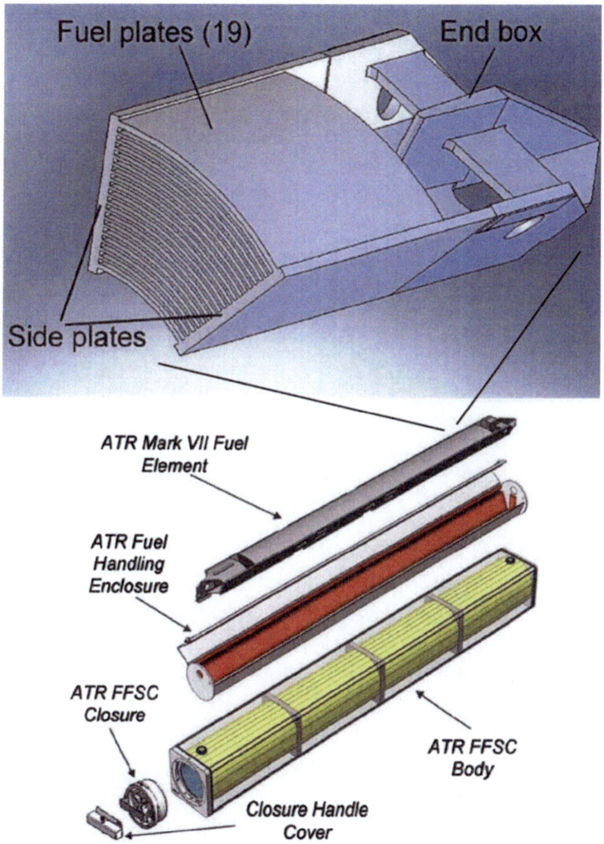

Fig. 3.13 ATR fuel element made up of 19 fuel plates [51]

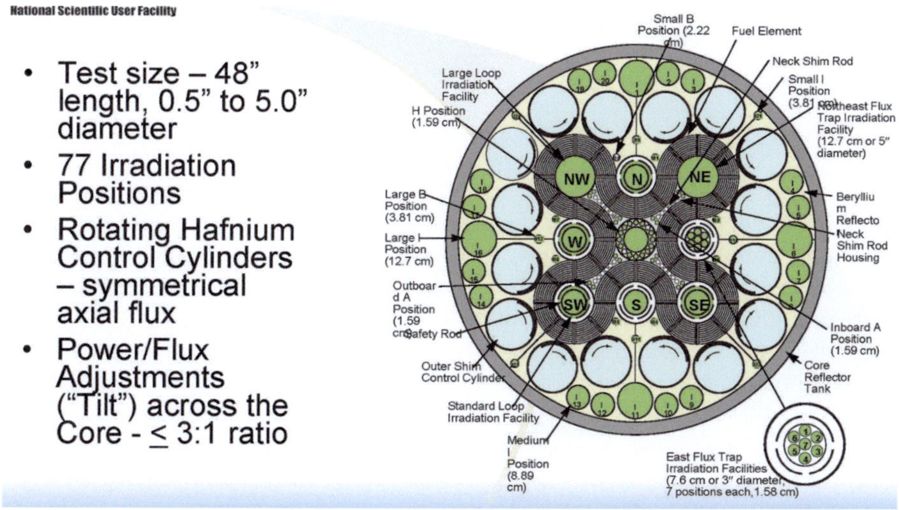

Fig. 3.14 Experimental test positions in the advanced test reactor [50]

Table 3.5 Maximum power deposition (W/cm^3) from various sources [1]

	Source type	Fast burst reactor	Transient test reactor	Pulsed TRIGA reactor	Steady state reactor	Radionuclide		
						^{210}Po	^{226}Ra	^{222}Rn
Surface	^{10}B	106	106	26.5	1.06			
	^{235}U	4,740	4,740	119	4.74			
	Radionuclide					0.042	0.263	
Volume	^{3}He	14,820	14,820	371	14.8			
	$^{10}BF_3$	3,500	3,500	87.4	3.5			
	$^{235}UF_6$	44,300	44,300	1,100	44.3			
	Uranofullerene (HEU)	44,300	44,300	1,100	44.3			
	Radionuclide							5.05

Flux from a fast burst reactor is 1×10^{17} n cm^{-2} s^{-1} in a 200 μs pulse, the flux from TREAT is 1×10^{17} n cm^{-2} s^{-1} in a 40 ms pulse, the flux from pulsed TRIGA is 2.5×10^{15} n cm^{-2} s^{-1} in a 10 ms pulse, and the flux from steady-state reactor 1×10^{14} n cm^{-2} s^{-1}

energy rapidly and also rapidly disassembles itself. It will be challenging to improve upon the pulse rise time characteristics of a fast burst reactor, but it may be possible to improve the k value with a new design as well as the peak neutron flux by having much better heat transfer characteristics.

3.9 Power Density

The achievable power density directly impacts the type of direct energy conversion device that is feasible. The maximum achievable power density from various types of radiation sources that are based on currently available technology is shown in Table 3.5.

3.10 Summary

As discussed in references [1, 7], the power density is a critical factor in producing chemicals from nuclear reactions, as discussed in references [1, 16, 53] it is an important factor in the production of electricity and as discussed in references [1, 6, 54], it is an important factor in driving lasers.

With regard to nuclear-pumping as applied to lasers, each laser system has a threshold pumping power density where oscillation begins. The issue with nuclear pumping is that the power density it can generate is limited to about 44.3 kW cm^{-3} with a pulsed fast burst reactor (Table 3.5). This may seem like a large power density, but the majority of high-power high-efficiency lasers require an even higher power density. Thus only a few unique laser systems have been able to operate with nuclear pumping (see Chap. 4). Of these, only a handful of these nuclear-pumped lasers exhibit an efficiency of 1 % or better. Furthermore there are only three of these nuclear-pumped lasers which have wavelengths in the ultraviolet to visible range. There is a good reason why so few nuclear-pumped lasers operate in ultraviolet to visible range since short wavelength lasers typically require a higher threshold power density than long wavelength lasers. The availability of so few efficient short wavelength lasers impact the use of nuclear-pumped lasers for applications of interest (discussed in Chap. 5) because these applications work best with short wavelength high-efficiency lasers.

Problems

1. What is the reaction rate of a thermal neutron beam (neutrons moving parallel with each other at a speed of 2200 m/s) entering a 5 μm thick ^{235}U metal film with an area much larger than the beam diameter normal to the film's surface.
2. If the film in problem 1 is interfaced with a gas plenum of thickness 5 cm and a pressure of 760 Torr. Calculate the average power density deposited in the gas for neutron fluxes of 5×10^{14} n cm^{-2} s^{-2}, 5×10^{15} n cm^{-2} s^{-2}, 5×10^{16} n cm^{-2} s^{-2}, and 5×10^{17} n cm^{-2} s^{-2}.
3. If a gas plenum of thickness 5 cm is filled with helium gas mixed with 10 % $^{235}UF_6$ at a total pressure of 760 Torr, calculate the average power density deposited in the gas for neutron fluxes of 5×10^{14} n cm^{-2} s^{-2}, 5×10^{15} n cm^{-2} s^{-2}, 5×10^{16} n cm^{-2} s^{-2}, and 5×10^{17} n cm^{-2} s^{-2}.
4. Derive an equation similar to Eq. 3.14 for a $^{235}UF_6$ fuel at various partial pressures of UF_6.
5. Derive an equation similar to Eq. 3.14 for a ^{235}U caged uranofullerene fuel at various uranofullerene densities.
6. Estimate the maximum possible power density in laser cell (driven with a ^{235}U surface source) placed in a test position of the Advanced Test Reactor operating that has an average thermal flux of 1×10^{15} n cm^{-2} s^{-1}.
7. What is the maximum achievable power density for a nuclear-pumped laser using the TREAT reactor as the neutron source? What is the integrated energy density over the pulse width of TREAT.
8. What is the maximum achievable power density for a nuclear-pumped laser using a spallation neutron source? What is the integrated energy density over the pulse width of the spallation source?
9. What is the maximum achievable power density for a nuclear-pumped laser using a best case Inertial Electrostatic Confinement (IEC) neutron source? What is the integrated energy density over the pulse width of an IEC?
10. Derive an equation similar to Eq. 3.14 for a ^{235}U thin film placed on the surface of a diamond solid-stated laser medium.
11. Derive an equation similar to Eq. 3.14 for a ^{235}U thin film fuel placed on the surface of a CsI scintillator crystal.
12. Discuss the possibility that uranium can be embedded in a scintillator crystal as a dopant. What are the pluses and minuses of doping a crystal with ^{235}U and exposing it to a large neutron source?

References

1. Prelas MA, Loyalka SK (1981) A review of the utilization of energetic ions for the production of excited atomic and molecular states and chemical synthesis. Prog Nucl Energy 8:35–52
2. Prelas M, Boody F (1992) A comparison of electrical, fusion-generated-ion, and fission-generated-ion ICF drivers. In: Miley G, Hora H (eds) Laser interaction and related plasma phenomena. Springer, New York, pp 67–78
3. Bethe HA (1979) The fusion hybrid. Physics Today, pp 44–55, May 1979. Available: http://large.stanford.edu/publications/coal/references/docs/bethe.pdf
4. Chung A, Prelas M (1984) Charged particle spectra from U-235 and B-10 micropellets and slab coatings. Laser Part Beams 2:201–211
5. Prelas M, Charlson E (1989) Synergism in inertial confinement fusion: a total direct energy conversion package. Lasers Part Beams 7:449–466
6. Prelas MA, Boody FP, Miley GH, Kunze JF (1988) Nuclear driven flashlamps. Laser Part Beams 6:25–62
7. Prelas MA, Romero J, Pearson E (1982) A critical review of fusion systems for radiolytic conversion of inorganics to gaseous fuels. Nucl Technol Fusion 2:143
8. Prelas M, Boody F, Zediker M (1984) A direct energy conversion technique based on an aerosol core reactor concept. In: IEEE International conference on plasma science, 84CH1958-8
9. Prelas MA (1981). In: U. Missouri (ed) Notorized notes dust core reactor and laser, p 7. doi:10.13140/RG.2.1.4119.9201
10. Prelas MA (1982) Nuclear powered space lasers: an evaluation of current technology. Nuclear-Pumped Laser Corporation, December 5, 1982. doi:10.13140/RG.2.1.1355.4405
11. Henry AF (1975) Nuclear-reactor analysis/Allan F. Henry. MIT Press, Cambridge, MA
12. Prelas MA, Jones GL (1982) Design studies of volume-pumped photolytic systems using a photon transport code. J Appl Phys 53:165–169
13. Voinov AM, Krivonosov VN, Mel'nikov SP, Mochkaev IN, Sinyanski AA (1991) Quasi-cw nuclear-pumped laser utilizing atomic transitions in argon. Sov J Quantum Electron 21:157
14. Prelas MA, Boody FP, Kunze JF (1986) A compact aerosol core reactor/laser fueled with reflective micropellets. In: Hora H, Miley GH (eds) Lasers and related plasma phenomena, vol 7. Plenum Press, New York
15. Ghosh TK, Prelas MA (2011) Energy resources and systems: volume 2: renewable resources. Springer, Dordrecht
16. Prelas MA, Weaver CL, Watermann ML, Lukosi ED, Schott RJ, Wisniewski DA (2014) A review of nuclear batteries. Prog Nucl Energy 75:117–148
17. Miley GH (1977) Direct nuclear pumped lasers: status and potential applications. In: Schwarz H, Hora H (eds) Laser interactions and related plasma phenomena, vol 4A. Plenum Press, New York, pp 181–229
18. Boody FP, Prelas MA, Anderson JH, Nagalingam SJS, Miley GH (1978) Progress in nuclear-pumped lasers. In: Billman KW (ed) Radiation energy conversion in space, vol 61. AIAA, New York, p 379
19. Prelas MA, Lecours MJ, Schlapper GM, Brugger RM (1980) Nuclear-pumped laser research at the University of Missouri. Trans Am Nucl Soc 34:810
20. Prelas M, Charlson E, Boody F, Miley G (1990) Advanced nuclear energy conversion using a two step photon intermediate technique. Prog Nucl Energy 23:223–240
21. Eerkens J et al (1965) Nuclear-pumped laser progress report, May 20, 1965, Northrop Space Labs Report NSL 66–64, Northrop Corporation
22. Eerkens JW (1967) Nuclear-pumped lasers. Northrop Corporation. Technical session VII-D, in technology review for AF systems command, USAF-RTD, pp 23–24
23. Eerkens J (1967) Research on Direct Nuclear Pumping of Gas Lasers (DNPGL). Northrop Corporation, October 1966, Report NCL-68-13R
24. Eerkens J et al (1968) Nuclear-pumped noble-gas ion-laser experiments, December 1968, Northrop Corporation.
25. Ebert PJ, Terderber LJ, Koehler HA, Kuckuck RW, Redhead DL (1974) Xenon pumped by gamma rays. Presented at the IQEC-8, San Francisco

26. Lyons PB, Clarke JS, Metzger DS (1974) Gamma initiated HF laser. Presented at the IQEC-8, San Francisco
27. McArthur DA, Tollefsrud PB (1975) Observation of laser action in CO gas excited only by fission fragments. Appl Phys Lett 26:187
28. Voinov AM, Dovbyshov LE, Krivonosov VN, Mel'nikov SP, Podmoshenskii IV, SinyanskiiI AA (1979) Nuclear-pumped IR lasers using ArI, KrI, and XeI transitions. Soviet Phys Technol Phys Lett 5
29. Rowe MJ, Liang RH, Schneider RT (1981) Nuclear-pumping CO_2. In: Proceedings of 1981 IEEE international conference on plasma science, Ottawa, Canada, p 149
30. Eckstrom DJ, Lorents DC, Nakano HH, Rothem T, Betts JA, Lainhart ME (1979) The performance of Xe2* as a photolytic driver at low e-beam excitation rates. In: Topical meeting on excimer lasers
31. Walters RA, Cox JD, Schneider RT (1980) Trans Am Nucl Soc 34:810
32. Prelas MA (1985) Excimer research using the University of Missouri Research Reactor's Nuclear-Pumping Facility, National Science Foundation. doi:10.13140/RG.2.1.3714.7366
33. Baldwin GC (1981) On vacuum ultraviolet light production by nuclear irradiation of liquid and gaseous xenon. Unpublished report, Los Alamos National Laboratory. Available: http://www.osti.gov/scitech/biblio/6416823
34. Miley GH, Boody FP, Nagalingham SJS, Prelas MA (1978) Production of XeF(B-X) by nuclear-pumping. APS Conf Bull Am Phys Soc 24:117
35. Boody FP, Miley GH (1979) Data on XeBr* which indicated a 15% efficiency, Nuclear Engineering Department, University of Illinois (unpublished)
36. Boody FP, Prelas MA (1983) Photolytic dual-media nuclear pumping of excimer lasers. AIP Conf Proc 100:349
37. Prelas MA (1979) Nuclear-pumped photolytic energy focus, November, University of Missouri Patent Disclosure, #80-P-UMC-020. doi:10.13140/RG.2.1.4169.0728
38. Prelas MA (1981) A potential UV fusion light bulb for energy conversion. Bult Am Phys Soc 26:1045
39. Galambos JD, Osen DK (1997) Parameter and cost model for spallation neutron source studies. In: Particle accelerator conference, 1997. vol 3, pp 3851–3853
40. Watterson JIW (2000) A review of accelerator based neutron sources and their applications. IAEA, Vienna, Austria IAEA-TECDOC-1153, May 2000
41. Miley GH (2013) Life at the center of the energy crisis: a technologist's search for a black swan. World Scientific, Hackensack
42. Oak_Ridge_National_Laboratory (2013) Spallation neutron source. Available: http://neutrons.ornl.gov/facilities/SNS/index.shtml. September 22, 2014
43. Meyer R, Smith Z, Prelas M, Loyalka S (2008) Ion flow convergence in spherical inertial electrostatic confinement devices. Phys Plasmas 15:022105–022105-13
44. Oak_Ridge_Associated_Universities (2014) Cf252 sources. Available: http://www.orau.org/ptp/collection/Sources/cf-252.htm. September 22, 2014
45. Ghosh MAPTK (2009) Nuclear. In: Ghosh T, Prelas M (eds) Energy resources and systems. Springer, Dordrecht, pp 453–647
46. Los_Alamos_National_Laboratory (1953) Lady Godiva at TA-18. Available: http://www.lanl.gov/newsroom/photo/history.php. September 22, 2014
47. Idaho_National_Laboratory (2014) Transient reactor test facility. Available: http://www.inl.gov/research/transient-reactor-test-facility/. September 22, 2014
48. General Atomics (2014) TRIGA reactors. Photograph provided by the nuclear, plasma and radiological engineering Department from its archives, University of Illinois at Champaign-Urbana, 2015. Available: http://www.ga.com/triga. September 22, 2014
49. Prelas M, Miley G (1981) Dynamics of the nuclear and electrically pumped 1.45-μm atomic carbon laser in mixtures of helium +CO and helium +$CO_{2.}$ Dynamics of the nuclear and electrically pumped 1.45-μm atomic carbon laser in mixtures of helium +CO and helium +$CO_{2.}$ Nucl Technol/Fusion 1
50. Idaho_National_Laboratory (2014) Advanced test reactor. Available: http://www.inl.gov/research/advanced-test-reactor-research/. September 22, 2014

51. deBoisblanc DR, Leonard BH (1964) Advanced test reactor, US Patent 3149044 A
52. Ghosh T, Prelas M (2009) Energy resources and systems: volume 1: fundamentals and non-renewable resources. Springer, Dordrecht
53. Prelas MA, Boody FP, Miley GH, Kunze JF (1988) Nuclear driven flashlamps. In: Lasers and particle beams, vol 6, pp 25–62
54. Prelas M, Watterman ML, Wisiewski DA, Neher JA, Weaver CL (2014) A review of nuclear pumped lasers and applications (Asteroid Deflection). In: 121st ASEE 2014 annual conference, Indianapolis, Indiana. Available: http://www.asee.org/public/conferences/32/papers/10774/view

Chapter 4
Nuclear-Pumped Lasers

Abstract Nuclear-pumped laser research has been constrained by the available power density from the various "pumping" schemes used. This section looks at the history of nuclear-pumped laser research and describes the 50 or so laser systems which have worked using the nuclear-pumping method. It also addresses the mechanisms for these lasers as well as the current understanding of why these specific laser systems worked.

Keywords Nuclear pumped lasers • Mechanisms • History

4.1 Introduction

High power, high efficiency lasers with short wavelengths (less than 1000 nm) typically require high pumping power densities (greater than 100 kW/cm^3) [1]. - Nuclear-pumping methods are able to achieve high energy density (up to 46 J/cm^3) but the pulse width of the energy delivery system is limited due to the pulse width of the reactors that deliver the neutrons to drive the reactions that power the laser (see Chap. 3 Table 3.5). Thus a fast burst reactor, which has the shortest pulse width, is limited to a power density of about 46 kW/cm^3. In contrast, an electron beam can deliver its energy in a much shorter pulse width (a microsecond or so) thus achieving power densities on the order of 1 MW/cm^3. Hence NPLs must be based on specialized high power/high efficiency lasers which can operate at the moderate power densities available from direct pumping by nuclear reactions or at the boosted power densities available from indirect pumping methods such as the energy focus.

Nuclear-pumped lasers, as discussed in prior chapters and reiterated here, have utilized the following neutron-initiated reactions to produce charged particles to pump the laser,

$$^{3}\mathrm{He} + \mathrm{n} \rightarrow \mathrm{T} + \mathrm{p} + 0.76\ \mathrm{MeV} \tag{4.1}$$

The tritium ion has a kinetic energy of 0.19 MeV and the proton has a kinetic energy of 0.57 MeV.

M. Prelas, *Nuclear-Pumped Lasers*, DOI 10.1007/978-3-319-19845-3_4

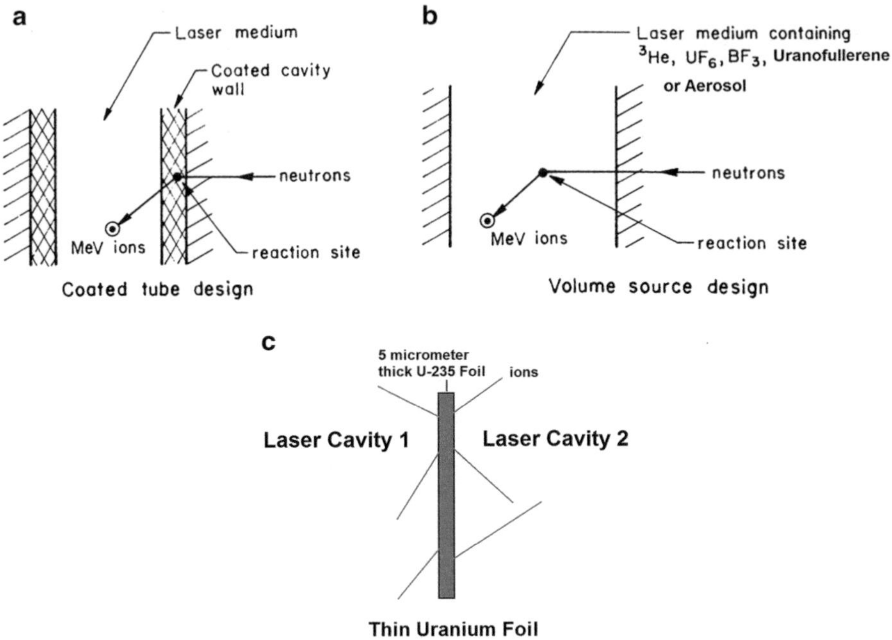

Fig. 4.1 Nuclear-pumping sources are of two forms: (**a**) a surface source where the high energy ions are created in a coating on the walls of the cell, (**b**) a volume source where a gaseous fuel, uranofullerene or aerosol fuel is used to create the high energy ions that pump the laser medium or (**c**) a thin uranium foil suspended between two separate laser cavities [1]

$$^{10}\mathrm{B} + \mathrm{n} \rightarrow {}^{7}\mathrm{Li} + {}^{4}\mathrm{He} + 2.35\ \mathrm{MeV} \tag{4.2}$$

The lithium-7 ion has a kinetic energy of 0.85 MeV and the helium-3 ion has a kinetic energy of 1.5 MeV.

$$^{235}\mathrm{U} + \mathrm{n} \rightarrow \mathrm{ff_h} + \mathrm{ff_l} + \mathrm{vn} + 210\ \mathrm{MeV}(\text{with 165 Mev in}\ \mathrm{ff_h}\ \text{and}\ \mathrm{ff_l}) \tag{4.3}$$

where the 165 MeV is the charged particle recoil energy (not total energy produced in fission), ff_l is the light fission fragment, ff_h is the heavy fission fragment, and v is the statistical average number of neutrons released per fission (see Chap. 2).

Thermonuclear Explosion → gamma rays

Of these sources, only fission promises high enough system efficiencies for applications. As is shown in Fig. 4.1, the uranium can be present in the system as: (a) a surface source, (b) a volume source; or (c) a thin foil suspended between two laser cavities (to maximize the capture of fission fragments in the laser media).

Thin coatings of B-10 or U-235 on surfaces are fabricated by electroplating (flat surfaces) or by using a colloid made with powder and a fluid which is baked on (curved surfaces). Surface sources have relatively low efficiency since nuclear reactions are isotropic. (50 % of the ions are emitted in the wrong direction and are immediately absorbed by the wall. Of the remaining 50 %, less than half of the ion energy will eventually escape the coatings.) Therefore surface sources, although they have the advantage of being chemically inert, have low efficiencies (on the order of 20–30 %) and can only be scaled to volumes on the order of the ion range in the gas. It is also possible to create thin foils of the fuel (e.g., ^{235}U) coated on a membrane (made from carbon nanotubes for example) which can be anchored between two laser cavities (Fig. 4.1c) so that the ions emitted by the fuel excite the laser media on both sides of the foil. The suspended thin foil design is capable of ion transport efficiencies of about 50–60 %. Several designs of reactors using a surface source (Fig. 4.1a) interfaced with a laser will be discussed in Chap. 5 [1, 2]. Surface sources are also useable for an energy focus [3] provided that the foil is coated with a thin layer of reflector material to form a waveguide.

Volume sources (Fig. 4.1b) can produce a much a greater energy release because the bulk of the charged particle energy is absorbed in the laser volume and the volume is not constrained in size to the range of the charged particles. Until 1981, UF_6 was the only volume type fission source considered. UF_6 vapor however produces strong quenching of the chemical kinetics of most laser systems [1, 2, 4–7] and cannot be used for most fluorescence production applications because of its very high absorption at wavelengths shorter than 420 nm (Fig. 4.2). An aerosol core reactor was studied in 1981 and can provide volume type source advantages without the major drawbacks of UF_6 [8–10]. Calculations by Chung and Prelas suggest that 50–80 % of the energy generated in a micropellet will be available to pump the fluorescer for an indirect driver (See Fig. 4.3) [11]. Another major advantage of the aerosol concept, in addition to the efficiency, is that the micropellets are essentially inert and will not affect the chemical kinetics of the reactant being "pumped". Finally, the aerosol concept can be used with structural materials such as silicon carbide which should allow for high operating temperatures. This last advantage implies that other energy producing cycles can be added to make the system even more efficient [2].

A gaseous core-like reactor concept using uranofullerene (uranium 235 atom(s) trapped in a C60 cage seen in Fig. 4.4) as a fuel has also been suggested [12]. A reactor fueled with nanoparticle sized uranofullerenes would be chemically inert, and would interface well with laser media or flashlamp media. The uranofullerene nanoparticles would be suspended in the gaseous laser medium or flashlamp media and would have gas-like properties in the gas mixture.

As will be discussed in Chap. 6, some of the possible applications of nuclear-pumped lasers are space based. The main problem with power sources for space-based lasers is the required energy that must be stored by the fuel (which directly influences the mass of the laser system). In Table 4.1, the energy stored in 1 kg of various fuel sources is shown. Hydrogen has one of the highest specific energies available in a chemical energy storage system and is used here as an illustration of the maximum expected specific energy for chemical energy storage. The energy stored in the fuel is an important factor for a deep space mission where launch mass

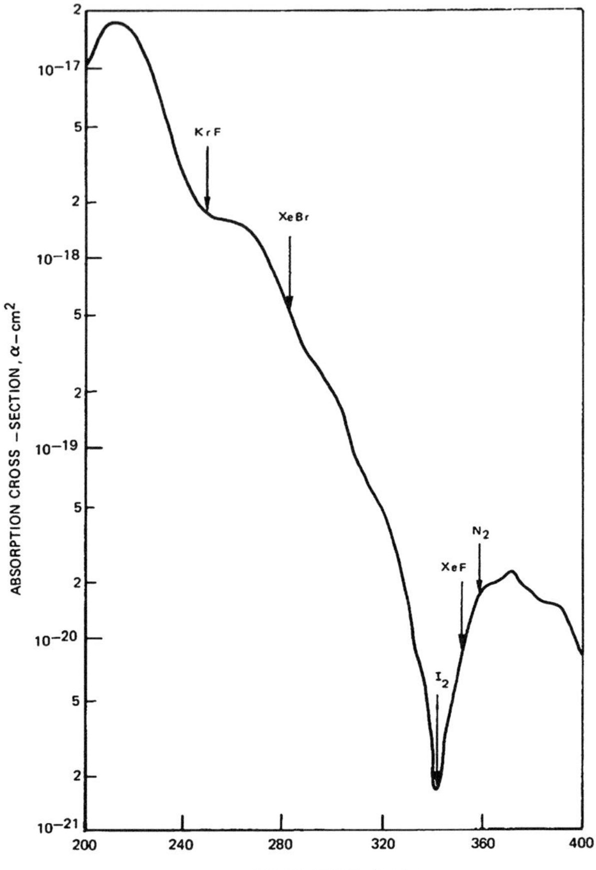

Fig. 4.2 UF_6 photoabsorption spectrum [4]. Shown are the wavelengths for the krypton fluoride excimer laser, the xenon bromide excimer laser, the molecular iodine laser, the xenon fluoride excimer laser and the molecular nitrogen laser

directly impacts the cost. For this example let us suppose that the mission is to use a nuclear-pumped laser to deflect a 1.4 km asteroid. The energy necessary to deflect the asteroid is on the order of 1.68×10^{15} J (as will be discussed in Chap. 6). To calculate a minimum fuel mass, assume that there is a 100 % efficient energy conversion system. The mass of the required fuel at minimum would be 1.87×10^{-2} kg of antimatter, 2.92 kg of Deuterium (D) and Tritium (T) for fusion, 1.17×10^{1} kg of U-235, 7.63×10^{2} kg of Pu-238 or 1.17×10^{7} kg of hydrogen (see Table 4.2). However, the efficiency of converting energy from a fuel to a laser beam is not even close to being 100 %.

A more realistic example would be that of an electrically pumped laser which uses electrical current from a fuel cell powered by hydrogen and oxygen. The fuel cell efficiency is about 50 %. The power transmission and conditioning circuits are about 80 % efficient. The efficiency of an e-beam driven XeF laser is about 1 %. Thus the minimum mass of the hydrogen fuel needed for this mission is about 2.3×10^{9} kg (i.e., Fuel Mass $= 1.17 \times 10^{7}$ kg/(0.5*0.8*0.01) $= 2.3 \times 10^{9}$ kg) If the

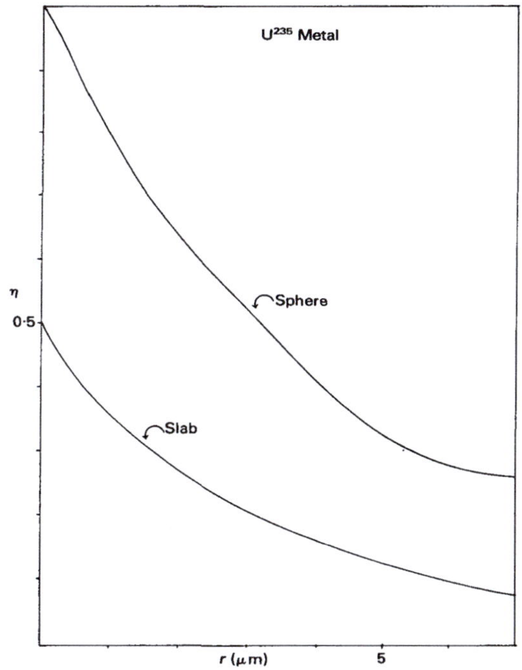

Fig. 4.3 Fission fragment transport efficiency in microspheres and "half-cell" slabs of various thicknesses [11]

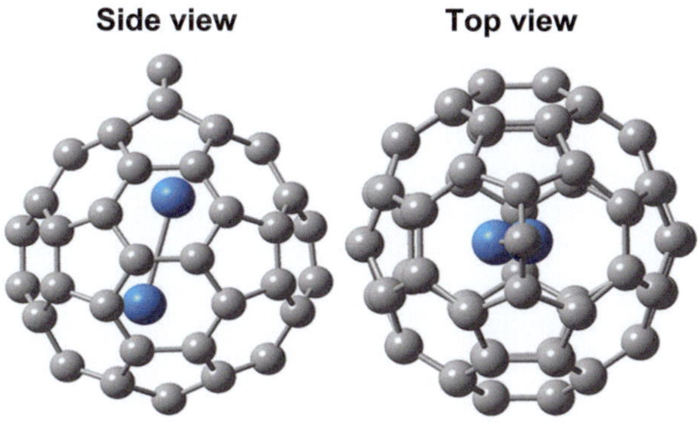

Fig. 4.4 Illustration of U2@C61 [13]

Table 4.1 Energy storage capability for various fuels

Storage type	Specific energy (MJ/kg)
Antimatter	8.99×10^{10}
Deuterium-tritium fusion	5.76×10^{8}
Uranium-235	1.44×10^{8}
Pu-238 α-decay	2.2×10^{6}
Hydrogen	143

Note that hydrogen is representative of chemical energy sources

Table 4.2 The total energy needed to deflect an asteroid of a given diameter and the minimum mass of U-235 or hydrogen needed to store that amount of energy

Laser power (W)	Asteroid diameter (km)	Lp*t (J) (t = 180 day)	Mass U235 (kg)	Mass H_2 (kg)
3.94×10^4	0.1	6.13×10^{11}	4.25×10^{-3}	4.28×10^3
4.92×10^6	0.5	7.66×10^{13}	5.32×10^{-1}	5.35×10^5
3.94×10^7	1	6.13×10^{14}	4.25	4.28×10^6
3.15×10^8	2	4.90×10^{15}	3.40×10^1	3.43×10^7
1.06×10^9	3	1.65×10^{16}	1.15×10^2	1.16×10^8
2.52×10^9	4	3.92×10^{16}	2.72×10^2	2.74×10^8
4.92×10^9	5	7.66×10^{16}	5.32×10^2	5.35×10^8
8.51×10^9	6	1.32×10^{17}	9.19×10^2	9.25×10^8
1.35×10^{10}	7	2.10×10^{17}	1.46×10^3	1.47×10^9
2.02×10^{10}	8	3.14×10^{17}	2.18×10^3	2.19×10^9
2.87×10^{10}	9	4.47×10^{17}	3.10×10^3	3.12×10^9
3.94×10^{10}	10	6.13×10^{17}	4.25×10^3	4.28×10^9

mass of the fuel cell, the mass of the oxygen, the mass of the power circuits, the mass of the laser, the mass of the radiator, the mass of the optics and the mass of the laser gas were also included, the total mass of the system would be significantly larger. In contrast, the amount of U-235 needed to supply the energy is so small, that it is not nearly enough mass to make up the fuel for a critical nuclear reactor core. The minimum mass of uranium for a critical nuclear reactor/laser system is about 10 kg. The other contributions to system mass are the laser cavity, the laser medium, the reactor control system, the radiator (for heat transfer in space) and the optics. This point is made to illustrate that a system based on nuclear fuels will have some mass advantages for use in space.

Estimating the mass of a NPL system is complex because of the efficiencies that govern the conversion of the energy stored in the nuclear fuel into laser energy. The choice of the laser is important both for the laser efficiency as well as the wavelength. The unfortunate fact is that there are not a lot of nuclear-pumped lasers to choose from due to the limited power density that is feasible from the nuclear reactors that drive the nuclear reaction. It took 11 years from the time nuclear-pumping was initially proposed in 1963 to the time that gamma rays from a thermonuclear explosion were used to pump a xenon excimer laser and a hydrogen fluoride laser in 1974 (Table 4.3). About a year later, the carbon monoxide NPL was reported which used a fast burst nuclear reactor as the neutron source, thus representing the first NPL pumped by a nuclear reactor. About 50 NPLs have been discovered to date (Table 4.3). The main characteristic of these NPLs is that they have low threshold pumping power densities. The highest efficiency achieved thus far is about 3 % as shown in Table 4.4 (it should be noted that advanced NPL concepts, to be discussed in Chap. 5, can potentially achieve efficiencies greater than 10 %). There have been a number of experiments in the Former Soviet Union aimed at finding high efficiency NPLs and these experiments used internal mirrors

Table 4.3 Summary of NPLs that have been discovered to date [15]

Laser [gas mixture]	Wavelength (nm)	Group	Pumping source	Threshold power density (W/cm^3)	Efficiency %
Xe excimer [16]	170	LLNL	Gamma rays from a thermonuclear explosion	$\sim 1 \times 10^6$	Unknown
HF [17]	2500–3000	LANL	Gamma rays from a thermonuclear explosion	$\sim 1 \times 10^6$	Unknown
CO [18] [$He + N_2 + CO$]	5100–5600	Sandia	Fission products from ^{235}U film	1.86×10^3	1
Xe [19] [He + Xe]	3500	LANL	Fission products from ^{235}U film	1.9	2×10^{-5}
N [20] [$Ne + N_2$]	860, 939	U of Illinois	Products from ^{10}B film	4.5	5×10^{-5}
Hg^+ [21]	615	U of Illinois	Products from ^{10}B film	4.5	9×10^{-5}
Ar [22] [He + Ar]	1790, 1270	NASA	Products from ^3He gas	9.8×10^2	1.0×10^{-3}
Xe [23] [He + Xe]	2026, 3508, 3652	NASA	Products from ^3He gas	1.6×10^2	1.57×10^{-2}
Kr [24] [He + Kr]	2520, 2190	NASA	Products from ^3He gas	5.2×10^2	4×10^{-5}
Cl [24] [$He + Cl_2$]	1586	NASA	Products from ^3He gas	1.1×10^2	2×10^{-4}
C [25] [He + CO or CO_2] [Ne + CO or CO_2] [$Ar + CO_2$]	1450	U of Illinois	Products from ^{10}B film	0.04	6×10^{-3}
Ar [6, 7, 26] [He + Ar]	2870, 2397, 2031, 2021, 2010, 2060, 1190, 1150	USSR	Products from ^{235}U film	3.7×10^2	2×10^{-2}
Xe [6, 7, 26] [He + Xe, Ar-Xe, He-Ar-Xe]	30,110, 2810, 2650, 2630, 2480, 2030	USSR	Products from ^{235}U film	0.07	1.33
Cd [27] [He + Cd]	534, 538, 441.6	USSR	Products from ^3He gas	044.	8×10^{-3}
CO_2 [28] [N_2 transfer to $CO_2 + He$]	10,600	U of Florida	Products from ^{10}B Film	2.7×10^1	1
Xe [29, 30] [He + Ar + Xe] [Ar + Xe]	1730, 2030	Sandia	Products from ^{235}U film	~10	2

(continued)

Table 4.3 (continued)

Laser [gas mixture]	Wavelength (nm)	Group	Pumping source	Threshold power density (W/cm^3)	Efficiency %
Nd^{3+} with the $^4F_{3/2} \rightarrow {}^4I_{11/2}$ transition in Nd:Yag * [14] [Pumped with a Xe and Kr flash lamp]	1064	U of Missouri	Products from products from ^{3}He gas	67	~0.2
Nd^{3+} with the $^4F_{3/2} \rightarrow {}^4I_{11/2}$ 4 m fiber 50 μm diameter [7] [Pumped with a Xe flash lamp]	1060	VNIIEF	Fission fragments from ^{235}U	Unknown	Unknown
Nd^{3+} with the $^4F_{3/2} \rightarrow {}^4I_{11/2}$ 4 m fiber 50 μm diameter [7] [Pumped with CsI (Tl) crystal]	1060	VNIIEF	Gamma rays	Unknown	Unknown
Ne [31] [He + Ne + H]	585.3	U of Illinois	Products from ^{3}He gas	0.001	0.1
I [32] [XeBr flashlamp Pumped C_3F_7I].	1310	U of Illinois	Products from ^{3}He gas	~0.4	0.6
Kr [6, 7], He-Kr, He-Ne-Kr, Ne-Kr	3070, 2860, 2630, 2520, 2190, 1870, 1780	USSR	^{235}U fission fragments	~0.9	0.6
Ar [6, 7], He-Ar, Ne-Ar, He-Ne-Ar	2870, 2400, 2310, 2210, 2100, 2060, 1790, 1690, 1270, 1190, 1150	USSR	^{235}U fission fragments	~4.2	0.6
Ne [6, 7], He-Ne-Ar, Ne-Kr	585.3, 703.2, 724.5	USSR	^{235}U fission fragments	14	0.02
Zn [6, 7], He-Zn	747.9	USSR	Products from ^{3}He gas	~1.48×10^3	0.05
O [6, 7], He-Ne-O_2	2653, 2760	USSR	^{235}U fission fragments	~14.2	0.05
Cl [6, 7], He-Ne-CCl_4	1587	USSR	^{235}U fission fragments	~0.94	0.2

(continued)

Table 4.3 (continued)

Laser [gas mixture]	Wavelength (nm)	Group	Pumping source	Threshold power density (W/cm^3)	Efficiency %
N [6, 7], Ne-N_2,	938.7, 904.6, 862.9,859.4	USSR	^{235}U fission fragments	~56.4	0.05
C [6, 7], Ne-CO, Ne-CO_2, Ne-Ar-CO_2	940.6, 833.5	USSR	^{235}U fission fragments	~17.1	0.3
N_2^+ [6, 7]	391.4, 427.8	USSR	^{235}U fission fragments	~10	0.3

*Nd:YAG small signal gain of 0.051 cm^{-1} measured. The calculated laser power was 62 W
Note: The nuclear-pumped laser used in the USSR did not have windows cut at Brewster's angle. These lasers used internal mirrors in order to minimize cavity losses
Note 2: The 1963–1968 work of Northrop researchers in which it was claimed that lasing was observed in oxygen at 800, 840, 890 and 1000 nm, Ne at 1150, 1520 and 2100 nm, unidentified lines a 340 and 440 nm and a possible line in Xe was not included in this table because these laser lines were not verified by other researchers [33–36]

Table 4.4 A listing of Falcon nuclear-pumped lasers

Lasing atom	Wavelength (nm)	Intrinsic efficiency (%)
Xe	3370	>0.1
Xe	2650	>0.1
Xe	2630	>0.1
Xe	2032	2.0
Xe	1733	2.5
Ar	1792	1.4
Ar	1291	1.1
Ne	725	0.10
Ne	703	0.05
Ne	585	<0.1

Intrinsic efficiency is defined as the ratio of laser light energy to total energy absorbed by the gas [37]

in order to eliminate losses through Brewster angle windows (Table 4.5). All of the NPLs found thus far, with the exception of one solid-state laser, were gas lasers. Gas lasers have notable advantages. For one, they are not susceptible to radiation damage. However, they also have disadvantages as will be discussed in Chap. 5. Due to the relatively low atomic densities of a gaseous medium, a gas laser will have a comparatively large volume thus leading to systems which will have large dimensions and volumes (greater than 5 m^3) that may require assembly in space

Table 4.5 IR nuclear-pumped lasers in Xe, Kr or Ar which operated at the Russian institute (VNIIEF) using the VIR reactors or the TIBR reactors

Laser	λ, μm	Mixture	p, atm	Laser power watt	η_l, %	$\Phi_{th}\cdot 10^{14}$, cm^{-2} s^{-1}	$\Phi_{max}\cdot 10^{15}$, cm^{-2} s^{-1}	Reactor
Xe	1.73	He-Xe	2	80	0.2	4.0	1.1	VIR-2
	1.73	Ar-Xe	0.5	500	2.2	0.1	1.1	VIR-2
	1.73	He-Ar-Xe	1	410	2.5	0.1	1.1	VIR-2
	2.03	He-Ar-Xe	1	700	1.0	0.6	2.5	VIR-2M
	2.03	Ar-Xe	0.25	–	–	0.015	2.5	VIR-2M
	2.48	He-Xe	1	–	–	–	26	TIBR-1M
	2.65	He-Xe	5	2000	0.8	10	26	TIBR-1M
	2.63; 2.81	Kr-Xe	0.35	150	0.2	0.03	2.5	VIR-2M
	2.63; 2.65	Ar-Xe	0.5	230	2.0	0.1	1.1	VIR-2
	2.65	He-Xe	3	250	1.2	0.6	1.1	VIR-2
	2.65	He-Ne-Xe	1.5	520	0.6	1.2	2.5	VIR-2M
	2.65	He-Ar-Xe	1	320	0.5	0.1	2.2	VIR-2M
	2.65	Ne-Ar-Xe	1.1	570	0.6	0.4	2.5	VIR-2M
	3.11 (or 3.37)	Ar-Xe	0.5	–	–	–	1.1	VIR-2
	3.51	Xe	0.6	90	0.1	0.8	2.5	VIR-2M
Kr	1.78	He-Kr	2	50	0.3	0.6	1.1	VIR-2
	2.52	He-Kr	2	110	0.6	2.6	1.1	VIR-2
	3.07	He-Kr	2	40	0.2	1.7	1.1	VIR-2
	1.78; 1.87; 2.19	He-Kr	2	120	0.2	0.29	2.5	VIR-2M
	1.78; 1.87; 2.19	He-Ne-Kr	1	140	0.2	0.19	2.5	VIR-2M
	2.52; 3.07	He-Kr	2	420	0.6	3.2	2.5	VIR-2M
	2.19; 2.86; 3.07	Ne-Kr	1	100	0.1	0.68	2.5	VIR-2M
	2.19; 2.52 2.86; 3.07	He-Ne-Kr	1	460	0.6	1.2	2.5	VIR-2M
	2.63	Kr	0.25	2	–	0.2	2.5	VIR-2M

Ar	1.15; 1.19	He-Ar	2	250	0.1	100	26	TIBR-1M
	1.27	He-Ar	1	10	–	5.0	2.1	VIR-2M
	1.69; 1.79	He-Ar	2	390	0.6	3.2	2.1	VIR-2M
	1.69; 1.79	Ne-Ar	1	140	0.2	0.9	2.1	VIR-2M
	2.10	He-Ar	2	220	0.3	6.3	2.5	VIR-2M
	2.06; 2.10; 2.21	Ne-Ar	1	110	0.15	1.2	2.5	VIR-2M
	2.40	He-Ar	1	–	–	–	26	TIBR-1M
	2.40; 2.87	He-Ar	2	190	0.25	8.4	2.5	VIR-2M
	2.21; 2.31; 2.40	Ne-Ar	1	60	0.1	1.3	2.5	VIR-2M
	2.31; 2.40; 2.87	He-Ne-Ar	1.3	140	0.15	2.0	2.5	VIR-2M
Ne	0.5853	He-Ne-Ar	–	140–	0.1	4	130	EBR-L
	0.7032, 0.7245	Ne-Kr	–	30	0.02	120	130	EBR-L
Cd	0.5337, 0.5378	^3He-Cd	0.53	0.1	–	3	130	EBR-L
	0.4416	^3He-Cd	1.8	1000	0.4	7	130	EBR-L
Zn	0.7479	^3He-Zn	1.1	60	0.05	100	2.5	VIR-2M
O	2.653,2.760	He-Ne-O_2	2	30	0.05	3.0	2.5	VIR-2M
Nd^{3+}	1060	Xe NDF		200	?			VIR-2M
	1060	CsI(Tl)NDF		4,000	?			BR-1
Cl	1.587	He-Ne-CCl_4	2	130	0.2	0.2	2.5	VIR-2M
N	0.8594,0.8629	Ne-N_2	0.25	5	0.01	12	2.5	VIR-2M
	0.9046	Ne-N_2	0.25	20	0.05	15	2.5	VIR-2M
	0.9387	Ne-N_2	0.4	0.4	–	13	2.5	VIR-2M
C	0.8335	Ne-CO	0.75	6	0.01	3.6	2.5	VIR-2M
	0.9406	Ne-CO	0.75	36	0.05	7.6	2.5	VIR-2M
	0.9406	Ne-CO_2	0.4	4	0.01	3.4	2.5	VIR-2M
	0.9406	Ne-Ar-CO_2	0.75	120	0.3	3.7	2.5	VIR-2M
N_2^+	391.4	He-N_2-H_2	3.5–5	10	0.3	10	50	BARS-6
	427.8	He-N_2-H_2	3.5–5	10	0.2	10	50	BARS-6

The laser pressure is p in atmospheres, η_l is the laser efficiency, Φ_{th} is the threshold thermal neutron flux, Φ_{max} is the peak thermal neutron flux [6]

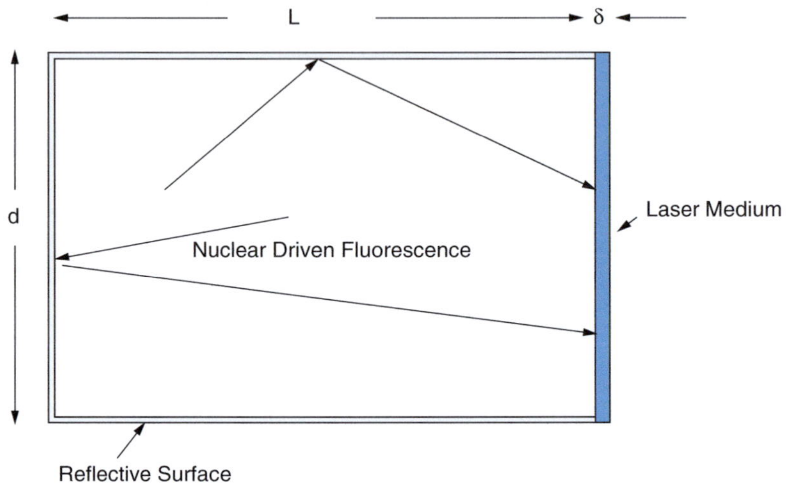

Fig. 4.5 An illustration of the principle of the energy focus concept [38] (Note that L is length of the nuclear driven fluorescence chamber, d is its diameter and δ is the thickness of the photolytic laser medium)

through the launch of multiple modular components [9]. On the other hand, solid-state systems can be susceptible to radiation damage but can have a relatively small volume (less than 1 m^3) which may be launched in a single payload [14].

It is desirable to design an efficient short wavelength NPL that operates at a steady-state high power level. The key to such a design is a system that is able to provide a high enough pumping-power density to drive the laser while allowing steady-state operation of the reactor. This is a difficult task given that efficient short wavelength lasers typically require high pumping-power densities while a reactor core typically needs to operate at a much lower power density (<0.1 kW/cm^3) to keep the fuel from melting. This is the reason that so few nuclear-pumped lasers have been discovered (about 50 or so) to date. The lasers in Tables 4.3 and 4.5 are some of the few that can operate at low power densities.

A method to boost the pumping power density of an NPL was developed in 1979 [38, 39] called the energy focus. The elements of the energy focus are that a nuclear-driven fluorescer produce photons from spontaneous emission and a non-imaging concentrator is used to focus the photons into the laser medium (Fig. 4.5). The energy focus reflects the photons from a large volume nuclear driven fluorescer to a small volume absorber (laser medium) [3, 9, 10, 40]. In Fig. 4.5, a simplified design is shown for a box lamp/reflector of length L with an absorber (laser) at the opening which has a thickness of a few photon mean free paths (δ). The power density in the laser will be much greater than the power density in the lamp (Eq. 4.4).

Table 4.6 Parameters of reactors that are typically available for research today

Reactor type	Peak neutron flux Neutrons/cm^2-s	Full width half maximum (ms)	References
TREAT	1×10^{18} fast neutrons	66	[42]
	1×10^{17} thermal neutrons		
Fast burst	1×10^{19} fast neutrons	0.2	[1, 43]
	1×10^{17} thermal neutrons		
TRIGA	2×10^{16} fast neutrons	10	[1]
	2×10^{16} thermal neutrons		
Steady-state research reactor	4×10^{14} fast neutrons	∞	[1]
	5×10^{14} thermal neutrons		

$$P_{laser} = P_{lamp}\eta_f\eta_{tr}\frac{L}{\delta} \tag{4.4}$$

where P_{laser} is power density of the laser, P_{lamp} is power density in the lamp, η_f is the fluorescence efficiency and η_{tr} is the photon transport efficiency.

If the reflector were perfect with no photon transport losses, then by using values for η_f of 0.5, η_{tr} of 1, L of 10 m, δ of 10×10^{-6} m (for a solid-state material), and P_{lamp} of 14.8 W/cm^3 (for a steady-state reactor using ^{3}He-see Table 3.5), it is found that the power density absorbed by the laser is,

$$P_{laser} = 14.8 \times 0.5 \times \frac{10}{1 \times 10^{-5}} = 7.4 \times 10^6 \ W/cm^3 \tag{4.5}$$

Reflector materials are not perfect but it was demonstrated that a right cylinder using a metallic reflector (e.g., aluminum for UV to visible light or sliver for visible light) makes a reasonably good waveguide [41]. The simplified waveguide shown in Fig. 4.5 will have a transport efficiency (η_{tr}) on the order of 90 % for a d/L ratio of 0.3 [41]. With a transport efficiency of 90 %, the power density absorbed by the laser using the simplified design will be about 6.67 MW/cm^{-3}. This type of power density is sufficient to drive any photolytic laser.

The energy focus is an advanced NPL concept that can be used to pump high efficiency gas lasers (such as the Iodine laser with a potential efficiency up to 10 %) or high efficiency solid-state lasers (such as GaN with a potential efficiency of up to 40 %).

There are a number of reactor based neutron sources that have been used in experiments as shown in Table 4.6. The sources are pulsed reactors (e.g., fast burst reactors or TRIGA reactors) and steady-state reactors. Table 4.6 shows the Full Width Half Maximum for the pulsed reactor sources as well as the peak thermal neutron flux.

The neutrons trigger one of the reactions shown in Eqs. (4.1), (4.2) and (4.3). If the reactant is Boron-10, He-3 or Uranium-235, there is an efficiency cost for the production of neutrons with the exception of uranium-235 which produces neutrons. Table 4.7 shows the theoretical maximum efficiency for a laser pumped by boron-10, He-3 and uranium-235. The low efficiencies for boron-10 and He-3 are due to the 400 MeV invested per neutron which is captured by the nucleus. Uranium-235 produces neutrons so it may be configured in a self-critical reactor configuration such that each fission reaction produces ions which "pump" the laser.

4.2 Thermonuclear Explosion

A thermonuclear bomb was used to drive a xenon excimer laser [16] and a hydrogen fluoride laser [17]. The gamma rays from the detonation excited the laser medium and the oscillation was able to occur before the optical cavity disintegrated [16]. The gamma rays can create power densities on the order of one Mega Watt per cubic centimeter.

The energy level of molecular xenon shown in Fig. 4.6, shows that the laser has an unbound lower laser level. It also shows that the system has a relatively high quantum efficiency. The Xe_2* excimer was modeled using either e-beams or ionizing radiation from nuclear reactions [44]. It was possible to benchmark the model with data taken from electron beam studies (note the reactions shown in Table 4.11). A study of xenon gas at 3 atm with a pump rate of 0.1 MW cm^{-3} was made. At an exposure time of 200 ns, the xenon excimer was found to have a fluorescence efficiency of 47.9 %, which agrees well with the results from experimental data of 48 % (Fig. 4.7). The dominant reactions in forming the xenon excimer are 1, 7, 11, 13, 14, 22 and 36 from Table 4.8 [44].

The xenon excimer laser requires high power densities ($>$100 kW cm^{-3}) but has excellent efficiency ($>$10 %) [46, 47]. The gamma rays from a thermonuclear explosion provides the excitation and ionization needed to pump the xenon excimer laser.

The hydrogen fluoride laser is an efficient laser that had been pumped by various types of electrical discharges [45, 48, 49]. The efficiency of the laser is 10 % or better with pumping power densities of on the order of hundreds of kilowatts per cubic centimeter. Thus the hydrogen fluoride laser was also a good candidate system to pump by thermonuclear explosion [17].

As discussed in Chap. 1, the high power density capability of a thermonuclear explosions makes it possible to pump a wide variety of lasers. The basis for the beryllium x-ray laser proposed by Edward Teller for the Strategic Defense Initiative was due the pumping power density of a thermonuclear explosion.

Table 4.7 Comparison of the absolute efficiency for the nuclear pumping source

	A	B	C	D	
Pump source	Charged particle energy produced (MeV)	Charged particle energy available (MeV)[a]	Nuclear energy produced by the source reaction (MeV)	Addition energy to produce neutron (MeV)	Pump source efficiency $\eta_{NP} = B/(C + D)$
^{10}B coating	2.35	0.5	2.35	400[b]	0.001
^{3}He gas	0.76	0.76	0.76	400[b]	0.002
^{235}U coating	165	40	200	None[c]	0.20
^{235}U dust	165	120	200	None[c]	0.6
$^{235}UF_6$ gas	165	165	200	None[c]	0.8
Uranofullerene	165	165	200	None[c]	0.8

[a]Only 50 % of the charged particles go in the right direction from a coating and self-absorption in the coating must be considered. An aerosol fuel particle can lose about 27 % of its energy through self-absorption. A gaseous form of UF_6 or uranofullerenes have minimal self-absorption

[b]^{3}He and ^{10}B reactions must be initiated by neutrons produced by an external fission reactor. About two fissions releasing ~400 MeV are required to produce one neutron captured by the ^{3}He or ^{10}B reactions

[c]^{235}U pumping sources can be self-critical so there is no energy expenditure to create a neutron for the reaction

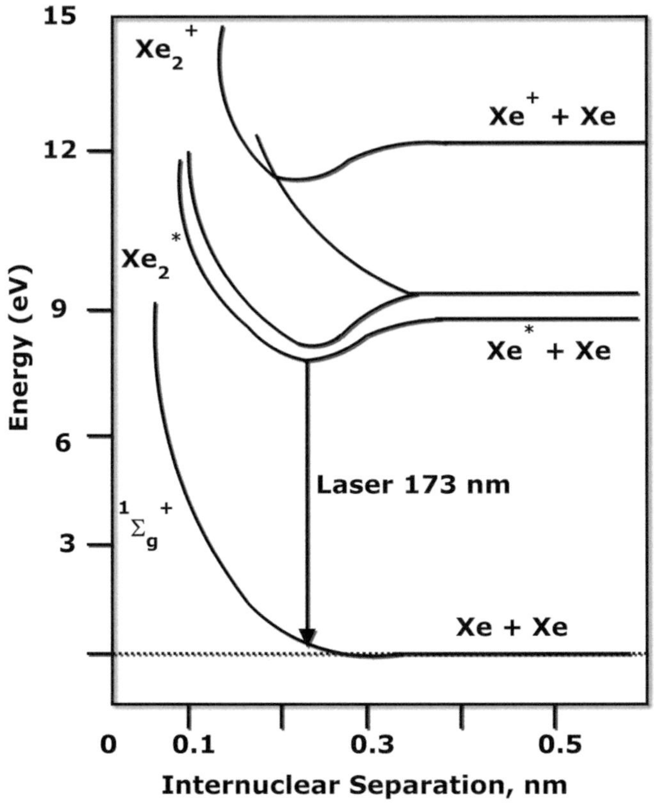

Fig. 4.6 Energy structure of the xenon molecule [45]

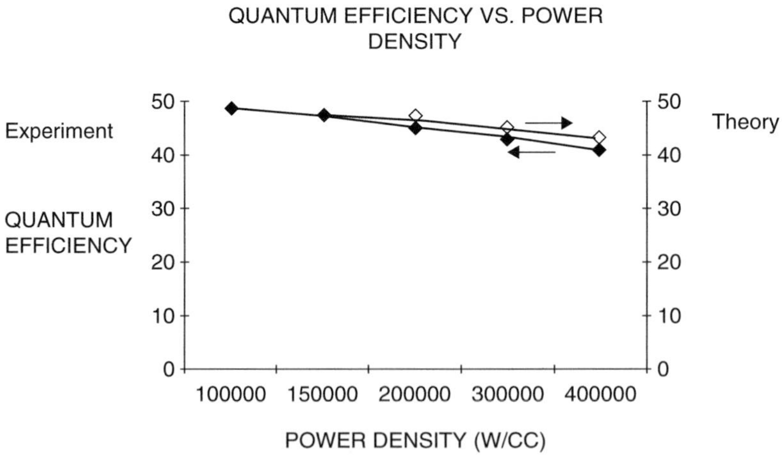

Fig. 4.7 Comparison of theoretical and experimental data for e-beam driven Xe_2^* fluorescence efficiency for 3 atm of Xe

Table 4.8 Reactions used in the Xe model [44]

#	Reaction	Rate constant (s^{-1}, $cm^{-3}\,s^{-1}$, or $cm^{-6}\,s^{-1}$)
1	$Xe^* + 2Xe \rightarrow Xe_2^*(1\Sigma) + Xe$	$1 \cdot 7 \times 10^{-32}$
2	$Xe^{**} + Xe \rightarrow Xe^* + Xe$	$1 \cdot 00 \times 10^{-10}$
3	$Xe^* + Xe \rightarrow Xe + Xe$	$3 \cdot 5 \times 10^{-15}$
4	$Xe^{**} + 2Xe \rightarrow Xe_2^{**} + Xe$	$1 \cdot 0 \times 10^{-31}$
5	$Xe_2^{**} + Xe \rightarrow Xe^* + 2Xe$	$1 \cdot 0 \times 10^{-11}$
6	$Xe_2^*(1\Sigma) + Xe \rightarrow Xe_2^*(3\Sigma) + Xe$	$1 \cdot 2 \times 10^{-13}$
7	$Xe^* + Xe^* \rightarrow Xe^+ + Xe + e$	$2 \cdot 0 \times 10^{-9}$
8	$Xe_2^*(1\Sigma) + Xe_2^*(1\Sigma) \rightarrow Xe_2^+ + 2Xe + e$	$5 \cdot 0 \times 10^{-10}$
9	$Xe^+ + 2Xe \rightarrow Xe_2^+ + Xe$	$2 \cdot 5 \times 10^{-31}$
10	$Xe_2^+ + e \rightarrow Xe^{**} + Xe$	$2 \cdot 3 \times 10^{-7}$
11	$Xe_2^*(1\Sigma) \rightarrow 2Xe + hv$	$2 \cdot 1 \times 10^{-8}$
12	$Xe^{**} \rightarrow Xe^* + hv$	$1 \cdot 5 \times 10^{-7}$
13	$Xe_2^*(3\Sigma) + e \rightarrow Xe_2^*(1\Sigma) + e$	$1 \cdot 8 \times 10^{-7}$
14	$Xe_2^*(1\Sigma) + e \rightarrow Xe_2^*(3\Sigma) + e$	$4 \cdot 9 \times 10^{-7}$
15	$Xe_2^{**} + e \rightarrow Xe_2^*(3\Sigma) + e$	5 · 0E-7
16	$Xe_2^{**} + e \rightarrow Xe_2^*(1\Sigma) + e$	2 · 0E-7
17	$Xe_2^*(1\Sigma) + e \rightarrow 2Xe + e$	1 · 3E-9
18	$Xe_2^+ + e \rightarrow Xe^* + Xe$	2 · 3E-7
19	$Xe^* + 2Xe \rightarrow Xe_2^*(3\Sigma) + Xe$	$4 \cdot 4 \times 10^{-32}$
20	$Xe_2^*(3\Sigma) + Xe_2^*(3\Sigma) \rightarrow Xe_2^+ + 2Xe + e$	$5 \cdot 0 \times 10^{-10}$
21	$Xe_2^*(3\Sigma) + Xe_2^*(1\Sigma) \rightarrow Xe_2^+ + 2Xe + e$	$5 \cdot 0 \times 10^{-10}$
22	$Xe_2^*(3\Sigma) \rightarrow 2Xe + hv$	$1 \cdot 0 \times 10^{-7}$
23	$Xe_2^*(3\Sigma) + e \rightarrow 2Xe + e$	1.3×10^{-9}
24	$Xe^+ + Xe + e \rightarrow Xe^* + Xe$	$1 \cdot 0 \times 10^{-26}$
25	$Xe_2^+ + e + Xe \rightarrow Xe_2^*(1\Sigma) + Xe$	$1 \cdot 0 \times 10^{-26}$
26	$Xe_2^*(1\Sigma) + e \rightarrow Xe_2^+ + 2e$	$5 \cdot 0 \times 10^{-9}$
27	$Xe_2^*(3\Sigma) + e \rightarrow Xe_2^+ + 2e$	$5 \cdot 0 \times 10^{-9}$
28	$Xe_2^+ + e + Xe \rightarrow Xe_2^*(3\Sigma) + Xe$	$1 \cdot 0 \times 10^{-26}$
29	$Xe^{**} + Xe \rightarrow Xe + Xe$	$1 \cdot 0 \times 10^{-15}$
30	$Xe_2^*(3\Sigma) + Xe \rightarrow Xe_2^*(1\Sigma) + Xe$	$4 \cdot 6 \times 10^{-15}$
31	$Xe^{**} + e \rightarrow Xe^* + e$	$8 \cdot 0 \times 10^{-7}$
32	$Xe_2^*(3\Sigma) + e \rightarrow Xe_2^{**} + e$	$3 \cdot 0 \times 10^{-7}$
33	$Xe^{**} + Xe^{**} \rightarrow Xe^+ + Xe + e$	$5 \cdot 0 \times 10^{-10}$
34	$Xe_2^{**} + Xe_2^{**} \rightarrow Xe_2^+ + 2Xe + e$	$5 \cdot 0 \times 10^{-10}$
35	$Xe^* + e \rightarrow Xe + e$	$1 \cdot 0 \times 10^{-9}$
36	$Xe^+ + 2e \rightarrow Xe^* + e$	$1 \cdot 0 \times 10^{-20}$
37	$Xe_2^{**} \rightarrow Xe^* + Xe + hv$	$1 \cdot 0 \times 10^{-8}$

4.3 Radioisotopes

A theoretical analysis of a system using a radioisotope as a pump source was made for a photolytic iodine laser using a nuclear energy focus [3]. The concept uses ^{220}Rn, a high activity isotope which is a noble gas (with a 55.6 s half life). The isotope is created in the thorium series from the decay of ^{224}Radium (which has a half-life of 3.66 years). It is feasible to place radium in a cell as a precursor for the production of radon. The two isotopes will be in secular equilibrium (meaning the activities of the two isotopes match) thus it is possible to derive excitation from both the radium and radon.

By using the radon isotope as part of a XeBr excimer fluorescer fill gas (a mixture of $Rn:Xe:Br_2$), the decay energy of radon is used to create the power density.

Assuming a 1000 T mixture of $He:Xe:Br_2:^{220}Rn$ at 7:2: 1.0.001, one finds that the power density absorbed by the photolytic laser medium is,

$$P_L = P_{Rn} 4.28 \times 10^2 \ Wcm^{-3} \tag{4.6}$$

where P_{Rn} = Rn pressure in Torr.

For 0.1 Torr of Rn gas,

$$P_L = 42.8 \ Wcm^{-3} \tag{4.7}$$

The average power deposition in the laser cell is about 162 W cm^{-3}. Therefore the laser power output was calculated to be,

$$P_{\text{Laser}} = 0.645 \text{ kW} \tag{4.8}$$

4.4 NPL Research

The nuclear-pumped lasers that have been discovered thus far originate either in the United States or Russia [6, 15]. These lasers have some common threads in their characteristics which make them viable. In order to overcome the limitations on power density that most nuclear-pumped lasers are subjected to (other than thermonuclear warheads used as a pump source), the lasers must have a means to efficiently channel energy to the upper laser level while quenching the lower laser level by collisional processes or short transition lifetimes in order to maximize gain. Many of the known nuclear-pumped lasers have specific traits or some combination of traits. In order to efficiently channel energy into the upper laser level, the majority of the known nuclear-pumped lasers depend on some feature other than a random distribution in the energy levels of the atomic or molecular system. Often times this manifests as some sort of resonance transition (such as the energy transfer of the first vibration level of nitrogen to the carbon dioxide upper laser level), or a favorable branching ratio in a process (such as recombination in the atomic xenon laser). Another trait which is complimentary is a relatively long lived upper laser level and a relatively short lived lower laser level. Sometimes there is a combination of resonance transitions with long lived upper laser level and short lived lower laser level (such as the atomic carbon laser). This section will examine the known nuclear-pumped lasers and will discuss what is known and what is believed to be true about the mechanisms and operational parameters of the laser.

Two of the known nuclear-pumped lasers [32, 50] have used the energy focus method [3] described above to improve the power deposition in a photolytically driven laser. Both the Nd:YAG and atomic iodine laser have favorable long lived upper laser level lifetimes and short lived lower laser level lifetimes.

4.4.1 Nuclear-Pumped Lasers

4.4.1.1 Atomic Rare Gas Lasers

The atomic rare gas lasers are a general class of laser based on the nd $\rightarrow$ (n + 1)p transitions in rare gases (where n = 5 for Xe, n = 4 for Kr, n = 3 for Ar and n = 2 for Ne). These transitions occur in the infrared and have been demonstrated by nuclear-pumping. These lasers have low power density thresholds and they have some of the highest efficiencies for known NPLs. The similarities in the structure of rare gas atoms leads to similarities in the mechanisms by which this class of laser operate. It is instructive to examine the energy level diagrams for Xe I (Fig. 4.8), Kr I (Fig. 4.9), Ar I (Fig. 4.10) and Ne I (Fig. 4.11) to begin to understand the mechanistic similarities. Each of these lasers uses a rare gas as a buffer majority species and the lasing medium a rare gas as minority species. The buffer gas is typically helium. The function of the buffer gas is to absorb the bulk of the energy transferred to the gas from the charged particle excitation source. As discussed in Chap. 2, when an energetic charged particle is stopped in a gaseous media, the primary excitation results in the formation of ions and metastable states (for rare gases the metastable states are n^1S and n^3S where n is 2 for He, 3 for Ne, 4 for Ar, 5 for Kr and 6 for Xe).

- The helium ion, the helium 2^1S and the helium 2^3S state have energies of 24.59 eV, 20.56 eV and 19.82 eV respectively.
- The neon ion, the neon 3^1S and the neon 3^3S state have energies of 21.56 eV, 16.8 eV and 16.62 eV respectively.
- The argon ion, the argon 4^1S and the argon 4^3S state have energies of **15.76** eV, 11.86 eV and 11.57 eV respectively.
- The krypton ion, the krypton 5^1S and the krypton 5^3S state have energies of **14.00** eV, 10.03 eV and 9.92 eV respectively.
- The xenon ion, the xenon 6^1S and the xenon 6^3S state have energies of **12.13** eV, 8.44 eV and 8.32 eV respectively.

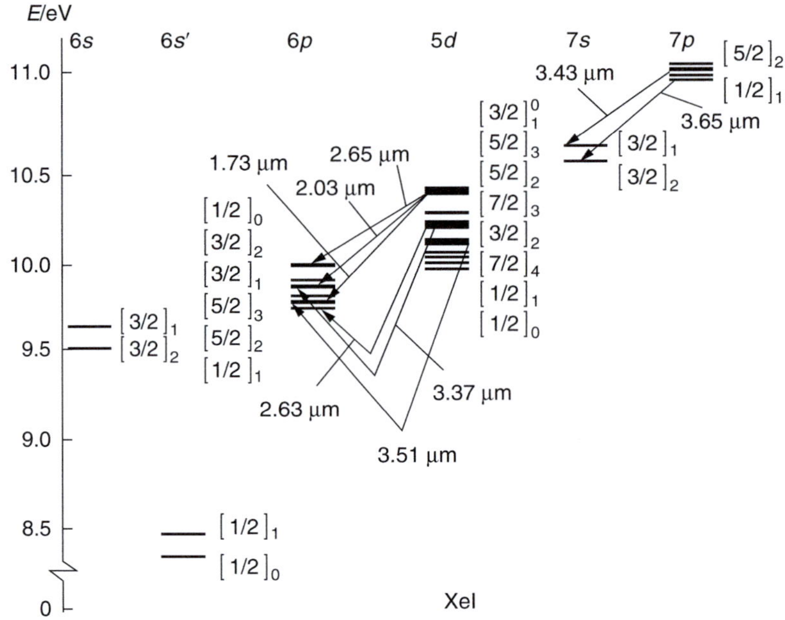

Fig. 4.8 Energy levels of prominent transitions in atomic xenon. The two important transitions for NPLs are the $5d[5/2]_2 \rightarrow 6p[5/2]_2$ at 1.73 μm with an efficiency of 2.5 % and the $5d[5/2]_2 \rightarrow 6p[3/2]_1$ at 2.03 μm with an efficiency of 2.0 % [51]

Fig. 4.9 Energy levels of prominent transitions in atomic krypton. The two important transitions for NPLs are the $4d[3/2]_2 \rightarrow 5p[3/2]_2$ at 2.52 μm with an efficiency of ~0.1 % and the $4d[1/2]_1 \rightarrow 5p[3/2]_2$ at 2.19 μm with an efficiency of <0.1 %

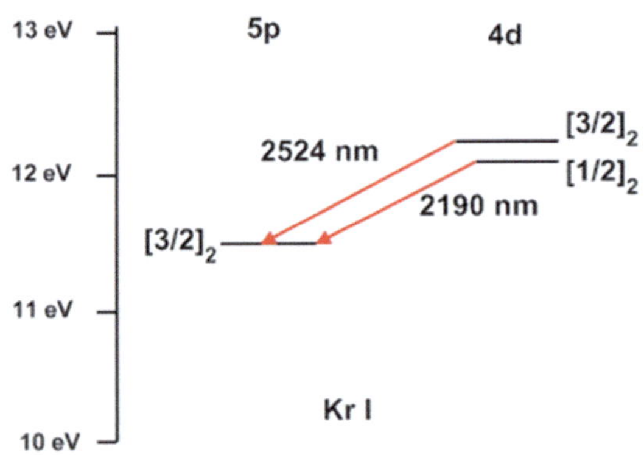

The buffer gas atomic density is generally very large compared to the lasing atom density (on the order of 1000–1). Thus the bulk of the energy deposited in the gaseous media goes to the buffer gas. The buffer gas ion and metastable states then transfer energy to the lasing atom by charge exchange or through penning ionization. The upper laser level is then populated by a variety of recombination reactions. The most efficient of the rare gas lasers are the xenon transitions $5d[5/2]_2 \rightarrow 6p$

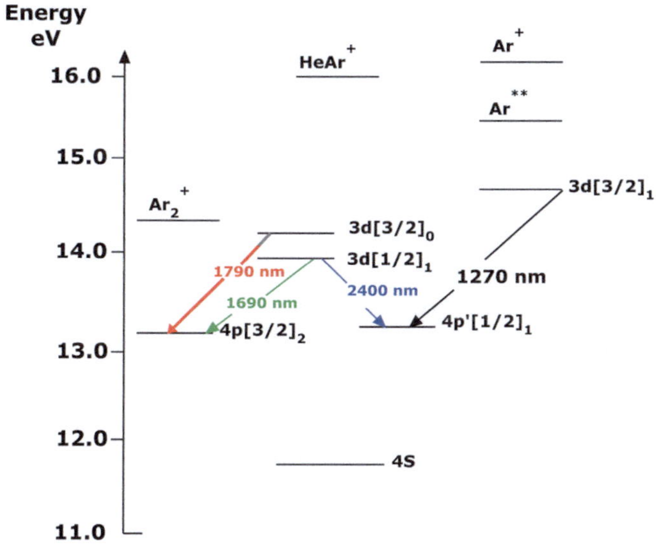

Fig. 4.10 Energy levels of prominent transitions in atomic argon. The most important transition for an NPL is the $3d[3/2]_0 \rightarrow 4p[3/2]_2$ at 1.79 μm with an efficiency of 0.5 % [52]

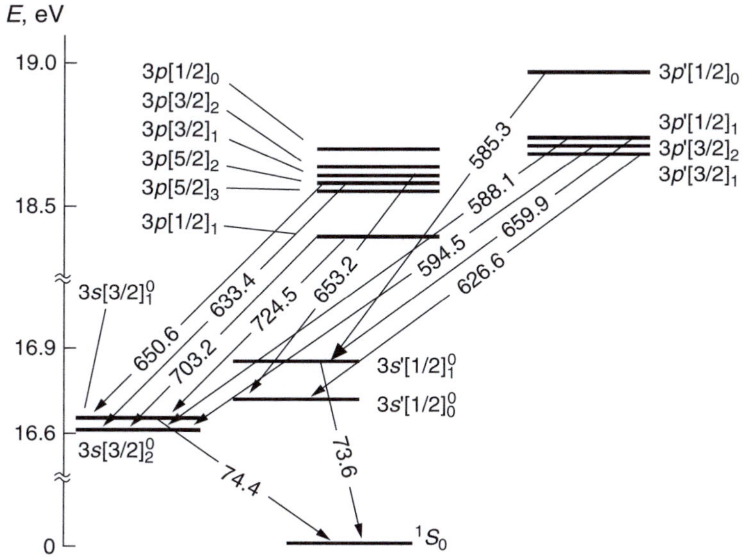

Fig. 4.11 Energy levels of prominent transitions in atomic neon. The most important transition for an NPL is the $3p[1/2]_0 \rightarrow 3\ s[1/2]_1$ at 585.3 nm with an efficiency of ~0.1 % [53]

$[5/2]_2$ at 1.73 μm with an efficiency of 2.5 % and $5d[5/2]_2 \rightarrow 6p[3/2]_1$ at 2.03 μm with an efficiency of 2.0 %. This efficiency is much higher than the next highest rare gas laser transition in argon $3d[3/2]_0 \rightarrow 4p[3/2]_2$ at 1.79 μm with an efficiency of 0.5 %. Thus for some reason the recombination process is more efficient for the atomic xenon transitions than for any of the other rare gas lasers. This reason shall be explored in the discussion that follows.

There is a general understanding of how the rare gas lasers work, but there are still questions that make this understanding less than perfect. A respectable body of modeling studies has been able to show trends but has not been able to fully explain all of the subtleties of the lasers performance. For example basics such as the threshold power densities are approximately predicted at under certain conditions but not under all conditions. Even though rare gas lasers that use a binary gas mixture depend upon similar ionization processes and recombination processes, the threshold power density have a large variation. For most of the rare gas lasers the threshold power density lies between 30 and 1000 W/cm^3. There is one exception for one line in neon which has a threshold power density of about 1 mW/cm^3 which is the lowest for any known nuclear-pumped laser. The variations in threshold power densities are not well understood. Another example is temperature dependence. It is know that elevated gas temperature will reduce the laser output, but this is not well understood. The incorporation of temperature dependence in models is still a work in progress.

The reason that models are not entirely accurate begins with the rate constants for the reactions that make up the model. There is a limited body of experimentally determined rate constants. When there is a lack of experimentally based data, the rate constants that are unknown are sometimes theoretically calculated [54], sometimes the rate constants are estimated based on comparisons to similar reactions with rate constants reported in the literature, sometimes the rate constants are guessed at with no basis, sometimes the rate constants are simply adjusted until the calculations from the model matches the operational characteristics of the laser (this happens more frequently than can be imagined). The body of experimentally verified rate constants with temperature dependence is very limited so this further complicates the modeling process. Often times rate constants are taken from a paper in the literature and the author does not research the origin of the rate constant. In summary there are a number of potential error sources in laser modeling.

There are approaches to mitigate the stream of errors that do occur in modeling. Due to the large number of possible reactions in typical plasma chemistry models, and the uncertainties associated with the rate constants for the reactions, it is possible through first-order sensitivity analysis to identify the major reactions, pressure effects, temperature dependence and rate constant uncertainties [44]. Several methods can be used: the Fourier Amplitude Sensitivity Test (FAST); the Direct Method (DM); the Green's Function Method (GFM); and the State Transition Matrix Method (STMM). The FAST and the DM focus upon the evaluation of sensitivity coefficients (i.e. the partial derivatives of the system solutions with respect to the system's parameters). In the FAST approach, each parameter is varied as a periodic function of a "search variable", s. Hence the system solutions are periodic in s and can be Fourier analyzed. Despite the success of the FAST approach, it still has several weaknesses: it is computationally intensive; and it yields time averaged sensitivity coefficients (impossible to explain transient effects). The DM solves the original set of system equations together with a large set of sensitivity equations. This approach has an advantage in that the sensitivity coefficients are local and therefore useful in transient analysis. However, the stiffness ratio of the system is very large (i.e. a total of $mn + n$ equations must be solved simultaneously where m = number of parameters and n = number of system equations). The Green's Function Method transforms the set of sensitivity equations into integral equations. For well-defined physical systems, the integrands are smooth. This reduces not only the stiffness ratio but also the computational time. The STMM approach follows the GFM except that the state transition matrix is solvable by methods developed by linear systems engineering.

The procedure for sensitivity analysis begins with the derivation of a set of first-order n-vector nonlinear differential equations [44],

$$\frac{dc}{dt} = g(C, u, t) \tag{4.9}$$

where C is a column vector of system solutions such as the laser intensity and species concentrations and u represents a column vector of m parameters such as the rate constants. For most physical systems, $m > n$.

The initial condition of $C(t)$ is given by

$$C(t = 0) = C_o \tag{4.10}$$

where C_o is an n-vector matrix of initial laser intensity and species concentrations. The first-order sensitivity coefficients are defined as, where y is an $n \times m$ matrix vector.

$$Y = \frac{dc}{du} \tag{4.11}$$

This equation can be transformed by first differentiating vector y with respect to time and interchanging some terms to give,

$$\frac{dy}{dt} = J(t)y + f(t) \tag{4.12}$$

where $f = \frac{dg}{du}$

f is an $n \times m$ vector

and, $J(t) = \frac{df}{dc}$

$J(t)$ is the Jacobian of the system and is an $n \times n$ vector. The initial condition y_{ij} depends on whether u_{ij} refers to an initial species condition or not. The Direct Method solves for Eqs. (4.9) and (4.12) directly. This system can be very stiff. To reduce the stiffness ratio, Eq. (4.12) can be transformed into an integral equation,

Table 4.9 Major reactions out of 37 from a plasma chemistry model for e-beam driven xenon excimer production are shown [44]

Reaction number	Sensitivity coefficient	Reaction
1	+0.177	$Xe^* + 2Xe \rightarrow Xe_2^*(1\Sigma) + Xe$
7	−0.095	$Xe^* + Xe^* \rightarrow Xe^+ + Xe + e$
11	−0.876	$Xe_2^*(1\Sigma) \rightarrow 2Xe + hv$
13	+0.120	$Xe_2^*(3\Sigma) + e \rightarrow Xe_2^*(1\Sigma) + e$
14	−0.110	$Xe_2^*(1\Sigma) + e \rightarrow Xe_2^*(3\Sigma) + e$
22	−0.134	$Xe_2^*(3\Sigma) \rightarrow 2Xe + hv$
36	+0.218	$Xe^+ + 2e \rightarrow Xe^* + e$

The normalized sensitivity coefficients are positive if the reaction is a source term or negative if the reaction is a loss term. The relative magnitude of the sensitivity coefficient is a quantitative measure of its impact on the production or loss on the excimer formation

$$y(t) = G(t,0)^* y_0 + \int_0^t G(t,u)^* g(u)^* du \tag{4.13}$$

where $G(t, 0)$ is solved from:

$$\frac{dg(t,0)}{dt} = J(t)^* G(t,0) \tag{4.14}$$

with $G(t = 0, 0) = I$ where I is the identity matrix.

$G(t, u)$ is obtained by solving the following system:

$$\frac{dg(t,u)}{du} = -G(t,u)^* J(u) \tag{4.15}$$

with $G(t, u = t) = I$ (solved backwards in time.)

$G(t, 0)$ and $G(t, u)$ are both state transition matrices with all elements having an absolute magnitude less than or equal to unity.

Since $G(t, 0)$ and $G(t, u)$ are stiff, the use of Gear's algorithm simplifies the numerical aspects of the solution. ·

To make the sensitivity coefficients comparable, the normalized sensitivity coefficients are defined as:

$$y_{ij} = \frac{dc_i}{c_i} \times \frac{u_i}{du_i} \tag{4.16}$$

such that the absolute value of $y_{ij} \leq 1$ for all i, j.

From the complete set of rate equations for the production of the $Xe_2^*(1\Sigma)$ excimer state, there were only 7 major reactions which dominated its production and quenching (See Table 4.9). Sensitivity analysis can improve the predictions of plasma chemistry models if used judiciously.

Table 4.10 Main plasma processes in He-B mixtures (B = Kr, Ar) [55]

No.	Process	Rate constant k	References
1	$He^* + B \rightarrow B^+ + e + He$	4.3×10^{-10} cm^3/s(Kr)	[56]
		3.0×10^{-10} cm^3/s (Ar)	[56]
2	$He^+ + 2He \rightarrow He_2{}^+ + He$	5.4×10^{-32} cm^6/s	[57]
3	$He_2{}^+ + B \rightarrow B^+ + 2He$	4.0×10^{-10} cm^3/s (Kr)	[58]
		2.5×10^{-10} cm^3/s (Ar)	[58]
4	$B^+ + B + He \rightarrow B_2{}^+ + He$	1.9×10^{-31} cm^6/s	[59]
5	$B^+ + e + e \rightarrow B^* + e$	$4.0 \times 10^{-9} Te^{-4.5}$ cm^6/s	[60]
6	$B^+ + e + He \rightarrow B^* + He$	1.7×10^{-21} $Te^{-2.5}$ cm^6/s	[61]
7	$B_2{}^+ + e \rightarrow B^* + B$	$3.7 \times 10^{-5} Te^{-0.55}$ cm^3/s (Kr)	[62]
		$3.0 \times 10^{-5} Te^{-0.61}$ cm^3/s (Ar)	[63]

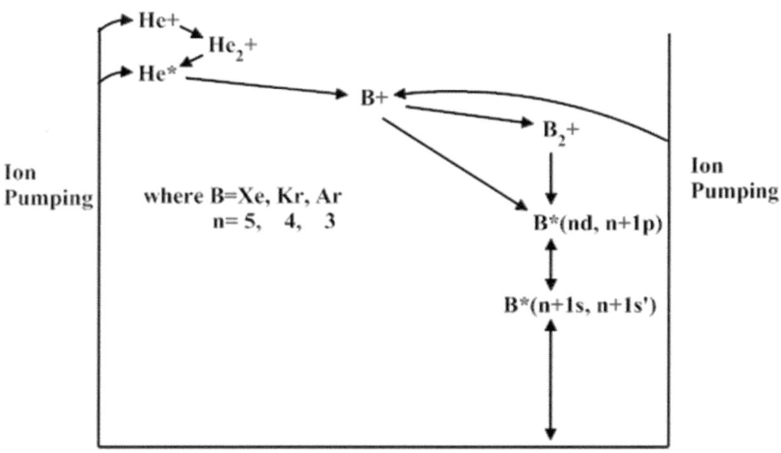

Fig. 4.12 Illustration of the plasma chemistry involved in two gas (He:B where B is Xe, Kr, or Ar) laser kinetics

The rare gas lasers have many similarities. First, the method by which the upper laser level is populated is similar. The rare gas lasers have a rare gas buffer (e.g., helium). The buffer gas is excited in the pumping process and produces ion and metastable states. The ion and metastable states transfer energy to the rare gas laser atom through penning ionization (process 1 in Table 4.9), through molecular ion forming reactions (processes 2 and 4 in Table 4.10) and through charge exchange reactions (process 3 in Table 4.10). The upper laser level is then populated by three body recombination (processes 5 and 6 in Table 4.10) and dissociative recombination (process 7 in Table 4.10). The plasma chemistry for binary gas mixtures is shown in Fig. 4.12.

The important factor in producing an efficient laser is how energy is channeled into the upper laser level and how the lower laser level is populated and quenched. The atomic xenon laser shown in Fig. 4.5 has particularly good efficiencies with

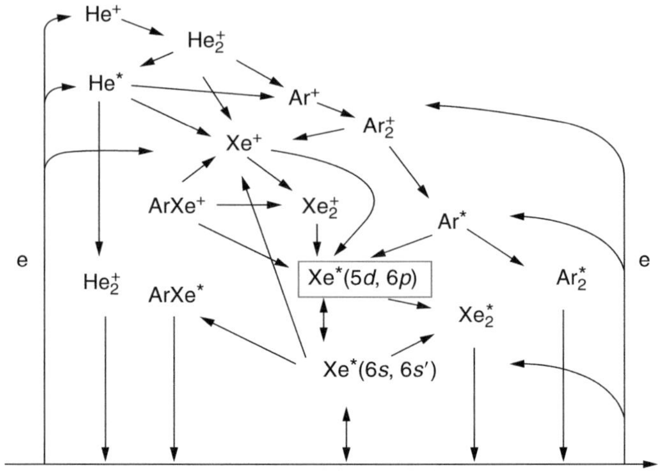

Fig. 4.13 Illustration of the complex plasma chemistry involved in the He:Ar:Xe laser kinetics [53]. The upper laser level (5d state) is shown in the box

nuclear pumping. The $5d[5/2]_2 \rightarrow 6p[5/2]_2$ transition at 1.73 μm has an efficiency of 2.5 % and the $5d[5/2]_2 \rightarrow 6p[3/2]_1$ transition at 2.03 μm has an efficiency of between 2 and 3 % [29, 30, 51]. The atomic xenon laser is optimal in mixtures of He:Ar:Xe at a ratio of 700:100:1. This addition of Ar to the mixture is very important in boosting the efficiency of the Xe laser (by a factor of 2–3). In a two gas laser medium (see Fig. 4.13), dissociative recombination and three body recombination are the major contributors to populating the upper laser level. In the three gas mixture, the process still depends upon penning ionization and recombination like other rare gas lasers based on two gas mixtures, but there appears to be unusually efficient dissociative recombination reactions with $ArXe^+$ and $Xe_2{}^+$ molecular ions Eqs. (4.17) and (4.18) that populate the 5d state [51]. The plasma chemistry of the three gas laser medium is more complex than a two gas laser (Fig. 4.13). In addition there are transfer reactions between Ar* and Xe (Eq. 4.19) and three body recombination reactions with the Xe^+ ion (Eq. 4.20) that also add to the population of the 5d state.

$$\mathrm{ArXe}^+ + \mathrm{e} \leftrightarrows \mathrm{Ar} + \mathrm{Xe}(5\mathrm{d}) \tag{4.17}$$

$$Xe_2{}^+ + e \leftrightarrows Xe + Xe(5\mathrm{d}) \tag{4.18}$$

$$Ar^* + Xe \leftrightarrows Ar + Xe(5\mathrm{d}) \tag{4.19}$$

$$Xe^+ + 2e \leftrightarrows e + Xe^* \tag{4.20}$$

The other process that adds to the efficiency of a laser is quenching of the lower laser level state. Quenching of the lower laser level occurs primarily by collisional de-excitation with heavy atoms (Eq. 4.21) and by spontaneous emission (Eq. 4.22).

$$\mathrm{Xe}(6p) + \mathrm{Ar}(\mathrm{Xe}) \leftrightarrow \mathrm{Xe}(6s') + \mathrm{Ar}(\mathrm{Xe}) \tag{4.21}$$

$$\mathrm{Xe}\left(6\,\mathrm{p}[3/2]_1\right) \rightarrow \mathrm{Xe}(6\mathrm{s}) \qquad \left(\mathrm{A}_{6\,\mathrm{p\text{-}6s}} = 2.78 \times 10^7 \mathrm{s}^{-1}\right) \tag{4.22}$$

It is important to note that collisional de-excitation and spontaneous emission will also quench the upper laser level (Eq. 4.23). However if the stimulated emission cross section and the cavity intensity are high enough, then stimulated emission will dominate the loss rate for the upper laser level. Thus the bulk of the energy flow through the upper laser level will end up as photon production through stimulated emission as discussed in Chap. 1.

$$\mathrm{Xe}\left(5\mathrm{d}[5/2]_2\right) \rightarrow \mathrm{Xe}\left(6\,\mathrm{p}[3/2]_1\right) \qquad \left(\mathrm{A}_{21} = 4.25 \times 10^5\ \mathrm{s}^{-1}\right) \tag{4.23}$$

The threshold pumping power density is also tied to the efficiency of populating the upper laser level and the quenching of the lower laser level. Parasitic processes, which act as competing channels for the energy flow, will cause an increase in the threshold pumping power density. The rare gas lasers have threshold pumping power densities on the order of 10 W/cm^3 for the 2.03 and 1.73 Xe lines, and between 37 and 980 W/cm^3 for other rare gas transitions. But the Ne transition 3p $[1/2]_0 \rightarrow 3\ \mathrm{s}[1/2]_1$ at 585.3 nm with an efficiency of ~0.1 % has a threshold pumping power density of 0.001 W/cm^3. This is the lowest threshold pumping power density of any nuclear-pumped laser which indicates that there are few parasitic processes which compete for the energy flow to the $3\mathrm{p}[1/2]_0$ state and that the quenching of the $3\ \mathrm{s}[1/2]_1$ state is efficient. But why would parasitic processes be so different for Ne versus Xe, Kr and Ar? The 585.3 nm neon nuclear-pumped laser has been reproduced [51]. As previously discussed the understanding of the wide variations of threshold power densities for rare gas lasers in a binary gas mixture can be improved especially for the extremely low threshold for the 585.3 nm line in neon.

Even though nuclear-pumped lasing on the $3\mathrm{s}_2 \rightarrow 2\mathrm{p}_4$ at 632.8 nm transition in neon was reported as having a threshold pumping power density of 0.0012 W/cm^3 [64], this claim has some serious issues. The cavity loss in this experiment was unusually high (about 50 %). It was shown that with even in the most optimistic assumptions, with the reported parameters, the 632.8 nm line in the neon laser could not have lased. The supposition used was that all of the power deposited was channeled to the upper laser level by directly pumping the lower laser level (i.e., a Pumping Scheme Quantum Efficiency (PSQE) of 100 % versus the actual theoretical quantum efficiency of about 1 %). With a PSQE of 100 % the laser still would not have enough gain to lase [65]. A response was published about the assertion that the He-Ne nuclear-pumped laser could not lase. The argument in this response was based on the premise that the absorption from all 3 s states to the upper laser level were source terms that were not considered and that if considered they would generate a higher gain [66]. The error in this argument is that the source terms from absorption were irrelevant. The model from reference [65] assumed the most optimistic conditions possible. The model assumed that the laser had no loss mechanism: i.e., all of the power density generated by the nuclear reactions was

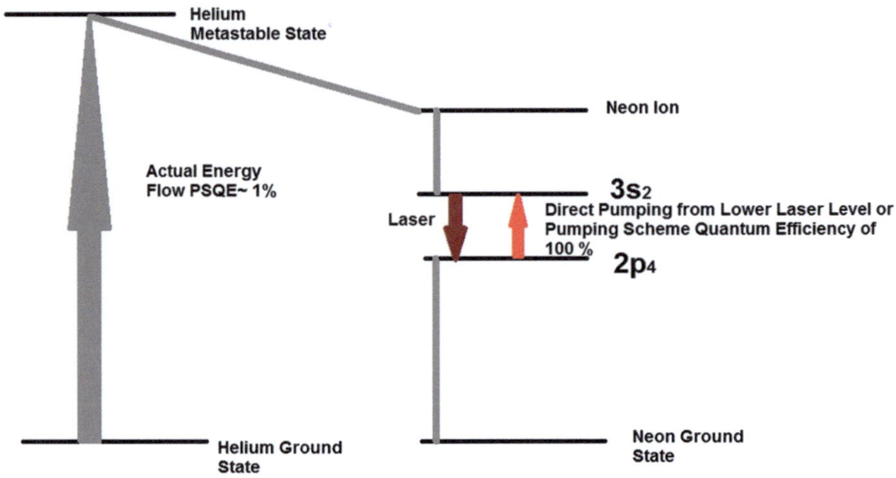

Fig. 4.14 This is an illustration of the concept of a perfect Pumping Scheme Quantum Efficiency (PSQE) versus a realistic pumping scheme based on energy transfer from the excited helium states to ionize neon and then recombination from the neon ion

directly channeled from the lower laser level to the upper laser level (Fig. 4.14). This assumption means that only the upper laser level was populated by the energy flow. The assumption assures that there would be no other states created. The assumption also assures that there would be no parasitic loss processes. The model optimistically assumed perfect quenching of the lower laser level by setting its density to zero thus making the calculated gain not only very optimistic but also giving the absolute maximum value of gain that is feasible. This does not mean that the 632.8 nm line in He-Ne can't lase. In fact with electrical pumping it works very well but at a much higher power density. The 632.8 nm Ne laser simply could not have worked with the reported parameters in this nuclear-pumped laser experiment. So the conclusion is that the observed laser was not the 632.8 nm line of Ne. As a point of interest, if the cavity losses had been less than the 50 % reported, it may have been feasible to observe the 585.3 nm Ne line because it's threshold power density is low enough (~1 mW/cm^3).

The scaling of the rare gas nuclear-pumped lasers to high power output has been demonstrated. In 1981 a multiple pass ^{3}He-Ar nuclear-pumped laser at 1.79 μm (Fig. 4.15) producing 1 kW of power was demonstrated [67].

The multi pass laser was constructed with a stainless steel frame with internal dimensions of 0.40 × 0.30 × 0.03 m. Bolted on both sides were aluminum cover plates covered by a 0.05 m thick polyethylene layer in order to moderate the fast neutrons from the fast burst reactor. Quartz windows at Brewster's angle were placed at opposite ends of the steel frame. Two flat gold mirrors were mounted on the internal walls of the steel frame to form the multi pass portion of the optical cavity. Terminating the end of the optical cavity were two 0.05 m diameter dielectric mirrors with a 10 m radius of curvature to form a stable oscillating

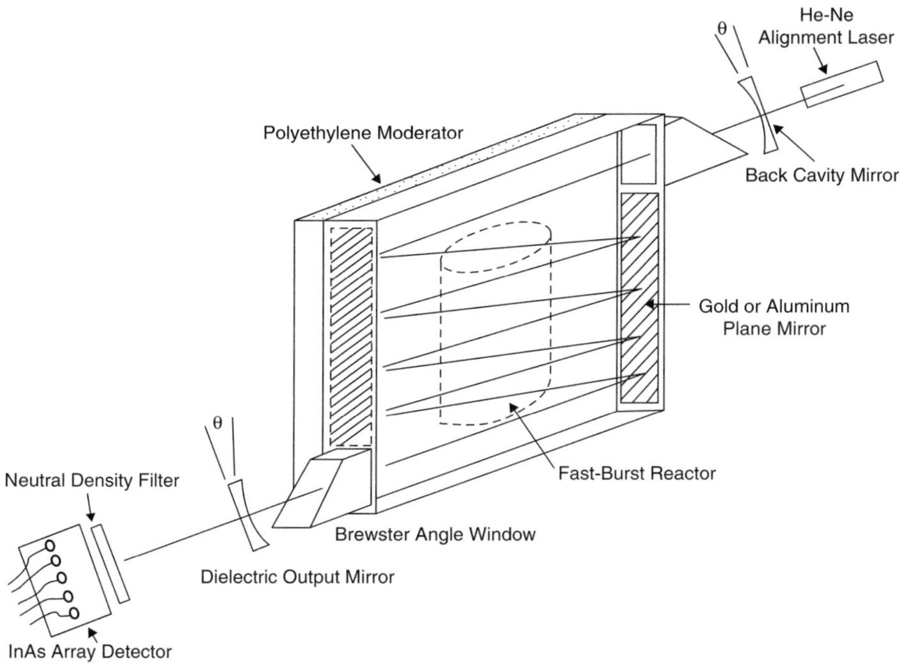

Fig. 4.15 Schematic of the multiple pass ^{3}He-Ar nuclear-pumped laser at 1.79 μm which produced over 1 kW of power [5]

system. The back mirror had a coating which provided the maximum possible reflectivity and the output mirror was had an 85 % reflectivity at 1.79 μm. The number of passes that the beam made through the gain medium was changed by varying the angle of these dielectric mirrors. The laser used a mixture of ^{3}He-Ar (with 1 % being Ar) at total pressures ranging from 80 to 290 kPa. Lasing was observed on the 1.79 μm line of Ar. The laser beam was viewed using a 13 element InAs detector array in the form of a cross. This detector configuration was able to determine the laser's intensity distribution. Figure 4.16a shows the peak power output of the laser as a function of peak thermal neutron flux for a total pressure of 80 kPa with a 2 % partial pressure of Ar. As can be seen in the figure, the power increase is linear with flux. Figure 4.16b shows the power output as a function of the number of passes in a gas mixture with a total pressure of 300 kPa with 0.7 % partial pressure of Ar and a flux of 4.3×10^{20} neutrons cm^2 s^{-1}. This shows that the power in the gain medium could be fully extracted with only 5 passes.

A basic nuclear-pumped laser design was used in many experiments in the USSR involving the VIR-2M reactor (Fig. 4.17a). The laser cavity was placed below the core of the reactor. This basic concept used internal mirrors in order to eliminate absorption in window materials. These basic experiments produced high power output as shown in Table 4.11. In addition to the validation of a high power nuclear-pumped laser, it has been shown that high energy output is possible with the

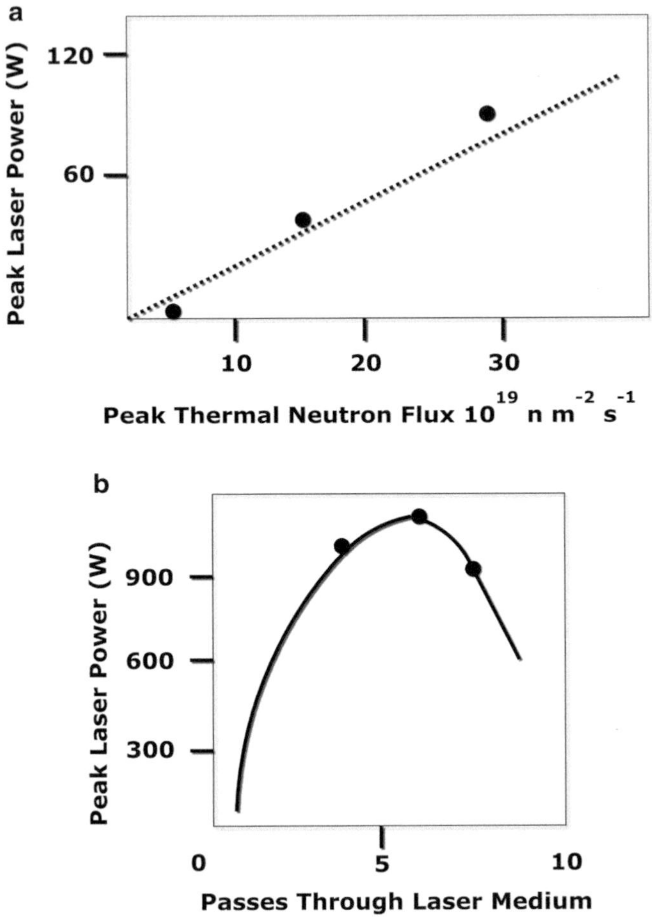

Fig. 4.16 Curve (**a**) shows the linear variation of laser power with thermal neutron flux. This shows that the laser directly scales with power density over the operating parameters of the fast burst reactor. Curve (**b**) shows the variation of power density with passes through the laser cavity. The optimum number of passes was five. This shows that relatively few passes were able to extract the power out of the gain medium [68]

nuclear-pumped Xe laser at 2.03 μm. A series of experiments utilizing a large laser cavity lined with uranium-235 coatings was pumped by a fast burst type reactor (Fig. 4.17b). This system had a laser output of over 500 J (although it was designed to produce a 1000 J or better) and a power output of 1.3 MW.

A series of experiments that demonstrated the highest reported nuclear-pumped laser energy output were run during the fall of 2011 [69]. An output of 526 J was the maximum energy achieved in these experiments using a He-Ar-Xe mixture with an atomic ratio of 700:100:1 (Table 4.11). The laser cavity was lined with ~3 mg cm^{-2} coating of highly enriched U_3O_8. Laser oscillation was achieved at 2.03 μm. A 0.4 ms pulsed reactor produced an estimated thermal neutron flux of about 3×10^{16} n cm^{-2} s^{-1}. The peak power of the laser reached 1.3 MW.

a

b

Fig. 4.17 (**a**) A typical set up for an NPL experiment in the USSR using a VIR-2M reactor [6, 7] (With permission of Springer Science+Business Media) (**b**) Diagram of a 500 J nuclear-pumped xenon laser (on the $5d[5/2]_2 \rightarrow 6p[3/2]_1$ 2.03 μm transition) with a peak power of 1.3 MW. The laser was placed in the proximity of a fast burst reactor. The Figure shows the laser assembly (*a*) and a close up of the 37 separate laser cavities which make up the laser assembly [69]

Temperature typically has detrimental effects on the output of rare gas lasers [51]. For mixtures of He:Ar:Xe (1.79 or 2.03 μm transitions) the laser output decreases as temperature increases. The reasons are several. First, there is a strong exponential dependence of the decay rate of the $ArXe^+$ ion with temperature. This impacts the formation of the upper laser level because of its dependence on the dissociative recombination of the $ArXe^+$ ion. The equilibrium of the excited gas shifts to the formation of the Xe^+ ion and the subsequent generation of the Xe_2^+ ion. The dissociative recombination of the Xe_2^+ ion lead to an increased source rate for lower laser level production. Temperature will also decrease the rate constant for dissociative recombination of the $ArXe^+$ ion which diminishes a source rate for the upper laser level. Finally, through collisions with hot helium, argon and xenon

Table 4.11 High energy laser experiments in Xe I and Ar I [69]

λ μm	Mixture	Pressure, bar	Resonator mirror reflectance (%)	Specific deposited energy J/cm^3	Energy/time, J/μs	Beam divergence, θ, mrad
2.03	He-Ar-Xe, 700:100:1	4	99	1	520 ± 104/400	3.9 ± 0.5
			85			
2.03	He-Ar-Xe, 700:100:1	4	99	1	241 ± 48/380	NA
			85			
2.03	He-Ar-Xe 700:100:1	4	99	1	526 ± 104/400	3.9 ± 0.5
			85			
1.79	He-Ar, 1140:1	3	99	0.71	39 ± 5/466	NA

atoms, the lower lying 6s and 6s' states become an enhanced source rate for population of the lower laser level.

Temperature has another deleterious effect- outgassing of water molecules from the walls of the laser cavity. Water is a poison which will further reduce the efficient energy flow to the upper laser level. One of the goals of nuclear-pumped laser technology is to develop a Continuous Wave (CW) or steady-state laser. In CW mode, the power deposited in the laser medium will cause the gas temperature to rise. Thus it can be surmised that a He:Ar:Xe Nuclear-Pumped Continuous Wave Laser (NPCWL) operating at elevated temperatures would have major problems maintaining the optimum conditions which allow it to operate at 3 % efficiencies in the pulsed mode. Going from a pulsed laser to a CW laser will be a huge technical challenge in many ways not least of which is the elevated gas temperature.

Using the methods discussed in Chap. 1, rate equations can be written for the binary mixture of He:Xe at a ratio of 1000:1 and can be solved. Based on the models of for a binary rare gas laser [52, 55], the energy transfer from the fission fragments create helium ions and helium metastable states,

$$ff + He \rightarrow He^+ + e \quad with \quad W = 30\left(\frac{eV}{ion\ pair}\right) \tag{4.24a}$$

$$ff + He \rightarrow He^* + e \quad with \quad W^* = 90\left(\frac{eV}{metastable}\right) \tag{4.24b}$$

The helium ions and helium metastable states transfer their energy to the background helium and xenon gases,

$$He^+ + 2He \rightarrow He_2^+ + He \quad with \quad k_3 = 5.4 \times 10^{-32} cm^6 s^{-1} \tag{4.25}$$

$$He^* + Xe \rightarrow Xe^+ + He \quad with \quad k_4 = 4.0 \times 10^{-10} cm^3 s^{-1} \tag{4.26}$$

The He_2^+ ion then undergoes a charge exchange with Xe,

$$He_2^+ + Xe \rightarrow Xe^+ + 2He \quad with \quad k_5 = 4.0 \times 10^{-10} cm^3 s^{-1} \tag{4.27}$$

The xenon ions formed in reactions 4.26 and 4.27 undergo the following reactions,

$$Xe^+ + Xe + He \rightarrow Xe_2^+ + He \quad with \;\; k_6 = 1.9 \times 10^{-31} cm^6 s^{-1} \tag{4.28}$$

$$Xe^+ + 2e \rightarrow Xe^* + e \quad with \;\; k_7 = 9.0 \times 10^{-8} cm^6 s^{-1} \tag{4.29}$$

Where, Xe^* is an excited state in Xe, with the estimated branching ratio to the X (5d[5/2]$_2$) state being 5 % (based on the reported efficiency of the He-Xe binary laser being 1 %).

$$Xe^+ + e + He \rightarrow Xe^* + He \quad with \;\; k_7 = 9.62 \times 10^{-21} cm^6 s^{-1} \tag{4.30}$$

$$Xe_2^+ + e \rightarrow Xe^* + Xe \quad with \;\; k_8 = 5.417 \times 10^{-5} cm^3 s^{-1} \tag{4.31}$$

The upper and lower laser levels then undergo spontaneous emission,

$$Xe^* \rightarrow Xe + hv \quad with \;\; A_{21} = 4.25 \times 10^6 s^{-1} \tag{4.32}$$

$$Xe^* \rightarrow Xe + hv \quad with \;\; A_1 = 2.78 \times 10^7 s^{-1} \tag{4.33}$$

The rate equations for this model can now be written beginning with the production of helium ions ($He^+ \, and \, He_2^+$) and helium metastable states (He^*). This model assumes that because helium is the major species in the gas mixture, the energy deposition from the fission fragments will flow primarily through helium. Thus the equations which govern helium ion production and helium metastable production are,

$$\frac{d[He^+(t)]}{dt} = \frac{S}{30} - 5.4 \times 10^{-32} [He^+(t)][He]^2 \tag{4.34}$$

where $S = Power Density\left(\frac{W}{cm^3}\right)/1.6 \times 10^{-19}\left(\frac{J}{eV}\right)$, it is assumed $[He^+(0)] = 0$ and that the He density is constant. The reason for this assumption is due to the fact that the initial atomic helium density is so large that that at the pumping power used, there is very little depletion of the initial pool of helium atoms.

$$\frac{d\left[He_2^+(t)\right]}{dt} = 5.4 \times 10^{-32} [He^+(t)][He]^2 - 4 \times 10^{-10}\left[He_2^+(t)\right][Xe] \tag{4.35}$$

where $\left[He_2^+(0)\right] = 0$

$$\frac{d\left[He^*(t)\right]}{dt} = \frac{S}{90} - 4.0 \times 10^{-10}\left[He^*(t)\right][Xe] \tag{4.36}$$

where it is assumed that the Xe density is constant and $\left[He^*(0)\right] = 0$

The xenon ions ($Xe^+ and Xe_2^+$) are formed through xenon's interactions with the helium ions, the xenon ion (Xe^+) and the helium metastable states. The rate equations are,

$$\frac{d[Xe^+(t)]}{dt} = 4 \times 10^{-10}\left[He_2^+(t)\right][Xe] + 4.0 \times 10^{-10}\left[He^*(t)\right][Xe] - 1.9 \times 10^{-31}[Xe^+(t)][He][Xe] - 9.0 \times 10^{-8}[Xe^+(t)][e(t)]^2 - 9.62 \times 10^{-21}[Xe^+(t)][He][e(t)] \quad (4.37)$$

where $[Xe^+(0)] = 0$ and the Xe density is assumed constant. The reason for this assumption is due to the fact that the initial atomic xenon density is so large that that at the pumping power used, there is very little depletion of the initial pool of xenon atoms.

$$\frac{d\left[Xe_2^+(t)\right]}{dt} = 1.9 \times 10^{-31}[Xe^+(t)][He][Xe] - 5.417 \times 10^{-5}\left[Xe_2^+(t)\right][e(t)] \quad (4.38)$$

where $\left[Xe_2^+(0)\right] = 0$

The xenon upper laser level ($[Xe_{ULL}(t)]$) rate equation is driven by the recombination of the xenon ions, stimulated emission and spontaneous emission. The rate equation is,

$$\frac{d[Xe_{ULL}(t)]}{dt} = 5.417 \times 10^{-5}\left[Xe_2^+(t)\right][e(t)] + 9.0 \times 10^{-8}[Xe^+(t)][e(t)]^2 - 9.62 \times 10^{-21}[Xe^+(t)][He][e(t)] - B_{21}I_\nu\frac{g(\nu)}{c}([Xe_{ULL}(t)] - [Xe_{LLL}(t)]) - A_{21}\left[Xe_{ULL}(t)\right. \quad (4.39)$$

where c is the speed of light (3×10^{10} cm s^{-1}) and $[Xe_{ULL}(0)] = 0$.

The xenon lower laser level ($[Xe_{LLL}(t)]$) rate equation can be written,

$$\frac{d[Xe_{ULL}(t)]}{dt} = A_{21}[Xe_{ULL}(t)] + B_{12}I_\nu\frac{g(\nu)}{c}([Xe_{ULL}(t)] - [Xe_{LLL}(t)]) - A_1[Xe_{LLL}(t)] \quad (4.40)$$

where $[Xe_{LLL}(0)] = 0$.

The laser intensity (I_ν) can be found by using a rate equation as discussed in Chap. 1,

$$\frac{dI_\nu(t)}{dt} = \Delta\Omega A_{21}[Xe_{ULL}(t)]\frac{c}{K} + B'_{21}I_\nu\left(\left[Xe_{ULL}(t) - \left[Xe_{LLL}(t)\right]\right]\right) - \frac{I_\nu(t)}{\tau_c} \quad (4.41)$$

where K is the number of photons per Joule, $B'_{21} = B_{21}\frac{g(\nu_0)}{c}\frac{ch\nu}{4\pi}$, the cavity lifetime is $\tau_c = \frac{2L_c}{c\ln\left(R_1R_2(1-L_i)^2\right)}$, L_c is the cavity length, R_1 is the reflectivity of mirror 1, R_2 is the reflectivity of mirror 2 and L_i are the diffraction losses.

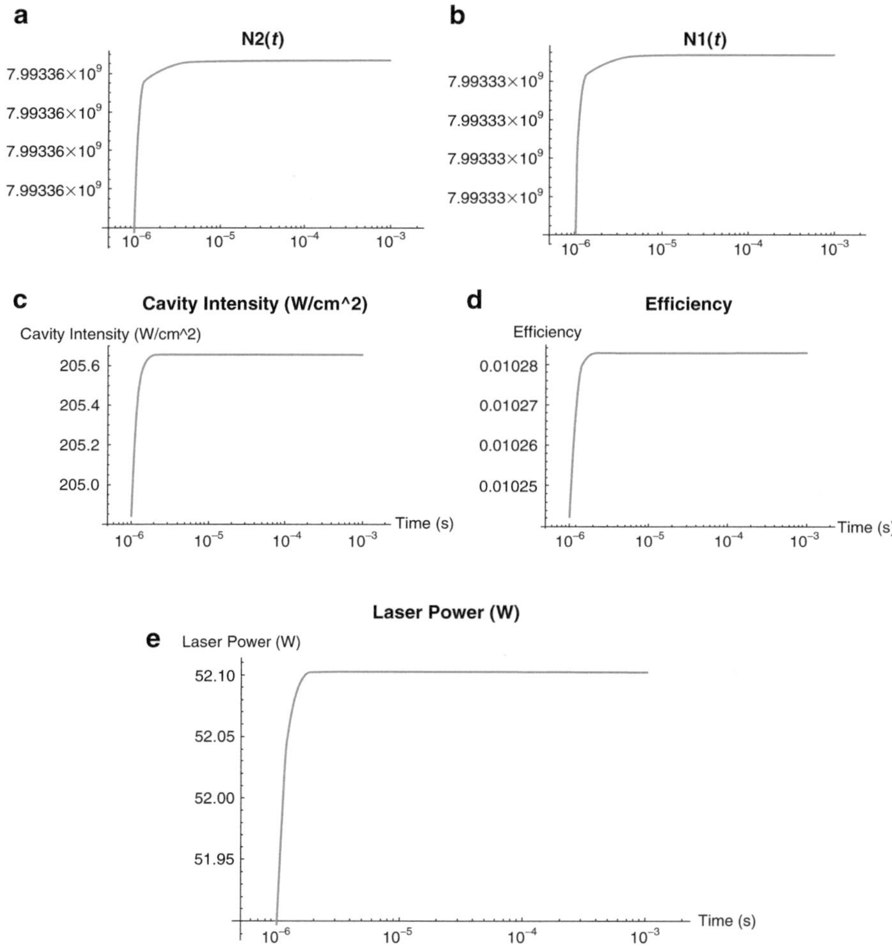

Fig. 4.18 The figure represents results for the atomic xenon laser at 2.03 μm with a pumping power density of 10 W/cm^3. Shown are (**a**) the upper laser level density change with time, (**b**) the lower laser level density with time, (**c**) the cavity intensity as a function of time (**d**) the laser efficiency as a function of time and (**e**) the laser power as a function of time. The cavity intensity saturates at 206.82 W/cm^2. With a 95 % reflective mirror, the laser power output is 52.1 W (with a cavity volume of 506.7 cm^3). The laser efficiency is 1.028 %

Equations (4.23), (4.24a), (4.24b), (4.25), (4.26), (4.27), (4.28), (4.29), (4.30), (4.31), (4.32), (4.33), (4.34), (4.35), (4.36), (4.37), (4.38), (4.39), (4.40), (4.41) can be solved using a program such as Mathematica as described in Chap. 1. The solution of these equations gives the results shown in Fig. 4.18 for the upper laser level density, the lower laser level density, the cavity intensity, the laser efficiency and the laser power output. The cavity length used in these calculations was 100 cm (typical of a nuclear-pumped laser experiment) with a cavity radius of 1.27 cm (also typical of a nuclear-pumped laser experiment). The cavity volume was 506.7 cm^3.

The mirrors used are a maximum reflectivity back mirror (100 %) and a 95 % reflective front mirror. Internal losses are 1 %. Recombination of the xenon ion is channeled into many states in the xenon atom. The fraction of recombination reactions that channel through the xenon upper laser level was assumed to be 5 %. This number was chosen because the resulting efficiency that was calculated approximately matches the experimentally measured efficiency for a binary He-Xe mixture.

The percentage of recombination reactions channeled through the upper laser level is an important concept to remember. This concept explains why the efficiencies of rare-gas lasers vary so widely despite the fact that rare gas lasers have similar mechanisms. It also explains why the tertiary He-Ar-Xe gas mixture is optimum for producing the high efficiency (~3 %) observed for the 2.03 μm xenon line. The recombination of the $ArXe^+$ ion channels a much higher percentage of the recombinations through the xenon upper laser level.

4.4.1.2 Molecular Carbon Dioxide Laser

A nuclear-pumped carbon dioxide laser was reported in 1981 after many years of research [28] and was met with skepticism. There are reasons for this skepticism. In Herwig's original nuclear-pumped laser concept from 1963, the carbon dioxide laser was suggested as the laser media [70]. Sandia National Laboratory's initial work in the early 1970s began with attempts to drive the carbon dioxide laser with nuclear reactions [5]. Sandia was not able to make carbon dioxide lase with nuclear-pumping alone, but was able to use a hybrid approach of combining nuclear-pumping with electrical pumping to make carbon dioxide lase. From these experiments, it was eventually determined that lasing was prevented by dissociation of the CO_2 molecule during irradiation. As discussed previously, the important factors in the success of a laser are the ways that the upper laser level is populated and how the lower laser level is quenched. The carbon dioxide laser is efficient because it uses a resonance transfer between the first vibrational level of nitrogen and the ground state of carbon dioxide to pump the laser. Molecular dissociation is a significant disadvantage in this laser because it channels energy away from the upper laser level, it depletes the ground state density of carbon dioxide and it depletes the nitrogen vibrational state. In essence to takes away the advantage of having a resonance process for transfer of energy into the upper laser level. In an electrically pumped laser, the vibrational levels of nitrogen are populated by collisional interactions between electrons and nitrogen. The nitrogen vibrational levels come into thermodynamic equilibrium with the electrons which have a temperature Te. Thus, for example, with an electron temperature of 0.5 eV (or about 5800 K), the fraction of nitrogen in the important vibrational energy levels from $v = 1–8$ (which can collisionally populate the upper laser level) is about 55 %. The electron temperature in an electrical discharge can be controlled so that the production of nitrogen vibrational energy levels from $v = 1–8$ is very efficient. The

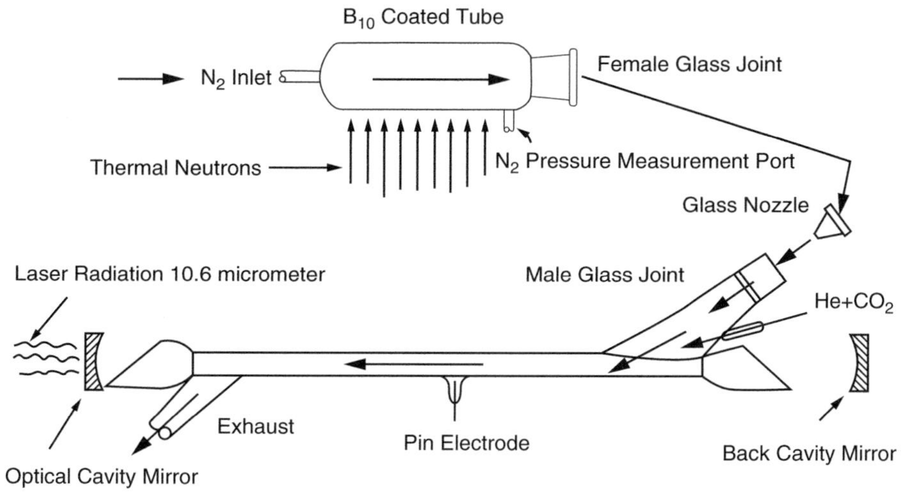

Fig. 4.19 An experimental diagram of the He-N_2-CO_2 nuclear-pumped transfer laser is shown. The transfer laser first used a fast burst reactor to drive the boron-10 capture reaction to excite nitrogen gas. The excited nitrogen gas then flowed to the laser cavity where it was mixed with He and CO_2 [68]

long lived nitrogen vibrational states can then be mixed with carbon dioxide to populate the upper energy level through a collisional resonance interaction and energy is transferred efficiently. Electron temperature is a direct function of the pumping power density so electrically-pumped lasers can control the electron temperature to optimize the formation of the vibrational levels of nitrogen. Nuclear-pumped lasers do not generate high enough electron temperatures to produce significant levels of the nitrogen vibrational states. This is the most likely reason for the early failures in nuclear-pumping a carbon dioxide laser.

The carbon dioxide laser experiment reported in 1981 was somewhat different in that it used a transfer scheme where nitrogen was nuclear-pumped and then transferred to a laser cavity where it was mixed with carbon dioxide (Fig. 4.19). The nitrogen was flowed through a pumping chamber coated with boron-10 exposed to the thermal neutron flux from a fast burst reactor. The nuclear-pumping both created ionization and excitation in the nitrogen gas. Typically the electron temperature is low (a few tenths of an electron volt in a nuclear-pumped cell) so the fraction of nitrogen molecules in the vibrational states 1 through 8 would be on the order of 10–15 %. There would be some recombination of ionized nitrogen that could also contribute to the key vibrational states as well. There should have been enough nitrogen in the correct vibrational states to drive a carbon dioxide transfer laser. The laser cavity was 0.80 m long and the gain medium was 0.50 m long with an inner diameter of 0.013 m. The back cavity mirror was 10 m radius of curvature gold-coated and the front mirror was a gold coated flat mirror with a small hole in its center. The gain medium had sodium chloride windows at Brewster's angle. The total mass flow rate was about $7.6 \times 10^{-4}\ m^3\ s^{-1}$. The peak neutron flux was of the order of 5×10^{16} neutrons cm^2 s^{-1} with a Full Width Half Maximum (FWHM) pulse of 200 μs. The pressures were 1 and 5 kPa at a mixing ratio of 1:3:8 (CO_2:N_2: He). The laser pulse was delayed about 30 ms after the reactor pulse. This delay corresponded to the time required for the nitrogen to flow from the excitation chamber to the gain cell. The laser pulse lasted 1 ms FWHM for 0.5 kPa and 3 ms FWHM for 1 kPa. The peak power output of the laser was 100 W and the power density in the reaction chamber system was around 50 W cm^{-3}. The efficiency of the laser was approximately 1 %. This efficiency is consistent with an excitation cell having an effective electron temperature of about 0.12 eV (or 1500 K—see the discussion below). The efficiency of the carbon dioxide laser is limited by the effective electron temperature. The nitrogen vibrational levels will come into equilibrium with the effective electron temperature through rapid collisional processes. Thus the nitrogen vibration levels can attain an effective equilibrium temperature of about 1500 K for a fast burst reactor. In order to reach the higher efficiency demonstrated by electrical pumping, the equilibrium temperature of the nitrogen vibrational levels would have to be increased substantially to about 0.5 eV (or 5800 K). This electron temperature is not achievable by traditional nuclear-pumping techniques.

There are alternative ways to use nuclear energy to drive high-power high-efficiency lasers without having subsystems that will increase the complexity, the size and the mass of the coupled reactor/laser (e.g., using a Rankine cycle to produce electricity which is then used to electrically drive the laser [71]). All NPLs to date have been based on the first step of ionizing radiation creating ion pairs. The efficiency of this step is typically 35–58 % dependent on the medium (e.g., solid, liquid, gas, type of gas etc.) and the type of ionizing radiation (e.g., ions). Thus the ionization efficiency is an upper bound for efficiency for this type of laser. In contrast, the energy from nuclear reactions can be converted to heat at nearly 100 % efficiency. Thus using a method which is based on thermal effects has some advantages.

The electrically pumped carbon dioxide laser has been extensively studied and modeled [72, 73]. The role of electrons in the electrically driven laser is well understood and has been successfully modeled using simplified energy balance equations and the assumption of a thermodynamic equilibrium within specific species of the laser like electrons and between species such as electrons and nitrogen vibrational levels or carbon dioxide vibrational-translational-rotational levels [72, 73]. The five and six temperature models accurately predict the behavior of the electrically pumped carbon dioxide laser [73]. The carbon dioxide nuclear-pumped laser used ions from the boron-10 neutron capture reaction to excite nitrogen gas. As discussed in prior sections of this book, interactions with energetic ions create ionization and excitation of the gas mixture. In addition, there is a population of electrons with a characteristic temperature that is a function of the power density. Thus the nuclear-pumped carbon dioxide laser has some similarities to electrically pumped carbon dioxide lasers but with a lower characteristic electron temperature.

In 1978 Basov and colleagues proposed that the carbon dioxide laser could be excited by thermal pumping and adiabatic cooling of the CO_2 gas mixture [72]. Several experiments testing this concept were successful [72, 74, 75]. In addition, similar thermal based concepts were demonstrated using gas dynamic lasers [76] and shock wave generated supersonic flow lasers [77]. These lasers pump the nitrogen vibrational levels by creating high temperatures and thus generating nitrogen vibrational levels by collisions with gas molecules to create a thermodynamic equilibrium between the gas temperature and the nitrogen vibrational level temperature. Thermal pumping methods do not have significant electron densities so electron collisions with nitrogen are not responsible for creating the nitrogen vibrational states. A model of the thermal process needs to calculate the effective temperature of the nitrogen vibrational levels.

A model for a high-power – high efficiency CO_2 laser thermally pumped by an extremely high temperature gas cooled reactor was developed by the author and his students [78]. The model starts with the use of an Ultra-High Temperature Reactor (UHTR) core which transfers heat to a nitrogen gas coolant. In order to generate the ultra-high temperatures needed for thermal pumping, technology developed for nuclear rockets was adapted for the purpose of creating a steady-state stream of ultra-high temperature nitrogen gas. This method of using an ultra-high temperature reactor to drive the carbon dioxide laser has a potential efficiency of between 8 and 16 %.

4.4.1.3 Molecular Carbon Monoxide Laser

After the nuclear-pumped laser group at Sandia concluded that radiation inhibited the operation of a carbon dioxide laser, their focus shifted to the carbon monoxide laser. Carbon monoxide lases on the vibrational-rotational transitions in the ground electronic state $X^1\Sigma^+$ (Fig. 4.20) [79]. The choice of a carbon monoxide laser was natural because of its high quantum efficiency of nearly 100 % as compared to 40 % for the carbon dioxide laser. The high quantum efficiency is due to the large number of lasing transitions from which the vibrational energy is extracted, as shown in Fig. 4.20, and the proximity of energy levels which reside close to the ground state. Also, an operating carbon monoxide laser is more efficient than any operational carbon dioxide laser (the highest efficiency obtained with a carbon monoxide laser is around 50 % [80, 81] while the highest efficiency carbon dioxide laser is about 33 % [72]).

The mechanism which creates the population inversion in the carbon monoxide laser is a combination of direct electron impact excitation of the lower vibrational levels of the CO states, followed by a rapid redistribution of the vibrational energy through inelastic collisions between the carbon monoxide molecules [82]. The carbon monoxide laser efficiency increases as temperature decreases and thus operates best at cryogenic temperatures [83].

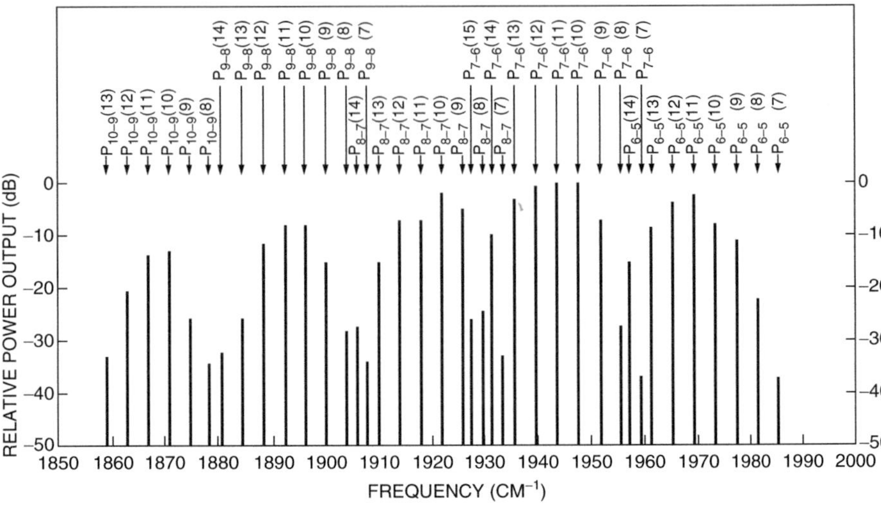

Fig. 4.20 The relative intensities of the lines which oscillate in the carbon monoxide laser as a function of frequency [79]

After trying to pump the carbon dioxide laser by both direct nuclear-pumping and a hybrid nuclear-pumped/electrically pumped system, a team from Sandia National Laboratory concluded that the carbon dioxide molecule was dissociated by radiation thus prohibiting the nuclear-pumping process from being effective. So the Sandia team began working with carbon monoxide. In 1974 the team demonstrated the first reactor-pumped laser using a cryogenically cooled carbon monoxide laser [18] as seen if Fig. 4.21. The axis of the laser apparatus was mounted about 20 cm from the axis of the SPR2 reactor. The interior surface of the laser cavity was coated with a 20 cm long $^{235}U_3O_8$ film and interacted with 100 Torr of pure CO gas at 77 K. The output mirror had either a reflectance of 96 or 98.5 % and the back mirror had a reflectance of 99 %. A CaF_2 window at Brewster's angle was used on the output side of the laser cavity. The Ge:Au optical sensor was placed 15 m away beyond the reactor shield (Fig. 4.22). The experiment had an issue with dust build up on the back mirror surface over time. The peak power of the laser was 2–6 W with the higher power level coming from the 96 % reflective mirror. The peak neutron flux from a fast burst reactor is about 1×10^{17} neutrons cm^{-2} s^{-1} with a 200 μs FWHM pulse width (Fig. 4.23). Lasing occurred on the vibrational bands of CO between 5.1 and 5.6 μm. A three dimensional Monte Carlo calculation yielded an average energy deposition in the carbon monoxide gas of about 200 J $liter^{-1}$. The laser efficiency was estimated to be 0.1–0.3 %.

The carbon monoxide laser was shown to work with a helium-3 and carbon dioxide mixture which was also notable for being the first demonstration of a volume pumped NPL [86]. The laser tube used in these experiments was a 2.5 cm inner diameter quartz tube cut at Brewster's angle on both ends. The middle 60 cm of the tube was cooled by flowing liquid nitrogen in PVC tubing thus maintaining a

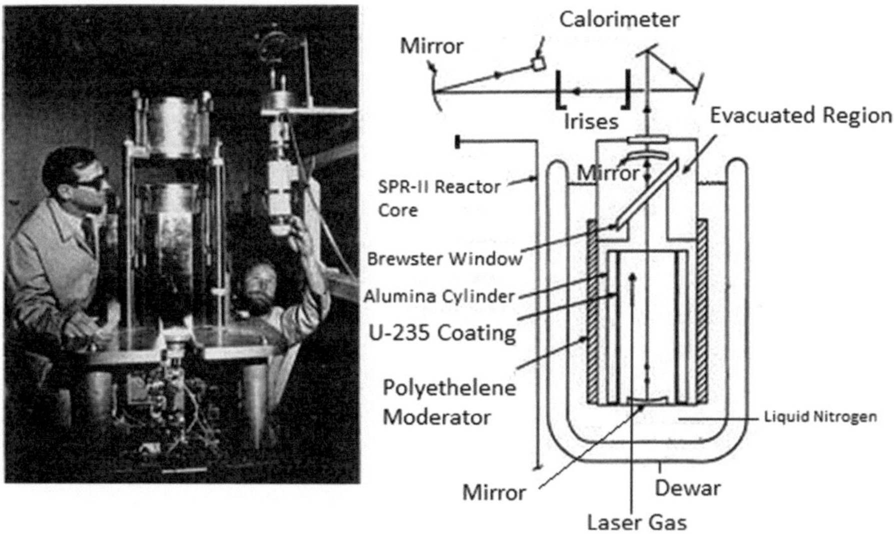

Fig. 4.21 Picture of the nuclear-pumped carbon monoxide laser which is cooled to liquid nitrogen temperatures [84]

wall temperature of 100 K and a gas temperature of about 150 K. The laser was pumped by the U.S. Army Pulse Radiation Facility (a fast burst reactor with a peak thermal neutron flux of approximately 3×10^{16} neutrons cm^{-2} s^{-1}). The gold coated back mirror had a 10 m radius of curvature and the front mirror was a gold coated flat with a 2 mm diameter hole in its center. The optical output was detected with an InSb sensor cooled to liquid nitrogen temperature. Nitrogen was added to the He-3 and CO mixtures at concentrations up to 10 %. A 5 % concentration of nitrogen was found to increase the power output to its maximum. The scaling of the laser with thermal neutron flux is not linear as would be expected but approximately proportional to the square of the thermal neutron flux. It was argued that this scaling probably rules out collisional radiative recombination as the main mechanism for populating the vibrational states of carbon monoxide. This behavior may be indicative of a process involving two ionized states in the production of vibrationally excited carbon monoxide. The reported efficiency for the $^3He:N_2:CO$ at a total pressure of 1,350 Torr (at a ratio of 1215:67.5:67.5) nuclear-pumped laser was of the order of 5 %.

Subsequent theoretical work on nuclear-pumped carbon monoxide was done in a study out of Russia [87]. Mixtures of He-CO and Ar-CO were examined. It was determined from the study that the energy transfer efficiency by fast ions into vibrational states of carbon monoxide were 1.5 times higher in the Ar-CO mixtures than with He-CO mixtures. In the Ar-CO mixtures about 18 % of the energy ended up in the vibrational states of carbon monoxide.

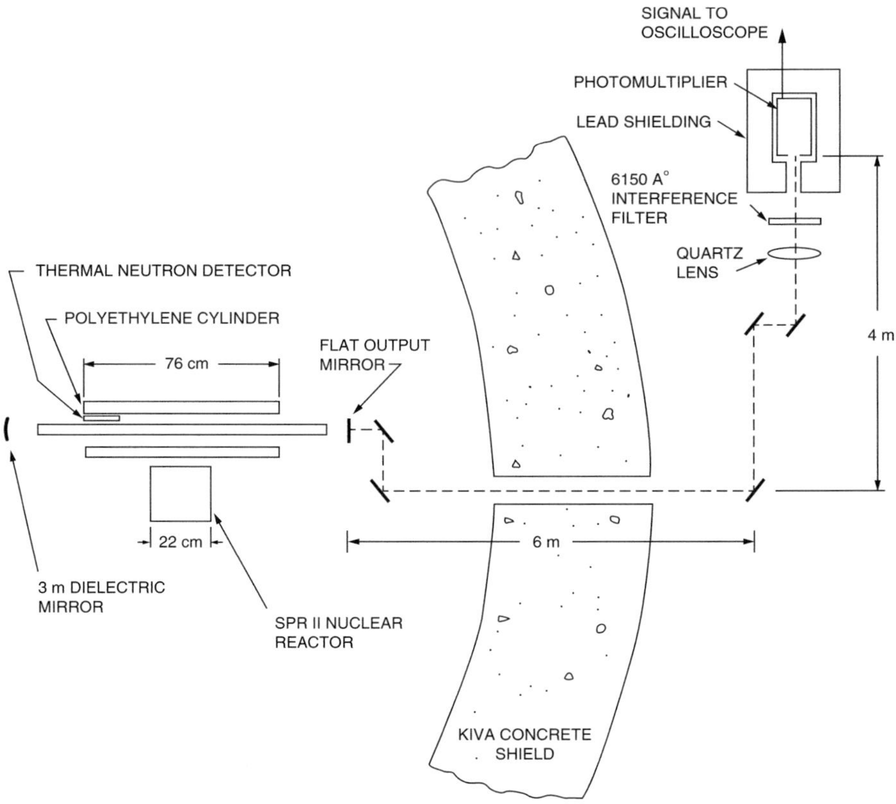

Fig. 4.22 Set up for a typical nuclear-pumped laser experiment using the SPR II fast burst reactor [4]. This figure is specifically for the mercury ion laser experiment at 615 nm

4.4.1.4 Atomic Nitrogen Laser

The atomic nitrogen laser utilized a pulsed TRIGA reactor as the neutron source and boron-10 coated on the inside of a cylindrical cell which surrounded the laser medium [20]. The laser's two lines (939.3 nm and 862.9 nm) oscillated simultaneously, had the lowest threshold power density of any nuclear-pumped laser prior to 1976, and had the shortest wavelength of any NPL prior to 1976. The laser is an impurity type laser which consists of a mixture of neon and nitrogen. The 939.3 nm transition lased at total pressures from slightly less than 75 Torr to slightly greater than 175 Torr while the 862.9 nm line lased from 75 to 375 Torr. The laser did not operate at pressures beyond 375 Torr because the lithium and alpha ions created by the ^{10}B – neutron reaction had lost most of their energy to the neon gas before reaching the centerline of the tube. Calculations indicated that at 450 Torr of neon, the alpha particle emitted from the boron-10 coating on the wall had a range equal to the centerline distance. The nitrogen was an impurity in the neon gas and also came from outgassing from the walls. It is believed that the nitrogen concentration

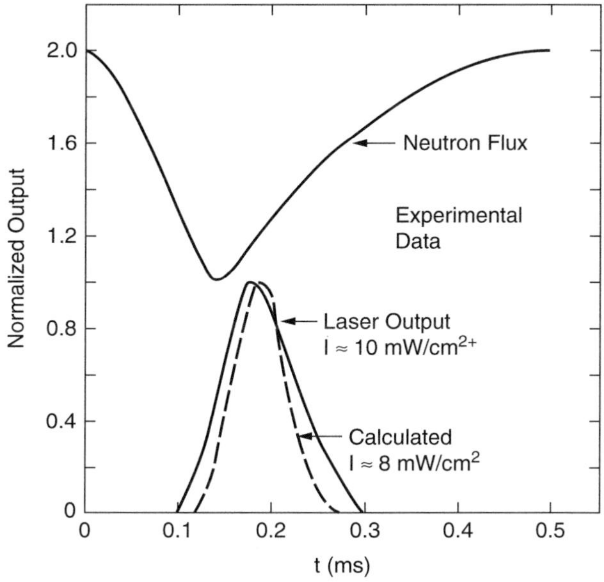

Fig. 4.23 Experimental data for the neutron flux profile from the SPR II fast burst reactor [85] (Note that the laser data at the bottom of the figure was for the atomic carbon laser on the 1.45 μm line)

was on the order of a few milli Torr. Since the Ne to N_2 ratio was on the order of 50,000 to one, the energy flowed to the neon atoms to produce neon ions and neon metastable states. The energy from these states was then transferred to the minority nitrogen species (Fig. 4.24). From the energy level diagram, it is clear that Ne^+, Ne_2^+ and the neon metastable states do not have enough energy to pump the nitrogen upper laser level directly. It was believed that the nitrogen laser is a recombination laser [88, 89]. Thus the process must occur in three steps in order for this to be true.

- The first step is to create an intermediate state such as the dissociation of the nitrogen molecule into a nitrogen atom.
- The second is to create a nitrogen ion through a process such as penning ionization.
- The third step is a recombination involving some form of the nitrogen ion to form the upper laser level.

It should be noted that mixtures of He:N_2 and Ar:N_2 do not lase. If recombination were indeed responsible for pumping the upper laser level, the ion and metastable states of He and Ar also require a three-step process to form the nitrogen ion. Thus the expectation is that nitrogen should lase with either a He or Ar buffer gas but does not. A possible explanation for why He and Ar buffer gases do not work centers on neon as a critical species in the laser's kinetics. Perhaps there is some undetermined resonance process or a complex ion that is formed (like NeN^+) which preferentially populates the $2P^0_{3/2}$or the $2D^0_{5/2}$ states through a recombination process. None the less, the mechanism is complex and questions still remain.

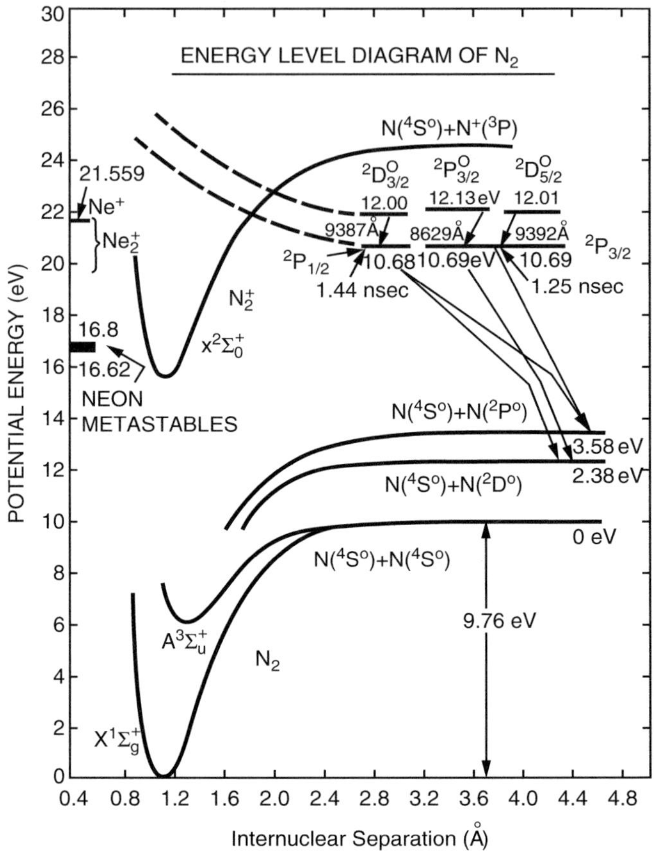

Fig. 4.24 Energy level diagram of the atomic nitrogen laser [88]

The nuclear-pumped atomic nitrogen laser was placed in the through port of the University of Illinois TRIGA reactor so that the laser centerline was coincident with the reactor core centerline as shown in Fig. 4.25. The detector system was separated from the laser by about 9 m of radiation shielding. The laser cavity consisted of a quartz tube cut at Brewster's angle which contained a 68 cm long by 2.5 cm inner diameter aluminum cylinder coated on the inner surface with boron-10. Two high reflectivity dielectric coated mirrors formed the optical cavity. The back mirror had a 2 m radius of curvature and was 99.9 % reflective at 844.6 nm while the output mirror had a 1 m radius of curvature and was 99.5 % reflective at 844.6 nm. The mirror separation was 87 cm. A small electric motor with chopping fan was mounted just in front of the back mirror such that it alternately blocked and unblocked the back mirror which caused the optical cavity to alternate between high gain and low gain. The University of Illinois TRIGA reactor produced a thermal neutron flux up to 1×10^{16} neutrons cm^{-2} s^{-1} with a 12 ms FWHM pulse (using a $4 pulse). The laser was pumped down by a vacuum system

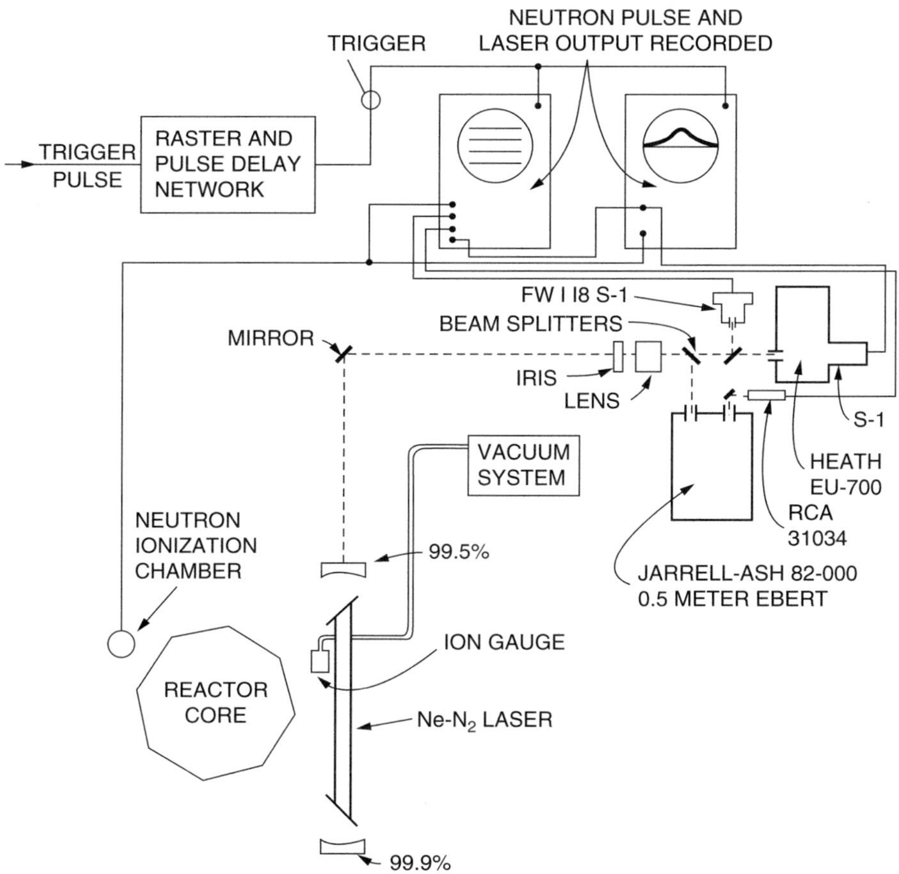

Fig. 4.25 Diagram of the experimental setup for the atomic nitrogen NPL [20]

(mechanical pump and diffusion pump combination) to background pressures of 4×10^{-5} Torr. The beam was split and one leg of the optical path went to a Jarrell-Ash 8200 monochromator with a RCA 31034 (Na-K-Sb-Cs: minimum wavelength 200 nm, peak wavelength 850 nm and maximum wavelength 930 nm) photomultiplier tube and the other leg went to a Heath EU-700 monochromator with an S1 photomultiplier tube (Ag-O-Cs: with a spectral range from 300 to 1200 nm).

Outputs from the two detectors were displayed on two Tektronix oscilloscopes. The S1 photomultiplier recorded the total light output while the RCA photomultiplier measured the laser output (Fig. 4.26). The top traces show the 862.9 nm and 939.3 nm laser outputs superimposed on the neutron pulse (measured with a BF_3 detector). The chopped signal indicates that the fan is blocking and unblocking the back mirror thus changing the cavity gain from high to low. The steep changes in signal are indicative of lasing.

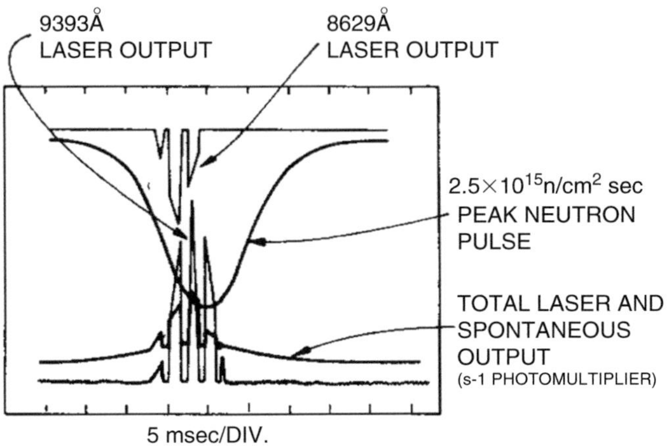

Fig. 4.26 Atomic nitrogen pulse characteristics for the 939.3 and 862.9 nm lines superimposed on the neutron pulse [20]

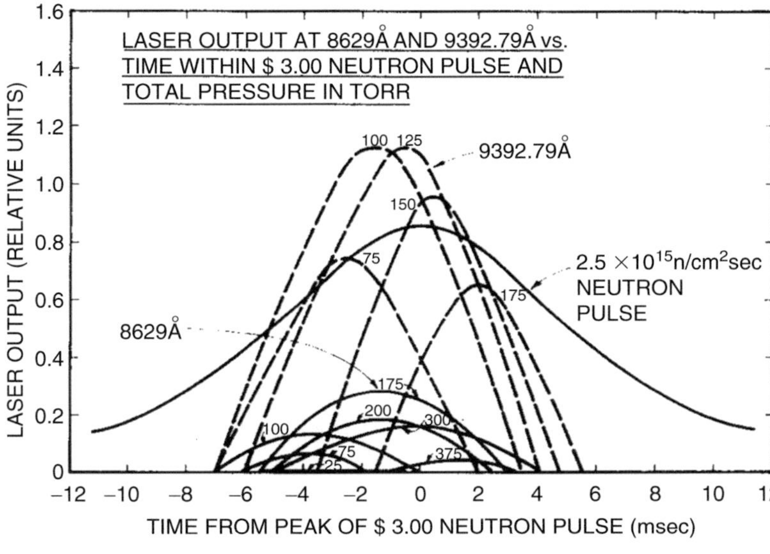

Fig. 4.27 Illustration of the time dependent behavior of the atomic nitrogen lines at 939.3 and 862.9 nm with pressure. The shape of the neutron pulse is superimposed [20]

Figure 4.27 shows the pressure and time dependent behavior of the 939.3 and 862.9 nm laser lines. The partial pressure of nitrogen came from the impurities in the neon as well as outgassing from the walls. It was assumed that the nitrogen partial pressure was on the order of a few milliTorr. The 939.3 nm line lased at pressures ranging from 75 Torr to about 175 Torr. The 862.9 nm line lased from 75 to 375 Torr. As previously discussed, it is believed that the 375 Torr runs were

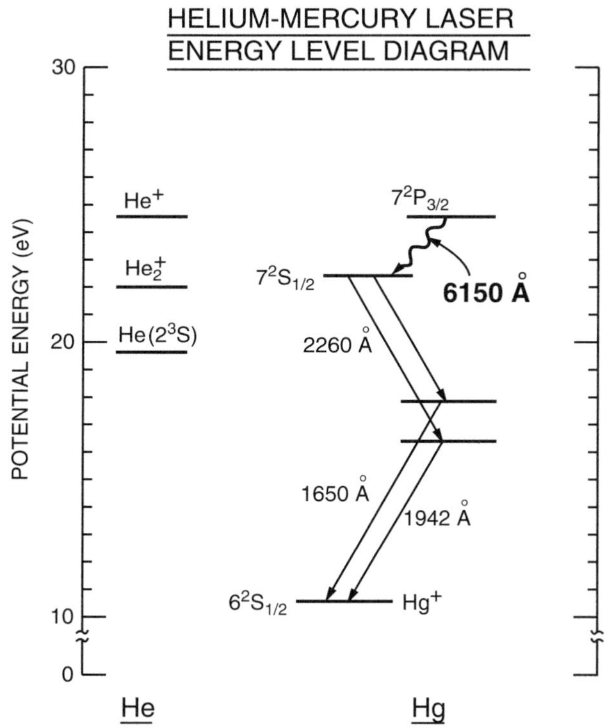

Fig. 4.28 The energy level diagram for mercury showing the mercury ion laser line at 615 nm (between the $7^2P_{3/2}$and the $7^2S_{1/2}$ states) [21]

near the point where the range of the alpha particles lost most of their energy before reaching the centerline of the laser cavity. Although, since there was no reliable way to measure or vary the partial pressure of nitrogen, the 375 Torr cut off could have also been due to changes in the partial pressure of nitrogen from outgassing.

The peak laser efficiency was found to be about 0.00014 % at a total pressure of 150 Torr which corresponded to 1.1 kW of power deposited in the neon gas and a laser power output of 1.5 mW.

4.4.1.5 Mercury-Ion Laser

The first visible nuclear-pumped laser was demonstrated in 1977 with the mercury ion laser line at 615 nm (between the $7^2P_{3/2}$and the $7^2S_{1/2}$ states) [21]. Figure 4.28 shows the energy level diagram for the mercury ion laser. Gain measurements were at first taken using the University of Illinois TRIGA reactor. The laser cavity was constructed of a 2.7 cm inner diameter Pyrex tube that was 86 cm long. High-temperature epoxy was used to seal thin Suprasil windows onto each end which was cut at Brewster's angle. A symmetrical hollow cathode and anode arrangement was employed in the laser cavity to allow low-pressure electrical operation of the laser for alignment purposes. The cathode length was a 60 cm and had an outer diameter

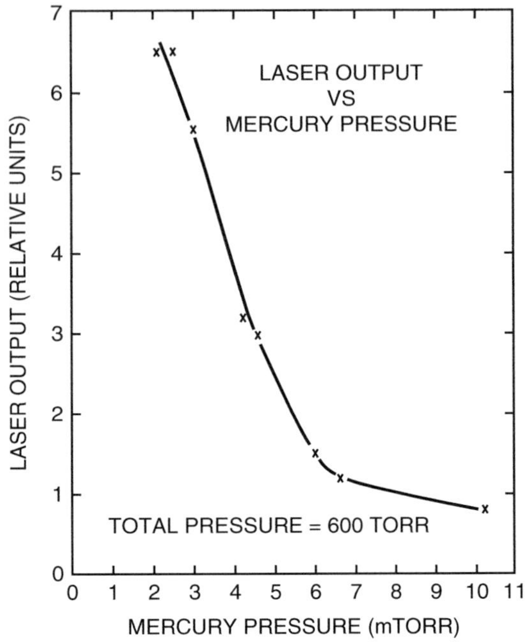

Fig. 4.29 The mercury ion laser output as a function of the partial pressure of mercury [21]

of 2.54 cm. This titanium tube was coated on the inside with a 4 mg cm^{-2} layer that was 66.8 % ^{10}B by weight. The thickness of the ^{10}B corresponds to the range of an alpha particle in boron. Heater tapes were wrapped around the laser to control the mercury partial pressure, and K type thermocouples were used to measure the temperature of the tube. Two reservoirs in the laser cavity contained a total of 79 mg of 67 % isotopic enriched ^{202}Hg. This specific isotope was used to increase the gain over that which was possible with non-enriched mercury. The mercury partial pressure was calculated from the temperature of the reservoir. The tube was mounted on rigid aluminum 1100 frame and placed next to the core of the TRIGA reactor for gain experiments and next to the SPR II reactor for the laser demonstration (Fig. 4.22). In the SPR II experiment, a 2.5 cm thick 76 cm long polyethylene sleeve was wrapped around the laser cavity in order to moderate the neutrons and create a thermal flux. The laser cavity used a 3 m curvature dielectric coated back mirror with a reflectivity of 99.9 % at 615 nm and a flat output mirror with a reflectivity of 99 % at 615 nm. A shielded GaAs photomultiplier tube was used as the detector and it was placed about 10 m (through a radiation shield) from the laser cavity as shown in Fig. 4.22.

Figure 4.29 shows the power output of the mercury ion nuclear-pumped laser as a function of mercury pressure. Mercury is an impurity in the helium gas medium (with helium pressures between 300 and 600 Torr). The mercury concentration was on the order of 3.3 ppm at the optimum conditions (600 T He and 2 mT Hg). The laser power increased as the mercury partial pressure decreased.

Table 4.12 A listing of TRIGA parameters used in the atomic carbon laser research [94]

ρ+	Φpeak	Pulse FWHM	Peak power deposition with ^{10}B (W cm^{-3})		
($)	(n_{th} cm^{-2} s^{-1})	(ms)	100 T He	400 T He	800 T He
1.80	5.0×10^{14}	20	0.8	3.1	6.3
2.25	1.5×10^{15}	17	2.4	9.5	19.0
3.00	2.5×10^{15}	12	3.9	15.8	31.5

ρ+ is the reactor reactivity measured in units of dollars

Several studies have shown charge transfer to be the dominant pumping mechanism in the mercury ion laser [21, 90]. The energy coincidence for the charge exchange reaction between He^+ and Hg can be seen in Fig. 4.28, as well as the mercury-ion upper and lower laser levels. At higher pressures, however, the formation of mercury and helium metastables, as well as the $He_2{}^+$ ion provide for potential energy loss channels. These species are thought to be responsible for the decrease in the laser output with increasing mercury partial pressure.

4.4.1.6 Atomic Carbon Laser

The atomic carbon laser is one of the best understood nuclear-pumped lasers. The laser was driven by the University of Illinois TRIGA reactor which is capable of producing a variety of pumping power densities based on the reactivity insert listed in column one of Table 4.12 [25]. The atomic carbon laser has a very low threshold pumping power density (40 mW cm^{-3}). The laser used binary gas mixtures consisting of a rare gas buffer (He, Ne or Ar) and a carbon donor impurity (CO or CO_2). The atomic carbon NPL line at 1,454 nm has operated at low threshold power densities in He + (CO or CO_2) [25], Ne + (CO or CO_2) [91, 92], and Ar + CO_2 [93] mixtures. As an Electrically Pumped Laser (EPL) [94] the atomic carbon system with He + CO mixtures required power densities greater than 90 W cm^{-3} while with mixtures of He + CO_2, and Ne + (CO or CO_2) required power densities greater than 900 W cm^{-3} for oscillation. Conversely, the atomic carbon NPL with He + (CO or CO_2) mixtures required a power density of about 1 W cm^{-3} and with Ne + (CO or CO_2) mixtures required a power density of about 9 W cm^{-3}. The Ar + CO_2 mixture by far had the lowest power density requirement, about 0.04 W cm^{-3} for the onset of oscillation. In the EPL with mixtures of He + CO, an efficiency (Laser Energy/Energy Deposited) of 2×10^{-5} % was seen, and with mixtures of He + CO_2 and Ne + (CO or CO_2) an efficiency of 5×10^{-7} % was observed [94]. [The EPL efficiency calculations utilized the maximum power deposited found by the product of voltage times current measurements without taking into account the energy losses (e.g., electrode phenomena)]. It is interesting to note that an EPL with mixtures of Ar + CO_2 did not lase at all. By comparison, the NPLs with mixtures of He + (CO or CO_2) and Ne + (CO or CO_2) were observed to have efficiencies of about 5×10^{-4} % and 1×10^{-5} % respectively. The efficiency of the Ar + CO_2 NPL was 9×10^{-3} % (Table 4.13).

Table 4.13 Comparison of parameters for the noble buffers gasses (He, Ne and Ar) and the donor gasses (CO_2 and CO) in the atomic carbon NPL [94]

	Buffer				
	He		Ne		Ar
Parameter	CO_2	CO	CO_2	CO	CO_2
Threshold (W cm^{-3})	~1–2	~1	10	9	0.04
Max delay (ms)	5	None	2	1	2
NPL efficiency (%)	1×10^{-4}	2×10^{-4}	1×10^{-5}	2×10^{-5}	1×10^{-2}
Max NPL efficiency (%)$^{+}$	3.4	4.3	3.4	4.3	3.4

$^{+}$Maximum theoretical NPL efficiency is defined as the available energy of the transition divided by the energy difference between the ULL and the ground state of the carbon donor

The atomic carbon laser has demonstrated unique characteristics apart from previously studied laser systems. The peak of the laser pulse is delayed from the peak of the neutron pulse by about 1–5 ms. This delay has been seen in all of the rare gas buffer and carbon donor mixture combinations with the exception of He + CO (Table 4.13) [95]. The absence of a time delay in He + CO leads one to conclude that the energy is not stored in the upper laser level since all of the mixtures studied oscillate on the same transition. As discussed in Chap. 1, the buffer metastable species actually transfers the energy to the carbon donor to populate the ULL and plays a crucial role in the delay as will be shown. The time delay, the low pumping power density of the NPL, the higher pumping power density of the EPL, the fact that the NPL operated with all combinations of buffer gas (He, Ne and Ar) and carbon donor gas (CO_2 and CO) with the exception of Ar + CO, and the EPL did not operate at all with the Ar buffer gas are all important factors in the determination of the laser's kinetics [85, 96–101]. Experiments in Russia with the atomic carbon nuclear-pumped laser confirmed lasing at 1,454 nm using mixtures of He (Ne, Ar) + CO and uranium foils. The power output obtained from the 1,454 nm line was 7 W [6]. In addition it was claimed that weak lasing on the 833.5 and 940.6 nm atomic carbon transitions were also observed.

The atomic carbon NPL is capable of operating over a wide variety of pressures utilizing the ^{10}B surface source. This phenomena becomes apparent by considering the three types of buffered atomic carbon NPLs. In the He buffered systems the atomic carbon NPL lased at total pressures ranging from 20 to 850 Twith carbon donor pressures ranging from 0.25 to 25 mT. The output power of the laser tended to increase as the He partial pressure was increased. As would be predicted by the range of alpha particles in He, the laser power output was still rising at 850 T when equipment limitations forced a halt to further pressure increases (the vacuum seals were not designed for pressures above one atmosphere). In the Ne buffered system, the total pressure varied between 75 and 600 T with carbon donor partial pressures ranging between 0.25 and 15 mT. The total pressure in the Ar + CO_2 system was varied between 20 and 350 T with the carbon donor partial pressures ranging between 0.25 and 150 mT (see Table 4.14).

Table 4.14 Laser power with respect to CO or CO_2 and rare gas buffer partial pressures [25, 91–95]

Buffer gas	Pressure (Torr)	Laser power output (mW/cm^2)									
		CO (mT)					CO_2 (mT)				
		1	3	6	15	20	1	6	15	30	100
He	50	12	11	9.5	0	0	5	2	0	0	0
	75	13	11	9.0	0	0	7	6	0	0	0
	200	20	20	18	0	0	10	9	3	0	0
	400	21	20	17	5	0	15	13	10	0	0
	600	22	21	19	0	0	18	14	11	0	0
Ne	150	0	0	1	1	0	0	2	0	0	0
Ar	50	0	0	0	0	0	2	5	2	1	0
	100	0	0	0	0	0	9	12	9	6	0
	200	0	0	0	0	0	19	24	18	15	9

As shown on Fig. 4.30, a laser cell with an 80-cm long 2.54 cm inner diameter (i.d.) Pyrex or quartz tube, containing approximately 70 cm of 2.5 cm outer diameter (o.d.) aluminum tubing (with a thin inner coating of ^{10}B), formed the active region of a 117-cm optical cavity. Two highly reflective 2.54-cm diameter mirrors (R ~ 99 %), with a radius of curvature of 3 m, were placed at both ends of the cavity. A small chopping fan mounted in between the back window and mirror provided a means of modulating the stimulated emission signal for the purpose of identifying gain or lasing on the transition. The cell was provided with a gas feed line in order to allow the pumping down of the system to a proper background pressure ($\sim 1 \times 10^{-5}$ Torr) as well as for providing a means of adding gases. Once the cell was connected to the vacuum and gas handling station it was then placed in the through port of the TRIGA reactor approximately tangent to the core (see Fig. 4.30).

Experiments were also conducted on the atomic carbon electrically pumped laser (EPL) using a positive column discharge. A cell of 3/8″ i.d. with a 60-cm active length was used (see Fig. 4.31).

4.4.1.6.1 He + (CO or CO_2) Mixtures

The atomic carbon laser on the $C(3p\ ^1P_1) \rightarrow C(3\ s\ ^1P_1{}^0) + 1{,}454$ nm transition was first discovered by Patel in 1963 using a positive column discharge with mixtures of He, Ne + (CO or CO_2) The optimum pressure for the carbon donor, (CO or CO_2), was ~10 mT and for the buffer gas He ~ 2 Torr. Experiments on the He + (CO or CO_2) EPL performed in reference [94] essentially duplicated the results of Patel.

The atomic carbon He + (CO or CO_2) system was made to lase using nuclear pumping and the ^{10}B surface source. It operated with carbon donor pressures of

Fig. 4.30 Experimental set of the atomic carbon nuclear-pumped laser using the University of Illinois TRIGA reactor [94]

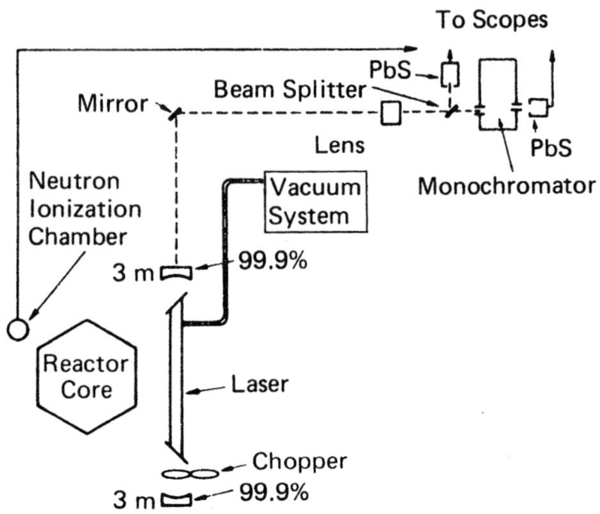

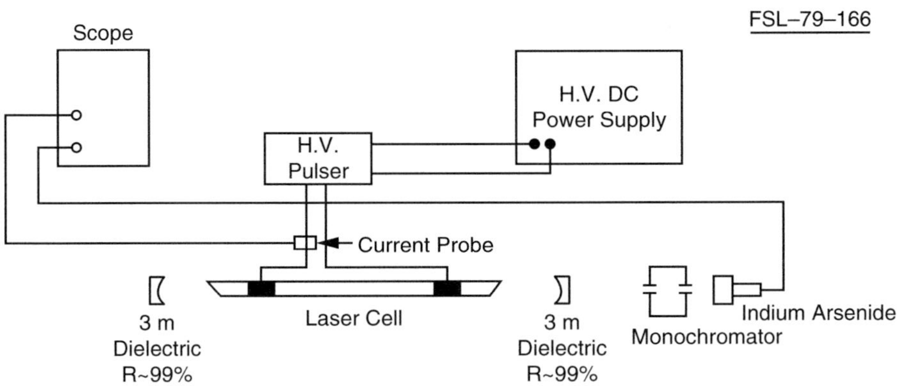

Fig. 4.31 Experimental set up of the atomic carbon discharge-pumped laser [94]

<25 mT and pressures of 20–850 Torr of He. The NPL was able to operate at high pressures because of the large transport distance of alpha particles in helium gas. A unique feature of the atomic carbon NPL with mixtures of He + CO_2 was the observation of up to 5 ms delays between the peak of the neutron pulse and the peak of the laser signal.

As can be seen on the energy level diagram in Fig. 4.32, a single collision with the metastable species of He can populate the upper laser level directly in the He + CO mixtures. Indeed, in the He + CO atomic carbon EPL it was shown that a single collision between He (2^3S) and CO ($v = 0$) is the dominant mechanism for populating the upper laser level [79]. In He + CO_2 mixtures, and in mixtures of Ne

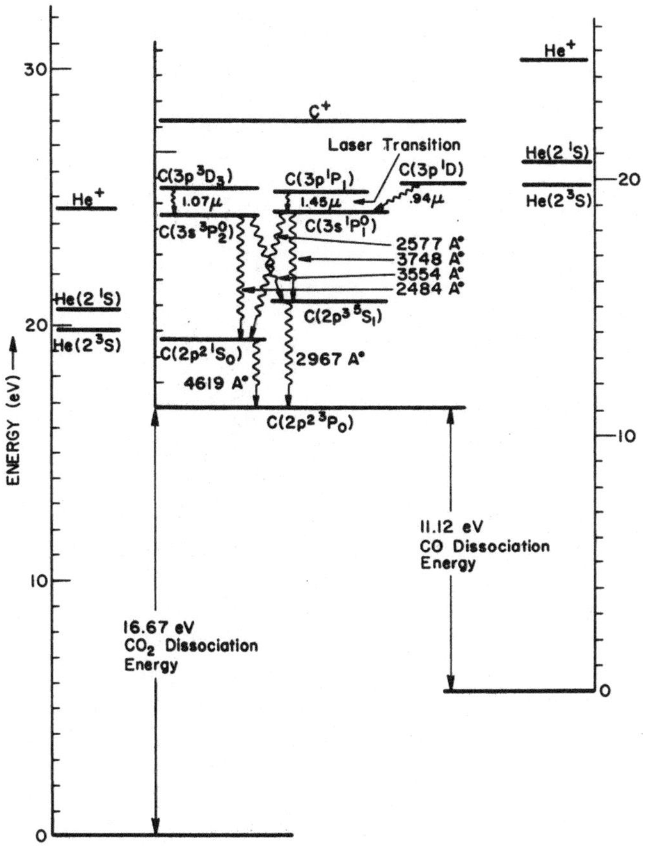

Fig. 4.32 Energy level diagram for the atomic carbon laser

or Ar with either CO or CO_2, at least two collisions with metastable species are required to populate the upper laser level. Figures 4.33 and 4.34 illustrate that the He + CO and He + CO_2 atomic carbon laser systems operated over a large range of pressures. Both the He + CO and He + CO_2 lasers tended to operate at higher power levels as the total pressure was increased. The power output of the He + CO system peaked at CO partial pressures of 2–3 mT, while the He + CO_2 system peaked at CO_2 partial pressures of 4–5 mT (Fig. 4.34). Power decreased in both systems as the carbon donor partial pressure was raised. The point where oscillation ceased was highly dependent upon the total pressure of the system. The stopping power of He for alpha radiation is strongly dependent upon pressure. All the energy is transferred to He in 2.54 cm at pressures of 800 T. The energy transferred to the He decreases at pressures lower than 800 T. This relationship is discussed in reference [94]. It was found that the delay phenomenon is also highly pressure dependent. The delay

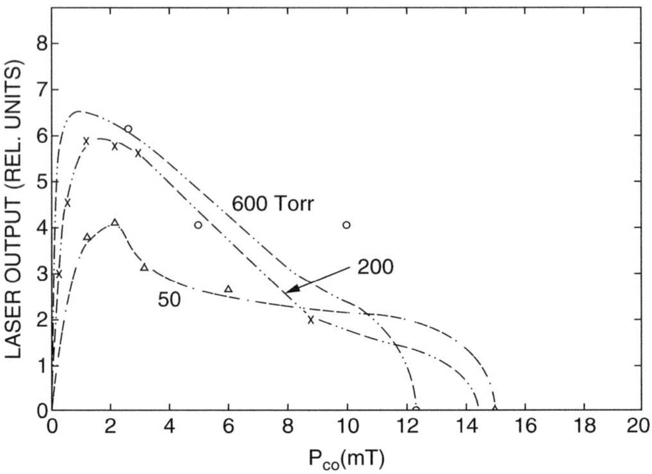

Fig. 4.33 Atomic carbon nuclear-pumped laser power output as a function of the partial pressure of CO and helium pressures of 600, 200 and 50 Torr

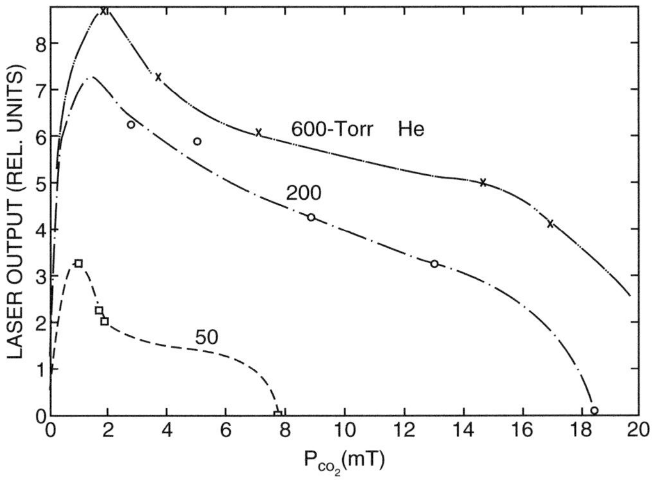

Fig. 4.34 Atomic carbon nuclear-pumped laser power output as a function of the partial pressure of CO_2 and helium pressures of 600, 200 and 50 Torr

tends toward zero as the pressure of He is increased and tends to become greater as the partial pressure of CO_2 is increased. This effect is shown in Fig. 4.35. Time dependent characteristics of the delay are a consequence of the kinetic dissimilarities between the He + CO and He + CO_2 systems. As seen in Fig. 4.36, the He + CO laser signal is coincident with the neutron pulse. In Fig. 4.37, the contrasting dynamics of the signal in the He + CO_2 laser can be seen: the laser signal is delayed by approximately 3 ms.

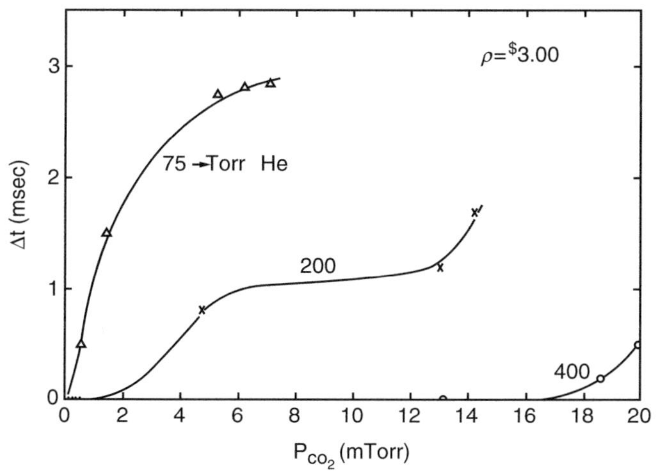

Fig. 4.35 The time delay between the thermal neutron flux and the laser power versus the partial pressure of CO_2 with He pressures of 75, 200 and 400 Torr

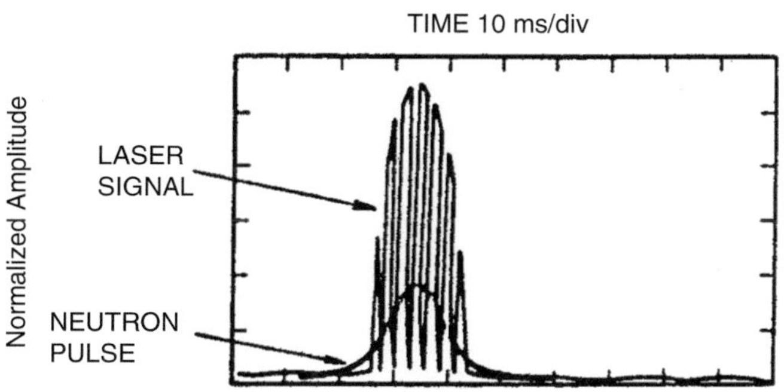

Fig. 4.36 The chopped signal of the He + CO nuclear-pumped laser caused by the high speed fan which blocks and unblocks the back mirror. The large ratio of the unblocked to blocked signal shows that the output is a laser

4.4.1.6.2 Ne + (CO or CO_2) Mixtures

Since the He + CO_2 laser demonstrated delays and was dependent upon a multiple collision process for the population of the upper laser level, Ne + (CO or CO_2) mixtures where multiple collisions would be necessary in order to populate the upper laser level were studied. The Ne + (CO or CO_2) systems had a high-power deposition threshold (900 W/cm) in a positive column discharge (EPL). By contrast the nuclear-pumped Ne + (CO or CO_2) mixtures oscillated 9–10 W cm^{-3} (threshold thermal neutron flux of ~9 × 10^{14} neutrons cm^{-2} s^{-1}). The Ne + (CO or CO_2) NPL did not begin oscillation until rather high partial pressures of the carbon donor were used (~5 mT). This behavior is unusual since the He buffered system required only background carbon impurities to lase. Further, the oscillation ceased at lower partial pressures of the carbon donor than in the He buffered systems (10 vs 15 mT). The

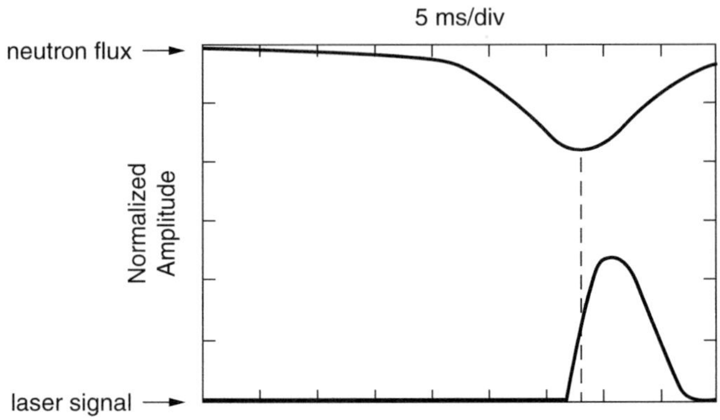

Fig. 4.37 The figure shows the signal from a He + CO_2 mixture which is delayed from the peak of the thermal neutron flux

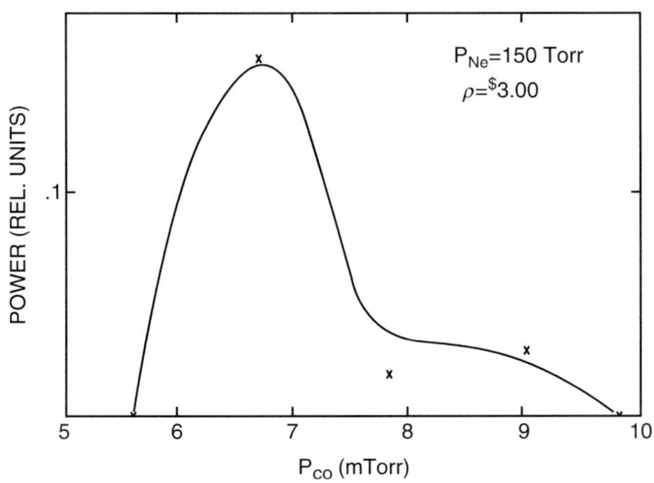

Fig. 4.38 Laser power variation with partial pressure of CO at a Ne pressure of 150 Torr

similarities of the Ne + (CO or CO_2) systems to the He + CO_2 systems, though, provide a key observation to explain why there is a delay in both Ne + CO and Ne + CO_2 mixtures.

The Ne + CO NPL pressure dependence is reflected in Fig. 4.38. The laser did not operate until the CO partial pressure was greater than 5.8 mT moreover the lasing cut off at a CO partial pressure of −10 mT. This contrasts sharply with the He + CO laser (Fig. 4.33), where a partial pressure equal to the background pressure of CO (0.25 mT) was sufficient to achieve oscillation in the He buffered system. The He gas used for these experiments contained less than 1 ppm CO as did the Ne gas. However, the He was 99.9999 % pure while the Ne was 99.995 % pure. It appears that this discrepancy in purity may have caused the necessity for higher partial pressures of CO in the Ne buffered NPL due to increased loss channels for

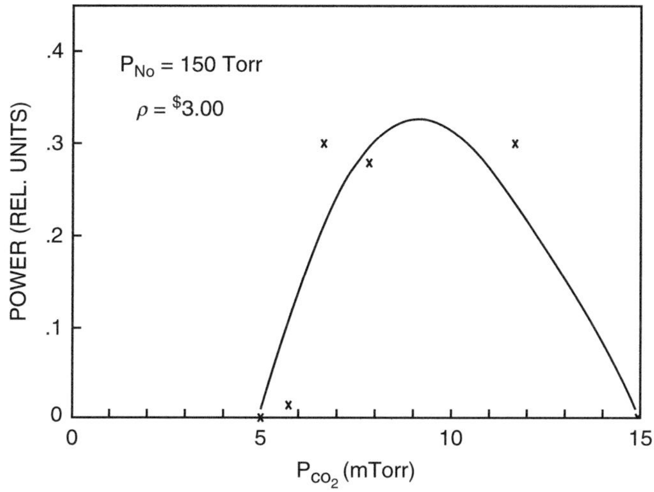

Fig. 4.39 Laser power variation with partial pressure of CO_2 at a Ne pressure of 150 Torr

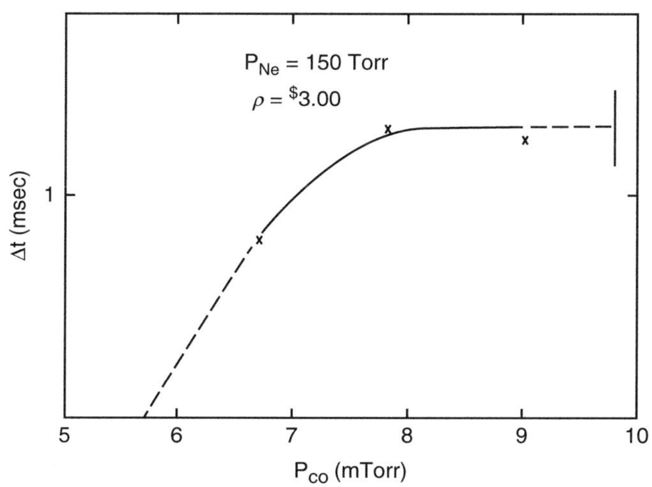

Fig. 4.40 Time delay variation with partial pressure of CO at a Ne pressure of 150 Torr

the metastable species through collisions with impurities other than CO or CO_2 (The normal background impurity is <3 mT (N_2+O_2) while the impurity in the Ne amounted to 5 mT (N_2, O_2, CH_4)). A partial pressure variation similar to that of the Ne + CO NPL was observed in the Ne + CO_2 NPL (Fig. 4.39). The minimum pressure of CO_2 needed in order to obtain oscillation is 5 mT while the pressure at which the NPL ceases to operate is ~15 mT. Pressure related delays have been observed in both Ne + CO and Ne + CO_2 mixtures. As seen in Fig. 4.40, a delay of up to 1.2 ms was evident in Ne + CO mixtures. Larger delays approaching 2 ms have been seen in Ne + CO_2 mixtures (Fig. 4.41).

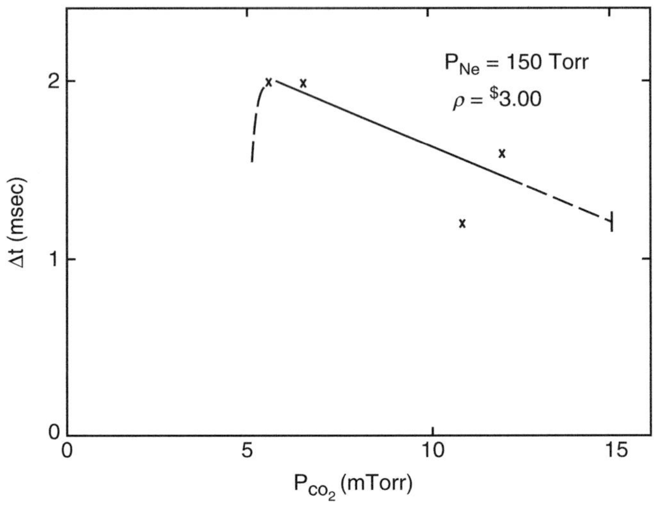

Fig. 4.41 Time delay variation with partial pressure of CO_2 at a Ne pressure of 150 Torr

4.4.1.6.3 Ar + CO or CO_2 Mixtures

The multiple collision process for the population of the upper laser level was investigated in Ar + CO or CO_2 mixtures. Interestingly, Ar + CO mixtures lased neither by electrical nor by nuclear pumping. The Ar + CO_2 mixtures have lased in the NPL but have not lased in the EPL. Moreover, slight additions of CO (3.5 mT) to the Ar + CO_2 NPL quenched the laser. The laser demonstrated a 15 % larger power output than did the He and Ne buffered NPLs. It also had the lowest lasing onset at a thermal neutron flux of about 4×10^{12} neutrons cm^{-2} s^{-1}. Thus it had a lower power deposition threshold with respect to the He buffer and the Ne buffer (0.04 W/cm^3 for Ar vs. 1 and 9 W/cm for He and Ne). The laser demonstrated delays of up to 1.8 ms which are comparable to the delays of the other multiple collision dominated systems (He + CO_2, Ne + CO and Ne + CO_2). A study of the pressure dependence of the Ar + CO_2 NPL established that the laser operated with higher peak laser power outputs than did the He and Ne buffered systems with the same peak power deposition. In Fig. 4.42, it can be seen that the Ar + CO_2 system is able to operate at relatively high partial pressures of carbon donor. (Partial pressures of CO_2 up to 150 mT were used with 200-T Ar to achieve oscillation). Like delays in the He and Ne systems, delays observed in the Ar + CO_2 system (Fig. 4.43) tended to increase as the partial pressure of CO_2 was increased. The maximum delay of 1.8 ms occurred with 200-T Ar + 150-mT CO_2.

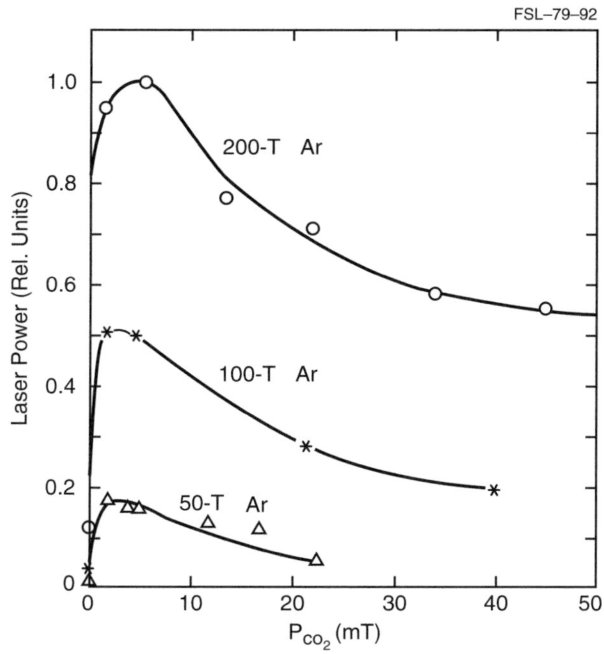

Fig. 4.42 Laser power variation with the partial pressure of CO_2 with Ar pressures of 50, 100 and 200 Torr

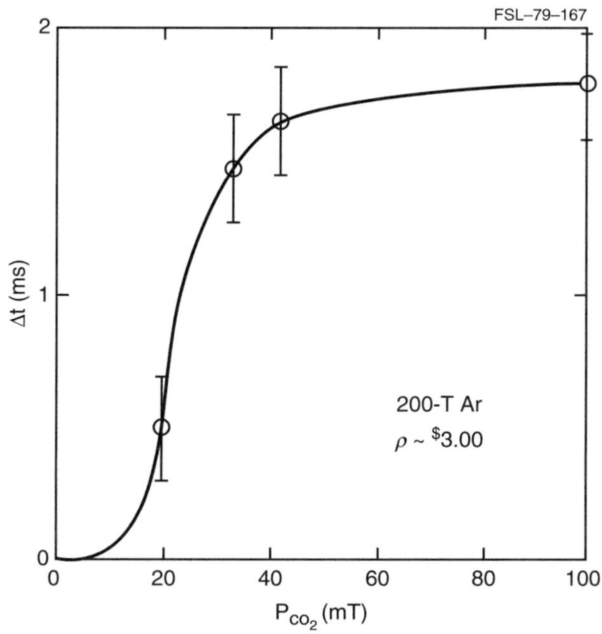

Fig. 4.43 Time delay variation with partial pressure of CO_2 at an Ar pressure of 200 Torr

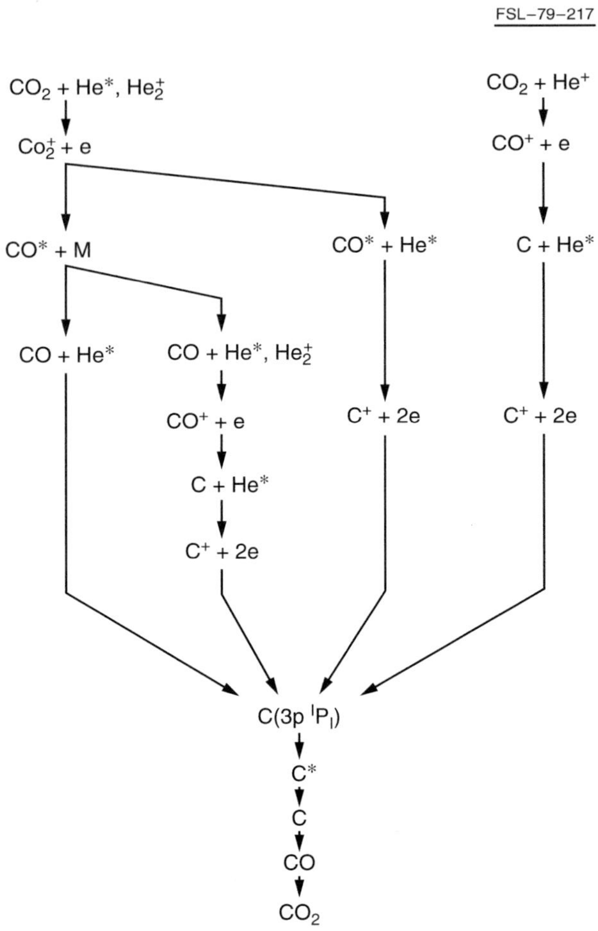

Fig. 4.44 The series of reactions used in the general He + CO_2 model for multiple collisions needed to populate the upper laser level. The first branch on the left has a resonance reaction that populates the upper laser level. The other three branches use a three body recombination with appropriate branching ratios to the upper laser level

4.4.1.6.4 Theory

In mixtures of He + CO_2, Ne + CO, Ne + CO_2 and Ar + CO_2 where a delay occurs, the donor molecule is dissociated into an intermediate state which in turn undergoes an interaction to populate the upper laser level. The process is either direct or cascading. Possible reactions that can take place are shown in Fig. 4.44.

Direct population of the upper laser level in the He + CO_2 mixtures can be accomplished by initial dissociation of CO_2 to form CO. Next, the helium metastable may directly populate the upper laser level through the following reaction:

$$\mathrm{He}\left(2^3\mathrm{S}\right) + \mathrm{CO} \rightarrow \mathrm{C}\left(3\mathrm{p}^1\mathrm{P}_1\right) + \mathrm{O} + \mathrm{He} \tag{4.42}$$

Cascading to the upper laser level may be achieved by the formation of C followed by three-body recombination and radiative relaxation. Three possible methods are illustrated in Fig. 4.44 with specific differences between them arising in the dissociation of CO_2 and the production of C^+.

Table 4.15 Lifetimes of species in the He + CO_2 nuclear-pumped plasma

Speed	#	Reaction	Rate constant ($cm^{-3}\ s^{-1}$ or $cm^6\ s^{-1}$)	Species	τ_{min} (s)
Fast	1	$CO_2^+ + e \rightarrow CO^* + O$	3.8×10^{-7}	CO_2^+	3×10^{-7}
	2	$CO^+ + e \rightarrow C + O$	8×10^{-8}	CO^+	1×10^{-6}
	3	$He^+ + 2He \rightarrow He_2^+ + He$	6.8×10^{-32}	He^+	1×10^{-6}
	4	$C^+ + 2e \rightarrow C + e$	8×10^{-21}	C^+	1×10^{-6}
	5	$He_2^+ + e \rightarrow He^* + He$	5×10^{-8}	He_2^+	2×10^{-6}
Intermediate	6	$He_2^+ + CO_2 \rightarrow CO_2^+ + 2He$	1×10^{-9}	CO_2	1×10^{-4}
	7	$He_2^+ + CO \rightarrow CO^+ + 2He$	5×10^{-10}	CO	2×10^{-4}
	8	$He^* + CO_2 \rightarrow CO_2^+ + e + He$	8.8×10^{-10}	CO_2	2×10^{-4}
Slow	9	$He^* + CO \rightarrow CO^+ + e + He$	1×10^{-10}	CO	1×10^{-3}
	10	$He^+ + CO_2 \rightarrow CO^+ + O + He$	1.6×10^{-9}	CO_2	1×10^{-3}
	11	$He^* + CO \rightarrow C_{ULL} + O + He$	6×10^{-11}	CO	1×10^{-3}
	12	$He^* + C \rightarrow C^+ + e + He$	5×10^{-11}	C	2×10^{-3}
	13	$CO^* + M \rightarrow CO + M$		CO^*	2×10^{-3}
After Pulse	14	$C + O + M \rightarrow CO + M$	1×10^{-3}	C	0.3
	15	$CO + O + M \rightarrow CO_2 + M$	1×10^{-34}	CO	0.3

Maximum concentrations used for these calculations are, [e] ~ 1×10^{13}, [He_2^+] ~ 1×10^{13}, [He*] ~ 1×10^{13}, [He^+] ~ 1×10^{12}, [He] ~ 3.22×10^{18}, [O] ~ 1×10^{14}, and [CO.] ~ 3.22×10^{13}. (Maximum values of [He*], [He^+], [He_2^+], and are taken for a pure He plasma and the maximum [O] was calculated by assuming all CO_2 is dissociated) [94]

Each of the four proposed channels in Fig. 4.44 has slow reactions which could account for the experimentally observed delay. Consider, for example, the dissociation of CO_2^- the rate constant for this reaction is ~8.20×10^{-10} cm^{-3} s^{-1} while the maximum metastable density is ~1.0×10^{13} cm^{-3} (i.e., no impurity species). From this information, a minimum e folding time for the CO_2 concentration can be calculated by setting up a simple rate equation:

$$He^* + CO_2 \rightarrow CO^+ + O + He \qquad \left(k = 8.20 \times 10^{-10} cm^{-3}\, sec^{-1}\right) \tag{4.43}$$

$$\frac{dCO_2}{dt} = k\left[He^*\right][CO_2] \tag{4.44}$$

$$[CO_2(t)] = [CO_2]_0 exp\left(-k\left[He^*\right]t\right) \tag{4.45}$$

The minimum decay constant (τ_{min}) of CO_2 is then:

$$\tau_{min} = \frac{1}{k\left[He^*\right]_{max}} = 0.174\ ms \tag{4.46}$$

As seen in Table 4.15, several species have τ_{min} values large enough to cause a delay on the millisecond time scale. The key species in these reactions are He^+, He_2 + and He*. Slight changes in the density of these species can affect the decay

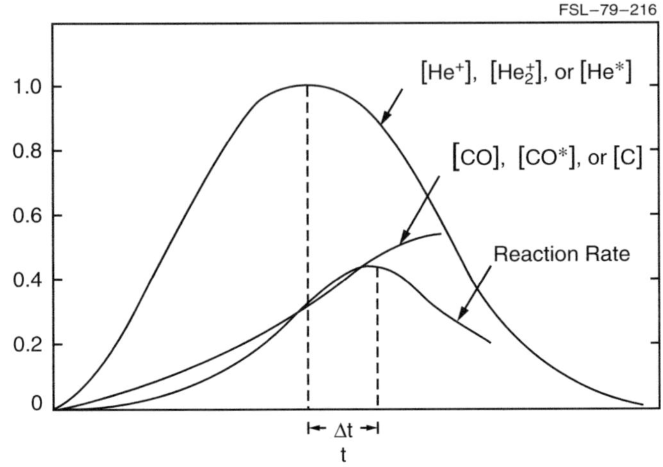

Fig. 4.45 This figure shows the relative density of primary species used in transporting energy to the carbon donor ($He+$, He_2+, He^*), and the main states of the carbon donor that lead to the production of the upper laser level (CO, CO^* and C). The reaction rate for creating the upper laser level depends on the energy transport species and the carbon donor species as show. The peak of the reaction rate is delayed by Δt from the peak of the energy transport species (which follow the neutron pulse)

constants, hence the delay. The addition of CO, CO_2 or other impurities (e.g., Ne, N_2, O_2, CH_4, etc.) to the plasma will reduce the metastable density.

Qualitatively, small additions of the carbon donor have a profound effect on the He metastable and ion densities. The model that was developed for the atomic carbon laser [94] demonstrates that by increasing the carbon donor density, the concentration of the metastable and molecular ions are reduced. This in turn causes the laser intensities to decrease at higher partial pressures of the carbon donor. Thus the buildup of CO, CO or C becomes slower, the pumping rate to the upper laser level is slowed, and a delay results (See Fig. 4.45).

The Ne and Ar buffered systems have scenarios similar to those of He in that the metastable concentration is important in the kinetics for dissociating the carbon donor. The upper laser level in the $Ne+CO_2$ system can be populated by two general methods as illustrated in Fig. 4.46. As shown in this figure, the Ne + CO system also has two general mechanisms. Population of the upper laser level in the $Ar+CO_2$ system is more difficult to pin down since the Ar + CO system doesn't work either as an NPL or EPL. Both Ar systems appear to have potential multistep mechanisms for C_{ULL} and C^+ production. A possible channel unique to the Ar $+CO_2$ system was proposed which involves a charge exchange between CO_2 and Ar followed by dissociative recombination to CO^{**} and direct excitation to C_{ULL} (See Fig. 4.47). In reference [89] it was suggested that the formation of an ArC^+ ion is a critical path followed by a recombination reaction with a high branching ratio to the atomic carbon upper laser level. This mechanism should also work for the Ar

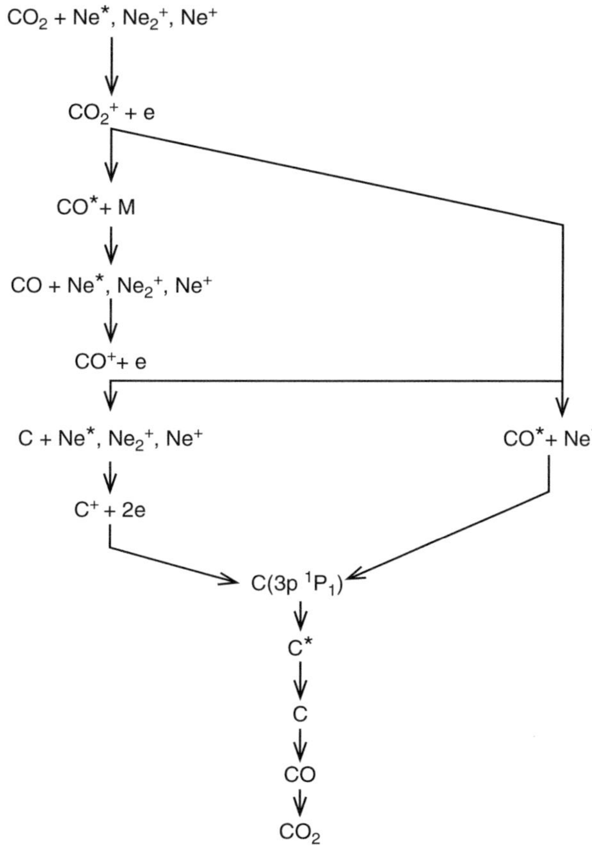

Fig. 4.46 Mechanisms for Ne + CO, CO_2 system which involve multiple collisions to populate the C_{ULL}

+ CO system and thus does not explain why Ar + CO does not lase either with nuclear pumping or with electrical pumping.

4.4.1.6.5 Model

To gain a better understanding of the delay, a model was developed. Studies highlighting the steps that could lead to a delay in the He + CO_2 system were undertaken since this system appeared to be the most straightforward and the most data was available for it [94, 97, 98, 101, 102]. Two simplifications were made. First, all fast processes were eliminated by allowing the reaction to proceed to the first slow step. The simplifications were verified [103].

$$e^{**} + He \rightarrow \text{intermediate steps} \rightarrow He^{*} + e \tag{4.47}$$

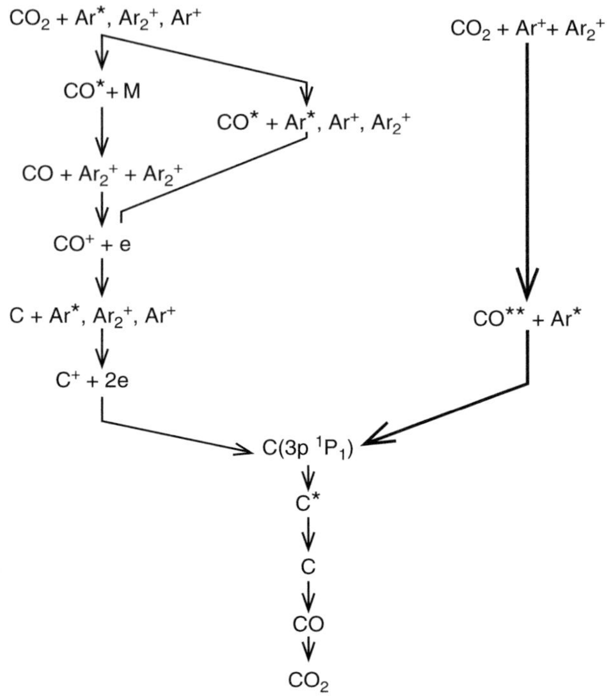

Fig. 4.47 Mechanisms for $Ar + CO_2$ system which involve multiple collisions to populate the C_{ULL}

$$He^* + CO_2 \rightarrow \text{intermediate steps} \rightarrow CO^* + O + He \tag{4.48}$$

$$He^* + CO \rightarrow \text{intermediate steps} \rightarrow C + O + He \tag{4.49}$$

$$He^* + C \rightarrow \text{intermediate steps} \rightarrow He + C \tag{4.50}$$

Second, cascading to the upper laser level was ignored because increases in the partial pressure of CO_2 show a decrease in recombination light (Figs. 4.48 and 4.49) experimentally [94]. This evidence leads to the conclusion that energy is not effectively channeled through the carbon ion [94]. The resulting model is shown in Fig. 4.50 [102]. The usefulness of the model rests on its ability to produce results similar to the experimental NPL data. Of particular interest are the experimentally measureable quantities (using the TRIGA reactor as a neutron source) that describe the peak power output of the laser, the laser's time dependence, and the delay. As it turns out, the calculated results agree with those of the experiment with surprising accuracy.

Table 4.16 shows that the calculated laser intensity was within 25 % of the measured laser intensity over pressures ranging from 50 to 800 Torr. The calculated and experimental power output with respect to the He and CO_2 partial pressures are shown in Figs. 4.51 and 4.52. In Fig. 4.51, the He partial pressure was varied between 50 and 800 Torr while the CO_2 partial pressure was held constant at 2 mT. Conversely, in Fig. 4.52, the CO_2 partial pressure was varied while the He partial

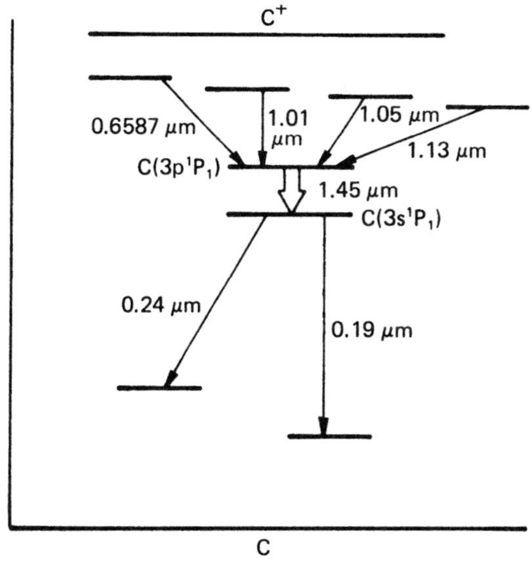

Fig. 4.48 Energy levels for lines cascading to the carbon upper laser level [102]

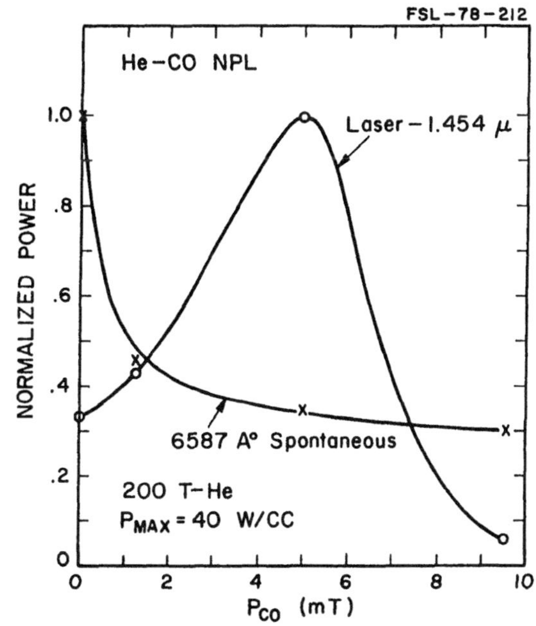

Fig. 4.49 Simultaneous measurements of the laser signal and the 658.7 nm emission in the He + CO NPL with 200-T He. Since the spontaneous emission does not follow the laser power, the cascading is not through this level. Similar measurements were made on the other lines which couple to the carbon upper laser level: 1,010, 1,050 and 1,130 nm lines. All of the lines behaved like the 658.7 nm line [94]. No other viable or near infrared lines coupled to the carbon upper laser level

pressure was held constant at 75, 200 and 800 Torr. (One can see the good correlation in either Figure between calculated and experimental values). Relationships between the calculated and experimental time dependence of the atomic carbon NPL signal are just as striking (Figs. 4.53, 4.54, 4.55 and 4.56). In Figs. 4.53 and 4.54, the pulse shape at 75-Torr He and at 3 and 5-mT CO_2 is

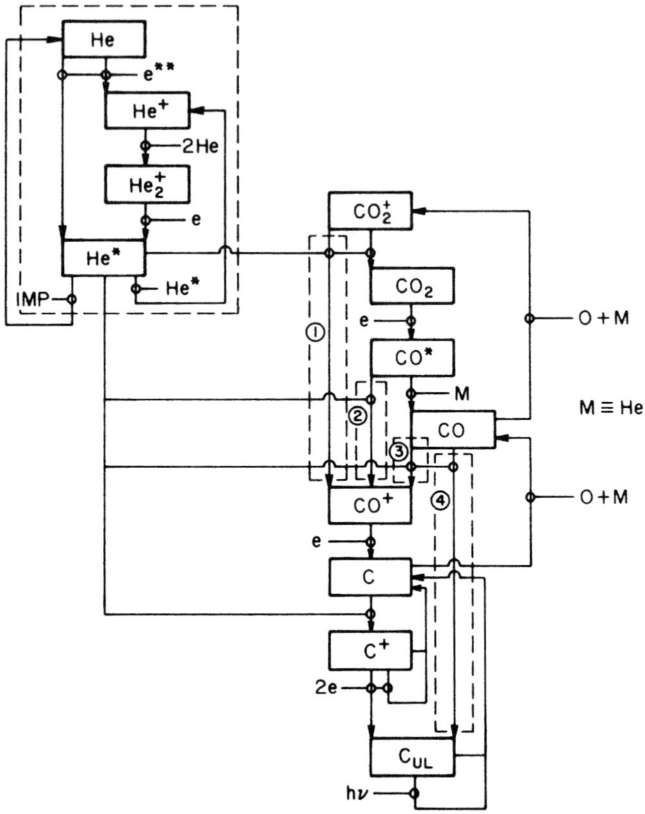

Fig. 4.50 Energy flow channels of the revised model which was used in the computer code [102]

Table 4.16 Laser intensity (from the model and measured) as a function of helium pressure holding the partial pressure of CO_2 at 2 mTorr

He pressure (Torr)	Laser intensity model (mW cm^{-2})	Laser intensity measured (mW cm^{-2})
50	5.9	5
100	7.8	8
200	18.2	15
400	26.1	19
600	31.0	22
800	30.1	25

shown. The laser power from mixtures of 200-Torr He and 2 and 10-mT CO_2 is illustrated in Figs. 4.55 and 4.56. Code generated and experimentally measured delays are seen in Figs. 4.57 and 4.58. Mixtures of 75 and 200-T He were tested at various partial pressures of CO_2 and the model generated results that matched experiments for the same conditions. Again, the correlation between experiment and calculation is good.

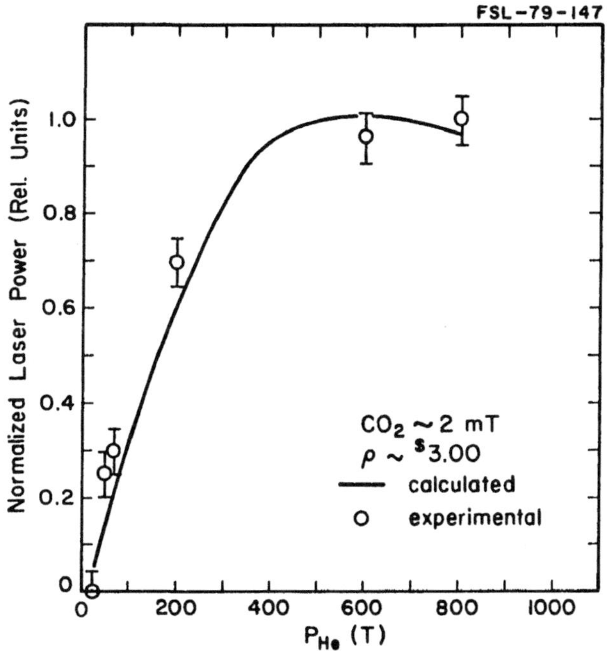

Fig. 4.51 Peak laser power holding CO_2 at 2 mT and varying the He pressure between 20 and 800 T. The calculated results are shown by the *solid line*, and the experimental results are shown by the *circles* [94] (Error bars were obtained from the high and low variation of data taken on different days)

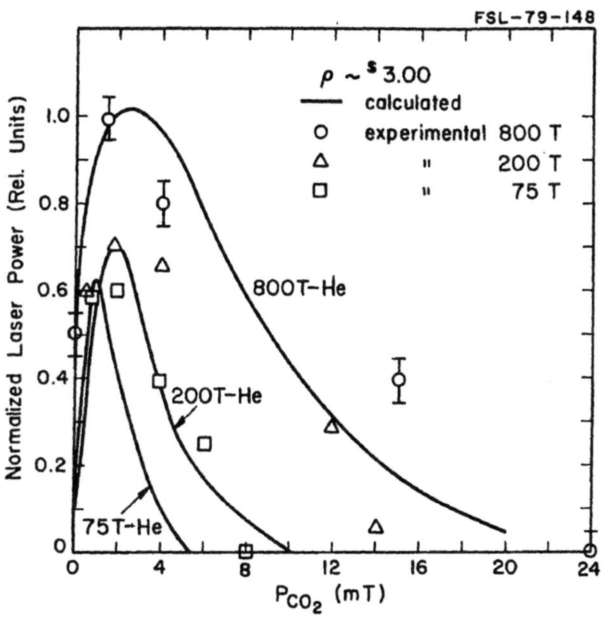

Fig. 4.52 Peak laser power holding the He pressure constant at 75, 200 and 800 T and varying the partial pressure of CO_2 between 0.25 and 24 mT [94]. The *solid line* was used to represent calculated results while the *circle*, *triangle* and *square* were used to represent experimental points at 800, 200 and 75-T He respectively (Error bars were used on 800-T He readings to represent high and low variation on data)

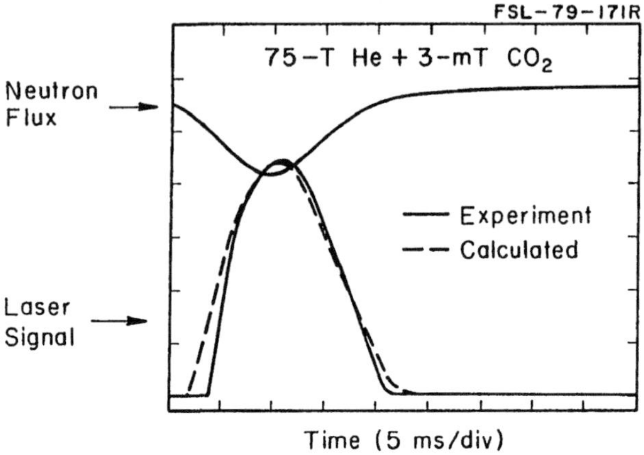

Fig. 4.53 Computer simulation of laser signal and excitation pulse with corresponding experimental output at 75-T He and 3-mT CO_2 [94]

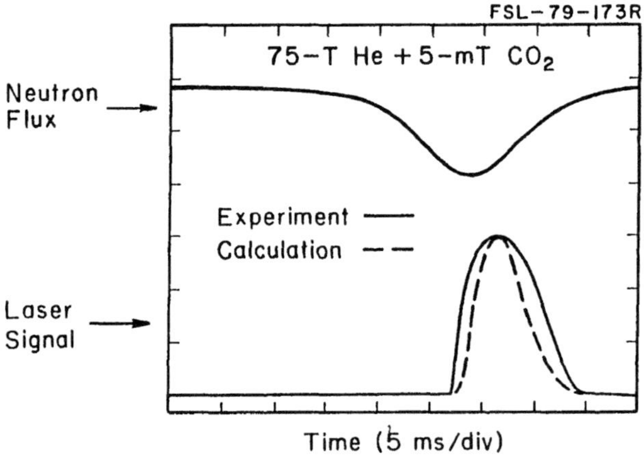

Fig. 4.54 Computer simulation of laser signal and excitation pulse with corresponding experimental output at 75-T He and 5-mT CO_2 [94]

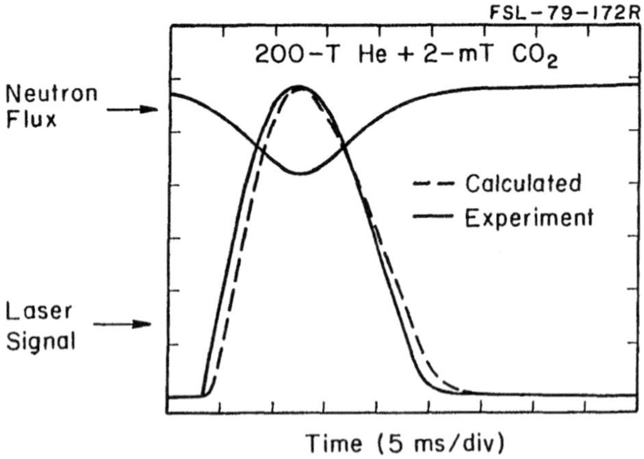

Fig. 4.55 Computer simulation of laser signal and excitation pulse with corresponding experimental output at 200-T He and 2-mT CO_2 [94]

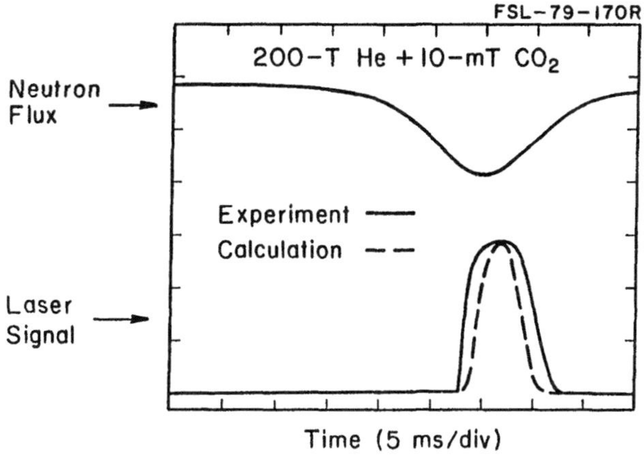

Fig. 4.56 Computer simulation of laser signal and excitation pulse with corresponding experimental output at 200-T He and 10-mT CO_2 [94]

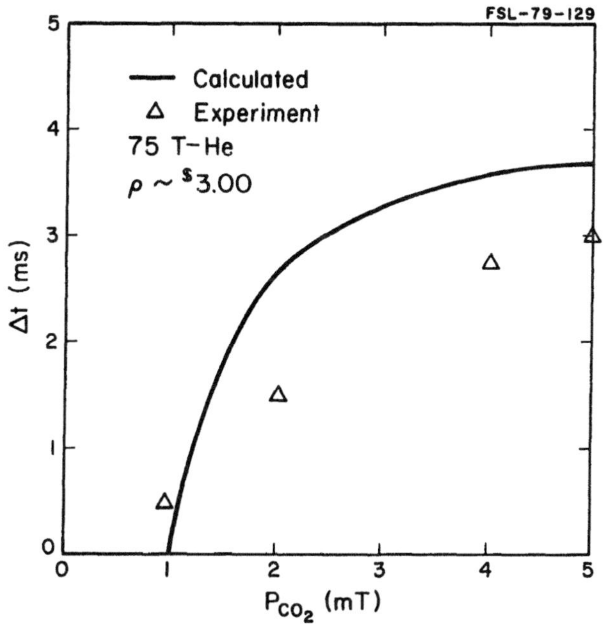

Fig. 4.57 Delay in He + CO_2 NPL at 75-T He and 1–5-mT CO_2. The calculated results are shown by the *solid line* while the experimental points are shown by *triangles* [94]

The code may also be used to model the fast burst reactor and electrically pumped atomic carbon lasers. Again, experimentally measurable quantities such as laser power and time dependence were compared to the calculations. Data for a fast burst reactor driven atomic carbon NPL is compared to the results generated in the model as shown in Fig. 4.59. The calculated laser signal correlates well with the experimental results. The peak power deposited in the fast burst system is higher than the peak power deposited in the TRIGA system (300 W cm^{-3} vs 30 W cm^{-3}). Thus, the He* density in the fast burst system is higher than that in the TRIGA. This

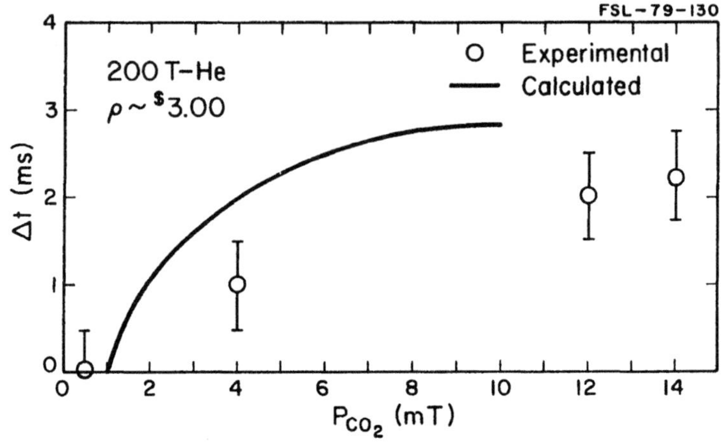

Fig. 4.58 Delay in $He+CO_2$ NPL at 200-T He and 1–15-mT CO_2. The calculated results are shown by the *solid line* and the observed results by the *circles* [94] (The error bars are due to the accuracy of reading the delay on data taken at 200-T He)

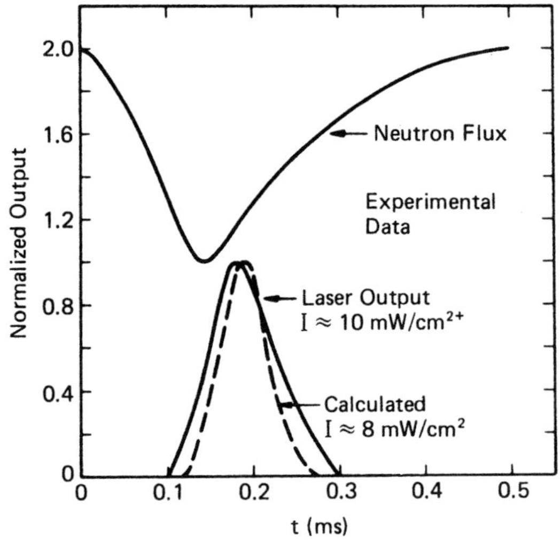

Fig. 4.59 Computer simulation of the laser signal in the fast burst reactor experiment and the corresponding experimental data for 400 Torr helium and 5 mTorr CO_2 [85]

implies that the CO density buildup is much faster in the fast burst, resulting in a relatively small delay (See Fig. 4.60).

The peak power deposition in the discharge pumped atomic carbon laser (EPL) is much greater than that in the TRIGA system (900 W cm^{-3} vs 30 W cm^{-3}). However, the EPL dumped the stored capacitor energy into the laser in about 2 μs. As shown in Fig. 4.61, the calculated laser power for $He+CO_2$ approximately matches the experimentally observed output. Note that the EPL operated in the afterglow. In Fig. 4.62, the densities of He, CO_2, CO, CO, C_{ULL} and I_{Laser} for the

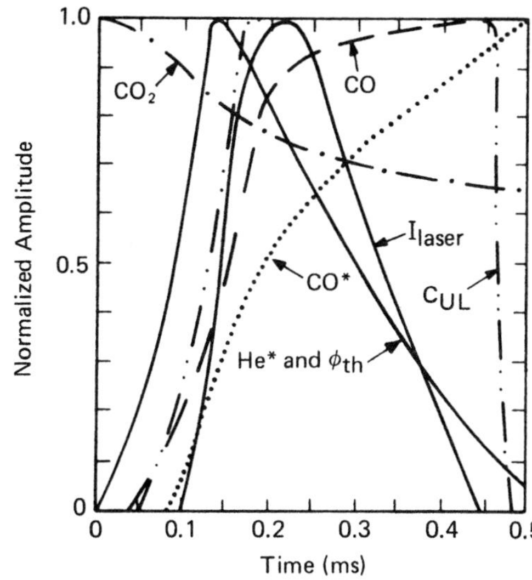

Fig. 4.60 The normalized values in a fast burst simulation for a mixture of 400 Torr helium + 5 mTorr CO_2. Normalizations are He* (7×10^{13} cm^{-3}), CO_2 (1.61×10^{14} cm^{-3}), CO* (8×10^{12} cm^{-3}), CO (3×10^{13} cm^{-3}) and C_{ULL} (4×10^{9} cm^{-3}), the flux Φ_{th} (6×10^{16} n cm^{-2} s^{-1}), and the laser intensity I_{laser} (8.3 mW cm^{-3}) [85]

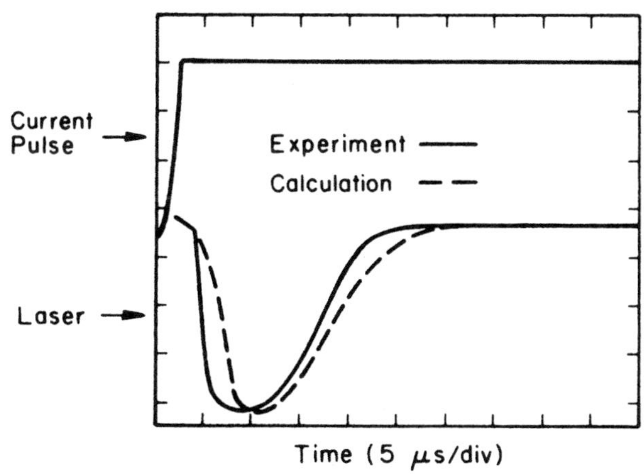

Fig. 4.61 Computer simulation of helium + CO_2 EPL with corresponding experimental output in 10 Torr helium and 1 mTorr CO_2 [85]

He + CO_2 EPL are shown as a function of time. The pulse is fast compared to the rate of change of the species. As a consequence, the CO density builds up to a peak value near 14 μs after the peak of the pulse while the He* density decays from an initial value of 2×10^{13} cm^{-3}. Calculated laser output is compared to the experimental laser output and the calculated densities of He*, CO, C_{ULL} and I_{Laser}, are plotted as a function of time for the He + CO discharge pumped laser in Figs. 4.40 and 4.41. Notice that the He + CO discharge pumped laser has many of the same features as the He + CO_2 EPL:

1. Oscillation occurs in the afterglow of the pulse,
2. The He* density starts out high and decays, and
3. The laser signal is delayed on the order of 2–10 μs.

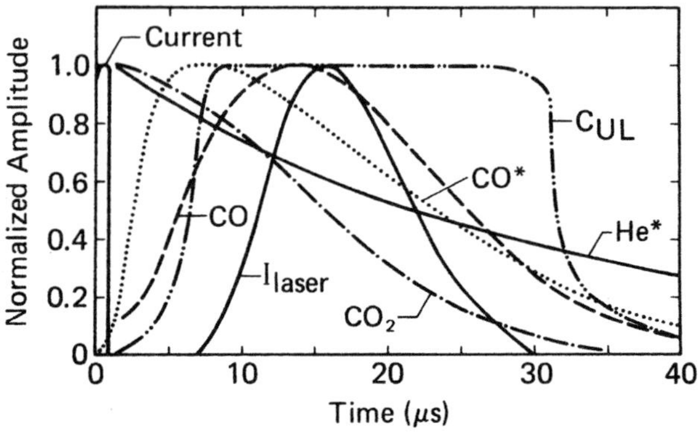

Fig. 4.62 Normalized values in an EPL simulation for the densities of He* (1×10^{13} cm^{-3}), CO_2 (9.90×10^{13} cm^{-3}), CO* (1.9×10^{12} cm^{-3}), CO (4.4×10^{12} cm^{-3}), and C_{ULL} (4.4×10^{9} cm^{-3}), the current and the laser intensity (21 mW cm^{-2}) in 10 Torr helium and 3.1 mTorr CO_2 [85]

Since the CO density is initially abundant, the laser commences oscillation once the C_{ULL} density builds up (~2 μs). The EPL typically had intensities on the same order as the NPL, however, the delays were considerably different. The reason for this difference is in the dynamics of He* production.

The atomic carbon nuclear-pumped laser has a low pumping power density which indicates that the process which populates the upper laser level is efficient (indicative of some type of resonance). Recombination is not the likely mechanism but rather a process which channels energy to the upper laser level.

In the USSR, the atomic carbon lased on the 833.5 and 940.6 nm lines using an experimental set up similar to that shown in Fig. 4.17a. The threshold pumping power density was about 17 W cm^{-3} [6]. The suggested mechanism was recombination.

4.4.1.7 Atomic Chlorine Laser

A chlorine nuclear-pumped laser (Fig. 4.63) was demonstrated at 1587 nm ($3d^4F_{9/2} \rightarrow 4p^4D_{7/2}$) using the fast burst reactor at the Aberdeen proving grounds [104] (Fig. 4.64). The laser used a mixture of $^3He + Cl_2$ (600 Torr total pressure with 0.008 % Cl_2). The threshold flux was about 7×10^{15} neutrons cm^{-2} s^{-1} and the peak output power was about 0.1 W. Laser action was quenched when the Cl_2 concentration reached 0.17 %.

Russian scientists also demonstrated lasing on the chlorine lines at 1587 and 2447 nm using mixtures of He-Ne-CCl_4 [6, 105]. The experimental work on the 1587 nm line used two 5 m radius of curvature 6 cm diameter dielectric mirrors which were 99.8 % reflective from 1400 to 1700 nm. The laser demonstrated 80 W of power output with a He/Ne/CCl_4 at a ratio of 250,000/250,000/1 at a total pressure of 2 atm (Table 4.17). For the 2447 nm line ($3d^4D_{7/2} \rightarrow 4p^4D_{7/2}$), two dielectric mirrors with a radius of curvature of 5 m and 6 cm diameter with 98 % reflectivity from 2400 to 3100 nm were used. The maximum power output was 5 W with a minimum threshold thermal neutron flux at 1×10^{15} n cm^{-2} s^{-1}.

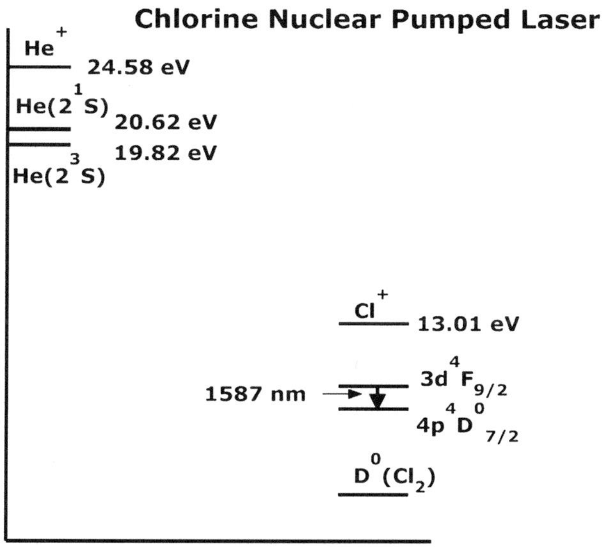

Fig. 4.63 Energy level diagram for the atomic chlorine nuclear-pumped laser

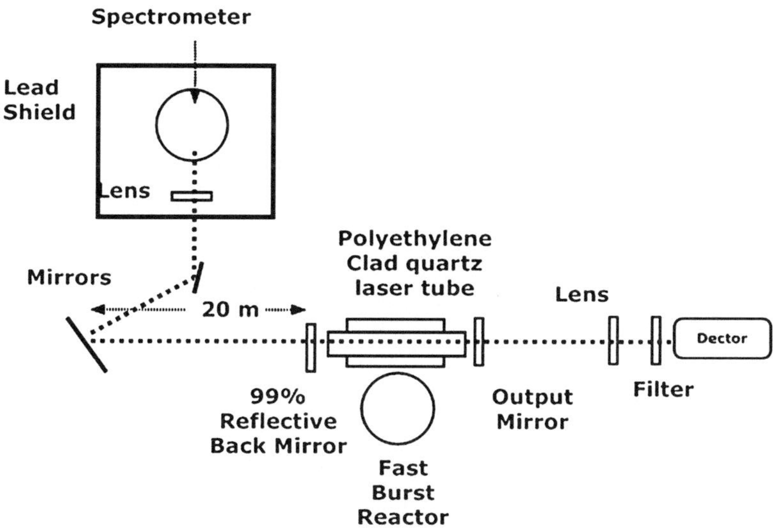

Fig. 4.64 A general experimental set up for the NASA nuclear-pumped laser work. The Cl NPL used two 2 m radius of curvature mirrors that were 98.4 % reflective at 1590 nm to form the optical cavity

Extensive kinetics studies were performed [106] in support of this work. The schematic of this model is shown in Fig. 4.65. The basic mechanism for populating the upper laser level are three body recombination of the Cl^+ ion, dissociative recombination of Cl_2^+ and CCl_2^+ and ion-ion recombination of CCl_2^{+*} and Cl^-. The optimal power output was estimated to be 160 W with an efficiency of 0.12 %.

Table 4.17 Experimental results for the chlorine laser at 1587 nm

Mixture	Total pressure (atm)	Output power (W)	Threshold thermal neutron flux (10^{14} n cm^{-2} s^{-1})
He/Ne/CCl_4	2	80	0.43
He/Ne/$CHCl_3$	2	40	0.56
He/Ne/CF_2Cl_2	2	8	0.46
He/Ne/$C_2F_3Cl_3$	2	10	2.2
Ar/CCl_4	0.4	2.5	18

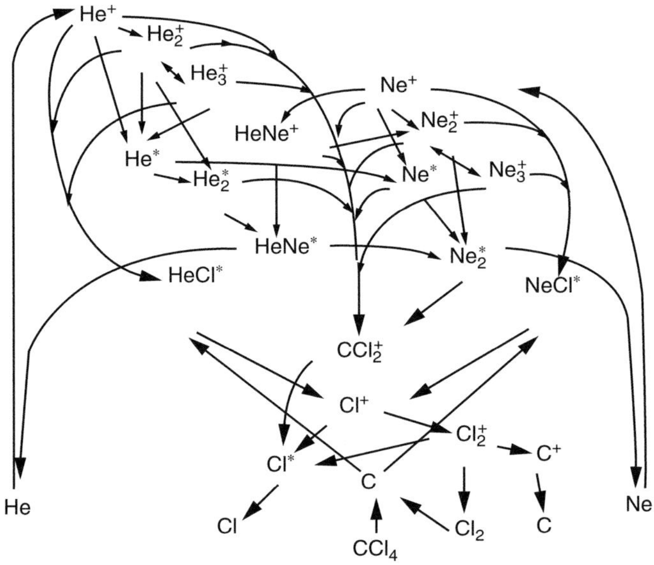

Fig. 4.65 Schematic of the atomic chlorine nuclear-pumped laser in mixtures of He + Ne + CCl_4 [106]

4.4.1.8 Atomic Oxygen

In the USSR atomic oxygen lased on the 2650 nm ($4d^5D^0 \rightarrow 4p^5P$); 2760 nm ($4p^5P \rightarrow 4s^5S^0$) lines using an experimental set up similar to that shown in Fig. 4.17a. Both lines lased using 5 m of radius of curvature dielectric mirrors, 6 cm diameter with 98 % reflectivity from 2400 to 3100 nm. The threshold pumping power density for the 2650 nm line was about 5 W cm^{-3} at a He/Ne ratio of 1 to 1 with the oxygen impurity coming from the rare gases (less than 10 m Torr) [6, 105]. The suggested mechanism was recombination. The 2760 nm line was observed for the first time ever. The 2760 nm line power output was 30 W using a He/O_2 ratio of 10,000/1 at two atmospheres with a threshold thermal neutron flux of 3×10^{14} n cm^{-2} s^{-1}. When Ne or Ar was added to the buffer gas mix, the laser did not work. Oxygen containing substances such as NO and CO also were used and led to lasing in He/Ne/NO(CO) mixtures with ratios of 10,000/10,000/1 with a power output of 8 W.

4.4.1.9 Neodymium YAG Photolytic Laser

Nd:YAG is a solid state laser medium which is photolytically pumped and is commercially available. The Nd:YAG laser emits at 1060 nm ($^4F_{3/2} \rightarrow {}^4I_{11/2}$) as seen in Fig. 4.66. It is an efficient laser due to its strong absorption band from 802 to 809 nm and efficient transition to the upper laser level ($^4F_{3/2}$) as shown in Fig. 4.67.

A nuclear driven rare gas flash lamp was used to pump a Nd:YAG rod (Table 4.18).

The first Nd:YAG nuclear-pumped laser experiment took place during the fall of 1987 [108] as part of the author's work with Idaho National Engineering Laboratory. The experiment used a reactor driven fluorescence source to pump the laser medium. The nuclear-driven lamp was designed using a ray tracing code called "concentrator." In this design it was necessary to develop a lamp which was capable of both concentrating and directing incoherent light generated in a volume source to a waveguide. A lamp was constructed which met this criteria (shown in Fig. 4.68). The lamp body was made of fused silica and was coated on the outside with silver which provided a reflective surface for visible photons. A reflective flat silver mirror was placed on the back of the lamp to reflect photons in the visible to infrared. The front window of the lamp was coupled to a Suprasil II lens (with the focal length optimized for the cross section of the fiberoptic bundle) and focused onto a fiberoptic cable which served as a waveguide to transport light from the nuclear-driven lamp to the laser medium.

At the peak of the TRIGA pulse, 27,000 W of power was deposited in the lamp volume. The photons created by the excitation of the gas were reflected by the walls of the lamp until they reached the front of the lamp. Here a Suprasil I lens focused the light onto the fiberoptic bundle. The light was collected by the fiberoptic cable and transported to the Nd:YAG crystal (Fig. 4.69). The Nd:YAG crystal (3 mm diameter cylinder that was 62.5 mm long) was coupled to the fiberoptic bundle through an illuminator shown in Fig. 4.70. The illuminator was placed on an optical bench (Fig. 4.71). Other optical components, detectors and a portable turbo pump-based high vacuum system were used along with a gas manifold.

Given that the total power deposited in the lamp was 27,000 W at the peak of the pulse, the nuclear-driven Kr fluorescer lamp deposited 570 W cm^{-3} in the Nd:YAG crystal at the peak of the pulse [108]. The measured 1064 nm emission (detector shown in Fig. 4.71) from the Nd:YAG crystal was 0.12 mW. This was slightly higher than the 0.1 mW predicted by the model used to design the experiment [50, 108]. The length of the laser rod was 6.25 cm and the diameter of the rod was 0.3 cm. The detector had an active area of 10 mm^2 and was located 5 cm from the face of the rod (Fig. 4.71). The transmission coefficient of the 1064 nm filter was 0.7. The total power of 1064 nm photons produced in the rod at the peak of the pulse

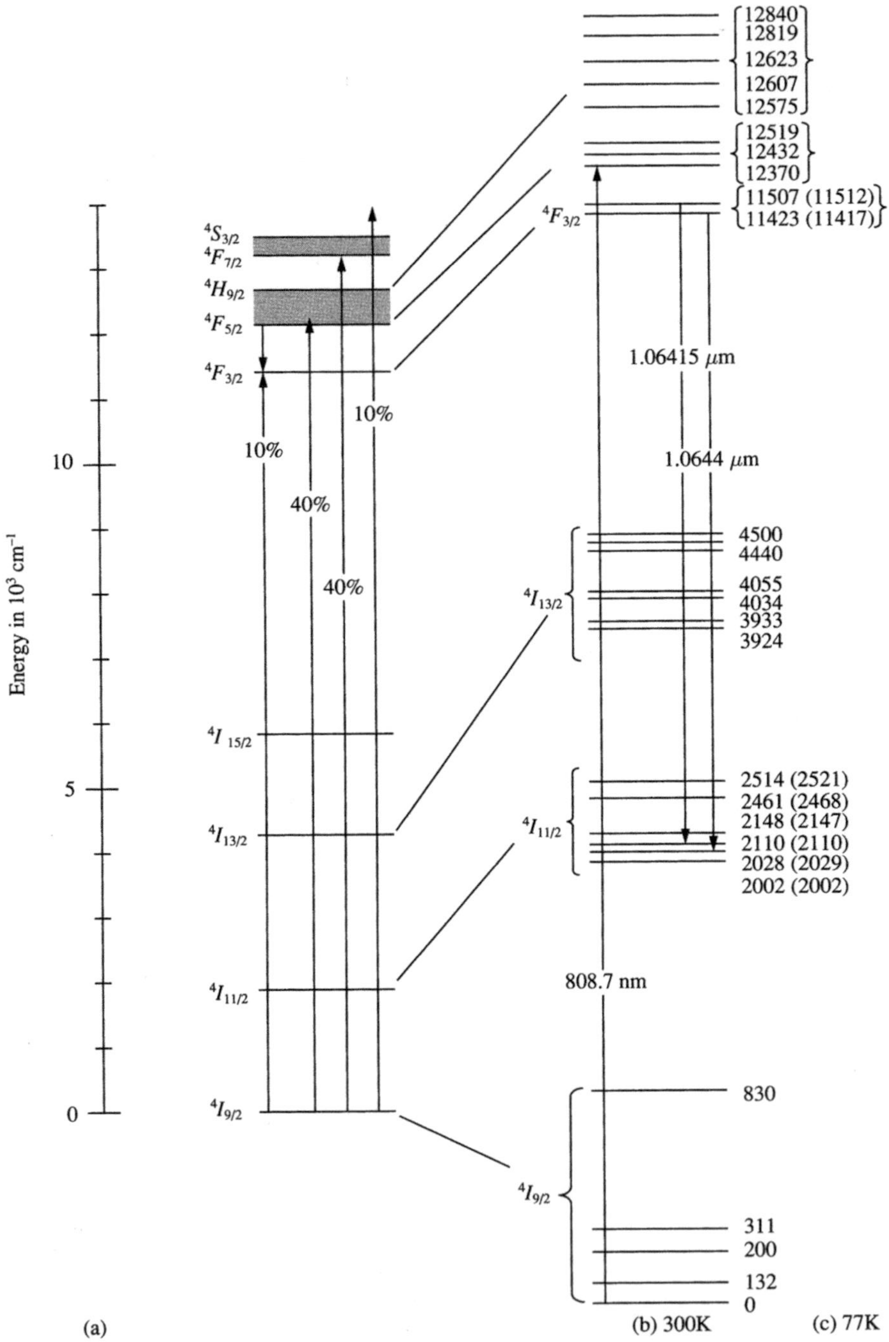

Fig. 4.66 Energy level for neodymium in a YAG crystal [107]. (**a**) YAG crystal structure showing the pumping routes the Nd+3 upper laser level. (**b**) Details of the spectral manifold for Nd+3 at 300 K. (**c**) Energy levels for a temperature of 77 K (Verdeyen, Joseph T., Laser Electronics, 3rd, ©1995. Printed and Electronically reproduced by permission of Pearson Education, Inc., New York, New York)

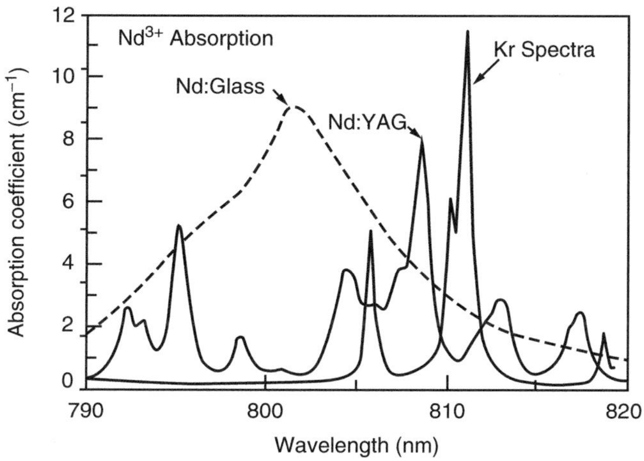

Fig. 4.67 Absorption spectra of the Nd:YAG and Nd:Glass lasers superimposed on the absolutely calibrated nuclear driven Kr lamp emission spectra

Table 4.18 Characteristics of a Nd:YAG rod [107]

Bulk parameters	Value
Chemical composition	Nd:$Y_3Al_5O_{12}$
Nd weight %	0.725
Nd atomic %	1
Density of Nd atoms	1.38×10^{20} cm^{-3}
Index of refraction at 1060 nm	1.81633
Scattering losses	0.002 cm^{-1}
Dominant transition	11.507 → 2111 cm^{-1} at 1064 nm
Fluorescent lifetime of $^4F_{3/2}$ state	255 μs
Fluorescent efficiency of $^4F_{3/2}$ state	99.5 %
Stimulated emission cross section	2.7–8.8 × 10^{-19} cm^2
Lower ;laser level lifetime ($^4I_{11/2} \rightarrow {}^4I_{9/2}$)	30 ns

can be calculated and was found to be 62.4 W. From the total emitted power of 1064 nm radiation from the rod, the upper laser level density was calculated and found to be 1.7×10^{17} cm^{-3}. The measured small signal gain for this configuration was 0.051 cm^{-1} (the predicted small signal gain for this experiment was 0.043 cm^{-1} [108]). The nuclear-flashlamp driven Nd:YAG rod had a significant amount of gain. This gain was more than enough to achieve lasing.

After the gain experiments that were completed in the fall of 1987, experiments to measure radiation induced emissions from the fiberoptic cable and the Suprasil I lens were conducted. In addition, the absolute calibrated efficiency from nuclear-driven ^{3}He:Ar, ^{3}He:Kr and ^{3}He:Xe flashlamps were measured as part of the research performed for a dissertation [109]. The fiberoptic cable was replaced by

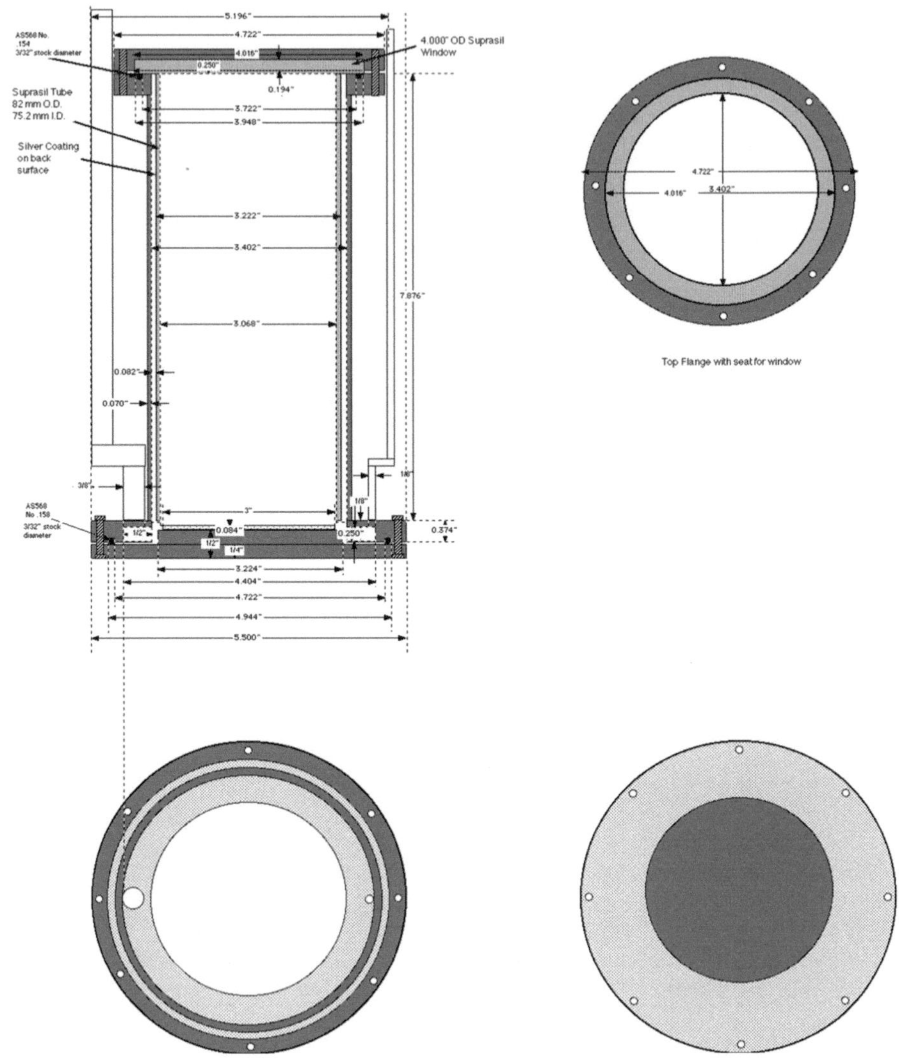

Fig. 4.68 Nuclear Driven Lamp design for rare gas oxide and mercury excimers along with inert gas line spectra

one that was flat on both sides. Using the experimental set up shown in Fig. 4.72, this new fiberoptic cable was placed in the through port near the core of the University of Illinois's TRIGA reactor. The cable was irradiated and the radiation induced fluorescence was examined (See Fig. 4.73). This radiation induced fluorescence (RIF) from the fiberoptic cable was measured over a wide range of power levels. It was noted that the spectral distribution remained constant at all power levels. This was tested by taking the spectral distribution digitally stored at one

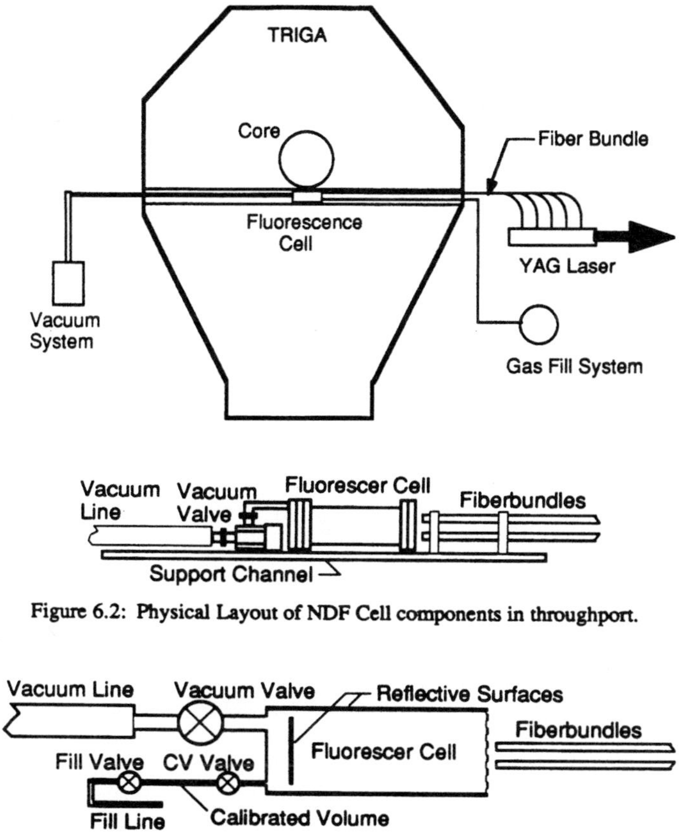

Fig. 4.69 Diagram of reactor driven fluorescer experiment. Details of the fluorescer cell are shown

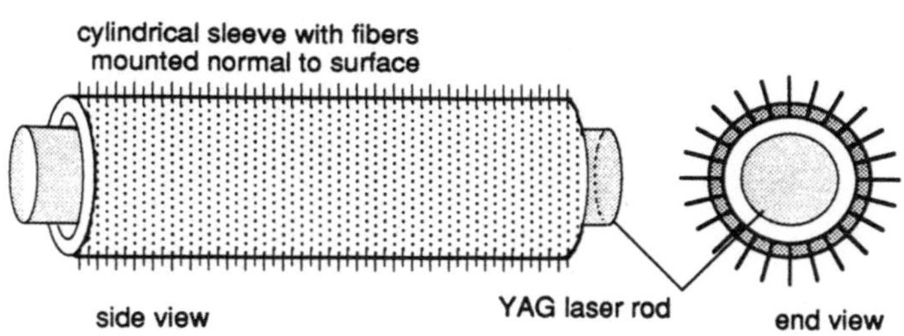

Fig. 4.70 Nd:YAG rod mounted inside of the illuminator

power level and multiplying it by a constant then subtracting it from a digitally stored distribution at a different power level. In all cases a constant was found in this subtraction operation such that the resulting curve was a line at the zero count level ±0.001 %. Thus, the radiation induced fluorescence noise could be digitally subtracted from signals coming out of the fiberoptic cable.

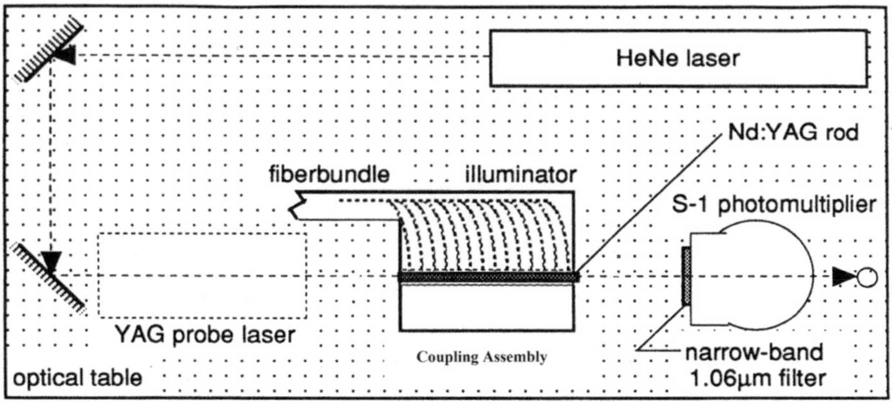

Fig. 4.71 Nd:YAG nuclear-pumped laser set up on optical table

Fig. 4.72 Experimental set up of fiberoptic radiation induced fluorescence experiment

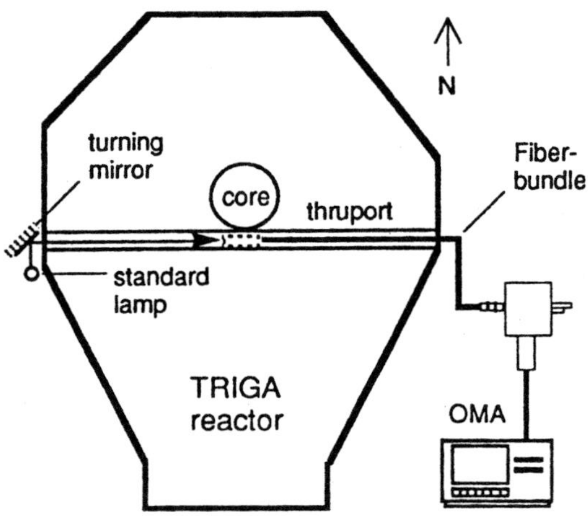

Using an Optronics Laboratory Model 246 M NBS traceable light source, spectra of the light source was taken at various stages of irradiation (See Fig. 4.74). From these spectra, it was possible to determine the fiberoptic attenuation at the wavelengths tested as a function of reactor power (Fig. 4.75). This data then provided a calibration for the radiation damage in the fiberoptic cable.

The experiment was repeated by attaching the fiberoptic cable to the lamp (under vacuum) using the experimental set-up of Fig. 4.72 to see the effects of radiation on the Suprasil I window (the lens was not used in these experiments). The window being made of fused silica as was the fiberoptic cable exhibited the same RIF as the cable. Thus the experiment was conducted with the lamp and attached fiberoptic cable and the lamp was filled with various mixtures of ^{3}He with Ar, Kr and Xe. The spectra taken from the lamp along with the calibration data for the optical path, the RIF correction data from Fig. 4.75, and the fiberoptic attenuation data from

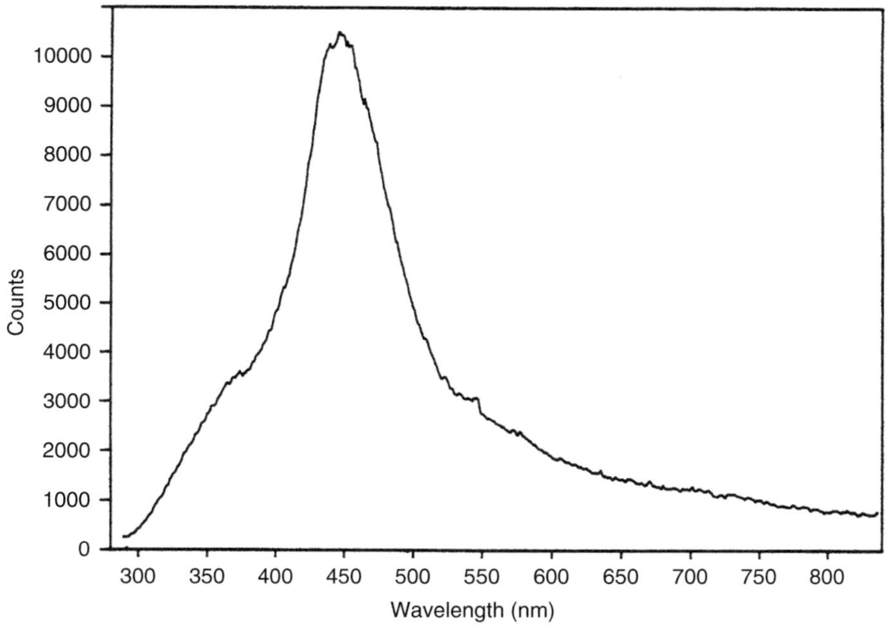

Fig. 4.73 Spectrum of radiation induced fluorescence in a high purity fused-silica fiberoptic cable

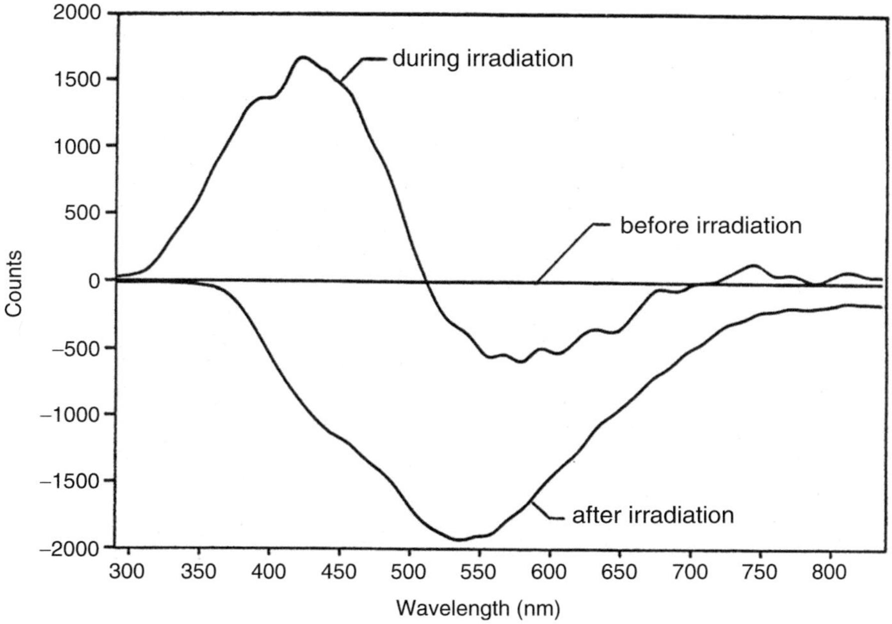

Fig. 4.74 Net lamp spectrum [lamp + RIF]-[RIF], before, at the peak of, and shortly after a $3 pulse

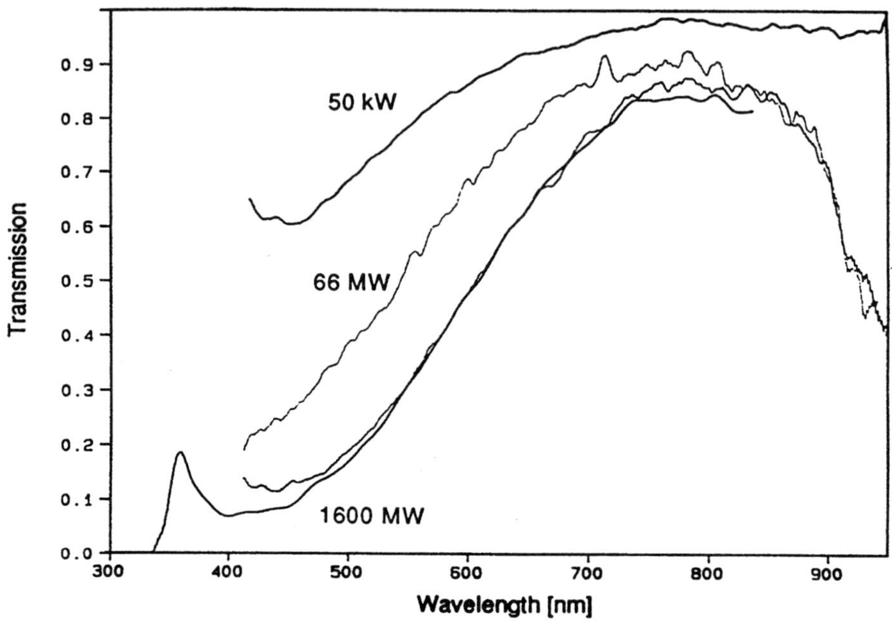

Fig. 4.75 Fiberoptic attenuation as a function of reactor power (Notice that this attenuation is shown for reactor power levels of 50 kW, 66 and 1600 MW)

Fig. 4.76, allowed the absolutely calibrated spectra from the lamp over the visible to near infrared range to be found. The line spectra from the rare gas lamps were cataloged. The results from nuclear-driven rare gas flash lamps were compared to results from electrically driven rare gas flashlamps. One example shows that the output intensity of a nuclear driven Kr flashlamp with a nuclear power input of 27 kW had a factor of two higher intensity line spectra than a 1.3 kW Kr arc lamp (See Fig. 4.76). The surface area of the Kr arc lamp was 1.4137 cm^2 while the surface area of the nuclear driven Kr lamp was 50.265 cm^2. Based on this comparison, and the fact that the Kr arc lamp releases 8 % of its energy in line spectra, the optical power of the nuclear driven Kr lamp was found to be around 3.7 kW [109].

The absolutely calibrated intensity data for the nuclear driven fluorescer lamp was in agreement with the work done in 1987 on pumping of a Nd:YAG laser with nuclear driven fluorescence. The intensity of the lamp showed that the power deposition in the crystal from the 1987 experiment would be about 500 W cm^{-3} which is in good agreement with the prediction [108].

It should be noted that at the end of the calibration studies for rare gas nuclear driven flashlamps, a laser experiment was attempted [109]. A window was used in the lamp rather than the focusing lens which was used in the 1987 experiments. This reduced the power transported by the fiberoptic cable to the Nd:YAG crystal significantly. There was insufficient power deposition in the crystal to observe any gain. This result was expected. The lens used in the 1987 experiment was

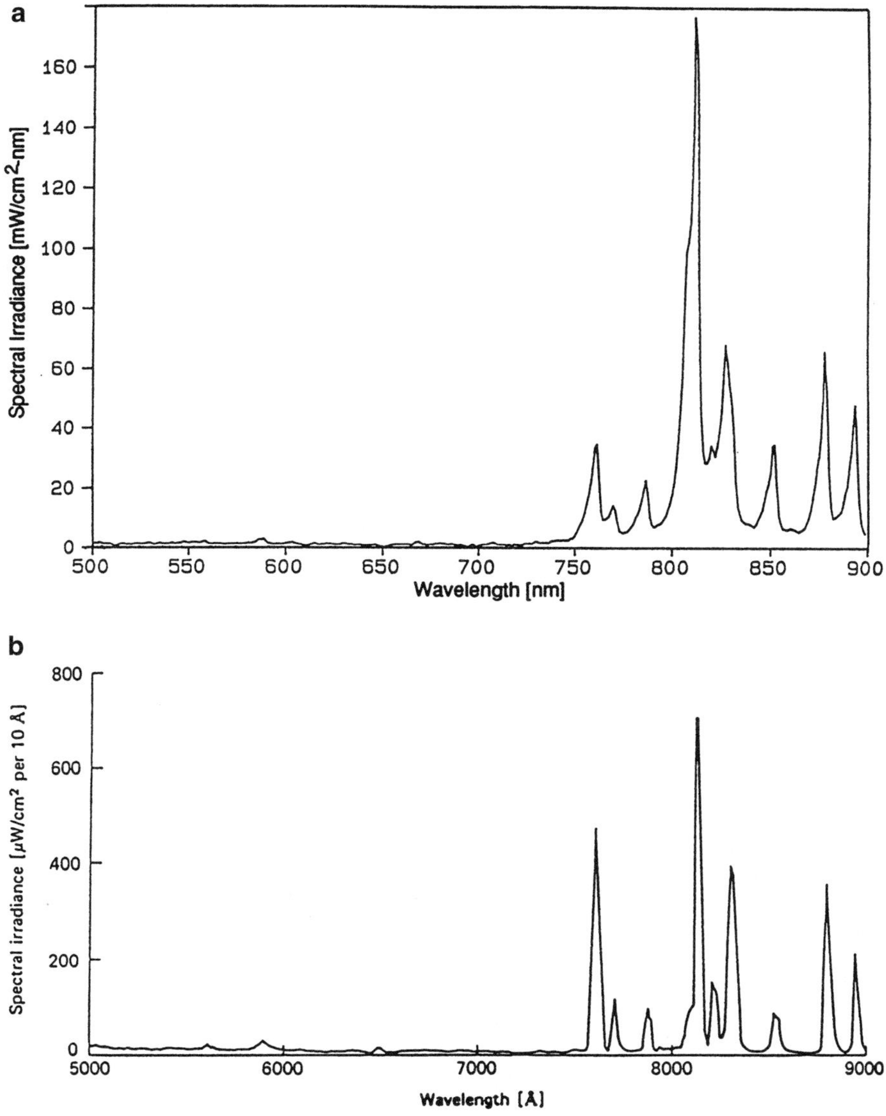

Fig. 4.76 Comparison of wide range, low resolution, Kr nuclear-driven fluorescer lamp absolute spectrum (**a**) [50 Torr Kr, 750 Torr ^{3}He, 30 Wcm^{-3}] to that of a Kr arc lamp (**b**) [6 mm bore, 50 mm long, 4 atm. Kr, 1.3 kW]

able to collect light from the full 50 cm^2 of the lamp's aperture. The 2.9 cm^2 fiberoptic bundle was only able to collect light from 5.7 % of the lamp aperture's surface. Thus, the coupling efficiency from the lamp to the laser was very low and the power density absorbed by the Nd:YAG crystal was only about 28 W cm^{-3}.

The main purpose of the above research project on solid-state lasers was to develop nuclear driven flashlamps with a good spectral matching efficiency with solid-state laser candidates so that reactor/nuclear-driven flashlamp photolytic laser system efficiencies greater than 1 % could be developed [110]. It was known from the beginning of the project that rare gas flashlamps would not have the optimum spectral matching efficiency but that they would be a good first step in developing research methodologies. In addition the rare gas lamps could be used to demonstrate the potential of the concept. One of the main goals of the project was to develop nuclear-driven alkali metal excimer flashlamps. They were a critical task item in the DOE funded research project because of their potential to boost lamp efficiency. It should be noted that a series of experiments were conducted with nuclear driven alkali metal excimer fluorescers (discussed in Chap. 1 and elsewhere [111, 112]) as part of this DOE funded research project on nuclear-driven flashlamps for pumping solid-state lasers.

Experiments in Russia using thin uranium coatings to drive a xenon fluorescer demonstrated lasing with 1–4 m long, 40–50 μm radius Nd doped phosphate glass fibers [7]. The fiber was coiled between two plates each coated with 5 mg cm^{-2} of metallic enriched ^{235}U. Xenon gas at 0.5 atm filled the space between the plates (Fig. 4.77). A neutron pulse from the VIR-2-M fast burst reactor was used as the driver. The peak energy from the laser was 2.5 J (Fig. 4.77). In addition to Nd glass fibers, the radioluminescence from a CsI(Tl) crystal driven by gamma rays from the center of the BR-1 pulsed reactor (60 μs pulse width) was used to pump a Nd:Glass laser. The highest laser power achieved was 4 kW [7].

4.4.1.10 Iodine Photolytic Laser

Photolytically driving the iodine laser with fluorescence from the XeBr excimer was proposed and demonstrated in the 1970s by the Lawrence Livermore National Laboratory [113, 114]. The laser uses the overlap of the XeBr excimer emission spectra with the n-C_3F_7I (hereafter called C_3F_7I) absorber (Fig. 4.78). Ultraviolet radiation (between 240 and 320 nm) photolytically dissociates C_3F_7I into an excited iodine $5^2P_{1/2}$ atom and a n-C_3F_7 free radical specie with nearly 100 % quantum yield (Eq. 4.51). The lifetime of the excited state of iodine is 130 ms and it is only quenched by impurities and I_2.

$$C_3F_7I + hv \rightarrow I\left(5^2P_{1/2}\right) + C_3F_7\cdot \tag{4.51}$$

Nuclear-pumped photolytic lasers were first proposed in 1978 in which the fluorescence efficiency of XeBr* was found to be of the order of 15 % [115]. A subsequent theoretical study indicated feasibility of the concept [116]. However, the true potential of the photolytic laser is due to the ability of the nuclear-pumped flashlamp to focus energy into the laser medium thus increasing the power density in the laser medium [38, 117]. An analysis of an iodine laser using an alkyliodide and XeBr* fluorescence (with a = 15.24 cm, r = 0.3175 cm and an average power deposition at a thermal neutron flux of 10^{14} = 1 W cm^{-3}) estimates that the

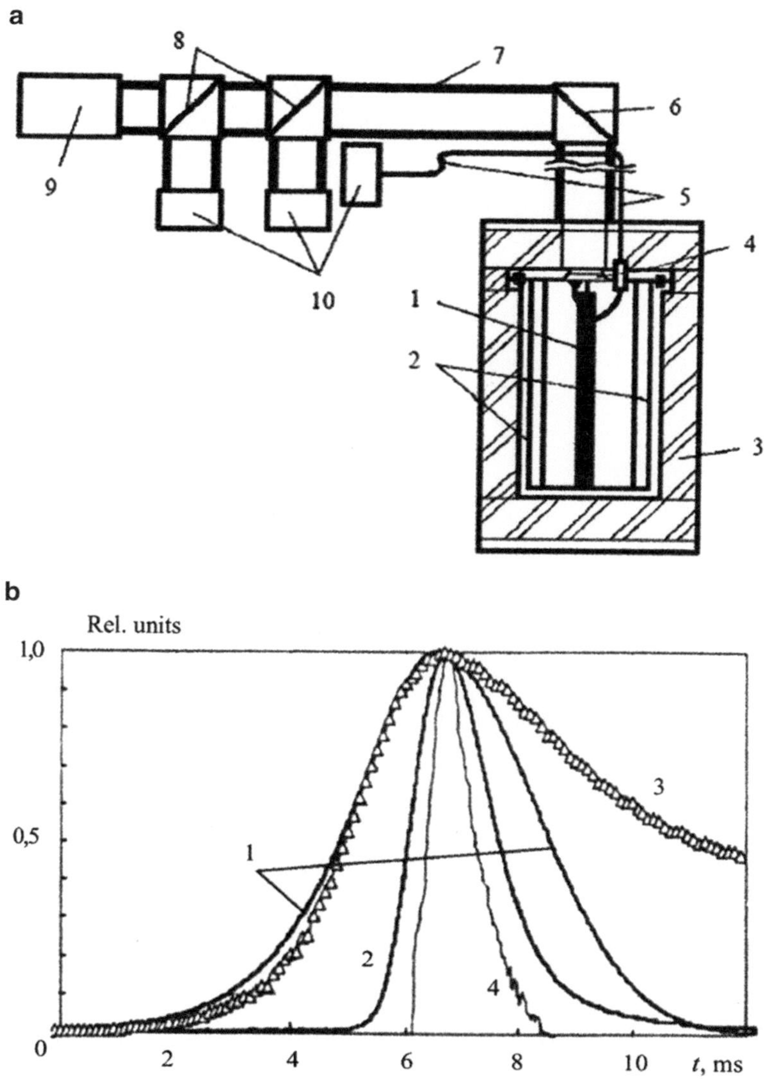

Fig. 4.77 The diagram of the Russian experiment with a nuclear driven xenon flashlamp pumped by thin uranium films and a neodymium fiber laser medium on *top* and a trace of the laser signal on the *bottom*. The *top* diagram are composed of (*1*) neodymium fiber laser; (*2*) plates with thin uranium coatings; (*3*) moderator; (*4*) optical connector; (*5*) quartz fiber; (*6*) rotary mirror; (*7*) aluminum pipe; (*8*) quartz plates; (*9*) power meter; (*10*) photodectors. The *bottom* diagram shows (*1*) neutron pulse; (*2*) the fluorescence from xenon; (*3*) gas pressure; and (*4*) laser pulse [7]

fluorescence intensity (I) on the laser cell for system without coupling is about 2.4 W cm^{-2} and with the nuclear energy focus (NEF) = 138 W cm^{-2} [117] (Fig. 4.79).

The concept of photon coupling between a volume source-driven fluorescer and a photolytic laser requires a reflector which concentrates the photons generated in

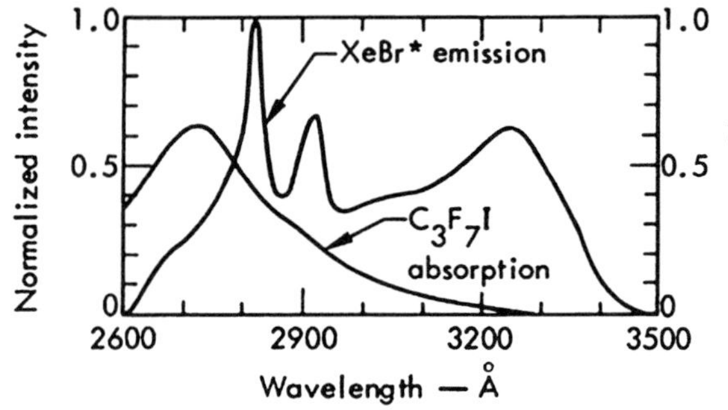

Fig. 4.78 Coupling of XeBr* emission coupled with the C_3F_7I absorber

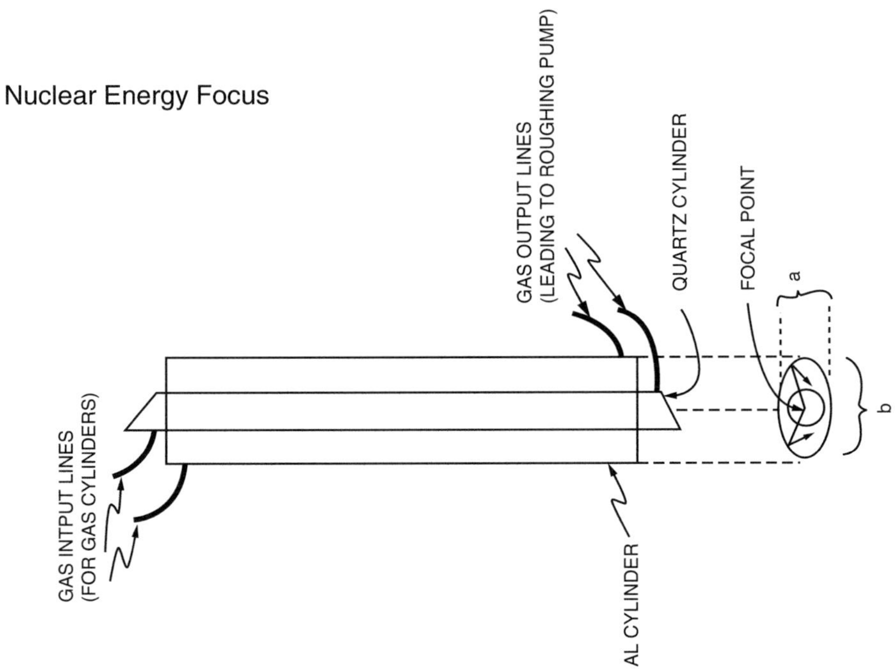

Fig. 4.79 Diagram for the Nuclear Energy Focus as applied to the XeBr* fluorescer and C_3F_7I absorber [117]

the nuclear driven fluorescer to the laser medium. Specifically, one can describe the average power deposition and the average intensity of and upon the laser medium as being,

$$\overline{P_{ds}} = \overline{P_{dl}}\eta_f\eta_{pl}{}^{V_l}/_{V_s} \tag{4.52}$$

$$\overline{I_s} = \overline{P_{dl}}\eta_f\eta_l{}^{V_l}/_{A_s} \tag{4.53}$$

Table 4.19 Reflectivity of aluminum and transmissivity of fused silica

Wavelength (nm)	10-mm fused silica (T)	Highly polished aluminum (R)
150	0.10	–
170	0.65	–
200	0.92	92.3
300	0.92	92.4
400	0.92	92.4
500	0.92	91.8
600	0.92	91.1
700	0.92	89.7
800	0.92	86.7
900	0.92	89.1
1000	92.3	94.0
1500	92.4	97.4
2000	92.4	97.8

where $\overline{P_{dl}}$ = power deposition in fluorescer (W cm^{-3}), η_f = fluorescence efficiency, η_{pl} = coupling efficiency for power deposition, η_I = coupling efficiency for intensity, V_l = volume of fluorescer, V_s = volume of laser, and A_s = surface area of laser. From Eqs. (4.49) and (4.50), it may be noted that $\overline{P_{ds}} \propto {}^{V_l}/_{V_s}$, and $\overline{I_s} \propto {}^{V_l}/_{A_s}$. Thus, in having a large ${}^{V_l}/_{V_s}$, or ${}^{V_l}/_{A_s}$, ratio, thus it is possible to design a system where $\overline{P_{ds}} \gg \overline{P_{dl}}$ and $\overline{I_s} \gg \overline{P_{dl}}$. This effect is called energy focusing.

In order to couple the fluorescer and laser efficiently, the surface of the fluorescer cell must be highly reflective to the fluorescence. Additionally, the number of reflections required before the photon reaches the laser should be reduced to a minimum for two reasons: (1) No reflecting material is perfect; as such, every reflection reduces η_{pl} or η_I; and (2) Since the fluorescer material must be self-absorbing to some degree, the more reflections it takes, the longer the distance the photon must travel and the higher the probability that self-absorption occurs.

In developing the model, a parabolic-type geometry was assumed for the fluorescer cell. The laser cell was placed at the focal point of the parabolic fluorescer cell and highly polished end plates were utilized (see Fig. 4.79). Aluminum was chosen as the reflector material and ultra-high purity fused silica was used for the highly transmissive laser cell. (The values of reflectivity and trasmissivity may be seen in Table 4.19) For the examples chosen in the modeling studies, the fluorescence cell utilized $^3He:Xe:Br_2$ and $He:Xe:Br_2:Rn$ mixtures. The efficiency of fluorescence generation from the XeBr excimer (282 nm) was assumed to be 11 % [118], even though there is evidence that this efficiency may be as high as 15 % for a nuclear-pumped system [2].

A 1000 Torr mixture of $^3He:Xe:Br_2$ at 7:2:1 was chosen for this model. The calculated efficiencies, intensities and power densities are shown for the NEF and for a system with no coupling in Table 4.20. The threshold intensity for the iodine

Table 4.20 Calculated efficiencies, intensities and average power deposition for the design with various neutron sources

Parameter	MURR	TRIGA	Fast burst
η_I	40.79	40.79	40.79
$\overline{I_s}$	1.16	28.9	1156.0
η_{pl}	37.02	37.02	37.02
$\overline{P_{ds}}$	3.73	93.1	3724.0
η_{nc}	2.10	2.10	2.10
$\overline{I_{nc}}$	0.10	1.51	60.0

Note: The threshold intensity for pumping the iodine laser is $\overline{I_s^{th}} = 7$ W cm^{-2}

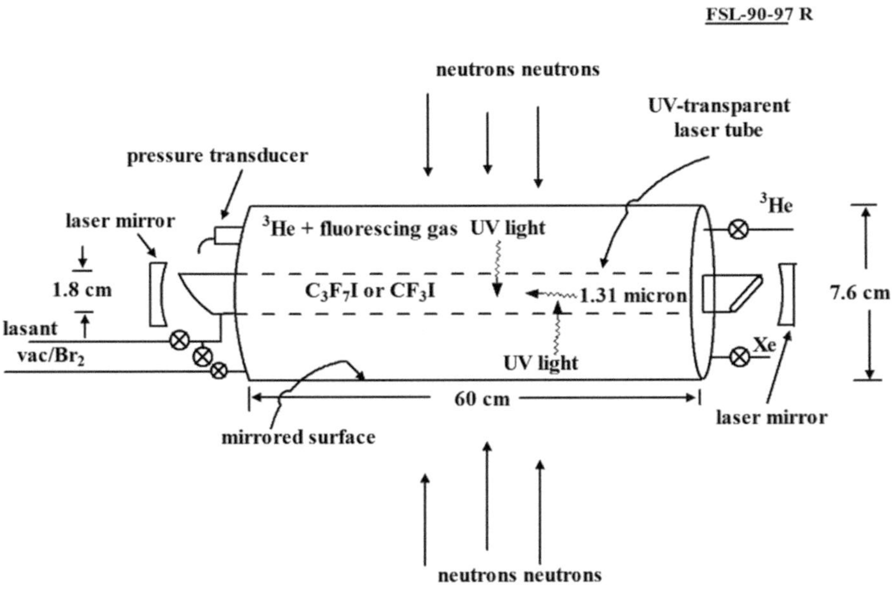

Fig. 4.80 Diagram of the nuclear-pumped iodine laser cell [32]

laser is $\overline{I_s^{th}} = 7$ W cm^{-2}. As can be seen in Table 4.20, the NEF using a TRIGA reactor and a fast burst reactor exceeds the threshold intensity. For a laser with no photon coupling, only the fast burst reactor exceeds the threshold intensity.

A reflective parabolic shape for the nuclear-pumped flashlamp cavity is optimum [40]. However as this reference shows, by changing the cavity shape to a reflective cylinder there would only be a loss of about 30 % of the absorbed photons in the laser medium. Due to the difficulty of building a reflective parabolic nuclear-pumped fluorescence cell, the experiment which demonstrated the nuclear-pumped iodine laser used a reflective cylindrical nuclear-pumped fluorescence cell [32] as shown in Fig. 4.80. The optical cavity was formed by using a back mirror with 99.7 % reflectivity at 1310 nm and a front mirror with 98 % reflectivity at 1310 nm.

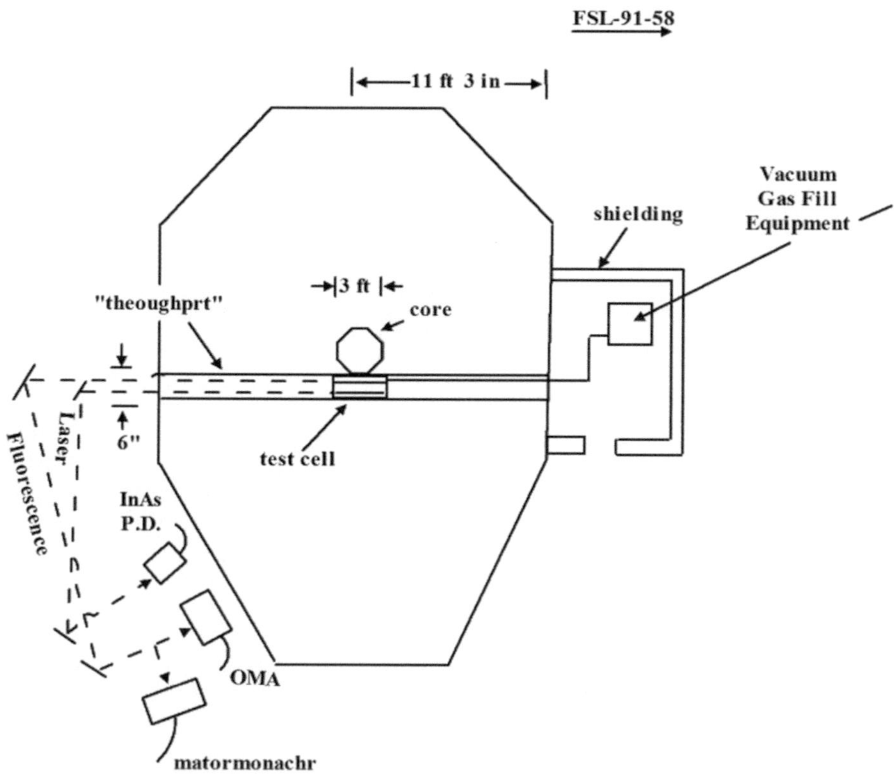

Fig. 4.81 Experimental set up of the nuclear-pumped iodine laser experiment in the through port of the University of Illinois TRIGA reactor [32]

The laser cell was placed in the through port of the University of Illinois TRIGA reactor as shown in Fig. 4.81. Research grade gases and chemicals were used and the $CHBr_3$ and i-C_3F_7-I were treated to remove dissolved gasses. The fluorescing mixture in the flashlamp used 0.5 Torr of $CHBr_3$, 200 Torr of Xe and 3 atm of ^{3}He. The fluorescence efficiency from the XeBr (B-X)* emission was measured as 1 ∓ 0.5 %. Figure 4.82 shows the trace of the laser output superimposed on the neutron pulse. The threshold pumping power was about 0.4 W cm^{-3} and compares well with an electrically flashlamp pumped iodine laser experiment at about 0.07 W cm^{-3}.

The peak laser output was about 20 mW with an efficiency of about 4.6×10^{-5} %. It was predicted that if the gas mixtures were optimized, the impurities removed and the flashlamp efficiency increased to 10 %, then the theoretical efficiency would be 0.6 %.

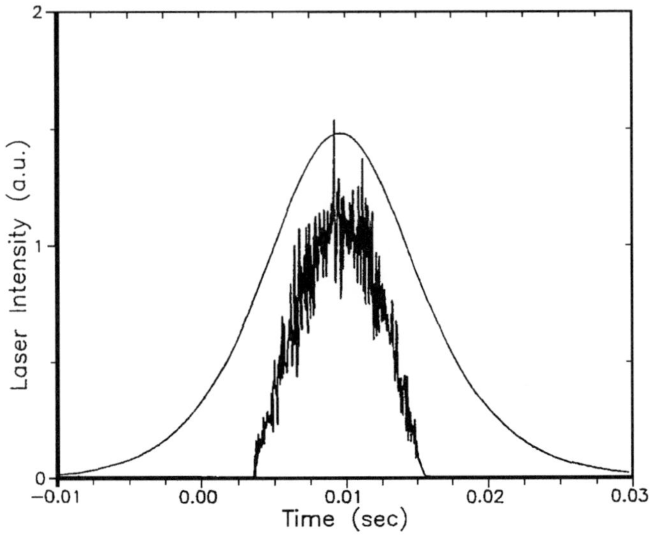

Fig. 4.82 1310 nm laser output from the nuclear-pumped iodine laser superimposed on the neutron pulse [32]

4.4.1.11 Molecular Nitrogen Ion (N_2^+) Laser

The molecular nitrogen ion laser (N_2^+) on the B→X transition on the 391.4 and 427.8 nm lines was pumped by fission fragments using the EBR-L and the BARS-6 reactors. The laser used mixtures of He-N_2-H_2 at total pressures between 3.5 and 5 atm. The partial pressures of nitrogen and hydrogen were on the order of 3–20 Torr. The laser initially operated on the 391.4 nm line using the EBR-L with a power output of 10 W and a laser efficiency of 0.01 %. In later experiments using the BARS-6 reactor, both the 391.4 nm and 427.8 nm lines lased. The efficiency of the 391.4 nm line was 0.3 % and the 427.8 nm line was 0.2 %. The threshold power density was 50–60 W cm^{-3} [6, 7]. The dominant mechanism for the population of the upper laser level is due to charge transfer ($He_2^+ + N_2 \rightarrow N_2^+(B) + 2He$) and the depopulation of the lower laser level due to $N_2^+(X) + H_2 \rightarrow N_2H^+ + H$.

4.5 Summary

A number of nuclear-pumped lasers have been discovered. Two of the lasers were driven by a thermonuclear explosion and the others used neutrons from a nuclear reactor. The main barrier which limited the total number of nuclear-pumped lasers is due to the nature of the energy conversion mechanism from neutron capture to the generation of charged particles. Pulsed reactors are relatively slow (200 μs for a fast burst reactor or 10 ms for a TRIGA type reactor) thus constricting the amount of power density that can be generated. Lasers, unfortunately, are power density hungry. Those nuclear-pumped lasers which have worked exhibit unusually efficient mechanisms for populating the upper laser level. These mechanisms either have resonance reactions or favorable branching ratios. A list of successful nuclear-pumped laser systems is shown in Table 4.21. The mechanisms which are believed to be relevant in populating the upper laser level and depopulating the lower laser level are shown for each laser.

There may be other viable nuclear-pumped laser systems in nature. For example the nuclear energy focus/photolytic laser systems remain largely unexplored. As described in prior sections, this is a viable method of boosting the pumping power density and can generate large power densities in the laser medium. Other areas which might prove fruitful are thermally pumped laser and directly driven solid-state lasers. Thermally-pumped lasers can take advantage of developments in material science which may make Ultra High Temperature Reactors possible. Semiconductor materials can stop fission fragments with 5–8 μm thus allowing the energy content of a fission fragment to be deposited in a small volume. This leads to very high power densities in the semiconductor material. Some of these advanced concepts will be discussed in Chap. 5.

Table 4.21 Summary of mechanisms of formation of the population inversion levels in NPL [6, 32, 50]

The active medium	Lasing wavelength (μm)	Population of the upper laser levels	Depopulation of the lower laser levels
IR lasers on transitions nd-(n + 1) p atoms B = Xe, Kr and Ar ($n = 5, 4, 3$ to Xe, Kr, Ar respectively)			
He-Xe, Kr, Ar; Ar-Xe; He-Ne (Ar)-Xe; Xe, Kr	26 lines in the 1.15–3.51 range	$B_2^+ + e \rightarrow B^*(nd) + B$;	"Quenching" in collisions with atoms A and B
		$AB^+ + e \rightarrow B^*(nd) + A$	
Visible laser at 3p-3s transitions of the atom Ne			
He-Ne-*M*; Ne-*M* ($M =$ Ar, Kr, Xe, H_2)	0.585; 0.703; 0.725	$Ne_2^+ + e \rightarrow Ne^*(3p) + Ne$	$Ne^*(3s) + M \rightarrow M^+ + e + Ne$
Laser using a mixture of inert gases with metal vapors			
He-Cd	ion Cd^+ (0.442; 0.534; 0.538; 0.807; 0.853)	$He^+ + Cd \rightarrow (Cd^+)^* + He$	Radiative decay and (or) collisional "Quenching"
		$He_2^+ + Cd \rightarrow (Cd^+)^* + 2He$	
		$He^* + Cd \rightarrow (Cd^+)^* + He + e$	
He-Cd	atom Cd (1.43; 1.65)	$Cd_2^+ + e \rightarrow Cd^* + Cd$	Radiative decay and (or) collisional "quenching"

(continued)

Table 4.21 (continued)

The active medium	Lasing wavelength (μm)	Population of the upper laser levels	Depopulation of the lower laser levels
He-Zn	ion Zn^+(0.748)	$He^+ + Zn \rightarrow (Zn^+)^* + He$	Radiative decay
		$He_2^+ + Zn \rightarrow (Zn^+)^* + 2He$	
		$He^* + Zn \rightarrow (Zn^+)^* + He + e$	
He-Hg	ion Hg^+ (0.615)	$He^+ + Hg \rightarrow (Hg^+)^* + He$	Radiative decay and (or) collisional "Quenching"
He-Xe-Hg-H_2	atom Hg (0.546)	$Hg_2^+ + e \rightarrow Hg^* + Hg$	"Quenching" molecules H2
Lasers on atomic transitions: C, N, O, Cl			
He(Ne, Ar)-CO, CO_2, N_2, O_2, Cl_2, CCl_4	C (0.833; 0.940; 1.45); N (0.859; 0.863; 0.905; 0.939); O (2.65; 2.76); Cl (1.59; 2.45)	The following processes of populating the upper levels:	
		(1) dissociative excitation:: $A^* + R_2 \rightarrow R^* + R + A(A^*$ – metastable inert gas atom, $R2$ – molecular gas); (2) - ion-ion recombination: $A^+(A_2^+) + R \rightarrow R^* + A(2A)$; 3 body recombination: $R^+ + e + M \rightarrow R^* + M$ ($M = e, A$);	
Laser on transitions of the molecular ion: $N_2^+(B \rightarrow X)$			
He-N_2-H_2	0.391; 0.428	$He_2^+ + N_2 \rightarrow N_2^+(B) + 2He$	$N_2^+(X) + H_2 \rightarrow N_2H^+ + H$
Laser vibrational transitions of the molecule CO			
CO	5.1–5.6	Generation mechanism is not understood	
Photolytic transitions			
Nd:YAG	1.06	$h\nu$ (100–810 nm) + $Nd^{3+} \rightarrow Nd_3^{+*}$	
I	1.31	$h\nu$ (282 nm) + $C_3F_7I \rightarrow I^*$	

Problems

1. Explain why only fission promises high enough system efficiencies for nuclear-pumped laser systems.
2. Contrast the pros and cons of surface sources, volume sources and nuclear-driven flashlamps coupled with photolytic lasers.
3. Discuss the pros and cons of an aerosol core reactor fueled with reflective aerosol particles.
4. Verify Table 4.1.
5. Contrast the pros and cons of using gases, liquids and solids as laser media.
6. Estimate the peak power density in the laser medium for a nuclear energy focus with the TREAT reactor as the neutron source. Assume the design parameters (with the modified lamp power density with the TREAT driver), photon transport properties and reflectivity used in Eq. (4.5).

7. In Sect. 4.4.1.2, the ^{220}Rn is used as the energy source for driving a XeBr lamp which in turn pumps the photolytic iodine laser. How can ^{224}Ra be incorporated in the lamp as a ^{220}Rn generator (a strategy which would give the radioisotope pumped laser a 3.66 year half-life)? Can the laser power be increased by maximizing the amount of ^{224}Ra in the laser?
8. Develop a 4 level laser model for the Xe $5d[5/2]_2 \rightarrow 6p[3/2]_1$ transition at 2.03 μm (see examples in Chap. 1).
9. Figure 4.17 shows a scale up experiment in which the Xe $5d[5/2]_2 \rightarrow 6p[3/2]_1$ transition at 2.03 μm produced an output greater than 500 J. Based on the experiment's design parameters, estimate the power density in the laser medium. It was reported that the laser had deleterious effects due to increasing gas temperatures in the laser cells. Explain how raising temperatures could reduce laser power output. How would these effects harm the design of a steady-state nuclear-pumped laser based on the 2.03 μm xenon transition?
10. Why was there a delay between the peak of the laser power and the peak of the neutron flux in the atomic carbon laser with mixtures of He+CO_2, Ne+CO, Ne +CO_2, and Ar+CO_2 but not He+CO and Ar+CO?
11. What is (are) the mechanism(s) for populating the upper laser level for the rare gas nuclear-pumped lasers, the nitrogen nuclear-pumped laser and the atomic carbon nuclear-pumped laser?
12. If the nuclear-pumped iodine laser were to be redesigned with a configuration like Fig. 4.5, could the efficiency of the laser be improved?

References

1. Prelas MA, Loyalka SK (1981) A review of the utilization of energetic ions for the production of excited atomic and molecular states and chemical synthesis. Prog Nucl Energy 8:35–52
2. Prelas MA, Boody FP, Miley GH, Kunze J (1988) Nuclear driven flashlamps. Laser Part Beams 6:25–62
3. Prelas MA, Jones GL (1982) Design studies of volume-pumped photolytic systems using a photon transport code. J Appl Phys 53:165–169
4. Boody FP, Prelas MA, Anderson JH, Nagalingam SJS, Miley GH (1978) Progress in nuclear-pumped lasers. In: Billman K (ed) Radiation energy conversion in space, vol 61. AIAA, New York, pp 379–410
5. Miley GH, McArthur DA, Deyoung RJ, Prelas MA (1989) Fission reactor pumped laser: history and prospects. In: Carlson JW, Behrens AD (eds) 50 years of nuclear power, national academy of science and NIST. American Nuclear Society, La Grange Park, IL, pp 333–342
6. Melnikov SP, Puning VT, Sinyanskii AA (2008) Gas lasers with nuclear pumping: physical processes and experimental techniques. Federal Agency for Education, Moscow Engineering Physics Institute, Moscow, Russia
7. Melnikov SP, Sizov AN, Sinyanskii AA, Miley GH (2015) Lasers with nuclear pumping. Springer, New York
8. Prelas MA, Boody FP, Zediker M (1984) A direct energy conversion technique based on an aerosol core reactor concept. IEEE International conference on plasma science, vol 84CH1958-8, p 8
9. Prelas MA, Boody FP, Zediker MS (1985) An aerosol core nuclear reactor for space-based high energy/power nuclear-pumped lasers. In: El-Genk MS, Hoover M (eds) Space nuclear power systems. Orbit Book Company, Malabar

10. Guoxiang G, Prelas MA, Kunze JF (1986) Studies of an aerosol core reactor/laser's critical properties. In: Hora H, Miley GH (eds) Laser interaction and related plasma phenomena. Springer, Boston, pp 603–611
11. Chung AK, Prelas MA (1984) The transport of heavy charged particles in a cylindrical nuclear-pumped plasma. Nucl Sci Eng 86:267–274
12. Mencin DJ, Prelas MA (1992) Gaseous like uranium reactors at low temperatures using C60 cages. In: Nuclear technologies for space exploration, vol 2, pp 403–433. NTSE '92 by American Nuclear Society, La Grange Park, ISBN 0894481746
13. Dai X, Cheng C, Zhang W, Xin M, Huai P, Zhang R et al (2013) Defect induced electronic structure of uranofullerene. Sci Rep 3:1341
14. Prelas MA (1990) Nuclear-driven solid-state lasers. In: Shay DG, Harris TM (eds) Proceeding of the international conference on lasers 89. STS Press, McLean
15. Prelas MA, Weaver CL, Watermann ML, Lukosi ED, Schott RJ, Wisniewski DA (2014) A review of nuclear batteries. Prog Nucl Energy 75:117–148
16. Ebert P, Ferderber L, Koehler H, Kuckuck R, Redhead D (1974) Amplified spontaneous emission in xenon pumped by gamma rays. Quantum Electron IEEE J 10:736–736
17. Lyons P, Clarke J, Metzger D (1974) Gamma initiated HF laser. Quantum Electron IEEE J 10:736–736
18. McArthur DA, Tollefsrud PB (1975) Observation of laser action in CO gas excited only by fission fragments. Appl Phys Lett 26:187
19. Helmick HH, Fuller JI, Schneider RT (1975) Direct nuclear pumping of a helium-xenon laser. Appl Phys Lett 26:327–328
20. DeYoung RJ, Wells WE, Miley GH, Verdeyen JT (1976) Direct nuclear pumping of a Ne-N_2 laser. Appl Phys Lett 28:519
21. Akerman MA, Miley GH, Mcarthur DA (1977) A helium mercury direct nuclear-pumped laser. Appl Phys Lett 30:409
22. Jalufka NW, DeYoung RJ, Hohl F, Williams MD (1976) Nuclear pumped ^{3}He-Ar laser excited by the ^{3}He(n, p)T reaction. Appl Phys Lett 29:189–190
23. Deyoung RJ, Jalufka NW, Hohl F (1977) Nuclear-pumped lasing of ^{3}He-Xe and ^{3}He-Kr. Appl Phys Lett 30:19
24. DeYoung RJ, Jalufka NW, Hohl F (1978) Direct nuclear-pumped lasers using the ^{3}He(n, p)T reaction. AIAA J 16:991–998
25. Prelas MA, Akerman MA, Boody FP, Miley GH (1977) Direct nuclear-pumped 1.45 μm atomic laser in mixtures of He-CO and He-CO_2. Appl Phys Lett 31:679
26. Voinov AM, Dovbyshov LE, Krivonosov VN, Mel'nikov SP, Podmoshenskii IV, SinyanskiiI AA (1979) Nuclear-pumped IR lasers using ArI, KrI, and XeI transitions. Sov Phys Tech Phys Lett 5
27. Mis'kevich AI, Dmitriev AB, Il'yashchenko VS, Salamakha BS, Stepanov VA, Gorodkov EM (1980) Lasing in Cd vapor pumped by products of the nuclear reaction ^{3}He(n, p)T. Sov Phys Tech Phys Lett 6
28. Rowe MJ, Liang RH, Schneider RT (1981) Nuclear-pumping CO_2. In: Proceedings of 1981 IEEE international conference on plasma science, Ottawa, Canada, p 149
29. Alford WJ, Hays GN (1989) Measured laser parameters for reactor-pumped He/Ar/Xe and Ar/Xe lasers. J Appl Phys 65:3760
30. Hebner GA, Hays GN (1993) Fission-fragment excited xenon/rare gas mixtures: laser parameters of the 1.73 μm xenon transition. J Appl Phys 73:3614–3626
31. Shaban YR, Miley GH (1993) Practical, visible wavelength nuclear-pumped laser. Laser Part Beams 11:559–566
32. Williams WH, Miley GH (1993) Lasing in atomic iodine with a nuclear-pumped XeBr* flash lamp. Appl Phys Lett 62:1724–1726
33. Eerkens J et al (1965) Nuclear-pumped laser progress report, May 20, 1965, Northrop Corporation, Northrop Space Labs Report NSL 66–64
34. Eerkens JW (1967) Nuclear-pumped lasers. Northrop Corporation, 23–24 January 1967, Technical Session VII-D, in Technology Review for AF Systems Command, USAF-RTD

35. Eerkens (1967) Research on Direct Nuclear Pumping of Gas Lasers (DNPGL). Northrop , First and Second Interim Reports on USAF Contract 33(615)-5000; Northrop Report NSL 66-141-1 Corporation, October 1966
36. Eerkens J et al (1968) Nuclear-pumped noble-gas ion-laser experiments. December 1968, Northrop Corporation
37. Felty JR, Lipinski RJ, McArthur DA, Pickard PS (1993) DOE reactor-pumped laser program. Department of Energy, Washington, DC, Report SAND–93-2719C; CONF-940142–11; ON: DE94006372; BR: GB0103012
38. Prelas MA, Lecours MJ, Schlapper GA, Brugger RM (1980) Nuclear-pumped laser research at the University of Missouri. Trans Am Nucl Soc 34:810–811
39. Prelas MA (1981) A potential UV fusion light bulb for energy conversion. Bult Am Phys Soc 26:1045
40. Javedani JB, Prelas MA (1986) Concentrating properties of simple two-dimensional geometries for isotropic light. In: Hora H, Miley GH (eds) Laser interaction and related plasma phenomena. Springer, Boston, pp 155–165
41. Boody FP, Prelas MA (1992) Efficient light transport using large diameter-to-length ratio hollow lightpipes. In: Proceedings of specialist conference on physics of nuclear induced plasmas and problems of nuclear-pumped lasers, vol 1. Institute of Physics and Power Engineering, Obninsk. doi:10.13140/RG.2.1.3433.5207
42. Walker J (1967) Nuclear reactors as research instruments. Rep Prog Phys 30:285
43. McArthur D, Schmidt T, Philbin J, Tollefsrud P (1977) In: Sandia national laboratory, SAND76-0584 (ed) Concepts for the construction of large reactor-excited laser systems, SANDIA National Laboratories, Albuquerque
44. Chung AK, Prelas MA (1987) Sensitivity analysis of Xe_2* excimer fluorescence generated from charged particle excitation. Laser Part Beams 5:125–132
45. Klein MB (1974). In: Hughes Research Laboratories (ed) Waveguide gas lasers. NTIS, Alexandria, VA, AFCRL-TR-75-0005
46. Aleksandr MB, Sergei IY (2006) On the possibility of pumping Xe_2* lasers and VUV lamps in the afterglow of a background-electron multiplication wave. Quantum Electron 36:1176
47. Collier F, Cottin P (1978) Efficient VUV xenon excimer laser ($\lambda = 172$ nm) AT 6 atm. Opt Commun 25:89–92
48. Panchenko AN, Orlovskii VM, Tarasenko VF, Baksht EH (2003) Efficient operation modes of a non-chain HF-laser pumped by self-sustained discharge, pp 303–310, International conference on lasers, applications, and technologies 2002: advanced lasers and systems, 303–310 (October 8, 2003). doi:10.1117/12.518087
49. Panchenko AN, Orlovskii VM, Tarasenko VF (2003) Efficient e-beam and discharge initiated nonchain HF(DF) lasers. Laser Part Beams 21:223–232
50. Prelas MA (1990) Nuclear-driven solid-state lasers. In: Lasers '89; Proceedings of the international conference, New Orleans, LA, Dec. 3–8, 1989 (A91-41326 17–36). STS Press, McLean, pp 263–269
51. Karelin AV, Simakova OV (2004) Complete optimisation of an electron-beam-pumped Xe laser emitting at 1.73, 2.03, 2.65, 2.63, 3.37, and 3.51 μm. Quantum Electron 34:29
52. Shon JW, Kushner MJ (1994) Excitation mechanisms and gain modeling of the high-pressure atomic Ar laser in He/Ar mixtures. J Appl Phys 75:1883–1890
53. Karelin AV, Tarasenko VF, Yakovlenko SI (2000) High-pressure atom and atomic-ion lasers. Laser Phys 10:827–844
54. Lawton S, DeTemple TA (1978) Near infrared gas lasers. In: AP Wright-Patterson AFB (ed). US air force, Ohio, Technical report AFAPL-TR-78-107, p 164
55. Melnikov SP, Sinyanskii AA (1994) The lasing mechanism of nuclear-pumped IR lasers at krypton and argon atomic transitions. Hyperfine Interact 88:165–171
56. Riola JP, Howard JS, Rundel RD, Stebbings RF (1974) Chemiionization reactions involving metastable helium atoms. J Phys B At Mol Phys 7:376

57. Deloche R, Monchicourt P, Cheret M, Lambert F (1976) High-pressure helium afterglow at room temperature. Phys Rev A 13:1140–1176
58. Bourne R, LeCalve J (1971) Effets d'additifs sur la postluminescence de He_2 dans l'hélium excité par une impulsion intense d'électrons. J de Phys 32:29–34
59. Dickinson AS, Roberts RE, Bernstein RB (1972) Ion-atom association reactions in the rare gases. J Phys B At Mol Phys 5:355
60. Mansbach P, Keck J (1969) Monte Carlo trajectory calculations of atomic excitation and ionization by thermal electrons. Phys Rev 181:275–289
61. Pitaevskii LP (1962) Vortex lines in an imperfect Bose gas. Zh Eksp Teor Fiz 42:1326
62. Shiu Y-J, Biondi MA (1977) Dissociative recombination in krypton: dependence of the total rate coefficient and excited-state production on electron temperature. Phys Rev A 16:1817–1820
63. Shiu Y-J, Biondi MA (1978) Dissociative recombination in argon: dependence of the total rate coefficient and excited-state production on electron temperature. Phys Rev A 17:868–872
64. Carter BD, Rowe MJ, Schneider RT (1980) Nuclear-pumped cw lasing of the ^{3}He-Ne system. Appl Phys Lett 36:115–117
65. Prelas MA, Schlapper GA (1981) Comments on "Nuclear-pumped cw lasing of the ^{3}He-Ne system". J Appl Phys 52:496–497
66. Schneider RT, Carter BD, Rowe MJ (1981) Reply to "Comments on nuclear-pumped cw lasing of the ^{3}He-Ne system". J Appl Phys 52:6980–6980
67. DeYoung RJ (1981) Kilowatt multiple path ^{3}He-Ar nuclear-pumped laser. Appl Phys Lett 38:297–298
68. Miley G (1984) Review of nuclear pumped lasers. In: Hora H, Miley G (eds) Laser interaction and related plasma phenomena. Springer, Boston, pp 47–72
69. Zagidulin AV, Bochkov AV, Mironenko VV, Sofienko GS (2012) A 500-J nuclear-pumped gas laser. Tech Phys Lett 38:1059–1062
70. Herwig L (1964) Preliminary studies concerning nuclear pumping of gas laser systems. Trans Am Nucl Soc 7:131
71. Day DA, Kennedy RG III (2010) Soviet star wars: the launch that saved the world from orbiting laser battle stations. Air&Space, January 2010. Available: http://www.airspacemag.com/space/soviet-star-wars-8758185/?no-ist
72. Cheo PK (1979) CO_2 lasers. In: Levine AK, DeMaria AJ (eds) Lasers, vol 3. Marcel Dekker, New York
73. Smith K, Thomson RM (1978) Computer modeling of gas lasers. Springer, Boston
74. Fein ME, Verdeyen JT, Cherrington BE (1969) A thermally pumped CO_2 laser. Appl Phys Lett 14:337–340
75. Makhov G, Wieder I (1967) Vibrational excitation of CO_2 by transfer from thermally excited nitrogen. Quantum Electron IEEE J 3:378–378
76. Gerry ET (1970) Gas dynamic CO_2 lasers. In: Meeting of the American Physical Society. Washington, DC, April 29
77. Kuehn DM, Monson DJ (1970) Experiments with a CO_2 gas-dynamic laser. Appl Phys Lett 16:48–50
78. Prelas M, White N, Wisniewski D, Boraas M, Kasiwattanawut H, Walton K, Correa S, De Castro J, De-La-Torre-Aguilar F, Knewtson T, Nelson S, Schutte J, Tchouaso MT, Watterman M (2014) Design of a space based very high temperature reactor thermally pumped carbon dioxide laser. Nuclear Science and Engineering Institute, University of Columbia, Columbia
79. Patel CKN (1966) Vibrational-rotational laser action in carbon monoxide. Phys Rev 141:71–83
80. Utkin YG, Goshe M, Adamovich IV, Rich JW (2006) Compact overtone band carbon monoxide laser. Opt Commun 263:105–110

81. Yu G, Baker HJ, Rodrigues NAS, Hall DR (1994) Compact high-efficiency carbon monoxide laser at 1 kW. Appl Phys Lett 65:2904–2906
82. Rich JW (1971) Kinetic modeling of the high-power carbon monoxide laser. J Appl Phys 42:2719–2730
83. Chen Chuan-Cheng (1974) The carbon monoxide laser: a review. The Rand Corporaton paper series. Available: http://www.rand.org/content/dam/rand/pubs/papers/2008/P5163.pdf
84. New Scientist (1974) A laser that runs on nuclear power. New Scientist, p 259
85. Prelas MA, Miley GH (1980) Dynamics of the nuclear and electrically pumped 1.45-μm atomic carbon laser in mixtures of helium + CO and helium + CO_2. Nucl Technol Fusion 1:402–413
86. Jalufka NW, Hohl F (1981) A direct nuclear-pumped ^{3}He-CO laser. Appl Phys Lett 39:139–142
87. Zherebtsov VA (1998) Nuclear pumping of a carbon monoxide laser. Tech Phys 43:818–823
88. Cooper GW, Verdeyen JT (1977) Recombination-pumped atomic nitrogen and carbon afterglow lasers. J Appl Phys 48:1170–1175
89. Karelin AV, Simakova OV (1998) Kinetics of the active media of nuclear-pumped lasers based on transitions in carbon and nitrogen atoms. Quantum Electron 28:307–312
90. Littlewood IM, Piper JA, Webb CE (1976) Excitation mechanisms in CW He-Hg lasers. Opt Commun 16:45–49
91. Prelas M, Anderson J, Boody FP, Miley GH (1977) A nuclear pumped laser using Ne-CO and Ne-$C0_2$ mixtures. In: 30th annual gaseous electronics conference, Palo Alto, CA. doi:10.13140/RG.2.1.3302.4483
92. Prelas M, Anderson J, Boody F, Miley G (1978) A nuclear-pumped laser using Ne-CO and Ne-CO_2 mixtures. In: APS Annual conference American Physical Society, College Park, MD, p 152
93. Prelas M, Miley G (1979) The Ar-CO_2 atomic carbon NPL. In: IEEE international conference on plasma science, New York, p 53. doi:10.13140/RG.2.1.4351.0248
94. Prelas MA (1979) Nuclear-pumping of the atomic carbon laser. PhD, Nuclear Engineering, University of Illinois, Champaign-Urbana, IL
95. Prelas M, Miley G (1978) Recent results with the atomic carbon laser @ 1.4539 micrometers. In: 31st gaseous electronics conference, Buffalo NY, p 117
96. Prelas MA, Anderson JH, Boody FP, Nagalingam SJS, Miley GH (1978) Nuclear pumping of a neutral carbon laser. In: Billman K (ed) Radiation energy conversion in space, vol 61. American Institute of Aeronautics and Astronautics, New York, pp 411–417
97. Prelas MA, Miley GH (1979) Kinetics of the atomic carbon nuclear pumped laser on the C (3p 1 P1) + C(3s 1 P1) + 1.454 μm transition with He + CO2 mixtures. In: Corcoran VJ (ed) International conference on lasers 79. Conference sponsored by the Society for Optical & Quantum Electronics, STS Press, MacLean, 1980. doi:10.13140/RG.2.1.3826.7362
98. Prelas M, Zediker M, Miley G (1980) Kinetics of the electrically-pumped atomic carbon laser. In: Annual conference of the American Physical Society, College Park, MD, p 111
99. Prelas M, Miley G (1980) Mechanisms of the atomic carbon nuclear-pumped laser @ 1.454 micrometers in He-CO_2 mixtures. In: Annual meeting of the American Physics Society, College Park, MD, p 111
100. Prelas M (1979) Nuclear pumping mechanisms in atomic carbon and in excimers. In: Nuclear-pumped lasers (SEE N80-13438 04-36). NASA Langley Research Center, Hapton, VA, pp 41–43
101. Prelas M (1979) Collisional processes in the He-CO_2 atomic carbon nuclear pumped laser 1.454 μm. Journal de Physique Colloques 40:C7-391–C7-392
102. Prelas MA, Miley GH (1981) Dynamics of the nuclear and electrically pumped 1.45-μm atomic carbon laser in mixtures of helium + CO advanced and helium + $C0_2$. Nucl Technol/ Fusion 1:402–413
103. Chung AK, Peck MS, Prelas MA (1991) Effects of helium buffer gas on the atomic carbon nuclear-pumped laser at low power densities. Laser Part Beams 9:829–839

104. Jalufka NW, Hohl F, DeYoung RJ, Williams MD (1978) Volumetric direct nuclear pumped laser, US Patent US4110703 A
105. Melnikov SP, Porkhaev VV (1995) Laser action on infrared chlorine and oxygen atomic transitions using uranium fission fragment pumping. In: Proceedings of SPIE 2619, International conference on atomic and molecular pulsed laser, 23 August 1995, pp 22–28. doi:10.1117/12.216920
106. Karelin AV, Simakova OV (1997) Kinetics of the active medium of a nuclear-pumped laser based on IR transitions in the chlorine atom. Quantum Electron 27:963–967
107. Verdeyen JT (1995) Laser electronics, 3rd edn. Prentice-Hall, Englewood Cliffs
108. Prelas MA (1987) Photolytic pumping of solid-state Nd_{3+} doped lasers using nuclear-driven flashlamps. Idaho National Engineering Laboratory, Idaho Falls, ID, Experimental design and testing of a nuclear-driven flashlamp to pump NdYAG, January 1987–December 1987. doi:10.13140/RG.2.1.1547.6329
109. Boody FP (1991) Nuclear-driven fluorescence pumped solid-state lasers. PhD, Nuclear Engineering, University of Missouri, Columbia, MO, USA
110. Prelas MA (1991). In: DOE/ER/13029-T3 (ed) Remote pumping of solid-state lasers with nuclear driven fluorescers. Department_of_Energy, Idaho Falls, ID, pp 1–35. doi:10.13140/RG.2.1.1285.4881
111. Lin LTS (1994) Microwave and nuclear excitations of Alkali metal vapors. PhD, Nuclear Engineering, University of Missouri
112. Prelas MA, Lin LT, Bahns JT (1994) Progress in nuclear driven flashlamps. In: Proceedings of the topical meeting on physics, safety, and applications of pulse reactors. Washington, DC, November 13–17, 1994 pp 155–162
113. Swingle JC, Turner CE, Murray JR, George EV, Krupke WF (1976) Photolytic pumping of the iodine laser by XeBr. Appl Phys Lett 28:387–388
114. Lawrence_Livermore_National_Laboratory_Energy_and_Technology_Review (1975) Lasers and laser applications. U.S. Department of Energy, 11/1975. Available: https://lasers.llnl.gov/multimedia/publications/pdfs/etr/1975_11.pdf
115. Miley GH, Boody FP, Nagalingham SJS, Prelas MA (1977) Production of XeF(B-X) by nuclear-pumping, APS conference. Bull Am Phys Soc 23(2):117
116. Wilson JW (1980) Nuclear-induced XeBr* photolytic laser model. Appl Phys Lett 37:695–697
117. Prelas MA (1980) Nuclear pumped energy focus. In: 1980 IEEE international conference on plasma science, vol 80CH1544-6. doi:10.13140/RG.2.1.4088.8801
118. Miley GH, Nagalingham SJS, Boody FP, Prelas MA (1978) Production of XeF(B) by nuclear-pumping. In: Corcoran VJ (ed) Proceedings of the international conference on lasers 78, Orlando, Florida. Society for Optical & Quantum Electronics, STS Press, Maclean, 1979

Chapter 5
Reactor and Laser Coupling

Abstract This chapter discusses the design of how nuclear reactors are coupled to lasers. The design is complex in that there are many ways to interface the reactor, which serves as a neutron source, to the laser medium. There are issues with the fuel interface to the laser that includes the phase, the type of fuel, how the fuel is put into proximity to the laser medium and the potential chemical interactions of the fuel with the laser medium. The coupling of the optics to the laser medium is also an important consideration and how radiation damage may impact the optical components, structural materials and subsystems. There are also cooling and control issues as well.

Keywords Space based nuclear reactors • Advanced nuclear fuels • Nuclear pumped lasers

5.1 Introduction

The design of nuclear reactor/nuclear-pumped laser systems is complicated and is dependent on a number of factors. One of the critical issues is the type of laser medium being used in the design. For example, the nuclear fuel/laser interface is dependent upon the phase of the fuel and the phase of the laser medium. The phase of the laser medium is limited. Two types of nuclear-pumped laser media have been demonstrated experimentally; solid-state and gas. The properties of the laser medium, such as threshold pumping power density, are critical. The nuclear fuel interface with the laser medium is an important consideration. The phase of the fuel, the fuel geometry, the type of fuel and the fuel characteristics play an important role. The range of the reaction products in the laser medium dictates the size of the laser cell and the reactor configuration. If the reactor is built from a series of unit cells where each cell is a stand-alone optical cavity/laser oscillator, then there could potentially be hundreds of laser beams being produced by the reactor. This complicates the output beam which combines all of the beamlets into one large beam. This problem means that attention must be paid to the phase matching of the beamlets in order to produce a Gaussian limited output beam. The thermal characteristics of the core are important since temperature does impact laser efficiency. Each has advantages and disadvantages which impact the design of a nuclear reactor/nuclear-pumped laser.

M. Prelas, *Nuclear-Pumped Lasers*, DOI 10.1007/978-3-319-19845-3_5

The phase of the laser medium and the fuel has eight different combinations where the laser medium or the fuel can be a gas, a liquid or a solid. There is even the possibility that the fuel can be in the plasma phase [1]. Each potential reactor/laser phase coupling combination is discussed in this chapter (with the exception of the plasma phase). There are also phase combinations that are not easily classified. For example it is possible to have a solid fuel (e.g., in the form of submicrometer to micrometer aerosol particles) that behaves like it is in the gas phase.

5.2 Engineering a Nuclear-Pumped Laser System

As described in Chap. 1, the steps in assessing a nuclear-pumped laser system are dependent upon many factors such as:

- The ionizing radiation used for excitation (e.g., ions, electrons, gamma rays and neutrons);
- The range of the ionizing radiation in the laser medium;
- The interface of the ionizing radiation source to the laser;
- The trigger that causes the emission of ionizing radiation (i.e., spontaneous emission such as that from a radioisotope or a neutron capture reaction where a nuclear reactor is used as the neutron source or possibly another type of neutron source such as a fusion reactor);
- The properties of the laser medium (such as photon absorption);
- The efficiency of the laser;
- Radiation damage issues which can limit the laser efficiency or laser lifetime;
- Operating temperature (which can impact the laser efficiency);
- The optical system which extracts the energy.

There are efficiencies for each step in the transfer of energy from the power source to the laser medium. The engineering details and power transfer in a laser system are complicated. For example in Fig. 5.1, the steps involved in delivering and extracting energy from an electrically excited laser is shown. The process begins with the generation of electricity at the power plant. Depending on the type of power plant, the efficiency of producing current and voltage could have a wide variation. For example, standard coal and nuclear power plants are about 33 % efficient while advanced power plants such as generation IV nuclear reactors will be 40 % or greater. Solar photovoltaics, a renewable energy source, on the other hand are on the order of 5 % efficient. These considerations determine the power plant efficiency (for this example $\eta_{Power} \sim 33$ % for a coal power plant). Once the electricity is generated, it is transmitted over high voltage lines to the laser. Then the voltage needs to be stepped down by transformers before the current reaches the wall plug. From the wall plug to the laser, the efficiency of the pumping method will have variations (e.g., electron beam pumping efficiency is ~90 %, AC or DC discharge efficiencies ~70–90 %, RF discharge ~60–90 %, microwave discharge ~90–100 %, photolytic pumping ~90–100 %, etc.). The transmission of power to

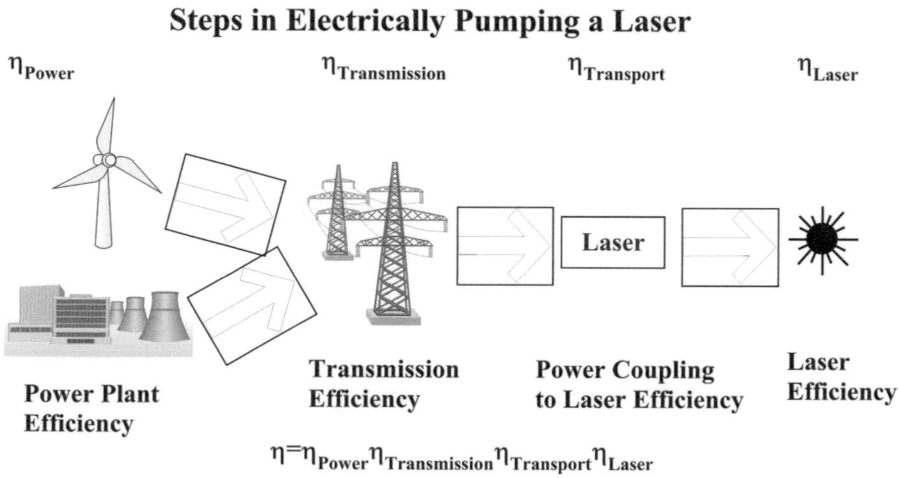

Fig. 5.1 The steps in flow of energy in an electrically pumped laser from power plant to beam output are shown. The efficiency for each step is illustrated and begins with the production of electricity from the power plant (η_{Power}), the transmission of the electricity to the wall plug ($\eta_{\text{transmission}}$), the efficiency of coupling electrical power to the laser medium ($\eta_{\text{transport}}$) and the efficiency of the laser (η_{Laser})

the wall makes up the transmission efficiency in Fig. 5.1 (for this example $\eta_{\text{transmission}} \sim 80$ %). Once the power is reaches the laser, one must consider the method of transporting electrons to the laser media like a DC discharge (for this example $\eta_{\text{transport}} \sim 70$ %). The conversion of the energy deposited in the laser medium to coherent radiation will also have wide variations depending upon the laser medium (for this example $\eta_{\text{Laser}} \sim 3$ % is used and is based on an excimer laser). The system efficiency of the electrically driven laser ($\eta = \eta_{\text{Power}}\, \eta_{\text{transmission}}\, \eta_{\text{transport}}\, \eta_{\text{Laser}}$) is around 0.55 %.

In contrast to the electrically driven laser, nuclear-pumped lasers have a simplified energy flow. The most basic type of nuclear-pumped laser uses thin fuel coatings (e.g., ^{235}U or ^{10}B) for the power producing portion of the laser (Fig. 5.2). The fuel interacts with a neutron flux produced by a nuclear reactor. In Fig. 5.2, the first step in the energy flow is the production of the neutron. As discussed in Chap. 4, the energy cost of producing a neutron can be high if the laser is decoupled from the reactor. If the laser is not part of a self-critical reactor/laser core, the energy cost of producing a neutron to trigger a capture event is approximately 400 MeV and this energy loss needs be taken into account when calculating the system efficiency. A laser decoupled from the reactor will waste 99.8–99.9 % of the energy from nuclear fission. If the fuel is part of a self-critical reactor/laser configuration, then the fission fragments are used to directly excite the laser medium. The fast neutrons produced from the fission reaction then trigger additional fission events. The efficiency of producing fission fragments from a neutron capture event is limited to 80 % because that is the fraction of energy carried off by the fission fragments in a fission reaction and is illustrated in Fig. 5.2 as the neutron to fission fragment efficiency ($\eta_{\text{neutron to ff}}$). The remaining 20 % of

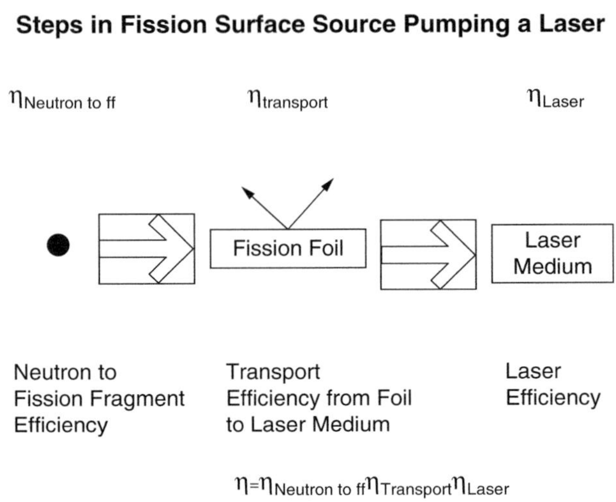

Fig. 5.2 The energy flow in nuclear pumping a laser with a solid-state fission source is shown. The first step begins with the neutron capture event and the efficiency of the nuclear reaction in producing energetic ions ($\eta_{\text{Neutron to ff}}$). The second step leads to the efficiency of the ions escaping the solid fuel medium to excite the laser medium ($\eta_{\text{transport}}$). The third step leads to the efficiency of the laser (η_{Laser})

the energy is carried off by neutrons, gamma rays and beta particles. If the uranium fuel is coated on a flat structural plate, then the fission fragment transport efficiency ($\eta_{\text{transport}}$) is about 25 %. The most efficient nuclear-pumped laser is the atomic xenon laser at 2030 nm with η_{Laser}. ~ 3 %. If a uranium coated plate is assumed then the system efficiency of the surface source nuclear-pumped laser ($\eta = \eta_{\text{Neutron to ff}}$ $\eta_{\text{transport}}$ η_{Laser}) is around 0.6 %.

A volume source driven nuclear-pumped laser has a minimum number of steps in the flow of energy to the laser as shown in Fig. 5.3. In this case the fuel is mixed in with the laser medium. The efficiency of producing fission fragments from a neutron capture event is limited to 80 % because that is the fraction of energy carried off by the fission fragments in a fission reaction and is in Fig. 5.3 as the neutron to fission fragment efficiency ($\eta_{\text{neutron to ff}}$). The remaining 20 % of the energy is carried off by neutrons, gamma rays and electrons. If the fuel is in a self-critical reactor configuration, then the fission fragments can be used to directly excite the laser medium. If the fuel is a fine aerosol suspended in the laser cavity then the fission fragment transport efficiency ($\eta_{\text{transport}}$) is about 80 %. If the fuel is UF_6 or uranofullerene mixed with the laser medium than the fission fragment transport efficiency ($\eta_{\text{transport}}$) is about 100 %. The most efficient nuclear-pumped laser is the atomic xenon laser at 2030 nm with η_{Laser}. ~ 3 %. If a uranofullerene is mixed with the laser medium, then the system efficiency of the surface source nuclear-pumped laser ($\eta = \eta_{\text{Neutron to ff}}$ $\eta_{\text{transport}}$ η_{Laser}) is around 2.4 %.

A nuclear-pumped flashlamp driven laser [2] has an energy flow as shown in Fig. 5.4. In this case the fuel is mixed in with the flashlamp medium. Photons are created in the flashamp medium and then are directed to the laser medium where

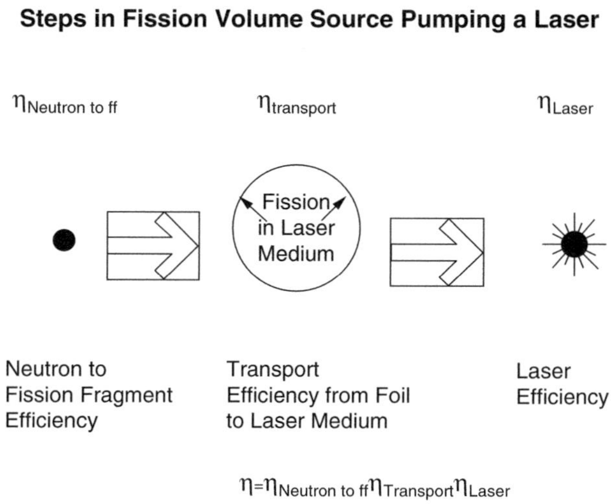

Fig. 5.3 The energy flow in nuclear pumping a laser with a volume fission source is shown. The first step begins with the neutron capture event and the efficiency of the nuclear reaction in producing energetic ions ($\eta_{\text{Neutron to ff}}$). The second step is the efficiency of the ions creating excitation in the laser medium -with a volume source this efficiency is 100 % ($\eta_{\text{transport}}$). The third step is creating the laser beam and the efficiency is indicated by the term η_{Laser}

they are absorbed. The first step is to produce fission fragments which begin with a neutron capture event to create excitation of the flashlamp medium. The efficiency of producing fission fragments is limited to about 80 % because that is the fraction of energy carried off by the fragments in a fission reaction. Figure 5.4 indicates the neutron to fission fragment efficiency ($\eta_{\text{neutron to ff}}$) as the first step. Assuming that the fuel is in a self-critical reactor configuration, then the fission fragments can be directly used to excite the fluorescer medium as shown in step 2. For example, if the fuel is a fine aerosol suspended in the laser cavity then the fission fragment transport efficiency ($\eta_{\text{transport}}$) is about 80 %. If the fuel is uranium hexafluoride gas or uranofullerene particles mixed with the laser medium then the fission fragment transport efficiency ($\eta_{\text{transport}}$) is about 100 %. The third step is production of the fluorescer photons, The fluorescence efficiency is defined as the fraction of deposited energy that goes into the production of photons (η_f) and is about 50 % for the xenon excimer or about 15 % for the XeBr excimer [3]. The fourth step is transporting the photons to the laser. The photon transport efficiency is defined as the fraction of photons produced by the lamp that are absorbed by the laser medium (η_{phTr}) and is about 80 % for a well-designed transport system. The fifth step is producing the laser. A highly efficient photolytic laser, such as iodine at 1315 nm, has an efficiency of η_{Laser}. ~ 30 % [4]. If a uranofullerene is mixed with the fluorescence medium, then the system efficiency of the nuclear-pumped laser ($\eta = \eta_{\text{Neutron to ff}}\ \eta_{\text{transport}}\ \eta_f\ \eta_{\text{phTr}}\ \eta_{\text{Laser}}$) is around 2.9 %.

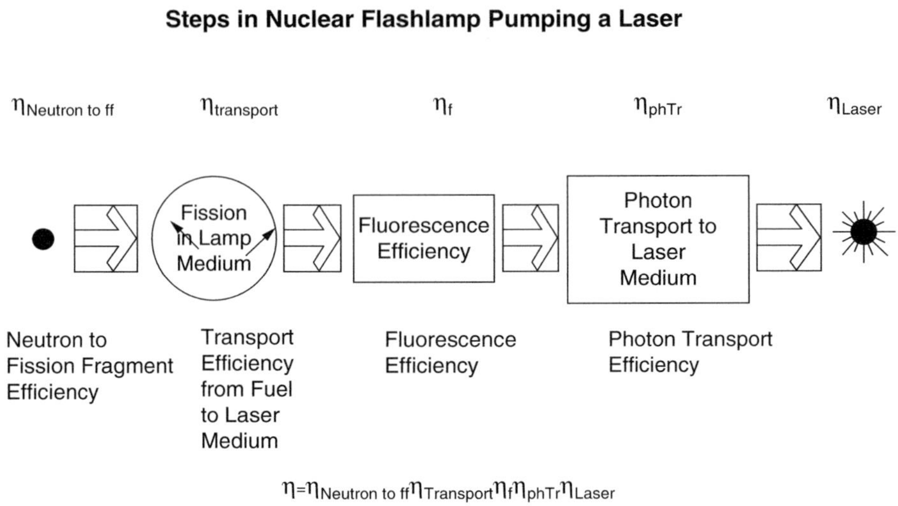

Fig. 5.4 The energy flow in nuclear pumping a flashlamp with a volume fission source and transporting the light to photolytically driving a laser is shown. The first step begins with the neutron capture event and the efficiency of the nuclear reaction in producing energetic ions ($\eta_{Neutron\ to\ ff}$), the second step is transporting the ions to the fluorescer medium -for a volume source fuel this efficiency is essentially 100 % ($\eta_{transport}$), the third step is the efficiency of the ions exciting the fluorescer to produce photons (η_f), the fourth step is transporting the photons to the laser medium (η_{phTr}) and the final step is the energy efficiency of conversion absorbed power to laser light (η_{Laser})

5.3 Laser Medium (Phase)

Like any material, the laser media phase can be a gas, liquid, solid or plasma. We will limit our discussion of laser media phases to gas, liquid or solid because lasers in the pure plasma phase are not common. Moreover, nuclear-pumped laser experiments have succeeded by using laser media in the gas or solid phases as described in Chap. 4.

5.3.1 Gas

Gaseous media are the most common phase used in nuclear-pumped laser research. As discussed in Chap. 4, numerous gaseous phase lasers have been demonstrated (e.g., numerous lines on rare gases, lines in nitrogen, lines in oxygen, lines in carbon and a line in Iodine). One of the most desirable types of gaseous lasers which have been pursued in nuclear-pumped laser research are excimer lasers. Experiments with the XeF excimer line at 351 nm showed initial promise in that gain (of ~ 0.01 %/cm) was measured in a nuclear-pumped laser experiment at the University of Illinois using a TRIGA reactor in 1978 [5–7]. Lasing was never achieved in subsequent studies at Sandia National Laboratory in the mid 1980's using a fast burst reactor (where a gain of 0.7 %/cm was observed) [8]. Excimer lasers require a high threshold power density as discussed in Chap. 1 and this is an important reason why these experiments did not lead to a laser.

5.3.2 *Liquid*

Liquid media are more complex given that liquid lasers typically have elements that are dissolved in a liquid solvent which are responsible for the laser transition. Liquids have advantages such as radiation damage resistance and are capable of operating at higher power density (due to enhanced cooling properties). Examples of a common liquid laser are dye lasers [9]. In early nuclear-pumped laser studies, a solution of $Eu^{3+}(TTA)_3TBPh$ in toluene was used [10]. The medium was excited by fission fragments from the fission of the ^{235}U isotope in the form of $UO_2(NO_3)_2$ which was dissolved in toluene. A pulsed reactor (1.17 ms pulse with peak thermal neutron flux of 5.6×10^{15} n cm^{-2} s^{-1}) was used for the neutron source and produced strong fluorescence at 615 nm. The solution (or gain medium) was placed in an optical cavity and a small signal gain coefficient of 0.033 cm^{-1} was measured. The medium did not lase due to high transient optical losses in the solution. What makes a liquid-liquid interface desirable is the direct interface of the fuel with the laser media thus making the energy transport process more efficient.

A series of experiments on the $Eu^{3+}(TTA)_32P$ compound were carried out in the Former Soviet Union using the TIBR reactor using various solvents (Ditolyl methane, α-Methylnapthalene, Benzol, Toluene, Xylene, Acetophenol, Acetone, Cyclohexane, Chloroform and Tetrachloride Carbon) [11]. Fluorescence studies of the $^5D_0$ line in Eu^{3+} were completed. A laser experiment was performed on a 30 cm long and 0.6 cm diameter cell filled with $Eu^{3+}(BTFA)_4HDPh$ with an acetone solvent using the neutrons and gamma rays from the TIBR reactor as the pump source. The laser cavity used silver coated mirrors (one with a transmission coefficient of 1.7 % for the output mirror). The cavity was cooled to 200 K by liquid nitrogen vapor. The laser exhibited what was believed to be superluminescence but the evidence did not indicate lasing. The failure to lase was attributed to high cavity losses.

Luminescent studies of inorganic liquids doped with europium and neodymium using fission fragments as the pump source and the VIR-2 pulsed reactor as the neutron source were performed in the Former Soviet Union [11]. The mixtures used included $POCl_3$-$SnCl_4$-$^{235}UO_2^{2+}$-$Nd^{3+}(Eu^{3+})$, $ZnCl_2$-$GaCl_3$-$^{235}UO_2^{2+}$-$Nd^{3+}(Tb^{3+}, Mn^{2+})$, $SOCl_2$-$GaCl_3$-$^{235}UO_2^{2+}$-Nd^{3+}, SO_2Cl_2-$GaCl_3$-$^{235}UO_2^{2+}$-Nd^{3+}, and D_2O-$^{235}UO_2^{2+}$-$Nd^{3+}(Tb^{3+})$. The studies provided information on the spectra, conversion efficiencies (up to 1.4 %) and lifetimes of the upper laser level of the medium, but lasing was not reported.

5.3.3 *Solid*

As discussed in Chap. 4, $Y_3Al_5O_{12}Nd^{3+}$ was excited by photons from a nuclear driven flashlamp (based on mixtures of ^{3}He:Ar, ^{3}HeKr and ^{3}He:Xe) [12]. These experiments demonstrated a very large absolutely calibrated output from a nuclear driven flashlamp within the absorption band of Nd^{3+} that was larger than the comparable spectral output from a xenon arc lamp. The emission from the $Y_3Al_5O_{12}Nd^{3+}$ crystal agreed with the model studies and indicated that the lasing threshold for the crystal was exceeded. Experiments in the Former Soviet Union were performed with a fiber neodymium laser pumped with a xenon flashlamp [11]. In this work the fiber was placed inside the lamp between two plates (10 cm by 6 cm) coated with a thin layer of uranium oxide-protoxide. The thickness of the metallic uranium layer was 5 mg/cm^2. The plates were separated by 2 cm. The fiber laser, 1–4 m long with a fiber diameter of 40–50 μm, was wound around a frame and placed symmetrically between the uranium layers. The neodymium was doped into phosphate glass. A xenon fill of 05 atm was used (creating a plasma temperature of 4,700 K). The maximum laser radiation produced in a single pulse was 2.5 J.

Experiments using a solid-state lamp based on a CsI(Tl) crystal was used to drive a neodymium laser [11]. The CsI(Tl) crystal used the BR-1 pulsed reactor to produce the luminescence. The maximum laser power output was 4 kW in a pulse with 60 microsecond duration.

A YAP:Tm^{3+} crystal (with wavelengths between 1.85 and 2.0 μm) was pumped by laser radiation from an argon nuclear-pumped laser (1.79 μm) and a xenon nuclear-pumped laser (1.73 μm) [11]. The laser crystal was 0.5 cm long with a volume of 4 mm^3. The nuclear-pumped lasers used the EBR-L reactor. The conversion efficiency of the argon or xenon laser emission to laser emission from the crystal was 4 %.

E-beams have been used to pump a number of semiconductor lasers (e.g., CdS, InSb, InAs, GaAs, etc.) [13]. The pumping of semiconductor lasers by ions generated in nuclear reactions was discussed by Watermann and Prelas [14]. The problem which occurs is the displacement of atoms within the semiconductor crystal by the high energy ions. Using a combination of self-annealing at high temperatures to help mitigate the displaced atoms and a limited exposure time, it was argued that reactor systems could be coupled with semiconductors to produce a limited duration reactor/laser design.

5.4 Phase of Fuel

The reactor fuel can be in the solid, liquid or gas phase. There are many permutations on how to couple a fuel to a laser medium and these will be discussed in Sect. 5.5. This section will focus on fuel in the gas, solid or liquid state.

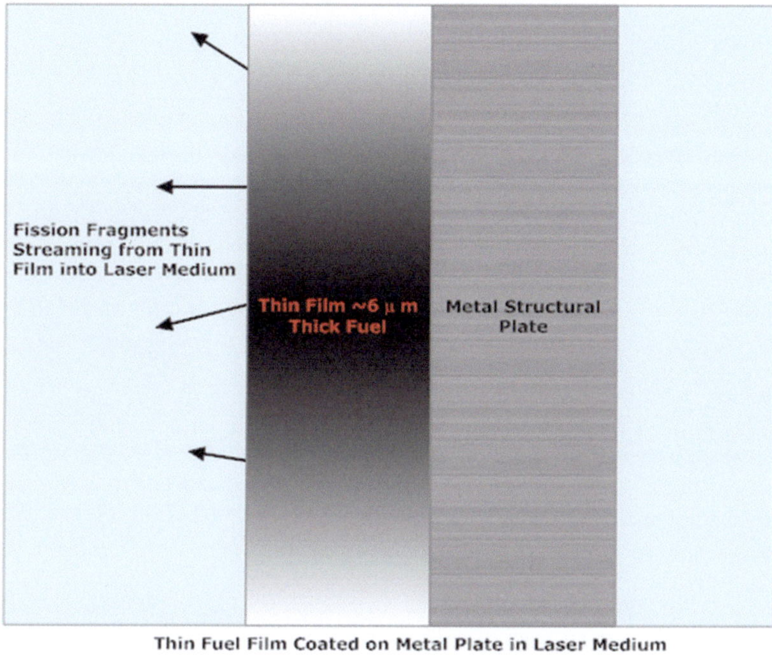

Fig. 5.5 A thin film of fuel (uranium metal, uranium oxide, uranium carbide or boron-10) coated on a metal structural plate

5.4.1 *Solid Phase*

Typically fuel (e.g., uranium 235) is in the solid phase. For example uranium can be in the form of uranium metal, uranium oxide or uranium carbide. The uranium is typically coated on a surface as a thin film (Fig. 5.5). The film is coated on a plate thus limiting the transport efficiency of the fission fragments to the laser medium as described in Sect. 5.2. In this configuration, half of the energy contained in the fission fragments goes into the plate. The remaining half of the energy is deposited in the layer of uranium (self-absorption) and in the laser medium (the transport efficiency is on the order of 20–30 %) [15, 16]. One possible way to improve the transport efficiency is to use the emission of fission fragments from both sides of the uranium layer. This can only happen if the uranium is coated on a very thin membrane made of a material such nano-carbon fibers woven into flat rectangular membrane (Fig. 5.6). Then, the fission fragments from both sides of the layer can be transported to a laser medium thus improving the efficiency (the transport efficiency for a membrane will be on the order of 50–60 %) [15, 17]. It is possible to further improve the geometrical coupling by coating thin fibers with fuel (Fig. 5.7).

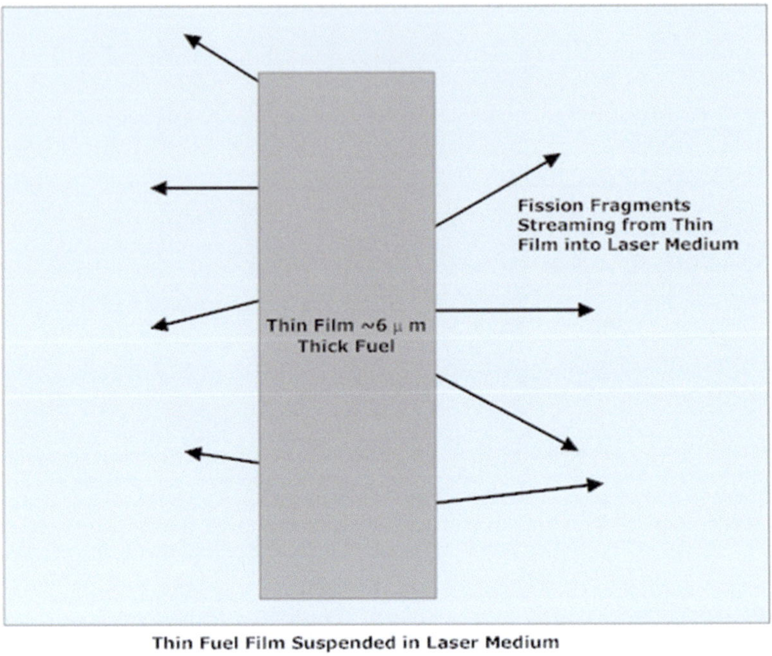

Fig. 5.6 A very thin membrane coated with fuel which allows fission fragments to escape from both the *left* side and *right* side to the laser medium

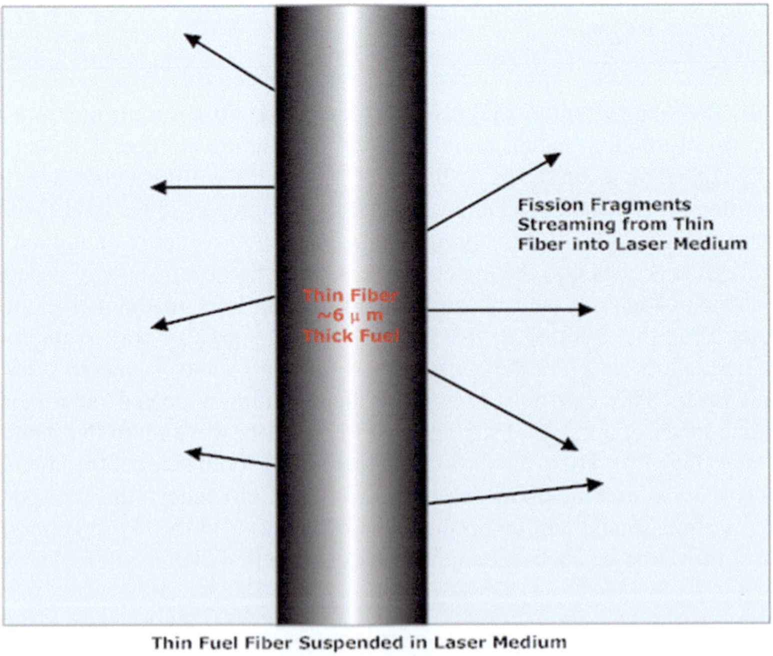

Fig. 5.7 A thin fiber with fuel suspended in the laser medium

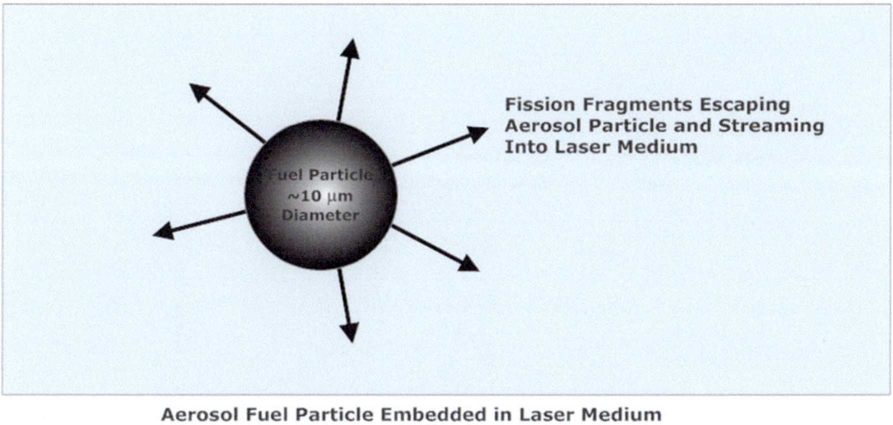

Fig. 5.8 A fuel particle in the form of an aerosol suspended in the laser medium [18, 19]

A thin fiber can improve the transport efficiency even more (60–70 %) [17]. A spherical geometry (e.g., aerosol) is the optimal route to improving the transport efficiency [18, 19]. If the fuel is in the form of an aerosol (Fig. 5.8), the transport efficiency would be on the order of 70–80 % [15, 19].

5.4.2 *Liquid Fuel*

Some forms of uranium salt (uranium sulfate ($U(SO_4)_2$) and uranium nitrate ($UO_2(NO_3)_2$)) can be dissolved in water to form an aqueous core reactor [2]. Uranium can also form liquids at high temperatures. An example being uranium dissolved in a fluoride salt to form molten NaF-ZrF_4-UF_4 [20].

5.4.3 *Gaseous Fuel*

Gaseous uranium, namely uranium hexafluoride (UF_6), has been of interest as a potential fuel for a nuclear-pumped laser almost from the beginning of the field [5]. With a gaseous fuel, the fission fragment transport efficiency would be 100 % (Fig. 5.9). In Chap. 1, some problems with uranium hexafluoride were discussed. These problems include corrosiveness and self-absorption of photons by the uranium hexafluoride molecule.

In 1990, when uranofullerenes were discovered, the possibility of incorporating them into a reactor as part of a gaseous-like core was proposed [21] (Fig. 5.10). A

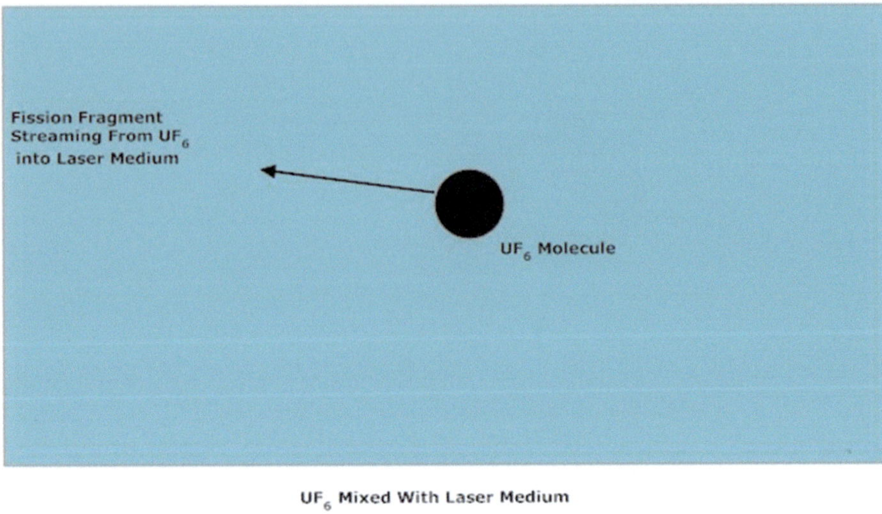

Fig. 5.9 Fission fragment transport from a UF_6 molecule to the laser medium

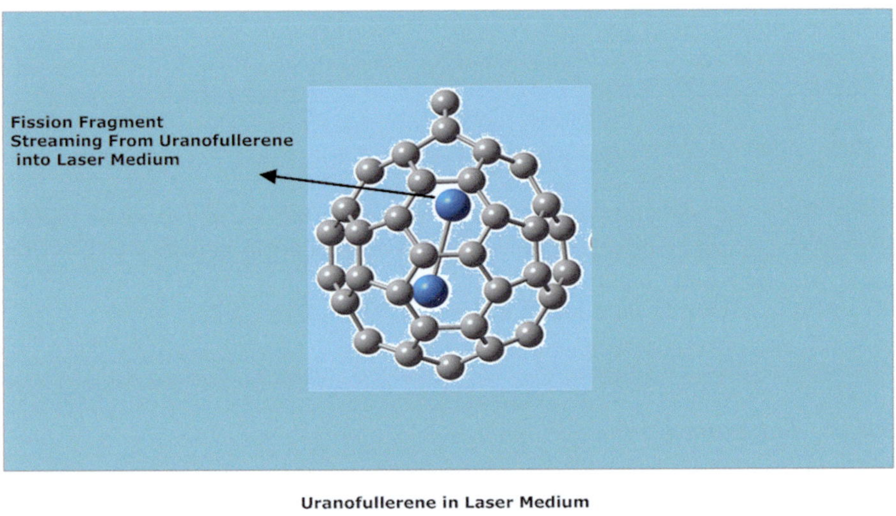

Fig. 5.10 Uranofullerene embedded in a laser medium showing a fission fragment escaping the uranofullerene and reaction with the laser medium

uranofullerene is a carbon-60 cage with two uranium atoms trapped in the cage (as discussed in Chap. 1). The uranofullerene is a large molecular-like structure which can behave like a gas. Much is unknown about the properties of uranofullerenes but there is some promise that the material will not be corrosive and may have limited self-absorption problems. The fission transport efficiency of a uranofullerene is 100 %.

5.5 Reactor/Laser Coupling

The coupling of the laser medium with the fuel is a critical component of nuclear-pumped laser design. It combines the need for high transport efficiencies with practicality. As will be discussed, only a limited number of combinations of laser medium phase and fuel phase are practical. The coupling can be direct, where the fission fragments directly excite the laser medium, or indirectly, where the fission fragments drive a flashlamp which provides photons for pumping a laser medium.

5.5.1 *Solid/Gas (Most Common)*

Historically, the easiest method of interfacing a solid fuel to a gaseous reactant has been to deposit a thin layer of fuel on a metal substrate and to form channels with the fuel/substrate components so that the gaseous reactant can flow through the channels in direct contact with the fuel. Since 50 % of the fission fragments interact with the substrate rather than the reactant and the remaining 50 % lose some energy travelling through the fuel, this technique is only 20–30 % efficient in transporting nuclear energy to the reactant [15]. One of the main advantages of this technique is that the solid fuel is passive to the kinetics of the gaseous reactant material [22]. A method of maintaining the inherent passivity of the solid-gas interface while improving the energy transport efficiency to the laser medium is to mix the gaseous reactant material with fuel in the form of micropellets [18]. Small spherical microspheres of fissile materials (^{235}UC, $^{235}UO_2$, and $^{235}U_3O_8$) have been shown to be between 60 and 90 % efficient in transporting fission fragment energy to a reactant [15, 19]. It has been shown that a reactor fueled with fissile micropellets can be constructed in the power reactor design presented by Martin 1964 [23]. Furthermore, a chemical processing system using a solid microsphere-gaseous reactant interface has been described in Conner and Davis 1962 [24]. Similarly, a method of placing micropellets in a porous catalytic material to interface with a gaseous reactant was presented by Fellows in United States Patent No. 3228848 [25]. The advantage of the catalytic surface is in the increased yield of chemicals: such as NO_2 from N_2O using microporous alumina [26]. The use of uranium aerosol particles as an excitation source for a coupled reactor/photolytic driven laser system was developed in 1981 [18].

The first nuclear-pumped laser discovered utilized a thin coating of $^{235}UO_2$ inside of a cylindrical tube that created the interface of the fuel to the laser medium-known as a surface source [27]. The passivity and inertness of this solid fuel-gaseous laser interface is a significant advantage which allows many types of laser media to be used. Another type of surface source uses a ^{10}B coating on the inside of a cylindrical tube and the $^{10}B(n, \alpha)Li$ reaction was used for excitation. The ^{10}B coating was originally developed for neutron detectors and was adapted to nuclear-pumped laser work due its simplicity. ^{10}B is not useable for a coupled reactor/laser design because it is a neutron sink which requires about two fission events for every $^{10}B(n, \alpha)Li$ reaction (400 MeV lost for every 2.35 MeV produced). The crucial results of the investigations of solid fuel-gaseous reactant interfaces are: (1) the solid-gas interface is passive to the chemical kinetics of the reactant; (2) a surface source is limited in efficiency; (3) Microspheres mixed with reactant are both passive and efficient; and (4) a neutron multiplying fuel, ^{235}U, is very important for the system efficiency of the coupled reactor/laser.

5.5.2 Solid-Liquid

In British Patent No. 770594 [28], published March 20, 1957, it is shown that a large number of chemical reactions may be initiated by fissioning uranium, which has been dispersed in solution or in very finely divided form throughout the reactants. This method of interfacing solid fission fragments with organic reactants in the liquid phase has produced molecular fragments which then recombine to form desired compounds. Examples of this process were cited: the conversion of methanol to ethylene glycol and formaldehyde, conversion of ethanol to mixed butanediols, conversion of acetic acid to succinic acid and of isobutane to iso-octane. The main advantage of a solid-liquid interface is in the increased LET which leads to higher local dose rates, and the increased local molecular density which leads to higher local reaction rates. There have not been any nuclear-pumped laser experiments based on solid/liquid interfaces.

5.5.3 Solid-Solid

Some work has been done on solid fuel/solid reactant interfaces. An example would be the irradiation of dry ice (solid CO_2) with fission fragments [29]. The results of these studies indicate that chemical processing in the solid phase of CO_2 is less efficient than the gaseous phase. The major advantage of a solid-solid interface is the higher local dose rates and the higher local reaction rates that are possible. No nuclear-pumped laser experiments have been reported using a solid-solid interface. However it is conceivable that crystals can be doped with ^{235}U atoms. Thus the embed uranium in the solid-state laser medium would produce a solid-solid interface.

5.5.4 Liquid-Gas

A liquid fuel interfaced with a gaseous laser medium is a very difficult combination thus no examples of this combination in radiolysis or nuclear-pumped laser research is known. One can imagine using droplets of a liquid form of fuel in a gas flow (laser medium). Dual phase systems work better in an indirect pumping scheme because of possible absorption and scattering losses of laser light.

5.5.5 *Liquid-Liquid*

Liquid core reactors have been operational as research reactors for some time [30]. Some of the potential applications of a liquid core reactor in chemical processing have been discussed in British Patent No. 770,594 [28], published March 20, 1957. Specifically this has been delineated in case 2 of this patent. The major advantage of the liquid core concept is in the high transport efficiency of the fission fragments to the reactant material in addition to the potentially high dose rates and high reaction rates. As described previously, a liquid-liquid interface was formed using a solution of $Eu^{3+}(TTA)_3TBPh$ in toluene [10] with dissolved $UO_2(NO_3)_2$. The experiment did not demonstrate lasing, but the principle of a liquid-liquid interface was used.

5.5.6 *Liquid-Solid*

There are no examples of liquid-solid interfaces. However one can imagine a solid-state laser medium suspended in an aqueous core reactor as a possible configuration. The system efficiency of this type of configuration has no advantages which would provide a motivation to develop such systems.

5.5.7 *Gas-Gas*

It was recognized early on that in order to build large scale energy conversion systems, an efficient method of interfacing the fuel to the reactant had to be developed. One of the earliest concepts was the gaseous core reactor in which a mixture of $^{235}UF_6$ gas and the reactant were combined [22]. The first demonstration of a gas-gas interface was with the He + Ar laser in ^3He gas was used rather than ^4He gas [31]. ^3He however, is a neutron sink (meaning 400 MeV has to be expended in a fission reactor in order to provide one neutron to trigger the reaction). Thus ^3He can never be used efficiently in a large scale system. This implied that a fissile fuel such as $^{235}UF_6$ (or some other gaseous neutron multiplying fuel) has to be used in a system using a gas-gas interface. Unfortunately, $^{235}UF_6$ is harmful to both the kinetic processes in the reactant and the optical properties of the medium [22]. The advantage in gas-gas interfaces is the high energy transport efficiency to the reactant material. Uranofullerenes have gaseous-like properties and may not be as active as $^{235}UF_6$. Thus uranofullerenes embedded in a laser medium may be a workable combination.

5.5.8 Gas-Liquid

There are no examples known of gas-liquid interfaces. One can imagine bubbling a gaseous fuel through a liquid. However, there are no advantages of a gas/liquid interface which could be sufficient to generate interest in such systems.

5.5.9 Gas-Solid

Gas/solid interfaces have not been studied. It is possible to envision a gaseous fuel flowing around a solid-state laser medium. There are no apparent advantages in using a gas/solid interface.

5.5.10 Indirect Coupling (Photolytic)

In systems where the phase of the fuel and the laser medium are different, there would be concerns that the scattering cross section for laser beam photons at the interface of the fuel embedded in the laser medium would be a significant loss factor in the gain of the laser medium. This disadvantage can be overcome by using an indirect coupling technique. Such a technique was described in the Aerosol Reactor Energy Conversion Systems (ARECS) [32, 33]. The ARECS used fuel in the solid phase in the form of an aerosol that was suspended in a gaseous medium which served as a nuclear driven fluorescer. The transport of photons through a dusty plasma (created by the interaction of fission fragments with the fluorescer gas) was studied [33–35]. It was shown in these studies that photons can be efficiently transported in a dusty plasma where aerosol particles represent absorption and scattering centers for the photons provided that the aerosol density and size distribution were in a specific range of values. It was also shown that in a self-critical reactor core, that the necessary aerosol density and size distribution to sustain criticality was within the useable parameter range to allow efficient transport of photons from the fluorescer gas to the laser medium.

Another type of indirect coupling involved studies of liquid scintillators with ^{235}U dissolved in the liquid have also been studied in the Former Soviet Union [11]. The scintillator served as a photon source and photons were transported out of the scintillator to a laser medium.

5.6 Reactor Criticality

A reactor is a complex system where neutrons produced in the fission of ^{235}U are born with energies in the MeV range (Fig. 5.11). The highest cross section for fission of ^{235}U occurs at neutron energies below 1 eV (Fig. 5.12). Thus the most advantageous situation occurs when the neutrons born in fission are "slowed down" to thermal energies. A reactor in which neutrons are "slowed down" to thermal energies is known as a thermal reactor. As discussed in reference [36], one can consider neutron production to occur in generations. Suppose that a thermal neutron triggers a ^{235}U fission event and, as discussed in Chap. 2, the statistical number of fast neutrons emitted is 2.44. The k value is defined as the number of thermal neutrons which remain as the neutrons slow down to trigger the next generation of fission events. When k is less than 1, the neutron population dies out in subsequent generations. When k is equal to one, the neutron population is constant in subsequent generations. When k is greater than one, the neutron population geometrically increases from generation to generation. In order for a nuclear reactor to operate, its k value must be larger than one. This allows the operator to use the insertion or removal of a neutron absorber (in the form of a control rod) to increase and decrease power levels. It gives the reactor excess reactivity (having the capability of exceeding k equal to 1) to overcome the formation of neutron absorbers (poisons) created in fission. The modeling of the process of neutrons being born at a high energy, moving through a collection of materials (fissile fuel, structural, coolant, absorbers, moderator, etc.) is both dependent upon the variables of energy, space and time. This is extremely complex to model. If the neutrons are in balance within the reactor ($k = 1$), the rate at which the neutrons of a given energy range leak out of a

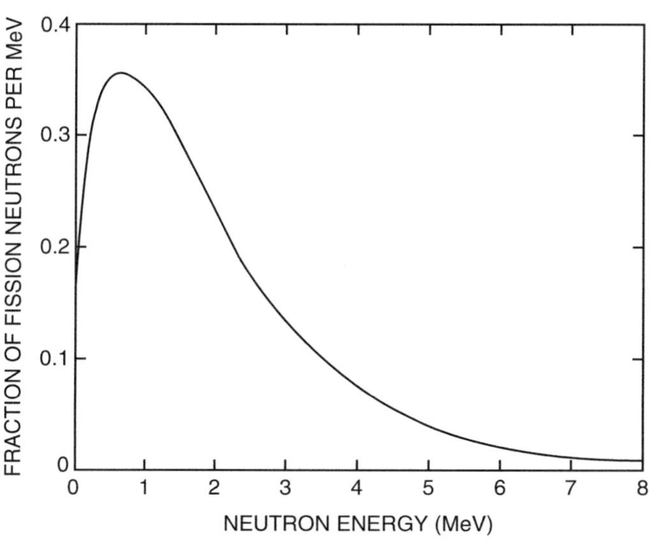

Fig. 5.11 Prompt neutron energy spectrum from the fission of ^{235}U [39]

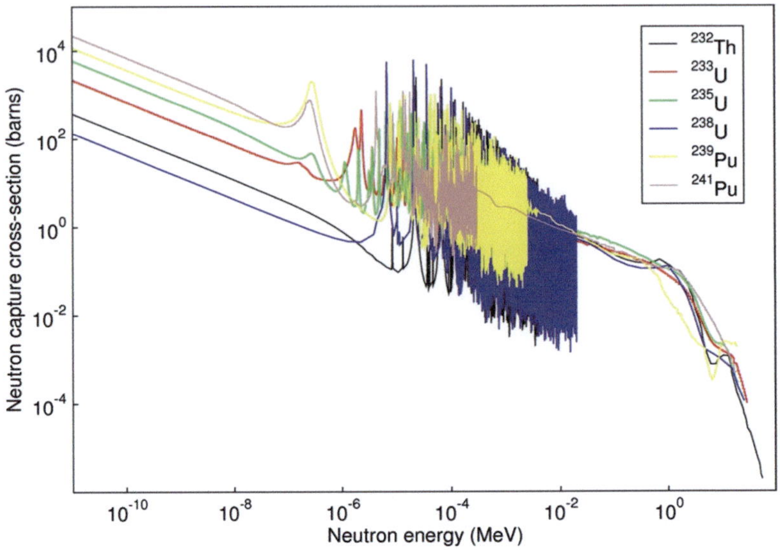

Fig. 5.12 Neutron capture cross sections for some the major actinides (Th-232, U-233, U-235, U-238, Pu-239 and Pu-241) [40]

differential volume should be equal to the rate at which they are added to that energy range within the differential volume (through fission and scattering) minus the rate at which they are removed by interactions (scattering out of the energy range or absorption) from the energy range and the differential volume. The mathematical statement for this balance is [37],

$$\nabla\vec{J}\left(\vec{r},E\right)+\Sigma_t\left(\vec{r},E\right)\Phi\left(\vec{r},E\right)=\int_0^{\infty}\left[\sum_j\chi^j(E)\nu^j\Sigma_f^j\left(\vec{r},E\right)+\Sigma_s\left(\vec{r},E'\rightarrow E\right)\right]\Phi\left(\vec{r},E'\right)dE' \tag{5.1}$$

where $\vec{J}\left(\vec{r},E\right)$ is the spatial and energy dependent neutron current density (neutrons per unit energy cm^2 s^{-1}), $\Sigma_t\left(\vec{r},E\right)$ is the spatial and energy dependent macroscopic total cross section (scattering, absorption etc.), $\Phi\left(\vec{r},E\right)$ is the spatial and energy dependent neutron flux, the fraction of neutrons emitted in the energy range dE is $\chi^j(E)dE$, the total number of neutrons produced from fission of isotope j is ν^j, $\Sigma_f\left(\vec{r},E\right)$ is the spatial and energy dependent macroscopic fission cross section, $\Sigma_s\left(\vec{r},E'\rightarrow E\right)$ is the spatially dependent macroscopic scattering cross section for scattering a neutron of energy E' to an energy E. A macroscopic cross section for an atom j is defined as, $\Sigma^j\left(\vec{r},E\right)=N_j\left(\vec{r}\right)\sigma(E)$ (where $N_j\left(\vec{r}\right)$ is the spatial distribution of the atom j and $\sigma(E)$ is the energy dependent cross section for the interaction of the atom j with a neutron).

Equation (5.1) is exact, but in order to solve the equation, a relationship between current density ($\vec{J}\left(\vec{r},E\right)$) and flux ($\Phi\left(\vec{r},E\right)$) needs to be found. An accurate approximation for this relationship comes from Fick's law [38] where,

$$\vec{J}\left(\vec{r},E\right) = -D\left(\vec{r},E\right)\nabla\Phi\left(\vec{r},E\right) \tag{5.2}$$

where $D\left(\vec{r},E\right)$ is called the diffusion constant.

The diffusion constant is approximated by the expression,

$$D\left(\vec{r},E\right) = \frac{1}{3\left[\Sigma_t\left(\vec{r},E\right) - \bar{\mu}_0\Sigma_s\left(\vec{r},E\right)\right]} \tag{5.3}$$

where $\bar{\mu}_0$ is the cosine of the scattering angle in the laboratory coordinate system averaged over all isotopes which make up the reactor core. ($\bar{\mu}_0 = \frac{\Sigma_j \bar{\mu}_0^j \Sigma_s^j\left(\vec{r},E\right)}{\Sigma_s\left(\vec{r},E\right)}$) [38].

The diffusion coefficient can be rewritten in terms of the transport cross section,

$$\lambda_{tr} = \Sigma_t\left(\vec{r},E\right) - \bar{\mu}_0\Sigma_s\left(\vec{r},E\right) \tag{5.4}$$

$$D\left(\vec{r},E\right) = \frac{1}{3\lambda_{tr}} \tag{5.5}$$

Substituting Eq. (5.2) into Eq. (5.1), leads to the energy dependent diffusion equation,

$$-\nabla D\left(\vec{r},E\right)\nabla\Phi\left(\vec{r},E\right) + \Sigma_t\left(\vec{r},E\right)\Phi\left(\vec{r},E\right) = \int_0^\infty \left[\sum_j \chi^j(E)\nu^j\Sigma_f^j\left(\vec{r},E\right) + \Sigma_s\left(\vec{r},E' \to E\right)\right]\Phi\left(\vec{r},E'\right)dE' \tag{5.6}$$

There are a number of criticality codes that can be used to solve the spatial and energy dependent diffusion equation. The codes can be based on stochastic methods or deterministic methods. Stochastic codes use methods such as Monte Carlo to track the discrete particle histories. Interaction probabilities are used to determine spatial and energy dependent scattering and absorption for these discrete particles. An example of a stochastic code is MCNPX [41]. A deterministic code uses either the transport equation or an approximation such as the diffusion equation. There are free codes such as DRAGON [42], those developed by national laboratories such as Denovo or XSDRNPM [43] and proprietary codes developed by nuclear vendors. In doing nuclear-pumped laser/reactor design, stochastic or deterministic codes would normally be used along with thermal codes and optical design codes.

It is possible to do some simplified analysis using Eq. (5.6) in its simplest form. It is possible to separate variables into those which are dependent upon space and those which are dependent on energy (e.g., $\Phi\left(\vec{r},E\right) = \Phi\left(\vec{r}\right)\Psi(E)$). In a thermal reactor, the neutron energy varies from the high energy at which the neutrons are born to the thermal energies which the neutrons slow down to. By integrating both sides of Eq. (5.6) over all possible neutron energies, the equation simplifies to the one group model. In the one group model the energy dependent coefficients are,

$$D\left(\vec{r}\right) = \int_0^\infty D\left(\vec{r},E\right)\Psi(E)dE \tag{5.7}$$

$$\Sigma_t\left(\vec{r}\right) = \int_0^\infty \Sigma_t\left(\vec{r},E\right)\Psi(E)dE \tag{5.8}$$

$$v\Sigma_f\left(\vec{r}\right) = \int_0^\infty v\Sigma_f\left(\vec{r},E\right)\Psi(E)dE \tag{5.9}$$

$$\Sigma_s\left(\vec{r}\right) = \int_0^\infty \Sigma_s\left(\vec{r},E\right)\Psi(E)dE \tag{5.10}$$

$$\Sigma_a\left(\vec{r}\right) = \Sigma_t\left(\vec{r}\right) - \Sigma_s\left(\vec{r}\right)$$

The one group form of Eq. (5.6) then becomes [44, 45],

$$-\nabla \cdot D\left(\vec{r}\right)\nabla\Phi\left(\vec{r}\right) + \Sigma_a\left(\vec{r}\right)\Phi\left(\vec{r}\right) = v\Sigma_f\left(\vec{r}\right)\Phi\left(\vec{r}\right) \tag{5.11}$$

A further simplification results if the reactor is homogeneous without a reflector ("bare"). Then there would be no spatial dependencies of the cross sections and diffusion coefficient. Thus Eq. (5.11) is rewritten as,

$$-\mathrm{D}\nabla^2\Phi\left(\vec{r}\right) + \Sigma_a\Phi\left(\vec{r}\right) = v\Sigma_f\Phi\left(\vec{r}\right) \tag{5.12}$$

Dividing Eq. 5.12 by – D,

$$\nabla^2\Phi\left(\vec{r}\right) + \frac{v\Sigma_f - \Sigma_a}{D}\Phi\left(\vec{r}\right) = 0 \tag{5.13}$$

Define

$$k_\infty = {}^{v\Sigma_f}\!/_{\Sigma_a} \tag{5.14}$$

$$L^2 = {}^{D}\!/_{\Sigma_a} \tag{5.15}$$

$$\nabla^2\Phi\left(\vec{r}\right) + \frac{k_\infty - 1}{L^2}\Phi\left(\vec{r}\right) = 0 \quad (5.16)$$

The equation can be rewritten by introducing the term geometrical buckling ($B_g^2 = \frac{\nu\Sigma_f - \Sigma_a}{D}$),

$$\nabla^2\Phi\left(\vec{r}\right) + B_g^2\Phi\left(\vec{r}\right) = 0 \quad (5.17)$$

It is traditional to define material buckling as ($B_m^2 = \frac{\nu\Sigma_f - \Sigma_a}{D}$). Geometrical buckling is found solving Eq. 5.17 for various reactor geometries while material buckling is based on the material properties. The geometrical buckling is equal to the material buckling when the core is critical,

$$B_m^2 = B_g^2 \text{ when the core is critical} \quad (5.18)$$

$$B_m^2 > B_g^2 \text{ when the core is supercritical} \quad (5.19)$$

$$B_m^2 < B_g^2 \text{ when the core is subcritical} \quad (5.20)$$

The average number of neutrons released per fission (ν) is energy dependent as well as being dependent upon the fissile isotope (^{233}U, ^{235}U and ^{239}Pu). The variation of ν with energy can be found empirically [45],

U-233 (where E is neutron energy in MeV):

$$\nu^{233} = 2.482 + 0.075E\,(0 \le E \le 1) \quad (5.21)$$

$$\nu^{233} = 2.412 + 0.136E\,(E > 1) \quad (5.22)$$

U-235 (where E is neutron energy in MeV):

$$\nu^{235} = 2.432 + 0.066E\,(0 \le E \le 1) \quad (5.23)$$

$$\nu^{235} = 2.349 + 0.15E\,(E > 1) \quad (5.24)$$

Pu-239 (where E is neutron energy in MeV):

$$\nu^{239} = 2.874 + 0.138E \quad (5.25)$$

The boundary conditions for Eq. (5.17) are that the derivative of the flux is zero at the centerline of the core (for symmetrical geometries) and the flux is zero at the extrapolated boundary. Because diffusion theory is an approximation of transport theory, it is necessary to create an extrapolated boundary were the flux goes to zero. As shown in Fig. 5.13, there is a difference between the exact flux and the flux from the diffusion approximation which the extrapolation length is used to minimize the error. The extrapolation length is defined as,

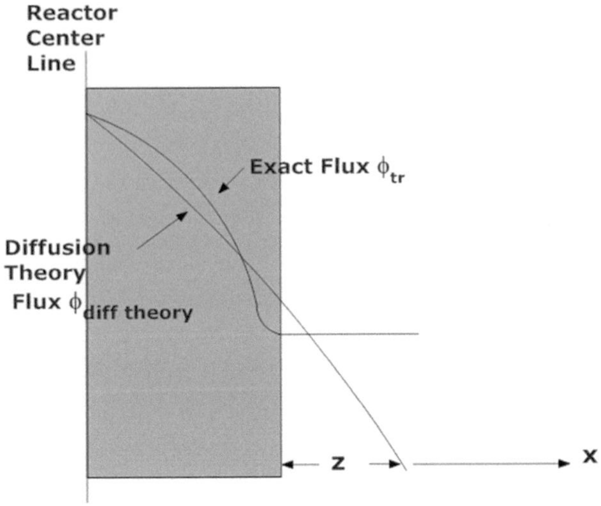

Fig. 5.13 An illustration of the extrapolation length as compared to the exact flux

$$z_0 = 0.7104\lambda_{tr} \tag{5.26}$$

where $\lambda_{tr} = \frac{1}{\Sigma_t - \bar{\mu}_0 \Sigma_s}$

The solution to Eq. (5.17) is dependent upon the geometry. If for example a right circular cylindrical geometry were chose as shown in Fig. 5.14, then Eq. (5.17) becomes,

$$\frac{1}{r}\frac{\partial}{\partial r} r \frac{\partial \Phi(r,z)}{\partial r} + \frac{\partial^2 \Phi(r,z)}{\partial z^2} + B^2 \Phi(r,z) = 0 \tag{5.27}$$

where r is the radial variable and z is the axial variable along the center line.

The boundary conditions are that the flux profile be symmetrical and that at the edge of the core,

$$\Phi(R + z_0, z) = \Phi\left(r, \pm\frac{H + z_0}{2}\right) = 0 \tag{5.28}$$

Using the method of separation of variable for this homogeneous partial differential equation [46], the solution will have the form,

$$\Phi(r,z) = \Phi(r)Z(z) \tag{5.29}$$

Substituting (5.29) into (5.27), the following differential equations can be separated,

Fig. 5.14 Right circular cylindrical reactor core of radius R and height H

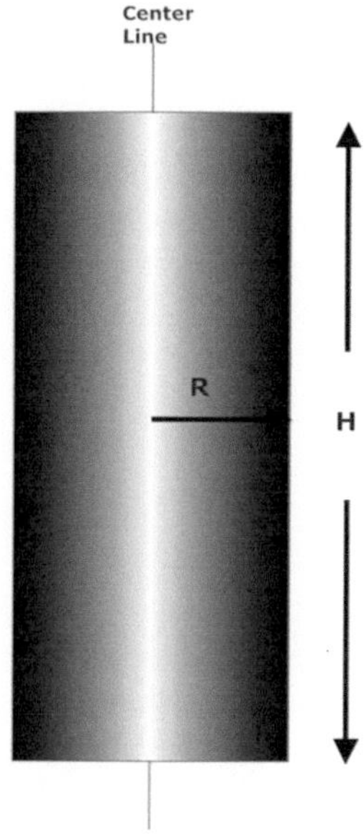

$$\frac{1}{r}\frac{\partial}{\partial r}r\frac{\partial \Phi(r)}{\partial r}+\alpha^2\Phi(r)=0\,\text{with}\,\Phi(R+z_0)=0 \tag{5.30}$$

$$\frac{\partial^2 Z(z)}{\partial z^2}+\lambda^2 Z(z)=0\,\text{with}\,Z\left(\pm\frac{H+z_0}{2}\right)=0 \tag{5.31}$$

The constants $\alpha^2+\lambda^2=B^2$. Each equation is an eigenvalue problem with the eigenfunction and eigenvalue solutions,

$$\Phi(r)=J_0\left(\frac{v_n r}{R+z_0}\right)\,\text{and}\,\alpha_n^2=\left(\frac{v_n}{R+z_0}\right)^2\,\text{for}\,\text{n}=1,2,3,\ldots \tag{5.32}$$

where v_n are the zeros of the Bessel function J_0.
and

$$Z(z)=Cos\left(\frac{n\pi z}{H+z_0}\right)\,\text{and}\,\lambda_n^2=\left(\frac{n\pi}{H+z_0}\right)^2\,\text{for}\,\text{n}=1,2,3,\ldots \tag{5.33}$$

The smallest value of B^2 occurs when n=1, thus the geometric buckling is,

$$B^2=\left(\frac{v_0}{R+z_0}\right)^2+\left(\frac{\pi}{H+z_0}\right)^2 \tag{5.34}$$

and the flux is,

$$\Phi(r,z) = AJ_0\left(\frac{v_0 r}{R+z_0}\right)Cos\left(\frac{\pi z}{H+z_0}\right) \tag{5.35}$$

where A is constant which is determined by the core power level.

There are some common geometries used in reactor theory for which the geometrical buckling and flux profile can be derived for bare homogeneous reactors using Eq. (5.17) (see Table 5.1) [45]. The nonleakage probability (P_{NL}) defines the probability that a neutron will not leak out of the reactor core. For a bare homogeneous reactor [45],

$$P_{NL} = \left(1 + L^2 B_g^2\right)^{-1} \tag{5.36}$$

A nuclear-pumped laser/reactor system can be approximated by a bare homogeneous core. As discussed previously, the fuel is interfaced to the laser medium by either using thin geometrical structures (such as thin layers) or by mixing the fuel with the laser medium as a gas (UF_6) or a medium with gaseous-like properties (aerosol or uranofullerene). In addition, the moderator structure and structural support materials are typically thin so that the material dimensions are typically less than the mean free path of a neutron in the material. Thus for practical purposes the reactor behaves like it is homogeneous. The core's infinite multiplication constant is,

$$k_\infty = {}^{\nu\Sigma_f}/_{\Sigma_a} \tag{5.37}$$

and the core's multiplication constant is,

$$k = k_\infty P_{NL} \tag{5.38}$$

The fuel interface to the laser medium needs to be chosen and a unit cell needs to be designed with consideration for the optical properties of the laser, the transport properties of the fission fragments, the neutronics of the core, the heat transport properties of the system and the mechanical integrity of the core. The decisions which have to be made are complex. For example, examine how one might design a nuclear-pumped laser/reactor based on the xenon laser (2030 nm). The typical configuration for successful experiments used a thin fissile fuel layer coated on a plate which is interfaced to the laser medium. A simple design is shown in Fig. 5.15. This design would flow the laser medium (since it is predominantly made up of helium) thus deriving cooling from the helium flow (like a High Temperature Gas Cooled Reactor). The optimum moderator to fuel ratio (atom density C/atomic density U) is on the order of 2000 (however there is little performance drop off over a range from 1000–10,000) [47].

As discussed in Chap. 2, the thickness of the UO_2 layer should be thin enough so that self absorption does not become an issue thus choosing t_3 to be 0.00025 cm. A unit cell has two coatings thus the total thickness of UO_2 is 0.0005 cm. The

Table 5.1 Geometric buckling and flux profiles for select bare homogeneous reactor geometries [45]

Geometry		$B_g^{\,2}$	Flux profile
Slab	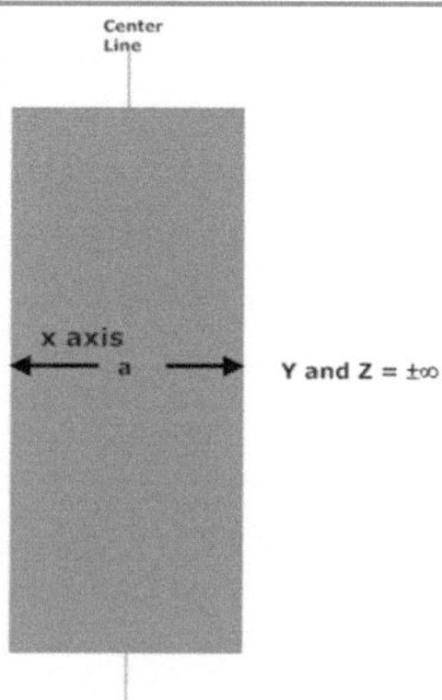	$\left(\frac{\pi}{a+z_0}\right)^2$	$\cos\left(\frac{\pi x}{a+z_0}\right)$
Infinite cylinder	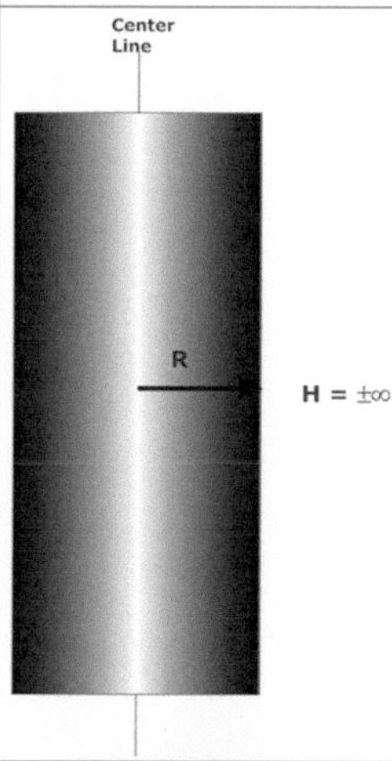	$\left(\frac{\nu_0}{R+z_0}\right)^2$	$J_0\left(\frac{\pi r}{R+z_0}\right)$
Sphere	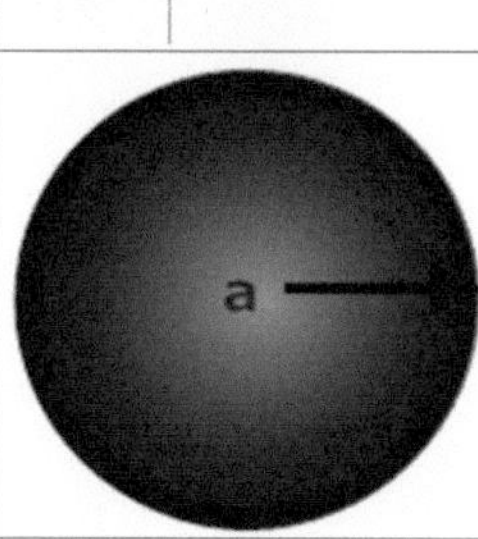	$\left(\frac{\pi}{R+z_0}\right)^2$	$r^{-1}\sin\left(\frac{\pi r}{R+z_0}\right)$

(continued)

Table 5.1 (continued)

Geometry		$B_g^{\ 2}$	Flux profile
Rectangular parallepiped	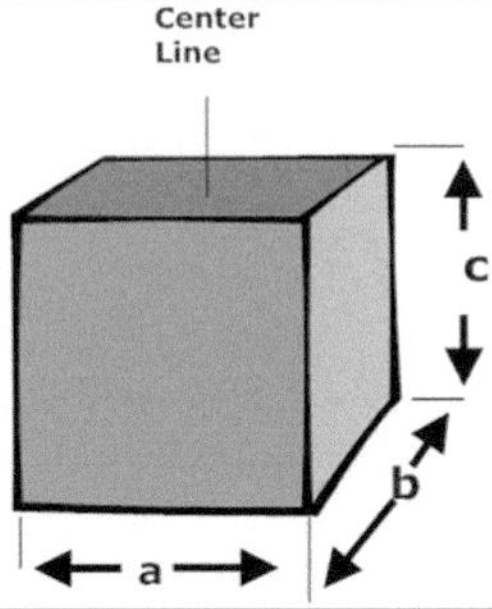	$\left(\frac{\pi}{a+z_0}\right)^2+\left(\frac{\pi}{b+z_0}\right)^2+\left(\frac{\pi}{c+z_0}\right)^2$	$\cos\left(\frac{\pi y}{b+z_0}\right)\cos\left(\frac{\pi z}{c+z_0}\right)\cos\left(\frac{\pi x}{a+z_0}\right)$
Finite cylinder	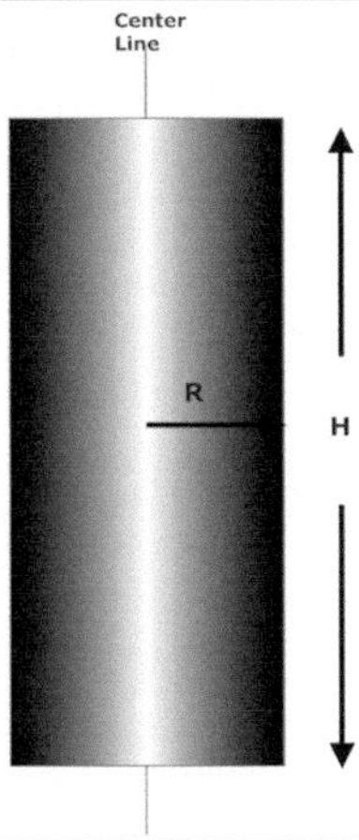	$\left(\frac{\pi}{H+z_0}\right)^2+\left(\frac{v_0}{R+z_0}\right)^2$	$\cos\left(\frac{\pi z}{H+z_0}\right)J_0\left(\frac{\pi r}{R+z_0}\right)$

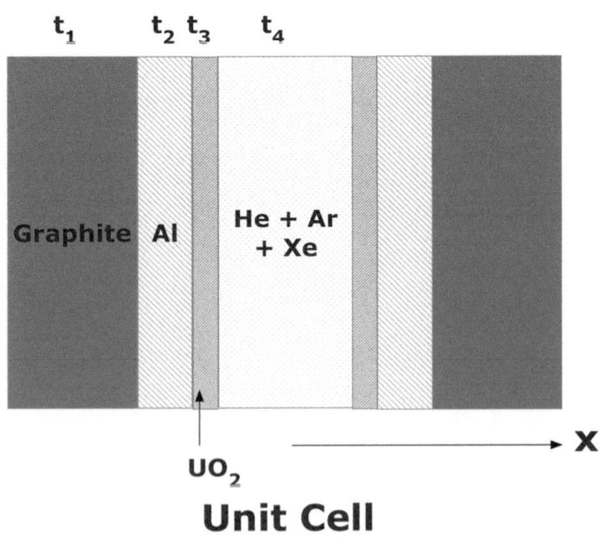

Fig. 5.15 A simplified cross section of a unit cell for a fuel/laser medium interface. The unit cell consists of a flowing gaseous laser medium (He +Ar+Xe) of thickness t_4 which also serves as a gaseous coolant. The UO_2 with a thickness t_3 is coated on an aluminum plate of thickness t_2. The unit cell also has a graphite moderator of thickness t_1

Table 5.2 Nuclear properties of atoms that make up the bare homogeneous core

Atom	Number density (atoms cm^{-3})	$1-\bar{\mu}_0$	σ_a (cm^{-2})	σ_s (cm^{-2})	σ_t (cm^{-2})	σ_f (cm^{-2})
C	8.03E+22	0.9444	4E–27	4.8E–24	4.8E–24	
Al	6.02E+22	0.9754	2.4E–25	1.4E–24	1.64E–24	
He	8.53E+19	0.8334	7E–27	8E–25	8.07E–25	
Ar	1.22E+19	0.9883	6.6E–25	1.5E–24	2.16E–24	
Xe	1.22E+17	0.9949	3.5E–23	4.3E–24	3.93E–23	
UO_2	2.23E+22	0.9972	7.6E–24	1.7E–23	2.43E–23	
O	4.46E+22	0.9583	2E–28	4.2E–24	4.2E–24	
U238	1.561E+21	0.9972	2.7E–24	1.1E–23	1.32E–23	
U235	2.0739E+22	0.9972	6.8E–22	1.7E–23	6.78E–22	5.77E–22

aluminum plate should be thick enough for structural integrity but minimized to avoid undo absorption. The aluminum plate thickness, t_2, was chosen to be 0.025 cm. The total thickness of aluminum in the unit cell is 0.05 cm. The graphite moderator should be of sufficient size so that the ratio of carbon atoms to uranium 235 atoms is between 1000 and 10,000. The graphite thickness, t_1, was chosen to be 0.5 cm. Thus the total thickness of graphite in the unit cell is 1 cm. Finally, the gas plenum should be thick enough to allow for the fission fragments to totally stop. The thickness of the gas plenum was chosen to be 1 cm. The optimum gas mixture, as discussed in Chap. 4, is 2,657 Torr He, 380 Torr Ar and 3.8 Torr Xe. To simplify the design process, the reactor core will be a cube with a side dimension equal to "a". The next step in the process is to determine the average atomic density of each atom that makes up a unit cell. The atoms that make up the unit cell are ^{235}U, O, C, Al, He, Ar and Xe. The one group properties for these atoms (from Appendix 4) are shown in Table 5.2.

Each component of the unit cell will have the same width and height and thus the only variable in finding the volume of each element is the thickness. The total thickness of the unit cell is 2.05 cm. This specific design has a moderator to fuel ration of 7,744. The macroscopic cross sections for the bare homogeneous core are calculated from Table 5.3. The average atomic fractions of the atoms in the core are calculated and shown in Table 5.4. With the macroscopic cross sections calculated in Table 5.3, the nuclear data of Table 5.2 and the average atomic fraction for the bare homogeneous core shown in Table 5.4, the diffusion coefficient, the transport meant free path, geometrical buckling, z_0, and core dimensions can be calculated for the just critical bare homogeneous core.

The macroscopic cross section is calculated by multiplying the microscopic cross section by the average number density for each atom in the core. The calculated values are,

Table 5.3 Macroscopic cross sections for the bare homogeneous reactor core

Atom	Thickness t_i (cm)	$\sum_i t_i$ (cm)	Volume fraction $(\frac{t_i}{\sum_i t_i})$	$\sum_a$ cm^{-1}	$\sum_s$ cm^{-1}	Σ_t cm^{-1}	Σ_f cm^{-1}
C	0.5	2.0505	0.243843	7.8322E–05	0.09399	0.093987	
Al	0.025	2.0505	0.012192	0.0001769	0.001028	0.001204	
He	1	2.0505	0.487686	2.912E–07	3.328E–05	3.357E–05	
Ar	1	2.0505	0.487686	3.9269E–06	8.925E–06	1.285E–05	
Xe	1	2.0505	0.487686	2.0824E–06	2.558E–07	2.338E–06	
UO_2	0.00025	2.0505	0.000122	2.0663E–05	4.541E–05	6.607E–05	
O	0.00025	2.0505	0.000122	1.0875E–09	2.284E–05	2.2838E–05	
U238	0.00025	2.0505	0.000122	5.1196E–07	1.998E–06	2.5103E–06	
U235	0.00025	2.0505	0.000122	0.001714	4.286E–05	0.001714	0.001459

Table 5.4 Atomic fractions of the atoms which make up the bare homogeneous core

Atom	Atomic fraction
C	0.961231211
Al	0.036031207
He	0.002042167
Ar	0.00029208
Xe	2.9208E–06
UO_2	0.000133471
O	0.000266942
U238	9.34298E–06
U235	0.000124128

$$\Sigma_a = 0.001976 cm^{-1} \tag{5.39}$$

$$\Sigma_f = 0.001459 cm^{-1} \tag{5.40}$$

$$\Sigma_s = 0.0.09512 cm^{-1} \tag{5.41}$$

$$\Sigma_t = 0.09698 cm^{-1} \tag{5.42}$$

$$\overline{\mu}_0 = 0.054686 cm^{-1} \tag{5.43}$$

$$D = 3.632 cm \tag{5.44}$$

$$z_0 = 0.71\lambda_{tr} = 7.7361 \,\mathrm{cm} \tag{5.45}$$

$$B_g^2 = 0.000436 cm^{-2} \tag{5.46}$$

$$3\left(\frac{\pi}{a + 3.873}\right)^2 = B_g^2 = 0.000436 cm^{-2} \tag{5.47}$$

$$a = 142.72 cm \tag{5.48}$$

The homogeneous core volume is 2.9071×10^6 cm^3 or 2.9071 m^3. This means that about 70 unit cells would be needed to make up a just critical core. To build in excess reactivity the core dimensions would have to be increased. The just critical nuclear-pumped laser/reactor would have 70 laser beams which would have to be phase locked and combined. The question of operating power density is an entirely different matter which is dependent upon core cooling. The gas temperature will also have an impact on the xenon laser efficiency as discussed in Chap. 4. In addition the use of control rods or other devices for control of the core is not discussed. The core dynamics is also important. The core may need to be pulsed in order to reach the necessary threshold power density for the laser medium. It is recognized that this analysis does not attempt to address these important issues.

The use of bare homogeneous core calculations in nuclear-pumped laser/reactor scale design studies can be utilized for rough sizing estimates. As a tool, the bare homogeneous reactor core approximation is fairly accurate due to the scale lengths of the fuel, the moderator, the structural components and the laser medium being less than a neutron mean free path in the core. A study using a discrete computer code to calculate the critical parameters was compared to the bare homogeneous core calculation shown above [48]. The study demonstrated that the bare homogeneous core calculation was within 5–10 % of the predicted value by the discrete code.

5.7 Reactor/Nuclear-Pumped Laser System Designs

As will be discussed, there are two basic strategies in designing a reactor/nuclear-pumped laser system. The first is use a critical reactor core as the neutron source and couple the neutrons to a subcritical nuclear-pumped laser assembly. This was the approach used in the one of the first conceptual designs of a reactor/nuclear-pumped laser system by Sandia National Laboratory (SNL). The second approach is to integrate the critical core with the nuclear-pumped laser. This section will look at the evolution of the concepts for reactor/nuclear-pumped laser systems design.

5.7.1 Sandia Fast Burst Reactor/Subcritical Nuclear-Pumped Laser and Gas Dynamic Carbon Monoxide Nuclear-Pumped Laser Designs (1974–1976)

One of the first reactor/nuclear-pumped laser concepts was developed by Sandia National Laboratory [5, 49–51]. These studies considered possible NPL systems based on the molecular carbon dioxide. A fast burst reactor coupled to a subcritical nuclear-pumped laser assembly was the basis of the design as shown in Fig. 5.16, where the fast-burst reactor is used as a primary neutron source. Neutrons from the reactor enter a surrounding subcritical uranium region which, in turn, produces fission fragments that escape into and excite the laser gas. Much like the design discussed in Sect. 5.6, the subcritical (laser "driver") region consists of laminated plates having a thin (~3 micron thickness) coating of uranium metal on neutron-moderator slabs of ~0.2 cm thickness. These slabs thermalize the neutrons to provide a better interaction with the uranium; they also serve as a heat sink and provide structural strength for the uranium. The thickness of the uranium coating is determined by the range of fission fragments (which is about 10 microns in uranium). With this design roughly 1/4 of· the fission fragments escape from the· surface. Half are born with velocities in the improper direction and thus are lost. Those emitted in the proper direction lose about half of their energy as they traverse the coating. Consequently, since about 80 % of the fission energy goes into charged particles, about one fifth of the energy produced by fission in the uranium subcritical region is deposited in the laser medium.

The difference between the energy storage design and the multiple optical cavity design in Fig. 5.16 involves the way that the subcritical excitation region is coupled to the laser cavity. If metastable states having sufficient lifetime are involved in the

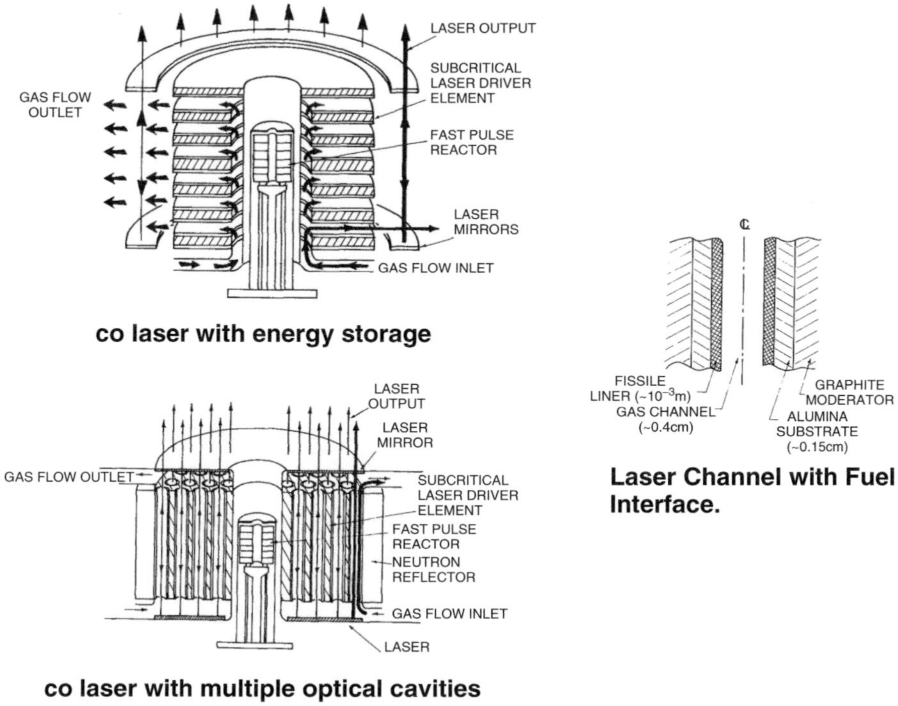

Fig. 5.16 A fast burst reactor coupled to a subcritical carbon monoxide nuclear-pumped laser. The CO laser where the excitation volume and laser cavity are separated, requiring long-lived laser states for energy storage and high gas flow rates is show on *top*. On the *bottom* is a combined excitation-laser region. Here gas flow rates are slow, depending only on cooling and gas renewal requirements for coupling a NPL driver to a fast burst reactor

laser mechanism, then the gas passes through the subcritical region, is excited and then flows into the laser cavity. This design has the advantages of removing the laser cavity from the radiation field and providing good cooling. However, excessive flow rates would be required except for only a few media that have quite long-lived metastable states (such as the nitrogen vibrational states discussed in Chap. 1).

To avoid the need for high flow rates, the multiple optical cavity design uses laser cavities which are directly superimposed on the subcritical region. In this case, flow along the cavity axis is employed using tubes with a uranium coating and construction similar to the plate design. Performance estimates involving four different cases of CO nuclear-pumped laser are summarized in Table 5.5. Based on coupling with a fast burst reactor, it is estimated that in one system 21 MJ could be deposited in the subcritical region, resulting in a 2.1 MJ laser. Due to the high efficiency of energy transfer to the vibrational states in CO by direct fission fragment excitation, it is anticipated that over 50 % of the energy deposited in the gas will be extracted as coherent light. Since 20 % of the energy released in the driver enters the gas, an efficiency of ~10 % for the nuclear-pumped laser is

Table 5.5 Characteristics of a CO nuclear-pumped laser assuming 13 MJ of energy from the fast burst reactor with a 100 microsecond pulse width [49]

	Driver dimensions			Graphite	Fuel				
Case	Inside radius (cm)	Outside radius (cm)	Height (cm)	Thickness (cm)	Atom density $\times 10^{19}$ (atoms cm^{-3})	Laser driver volume (liters)	E_{LD} (MJ)	E_{Gas} (MJ)	E_{Laser} (MJ)
1	30	85	150		5.36	2980	4.81	0.96	0.481
2	30	85	150	40	1.34	2980	7.69	1.54	0.769
3	30	85	150	40	5.36	2980	11.75	2.35	1.17
4	30	130	150	40	5.36	6362	21.04	4.21	2.10

possible. (Since an additional 1–3 times the driver energy is produced in the fast burst core, the system efficiency would approach 3.5 %) The relatively small size of this system demonstrates an important advantage of coupled reactor/nuclear-pumped laser systems. Thus, the 2-MJ laser-driver system would have a volume of 6 m^3 while the overall unit (including the reactor) is approximately double this size.

The pulse repetition rate of such a system would be set by cooling requirements of the fast burst reactor core. Without forced convection cooling, approximately an hour is required between pulses. Special designs might eventually reduce this time to minutes as illustrated by a repetitive pulsed reactor such as the TRIGA discussed in Chap. 3. A system without forced convection cooling should be adequate for some applications such as ballistic missile defense (as will be discussed in Chap. 6). For the cases shown in Table 5.5 the mean temperature rise of the driver section would only be 4–5°K although the peak (adiabatic) temperature in the uranium foil itself could approach 800°K.

An important aspect of the Sandia concept is that near term development seems quite feasible. The basic reactor core uses existing fast burst reactor technology. The subcritical driver system requires some new metallurgical techniques to assure good plating and proper moderator characteristics, but this seems relatively straightforward and similar uranium films were developed by Sandia for the original CO laser experiments. The CO laser-driver is also based on the Sandi CO nuclear-pumped laser which used a tubular design.

While the fast burst reactor/subcritical nuclear-pumped laser concepts employs uranium coated tubes, the same basic arrangement could easily be extended to a volume source. However, to take advantage of existing nuclear-pumped laser technology, it would be necessary to use ^3He with the noble gas. As NPL research progresses, however, more attractive mixtures such as $Ar/Xe/F_2$ $/UF_6$ can be envisioned.

Thus, using technology at hand, it should be possible to design and build a pulsed nuclear-pumped laser capable of delivering energies greater than 1 MJ per pulse. Further, the system would be self-contained and sufficiently compact for use in space or for high energy ground based laser systems.

Sandia National Laboratory scientists also explored more advanced nuclear-pumped laser systems. There are two possible routes to high energy nuclear-pumped lasers, namely: (1) replace the fast burst reactor with yet larger special purpose reactors capable of driving even larger nuclear-pumped laser volumes, or (2) combine the reactor with the laser-driver (i.e. design a self-critical nuclear-pumped laser driver). Since the latter potentially offers the best performance, an advanced system based on a gas dynamic laser was proposed. By employing a plate-type driver at high pulse repetition rates or in a steady-state mode, cooling becomes a critical path problem. Consequently, the gas-dynamic nuclear-pumped laser was studied [5] as shown in Fig. 5.17. The driver region, as detailed in Fig. 5.18, employs a laminate structure similar to that proposed for the earlier fast burst driver design. However, a sufficient number of plates are provided here to create a critical reactor which, depending on the design, could either operate in a

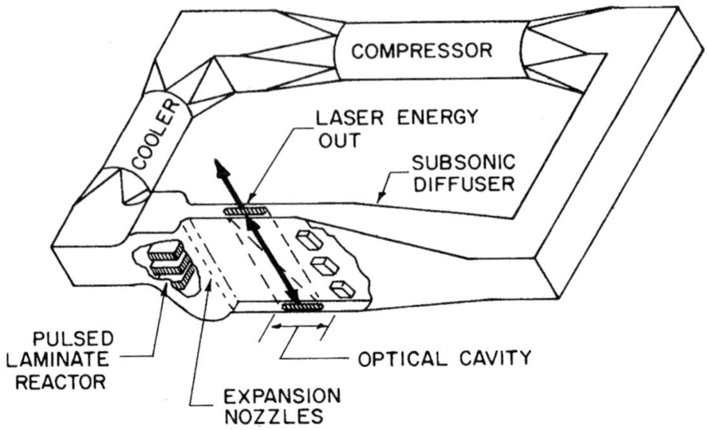

Fig. 5.17 A gas dynamic carbon monoxide nuclear-pumped laser concept [52]

pulsed or steady-state mode. Also, the plates are shaped to provide proper flow with an expansion nozzle region. Except for the nuclear reactor, the system is identical to a conventional closed-cycle gas-dynamic laser (i.e. the reactor region simply replaces the electric excitation region). Sandia scientists estimated that the operating conditions of Table 5.6 might be achievable with a carbon monoxide nuclear-pumped laser. Quasi steady-state laser powers levels of 60 MW were estimated. If, for example, the laser operated with a 2.5 s pulse, the output beam energy would be about 150 MJ.

The Sandia National Laboratory designs demonstrated the possibility that large scale nuclear-pumped laser systems could be built. This work was significant in that it spurred interest in nuclear-pumped lasers for the strategic defense initiative programs of the mid 1980s.

5.7.2 *NASA Uranium Hexafluoride Fueled Reactor/Nuclear-Pumped Laser Design (1979)*

NASA's nuclear-pumped laser group pushed for the use of gaseous uranium hexafluoride as the fuel mixed with a gaseous laser medium. They commissioned a study by the United Technologies Research Center for a conceptual design of a self-critical nuclear-pumped laser system [53]. The laser configuration consisted of a 3.2 m^3 lasing volume surrounded by a graphite moderator and a region of heavy water. The neutronics calculations indicated a critical mass of 4.9 kg of ^{235}U in the form of UF_6. The reactor was designed to be capable of operating in the steady-state mode with an average gas temperature of 600 K. The partial pressure of UF_6 was 0.34 atm. The reactor thermal power was 100 MW with an average power density in the laser medium of 31 W cm^{-3} and an average neutron flux of 7×10^{14} n cm^{-2} s^{-1}.

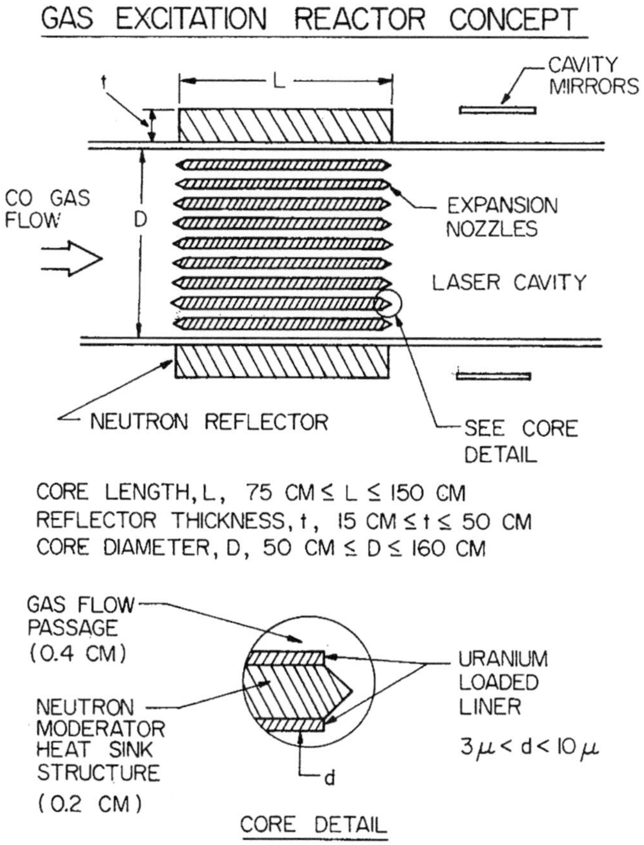

Fig. 5.18 The fuel/laser interface details for the gas dynamic carbon monoxide laser concept [52]. The core is self-critical

Table 5.6 Typical operating parameters for a carbon monoxide gas dynamic nuclear-pumped laser [52]

Parameter	Value
Laser power	60 MW
Energy density of gas	600 J/l-atmosphere
Flow length	1 m
Flow time	23 ms
Reactor temperature differential	360 K

The overall reactor/nuclear-pumped laser design is shown in Fig. 5.19 with a schematic shown in Fig. 5.20. The core is made up of a matrix of cells (Fig. 5.21) in which the fuel and laser medium flow in the center of each cell (Fig. 5.22). The length to diameter ratio of the core is about equal to one. Above and below the fuel matrix are regions which contain the laser optics (e.g., mirrors, beam splitters, etc.). A tank of heavy water is used as the reflector and moderator and it surrounds the

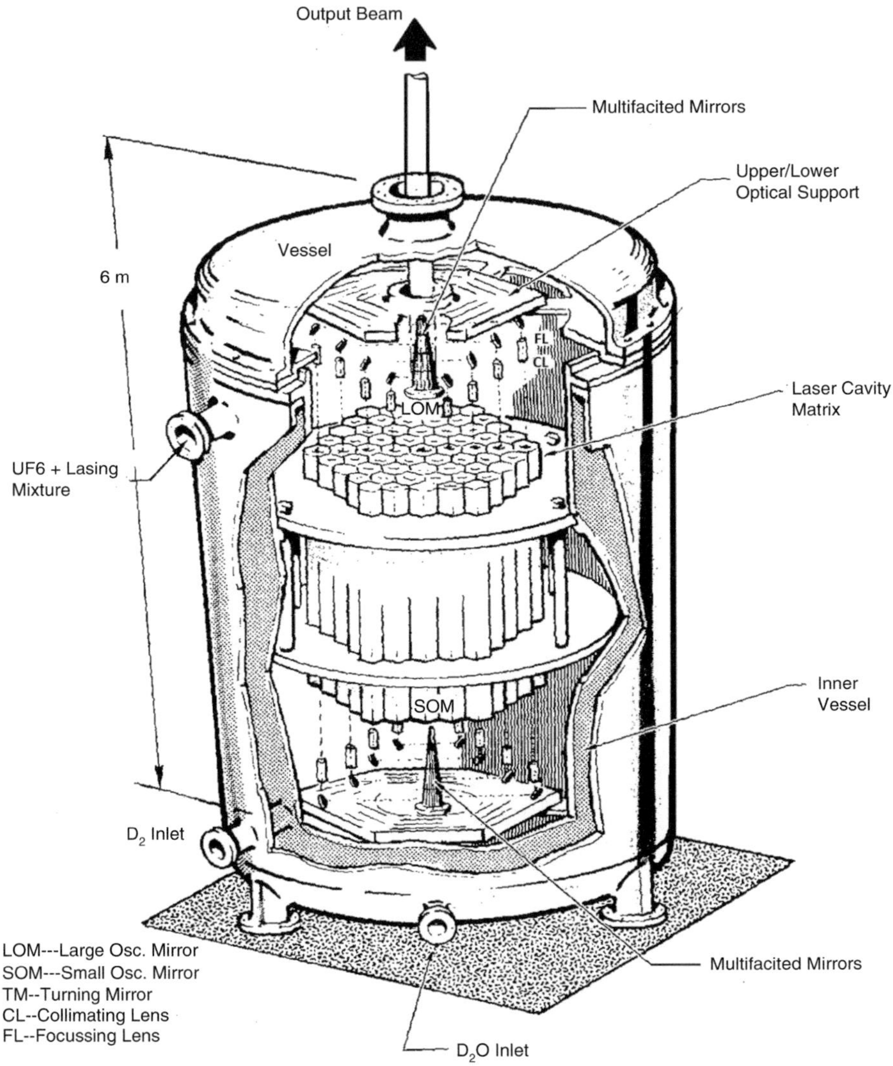

Fig. 5.19 The design of the uranium hexafluoride fueled self-critical reactor/laser is shown [53]

fuel matrix. At the top of the tank, there are extraction channels through which the laser beam passes through windows. These channels are filled with deuterium gas at a high pressure in order to reduce neutron leakage. A unit cell is hexagonal with a length of 200 cm. In the center of unit cell is a 20 cm diameter laser cell (Fig. 5.22). There is 30 cm between any two parallel surfaces within the unit cell. A minimum wall thickness for the graphite portion of the cell is 4.5 cm. These unit cells are packed in the reactor as shown in Fig. 5.21.

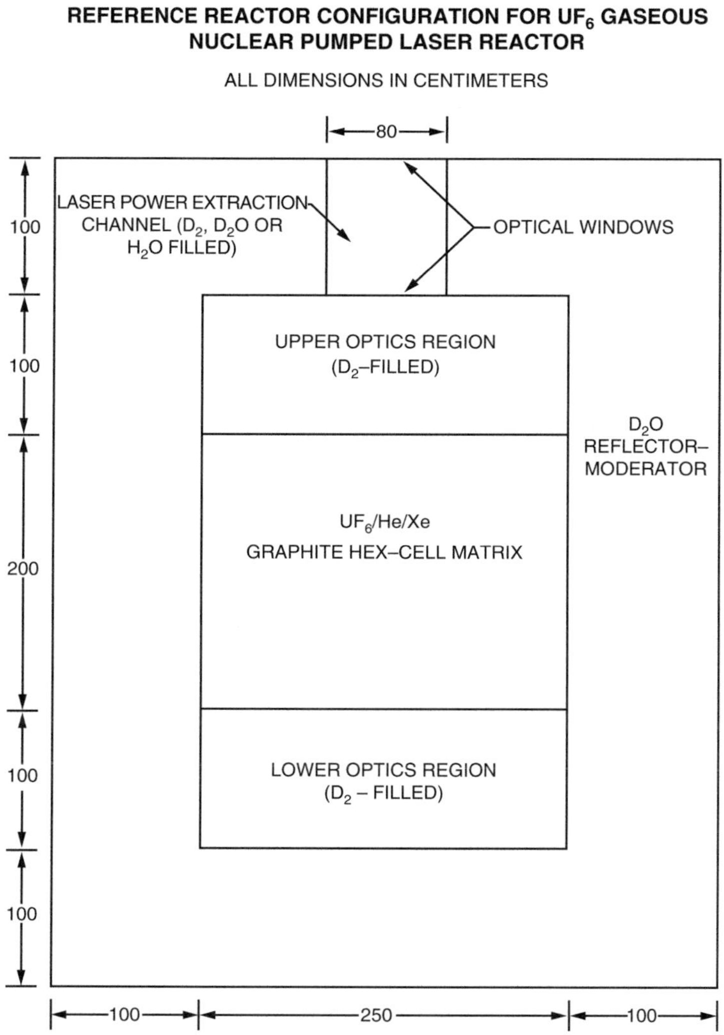

Fig. 5.20 The reference configuration for the uranium hexafluoride fueled reactor/laser [53]

The reactor design addresses the important issue of how to optically couple all of the laser cells to create a single Gaussian limited beam. In the center of the reactor there is a master oscillator that is optically coupled to the other laser cavities which act as single-stage amplifiers connected in parallel (Fig. 5.23). The approach of using a master oscillator and then using the laser cells as single stage amplifiers was also used for a nuclear-pumped solid-state laser as will be discussed in a later section.

For the neutronics calculations, the gas mixture within the laser cells was assumed to be a 10 to 1 ratio of helium to UF_6 (enriched to 93 %). The basis for

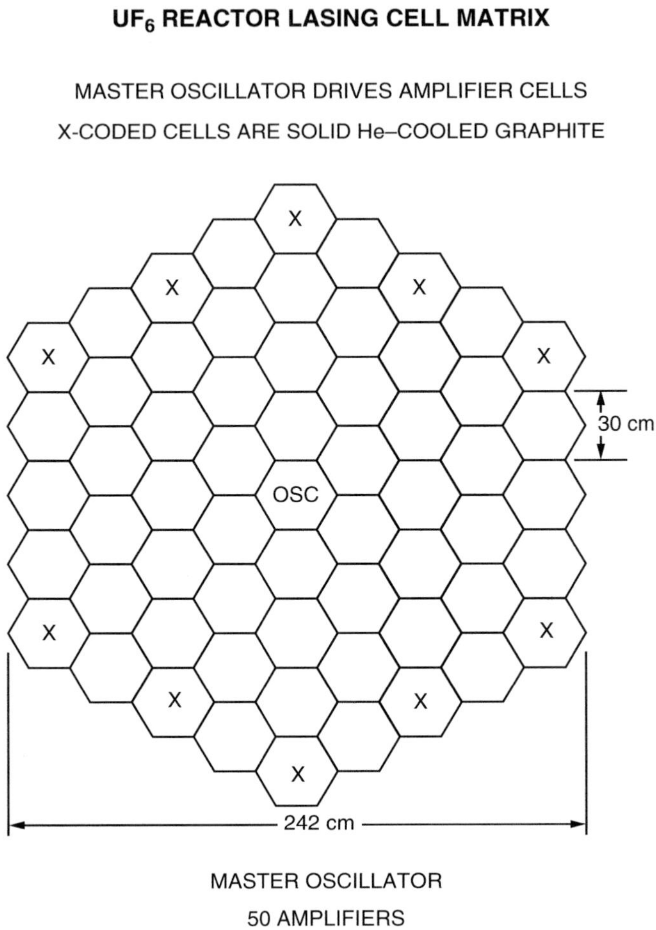

Fig. 5.21 The assembly of cells which make up the core of the uranium hexafluoride fueled reactor/laser [53]

this choice in gas mixtures came from experiments. The operating pressure was 10 atm and an operating gas temperature of 600 K was selected based on a parametric study. The fission fragments excite the gas mixture thus creating the required gain needed for laser operation. The range of fission fragments in this gas mixture at a pressure of 10 atm and an operating temperature of 600 K is about 1 cm. Thus 95 % of the fission energy is deposited in the laser medium.

The gas flow path is shown in Fig. 5.24. The gas flow enters through a slot at the peripheral of the wall which extends the entire cavity length. A vortex flow is then created in the laser cell. The bulk of the flow exits from the cell through a bypass flow exhaust duct through a perforated plate in the cylindrical wall of the laser cavity. Another portion of the flow spirals radially inward and then passes out of the

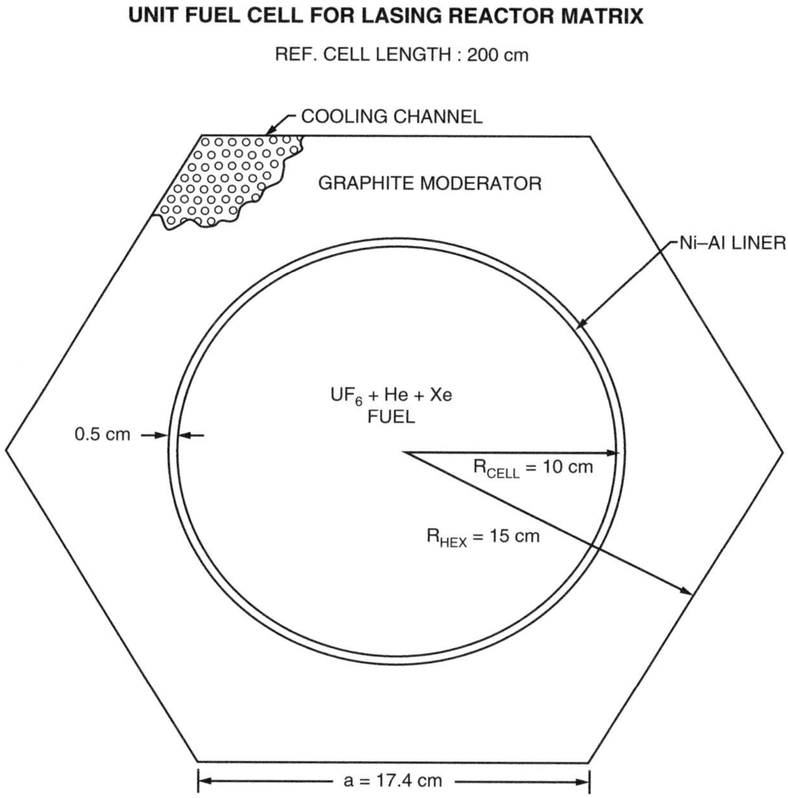

Fig. 5.22 The cross section of a cell in which the center is made up of the uranium hexafluoride gas plus laser medium. This gas mixture flows through the cell [53]

laser cell through an exhaust port at the ends of the laser cavity. An important feature of this design is that through variations of the injection flow rate, the temperature of the gas can be controlled so that a temperature gradient can be established. This temperature gradient causes a gradient in density which allows the energy deposition from fission to be spatially controlled. The critical mass was found as a function of UF_6 pressure (Fig. 5.25).

Experiments demonstrated that the xenon laser (2650 nm) can be pumped with fission fragments from UF_6 mixed in with the laser gas [54]. The efficiency of rare gas lasers is limited to about 2 % so even though this reactor design can work with rare gas lasers, it is desirable to find a more efficient laser. As discussed in Chap. 1, the optical absorption properties of uranium hexafluoride do limit which laser media can be used. Three laser candidate laser media which operate at suitable wavelengths that were mentioned in the study are the I_2 (342 nm), XeF (351, 353 nm), and N_2 (358 nm) lasers. The I_2 (342 nm) laser and the XeF (351, 353 nm) laser both can operate at efficiencies that exceed the rare gas lasers.

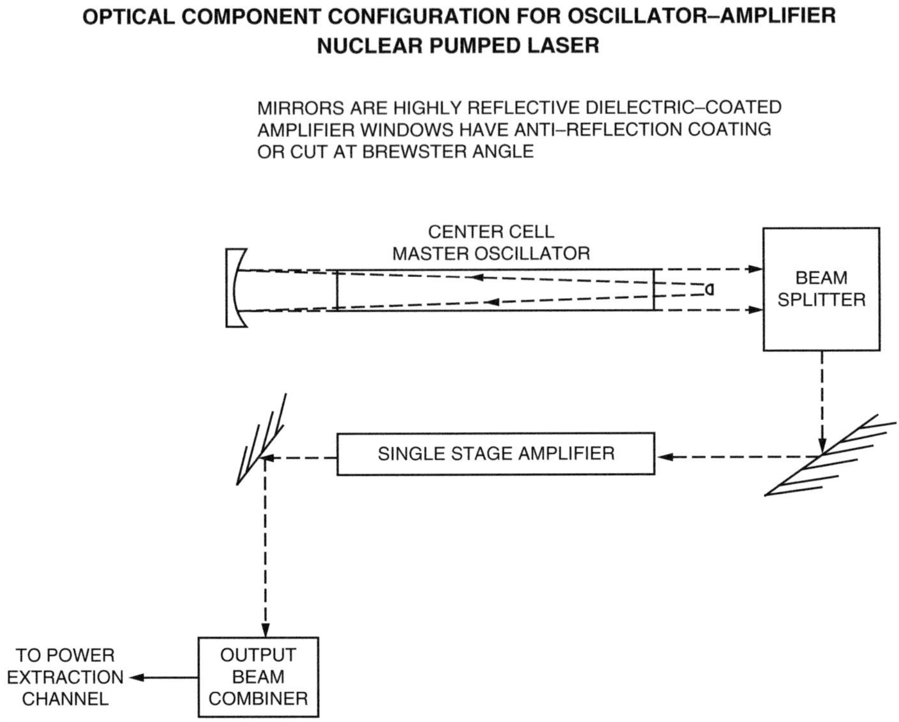

Fig. 5.23 Diagram of the master oscillator coupled to a single stage amplifier [53]

However, the power density threshold for these lasers is much higher than the rare gas lasers. Neither I_2 or XeF have been nuclear-pumped.

5.7.3 Reactor/Nuclear-Pumped Laser System Studies the University of Missouri (1982–1987)

The University of Missouri considered multiple designs of the nuclear reactor driver system with a laser medium driven by a volume type source (thermal NPL); an aerosol source (aerosol core NPL), and a surface source (conventional NPL). Reactor criticality computer codes were used (DISNL, XSDRN, PARET or IREKIN, and AMPX) for calculations. The feasibility of the above three types of high power/energy nuclear-pumped laser/reactor concepts were examined using criticality codes and neutron kinetics models [55, 56].

A thermal reactor/nuclear-pumped laser system cell uses a combination of solid fuel and moderator, but the coolant should serve another function as the laser medium. The fission energy is coupled to the laser medium by heat conduction. The carbon dioxide laser (a mixture of H_2O:N_2:CO_2) was chosen (see Chap. 1) and

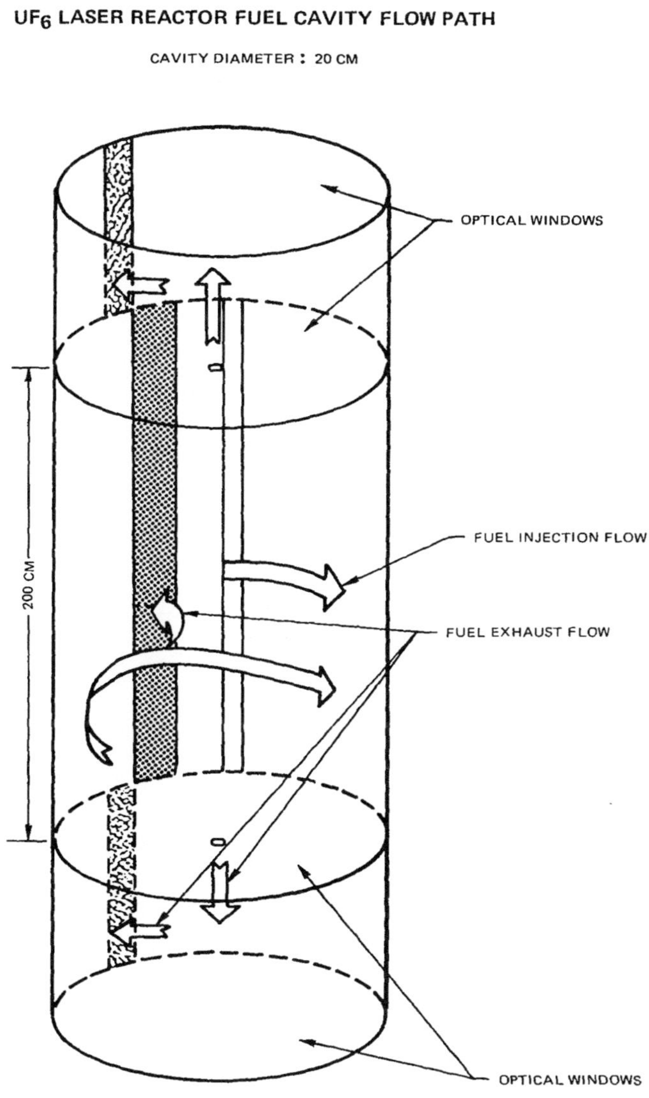

Fig. 5.24 The gas flow diagram for the reactor [53]

the laser gas mixture has good coolant properties [56]. Therefore this concept has a similarity to gas cooled reactors. The coolant operating temperature was chosen as 1400 K. In order to make the reactor more compact for use in space, beryllium and highly enriched UO_2 (93 %) can be used as a moderator and fuel respectively. The laser tube dimensions were chosen to be consistent with the range of fission fragments in the laser gas (tub radius of r_0=2 cm and a lattice radius of r=4 cm, and a zirconium refractory pressure tube thickness of 2 mm). Criticality calculations showed that a k_{eff}=1.2 was easily achieved with a small core volume (0.4 m^3)

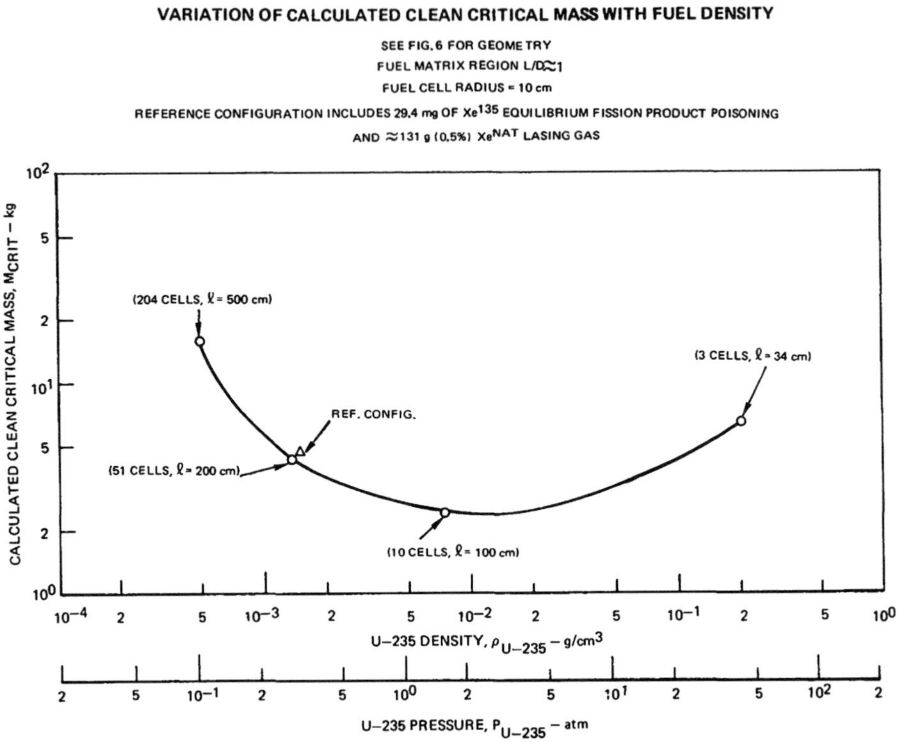

Fig. 5.25 Calculation of cell critical mass as a function of the U-235 density or pressure

and a berillium reflector thickness of 30 cm. If molybdenum is chosen as the refractory tube, it would be thinner than the zirconium tube and it would reduce the thermal neutron absorption. The advantages of molybdenum are seen in Fig. 5.26 which shows one lattice cell design with a molybdenum refractory coating (0.2 mm). The laser radius of 2 cm remains the same, but the lattice radius is 4.5 cm, the fuel clad is zirconium, the vacuum thickness is about 4 mm and both moderator and reflector have coolant channels which represent 5 % of the volume. When the core volume and the reflector thickness are about 0.4 m^3 and 30 cm respectively, the reactor multiplication factor, k_{eff}, change with the fuel meat (UO_2) thickness (see Table 5.7). These results show that small compact thermal reactor/nuclear-pumped lasers can be built with a total volume of around 2 m^3, a total mass of 3.3 Tonns and a total fuel loading of 8.4 kg (93 % UO_2) for 0.1 mm fuel thickness.

A self-critical reactor/nuclear-pumped laser system driven by a surface source was studied in 1981 (Fig. 5.27) [22]. The goal of this study was to develop a steady-state nuclear-pumped laser. A high flux steady-state nuclear reactor with a surface source is capable of producing a power deposition of about 4.74 W cm^{-3} (Chap. 4). The 2 % efficient xenon laser (at 2030 nm) requires a threshold power density of 10 W cm^{-3}. There are issues with using a thin layer of coated on a metal plate. The reactor will have a low average fuel density. This design used unit cells which had a

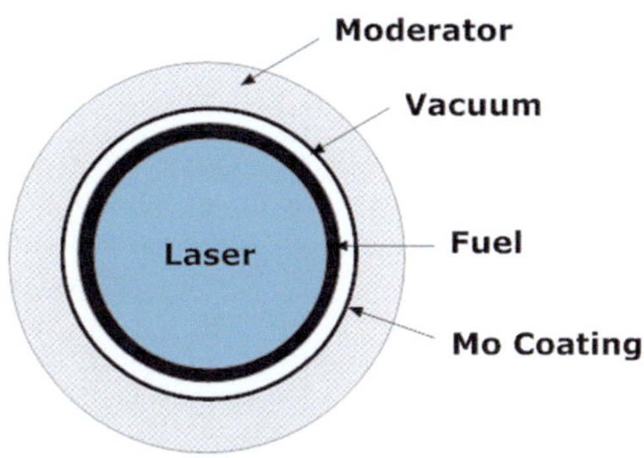

Fig. 5.26 The unit cell for the thermal reactor/nuclear-pumped laser [56]

Table 5.7 Multiplication factors for a reactor/nuclear-pumped laser system using a 0.2 mm molybdenum refractory coating [56]

Fuel thickness (mm)	0.05	0.1	0.3	0.5
Be/U^{235} ratio	3590	1793	593	351
k_{eff}	0.48	1.2	1.31	1.32

laser in the center of the cylindrical cell surrounded by a thin film of enriched uranium coated on the inside of a metal cylindrical sleeve. The metal sleeve was mounted in the middle of a graphite block (Fig. 5.27). The system was cooled by the laser gas mixture which consisted of 700 parts helium, 100 parts argon and 1 part xenon. The cooling properties of the gas mixture is dominated by the helium thus by flowing the gas mixture through the cell it serves as a coolant (e.g., like a high temperature gas cooled reactor). The study was valuable in that it identified a need for increasing the neutron flux by a factor of three in order to create a high enough power density to directly drive the xenon laser at 2030 nm. This in turn led to more extensive studies in the mid 1980s discussed below that used a number of reactor codes to model several types of low fuel density reactor/laser cores concepts and their operational characteristics. The main goal of these studies was to identify methods of boosting the power density of the reactor/laser core.

One of the reactor concepts that showed promise in boosting the power density of a reactor/laser core was the aerosol core reactor. The aerosol core reactor concept employs fuel in the form of an aerosol which leads to a low fuel density that remains optically transparent [19, 33, 57]. Both black (absorbs photons) and reflective aerosol fuel particles are considered. For black aerosol particles the fuel density must be limited to less than 1 mg cm^{-3} in order for photons to escape the gas (Table 5.8). Light water can't be used as a moderator and reflector and still have a critical reactor. Graphite can be used but the size of the core would be too large. In considering various combinations of fuels, moderators, reflectors and lattice cell, core and reflector sizes, a series of calculations were made. The result of these studies indicated that a k_{eff}=1.2 can be achieved. The core however was large due to the low average fuel density (R=2.226 m). One of the systems used U^{235}:Xe:F:

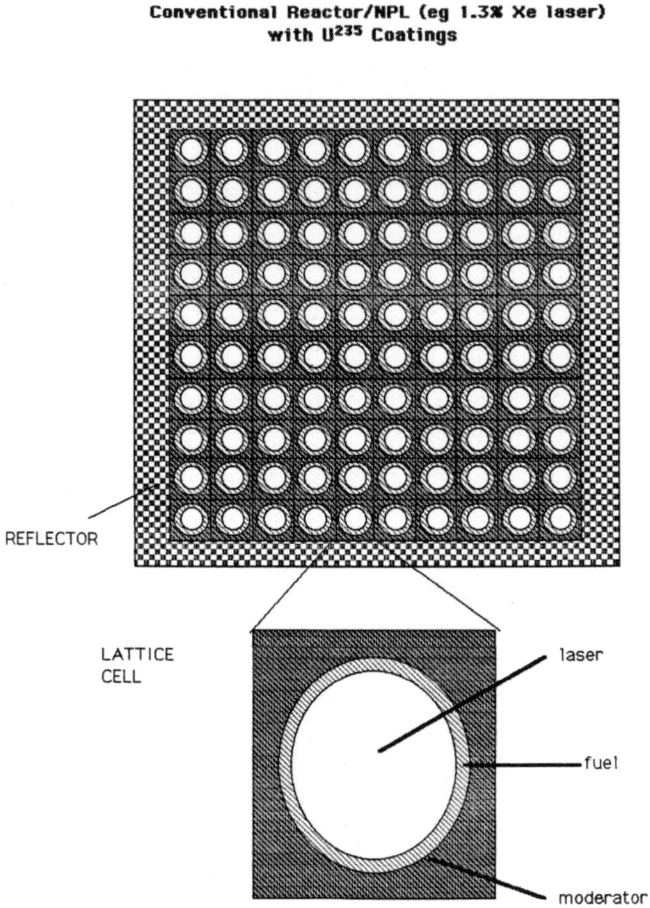

Fig. 5.27 An illustration for a conventional NPL/Reactor using a Xe laser. The fuel is a thin coating on the surface (surface source) [22]

Table 5.8 Parameters of a black aerosol and reflective aerosol fueled reactor

	R0(cm)	R1(cm)	^{235}U fuel density
Black aerosol	1.18	9.1	1 mg cm^{-3}
Reflective aerosol	2	4	5 mg cm^{-3}

D_2O as shown in Fig. 5.28. The minimum fuel density for $k_{eff} = 1$ is 0.3, 0.45 and 0.5 mg cm^{-3} for ^{239}Pu, ^{233}U and ^{235}U respectively. A reactor fueled with black aerosol particles has a volume of 46 m^3 and a mass of 20 Tonns [33].

A reactor fueled with a reflective aerosol fuel particle has a much higher allowed fuel density (5 mg cm^{-3}) while remaining optically transparent to photons [58]. With a higher fuel loading, the reactor becomes more compact. It was determined that the reactor can be controlled by pulsing the fuel density in the gas flow. Using a U or Pu:Xe:F:D_2O system with a fuel density of 5 mg cm^{-3}, a core volume of 1.58 m^3 and a reflector thickness of 140 cm was found. The total

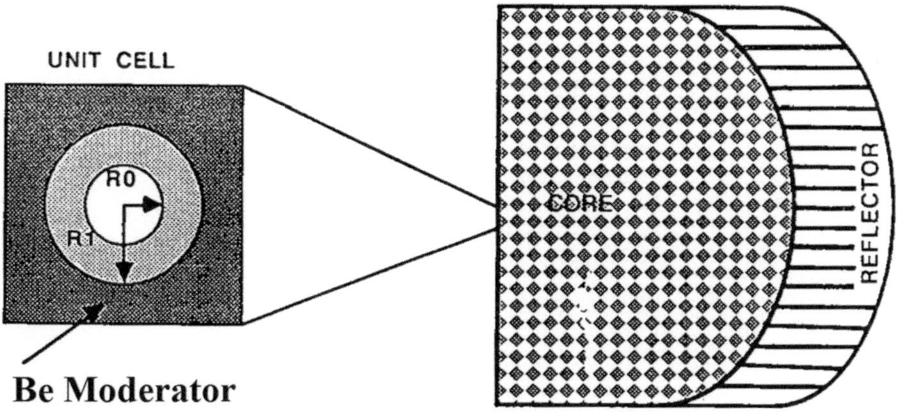

Fig. 5.28 A unit cell for an aerosol fueled reactor/nuclear-pumped laser

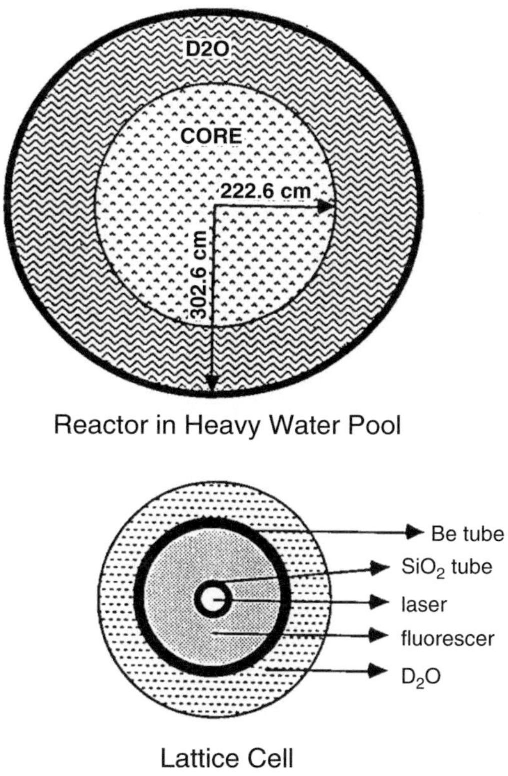

Fig. 5.29 Diagram of the U/Pu aerosol fueled Xe:F: D_2O reactor/laser

core volume was 52 m^3 and the keff = 1.427 for ^{239}Pu, 1.3138 for ^{233}U and 1.202 for ^{235}U. The core (Fig. 5.29) parameters are shown in Table 5.9. There are 136 cells which make up the core volume.

Table 5.9 Parameters of the reflective aerosol fueled reactor/flashlamp which pumps a xenon fluoride laser

Lattice cell	Laser tube radius (cm)	2
	Fluorescer tube radius (cm)	4
	Laser tube SiO_2 thickness (mm)	2
	Fluorescer tube thickness including 0.25 mm MgF coating (cm)	0.5
	D_2O moderator thickness (cm)	2
Materials	^{235}U:Xe:F:D_2O	
Cylindrical Reactor Size	Height (cm)	123
	Radius (cm)	64
	D_2O Reflector thickness (cm)	140
	Laser plus fluorescer volume (cm^3)	6×10^5
	Core volume (cm^3)	1.58×10^6
	Reactor volume (cm^3)	52×10^6
	k_{eff}	1.20

The neutron multiplication factor calculation assumes that the mixture of aerosol fuel and laser gas flows through the core at high speed. The pulse shape results from the combination of the fuel flow and the reactor kinetics. The pulse width of the neutron flux can be adjusted by varying the flow rate (100 m s^{-1}). The absorption macro cross section are much lower than that of a typical power reactor, thus the prompt neutron lifetime is relatively long for the low average fuel density that is characteristic of an aerosol core reactor. The total delayed neutron fraction for ^{235}U, ^{233}U and ^{239}Pu are 0.0065, 0.0026 and 0.0021 respectively. The largest fraction of delayed neutrons is with ^{235}U thus it is easier to control. The point reactor kinetics equation can be written as:

$$\frac{dn}{dt} = \left(\frac{\rho(t) - \beta}{\tau}\right) n(t) + \Sigma\lambda_i C_i(t) + S_n(t) \tag{5.49}$$

$$\frac{dC_i}{dt} = \frac{\beta_i}{\tau} n(t) - \lambda_i C_i(t) \tag{5.50}$$

where n(t) is neutron density, β is the delayed neutron fraction, $C_i(t)$ is the precursor concentration (which is assumed to homogeneous thus no spatial dependence), λ_i is the decay constant for the ith precursor (Table 5.10), $S_n(t)$ is the neutron source rate, $\rho(t) = \frac{k(t)-1}{k(t)}$ is the reactivity, β_i is the delayed neutron fraction for the ith group of precursors (i = 1, 2, 3, 4, 5 and 6) , $\tau = \frac{\ell}{k}$ is the mean generation time, ℓ is the prompt neutron lifetime.

The kinetic parameters of the reactor/laser fueled with reflective aerosol particles are shown in Table 5.11. If the reactivity feedback is ignored (approximately true for ^{239}Pu due to its small temperature coefficient) and the initial reactor power is assumed to be 1 kW, the core kinetics calculations were performed with PARET and generate some useful data on the dynamics of the pulsed reactor (see

Table 5.10 Delayed neutron data for ^{235}U

Group	Half life (s)	Decay constant (s)	Fraction β_i
1	55.72	0.0124	0.000215
2	22.72	0.0305	0.001424
3	6.22	0.111	0.001274
4	2.30	0.301	0.002568
5	0.61	1.14	0.000748
6	0.23	3.01	0.00273

Table 5.11 Kinetic parameters of a reactor/laser fueled with reflective aerosolized fuel

	^{235}U	^{233}U	^{239}Pu
Temperature reactivity coefficient α_T (1/°F) from 68 to 1400 °F	-1.95×10^{-4}	-1.99×10^{-4}	-6.95×10^{-5}
Effective delayed neutron fraction β_{eff}	0.71 %	0.31 %	0.27 %
Mean prompt neutron lifetime (ℓ) sec	2×10^{-3}	3×10^{-3}	1×10^{-3}

Table 5.12 Parameters for a pulsed aerosol core reactor fueled with reflective aerosol particles with an operating temperature of 1160 K [56]

Fuel	K_{eff}	Prompt neutron life (ms)	Pulse width (ms)	Peak power MW	Peak thermal flux $cm^{-2}\ s^{-1}$
^{239}Pu	1.42	1	50	1600	2.0×10^{16}
^{233}U	1.31	3	170	80	1.8×10^{15}
^{235}U	1.2	2	180	130	3.0×10^{15}

Table 5.12). These calculations show that ^{239}Pu is superior to the other fuels. Unfortunately the heavy water reflector assumed in the preliminary design leads to a large system mass (56,000 kg). The pulse characteristics of the reactor are shown in Fig. 5.30. The mass can be reduced substantially by using a 20 cm thick beryllium reflector (total volume of 3.61 m^3 and total mass of 6,500 kg). However the core reactivity reduces to 1.20.

A conventional nuclear-pumped laser design, such as that described in Sect. 5.6, using thin uranium coatings interfaced with the laser gas. The thickness of the uranium coating must be kept at a minimum to reduce self-absorption (i.e., 1–3 μm). The ranges of fission fragments in some materials common to a reactor/laser care are shown in Table 5.13.

This design has the same problem as an aerosol core reactor because the fuel thickness needs to be made thin (less than a mean free path of a fission fragment) in order to maximize the power deposition in the laser medium. For a low fuel density reactor, a moderator should be used as a structural material in order to avoid large neutron absorbers. If a beryllium moderator to fuel ratio is on the order of 3000/1, then the average fuel loading is going to be on the order of 1.2 mg cm^{-3}. This fuel loading is very similar to an aerosol core reactor fueled with black aerosol particles (volume of 46 m^3 and a mass of 20,000 kg) [56].

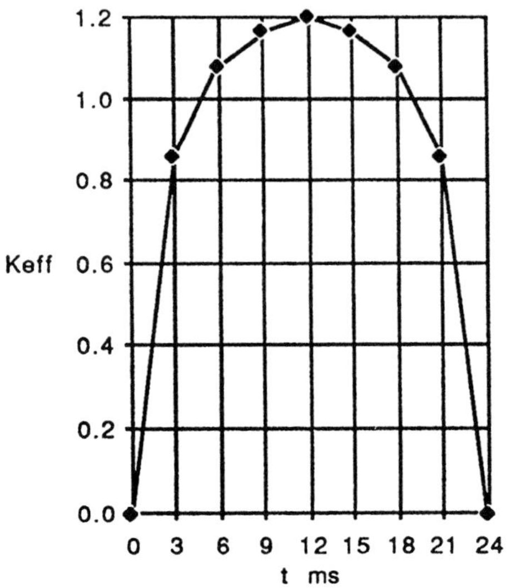

Fig. 5.30 Calculated pulse reactivity for a reactor/laser fueled with reflective aerosol fuel particles and an operating temperature of 1160 K. The flow rate is 100 m s^{-1}

Table 5.13 Ranges of fission fragments for uranium metal, UO_2 and U_3O_8 coatings and other materials in a reactor/laser core

	Heavy fission fragment	Light fission fragment
Energy (MeV)	67.5	98.7
Mass number	138.8	94.7
Range in Helium (cm)	2.3	2.87
Range in Neon (cm)	2.0	2.66
Range in Argon (cm)	1.94	2.39
Range in CO_2 (cm)	1.96	2.50
Range in U_3O_8 (cm)	1.04×10^{-3}	1.35×10^{-3}
Range in UO_2 (cm)	8.60×10^{-4}	1.19×10^{-3}
Range in U metal (cm)	4.67×10^{-4}	6.07×10^{-4}

Gas pressures are assumed to be 1 atm at 288 K

A nuclear-driven flashlamp fueled with a fissile aerosol to drive various kinds of lasers (iodine, excimer and Nd^{+3} doped solid state lasers) were studied and will be discussed in the next section [12, 32, 33]. A nuclear driven flashlamp fueled with UF_6 to produce fluorescence in the near infrared to drive Nd^{+3} doped solid state lasers was also examined as part of this group of reactor/laser designs with promising results [59]. The absorption properties of UF_6 in the near infrared are suitable for its use as fuel for a near infrared flashlamp to drive Nd^{+3} doped solid state lasers.

UF_6 decomposes to UF_4 and will plate out on walls. Such decomposition becomes significant above 1500 K, but can be stabilized (equilibrated) with the addition of some fluorine gas. So it is difficult to combine the UF_6 gaseous fuel and

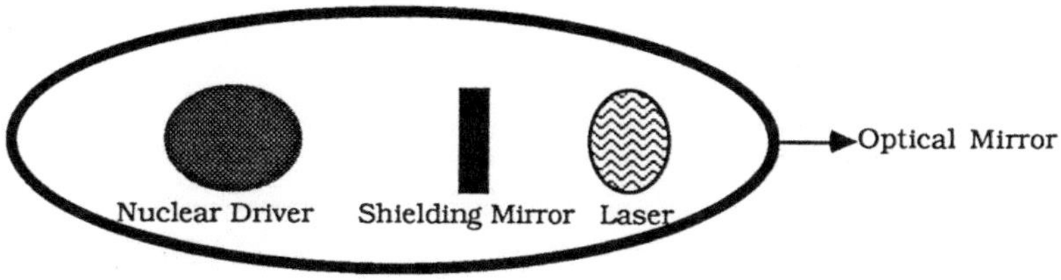

Fig. 5.31 Concept for the coupling a nuclear driven flashlamp to a photolytic driven laser

the gas laser medium as a coupled system without proper engineering considerations. In order to use UF_6 as the energy source and to protect the laser medium from the problems associated with fluorine it appears that separation of UF_6 and the laser medium is required. Therefore a nuclear flashlamp based on a gaseous UF_6 was designed and an artist's conception is shown in Fig. 5.31. In this concept the $^{235}UF_6$ is both the fuel and part of the fluorescer mixture. With a moderator it can make a self-critical nuclear-driven flashlamp. The ^{235}U fission fragments excite atomic and/or molecular levels in the gas mixture in the fluorescer cell which then emit photons (red or infrared for a Nd^{3+} doped solid-state laser). These photons are concentrated into a laser region to optically pump the solid-state laser. The nuclear driver is thus a gas core reactor that emits radiation in the visible or infrared that will "pump" the solid-state laser.

The neutronics of the UF_6 gaseous laser driver is examined through a series of calculations for different gas pressures, different temperatures and different moderator-to-uranium ratios have been performed. The codes that were used is the one-dimensional multigroup transport theory code XSDRN with 123 group cross section library. the one dimensional multigroup (19) diffusion code DISNEL and the reactor transient code PARET

The nuclear-driven flashlamp has a thin transparent tube which is needed to transport photons from the lamp to the laser (Fig. 5.32). The UF_6 gas pressure was limited between 0.5 and 2 atm. Three different temperatures, 323, 573, and 900 K, were chosen for the analysis. In order to optimize the moderator-to-uranium ratio, four volume fractions of beryllium were compared. These volume fractions are listed in Table 5.14. The fuel corresponds to typical high enrichment output from a gaseous diffusion plant and is known as Oralloy (93.2 % ^{235}U. 1 % ^{236}U, 0.4 % ^{234}U and 5.4 % ^{238}U), and is used for the U in the gaseous UF^6. From the ideal gas assumption, the number density of UF_6 at standard temperature and pressure is $2.69\times10^{19}\ cm^{-3}$ (about 10 mg/cm^3). Table 5.15 shows the number densities of UF_6 for the different cases, and the equivalent Oralloy metal (specific gravity 18.7) volume fractions. These number densities and volume fractions are used in the calculations. In the gaseous core the uranium number density is much lower than for solid fuels.

(The cross section correction factors (self-shielding factors) can be ignored for both the thermal and the resonance regions. Because S/V (surface to volume ratio) is about equal to 1 and Σ_a is about 10^{-3} for the gaseous core, the thermal

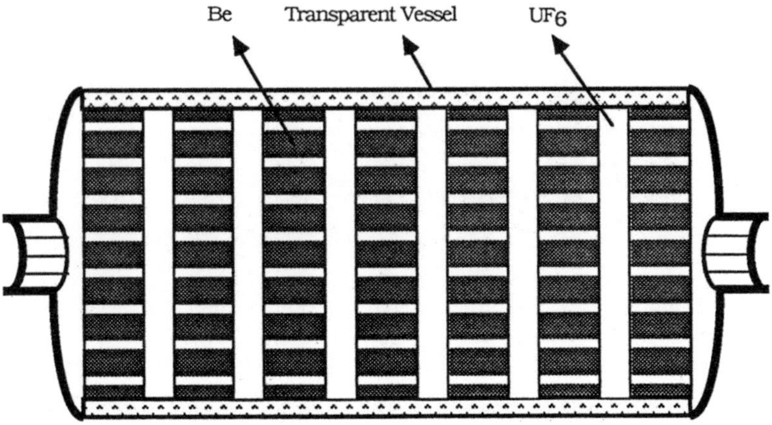

Fig. 5.32 A transparent UF_6 core reactor/nuclear-driven flashlamp

Table 5.14 Four volume fractions of beryllium selected for the UF_6 core reactor

Volume fractions (%)			
Core configuration	Be	SiO_2	UF_6
1	75	2	23
2	65	2	33
3	40	2	58
4	20	2	78

Table 5.15 Influence of pressure and temperature on the number density of UF_6 in the gaseous core

Pressure (atm)	Temperature (K)	Uranium atoms cm^{-3}	Oralloy volume fraction (18.7 g/cm^3)
0.5	323	1.136×10^{19}	2.377×10^{-4}
	573	6.403×10^{18}	1.340×10^{-4}
	900	4.076×10^{18}	0.853×10^{-4}
1.0	323	2.262×10^{19}	4.753×10^{-4}
	573	1.281×10^{19}	2.68×10^{-4}
	900	8.153×10^{18}	1.706×10^{-4}
2.0	323	4.544×10^{19}	9.506×10^{-4}
	573	2.562×10^{19}	5.360×10^{-4}
	900	1.631×10^{19}	3.412×10^{-4}

disadvantage factor, $F = \left(1 - {}^{1}\!/_{1+0.314\left(\frac{S}{V}\right)\Sigma_a}\right) \approx 1$. For similar reasons, the effective scattering cross section per resonance absorber for the gaseous core was on the order of 10^4–10^5 b (the path length in the moderator is much larger than in the fuel) and the resonance self shielding can be ignored. This is true for both the ^{235}U and the small (5.42 %) amount of ^{238}U present.).

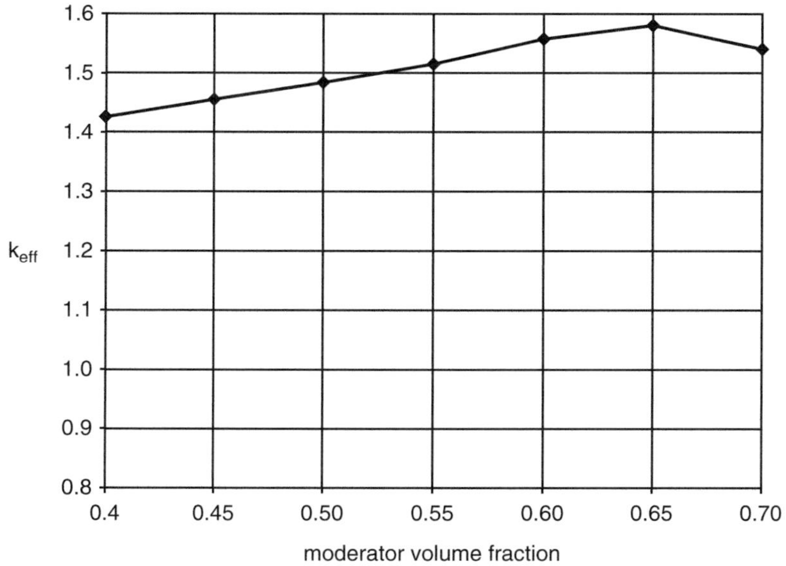

Fig. 5.33 The moderator volume fraction was varied and keff calculated to find an optimum value which turned out to be 0.65

The reactor size was chosen based on preliminary calculations. A cylindrical core of 2 m in diameters and 4 m long (core length 3.6 m plus two end reflectors, each 0.2 m thick) was chosen giving a total core volume of 12.5 m^3. Given the low partial pressure of UF_6 in the gaseous core, the calculations demonstrated that it is difficult to build a radial bare compact reactor with k_{eff} on the order of 1.20 (a number which is required for fast high power pulsed operation). For a bare cylindrical reactor, a pressure of 2 atm of UF_6 gas is needed to reach a k_{eff} of 1.20 at temperature of 900 K. This is a huge amount of UF_6 and with this concentration, the plasma kinetics of the reactor become a problem. It is thus desirable to operate at a much lower fuel density. This will require either a larger reactor core size or a D_2O radial reflector to attain a k_{eff} of 1.20. As a first step in reducing the core size, calculations of the optimization of the moderator (Be slabs in the core region) volume fraction is 65–70 % were done. Figure 5.33 from this study shows the variation of k_{eff} with moderator volume fraction.

Given that the optimized moderator volume was 65 %, this value was fixed for all subsequent calculations. Figure 5.34 shows the effect of temperature on k_{eff}. As can be seen in this figure, even at high temperatures ($T > 900$ K), k_{eff} is still on the order of 1.2. (It should be noted that the gaseous fuel shows a distinct advantage over the aerosol core reactor fueled with "black" aerosol particles, which needs a total volume of 52.7 m^3 for the ^{235}U:Xe:F:D_2O system to achieve a k_{eff} of 1.20 and still maintain optical transparency. Thus, the gaseous core reactor concept is relatively compact (12.6 m^3) compared to the "black" aerosol fueled system. The

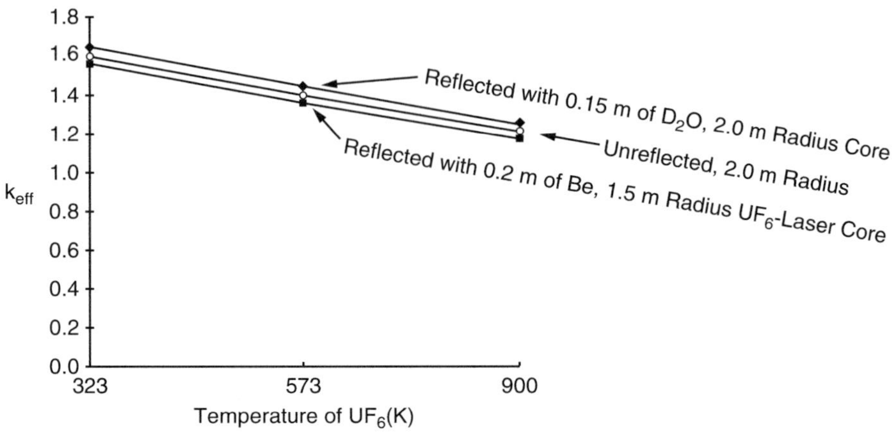

Fig. 5.34 Temperature effect on the multiplication factor

reason being that the fuel density for gaseous UF_6 at 2 atm and 573 K is 10 mg/cm^3, a factor of 10 higher than the 1 mg/cm^3 for the "black" aerosol core.)

The calculations assume that the thickness of the transparent vessel is 0.3 cm. This gives a hoop stress of about 7×10^7 N/m^2 with a pressure of 2 atm, which is well below the fracture stress of fused silica. The ultimate strength is as high as 1.4×10^9 N/m^2 for silica glass without surface flaws. A word should be said about the feasibility of using fused silica as containment for the UF_6. Fused silica is radiation resistant and maintains transparency at fairly high doses [60].

Placing a reflector of clear heavy water (15 cm thick) around the reactor produces a higher multiplication fact as shown in Fig. 5.34. The core size for this reflected configuration is the same as for the core of radial bare reactor (diameter of 2 m and height of 4 m), but the multiplication is about 4 % higher. Thus, a thin radial D_2O reflector does not add significantly to k_{eff}. Therefore, the flashlamp concept is shown to effectively reach the high multiplication factors needed with a design that does not require a radial reflector, thus avoiding the penalty that such a reflector would create in system mass and in light transmission losses (a few percent in the visible range).

Calculations were also made on a laser concept in which a clear path for light propagation along the axial (longitudinal) direction of the cylinder was created. In this case, a 65 % volume fraction of small rods of beryllium, oriented axially, were used as the moderator, the reactor core diameter was 1.5 m (instead of 2 m as used in all previous designs in this study) and the thickness of the radial beryllium reflector was 20 cm (see Fig. 5.35). The results show that a k_{eff} of 1.2 can be reached at a 900 K operating condition (Fig. 5.36).

Using perturbation theory for a UF_6 core reactor with a 65 % moderator volume fraction at 2 atm and 573 K, the mean prompt neutron lifetime was found to be about 4×10^{-4} s and the effective delayed neutron fraction, β_{eff}, is 0.71 %. This is a rather long prompt neutron lifetime for a configuration intended to be used in the

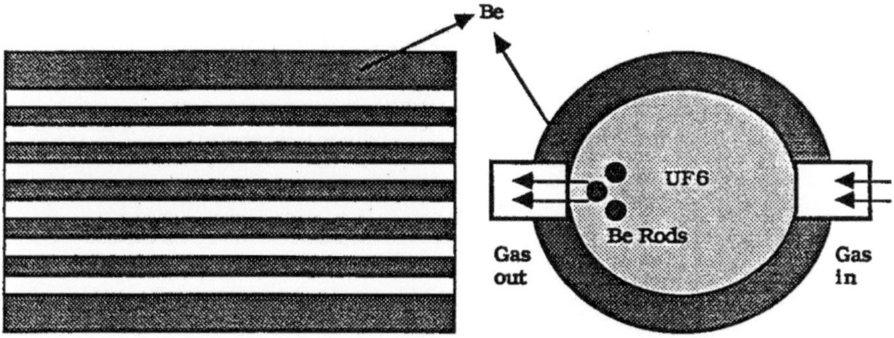

Fig. 5.35 Gas core nuclear-driven flashlamp with radial beryllium reflectors

Fig. 5.36 Supercritical pulse reactivity of a UF_6 fueled nuclear-driven flashlamp with radial beryllium reflectors

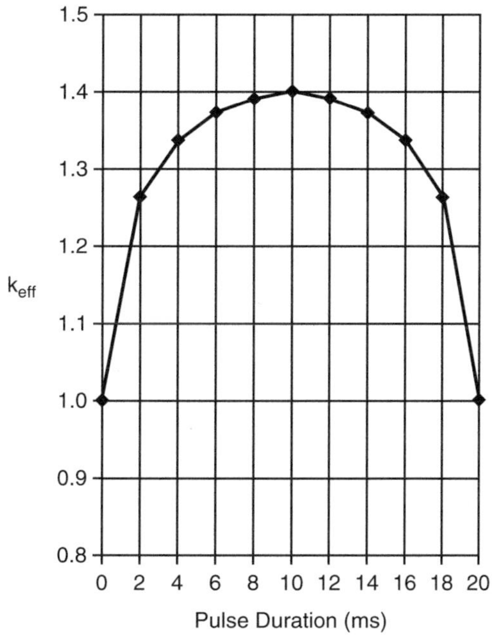

pulsed mode with high peak powers, hence short pulses. An approximate analysis of the expected pulse width was made so as to give an indication of what might be expected from this nuclear-pumped laser system. If the initial volume of the UF_6 gas moving through the core, envision it to be a "slug" of mass that is as long as the core (4 m), then that "slug" of mass is injected into the core at approximately the velocity of sound 320 m/s at room temperature. This "slug" of mass will cause the core reactivity to rise as it moves through the core and fall as it moves out of the core. The full width of this pulse of mass addition to the core will create a supercritical reactivity pulse. The PARET code is used for the point kinetics model calculation and the reactivity pulse according is found for the flow of the mass "slug" moving through the core. As shown in Fig. 5.36, the pulse width was about 20 millisecond. During this pulse the core power increased by a factor was 3.4×10^6 (i.e., if the initial power, Po, is 100 W then the peak power is 340 MW). The initial power (source) level can be tailored somewhat immediately prior to operation by taking advantage of the (delayed) gamma-neutron reaction in the beryllium. This estimate of pulse width was then used to determine the survivability of the UF_6 fuel in its molecular form. Significant dissociation begins at temperatures above 1500 K and must be avoided. The above pulse width, peaking at 1000 MW, develops a total energy of 2.0 MJ. If all of this energy is deposited in the UF_6 (840 J/kg-K specific heat), then the temperature rise of the 9 kg fuel slug will be 265 K, well below the dissociation level (assuming near room temperature initial conditions). The Doppler coefficient was calculated to be about 10^{-5}/K, thus giving a peak effect during the pulse of only 0.26 % of k_{eff}. The space-time dependent kinetics in this rather large reactor could be impacted by pressure waves and perturbations that are created by the temperature increase. The results given here are based on simplified assumptions and are meant only to show nominal feasibility of pulsing the UF_6 fueled core.

In space, heat must be rejected by radiators. This puts a constraint on the laser in that it should operate at as high a temperature as possible in order to reduce the size of the radiator. The systems that are necessary for a space based lasers are (Fig. 5.37):

- Fuel Source
- Energy Convertor
- Energy Storage
- Laser System
- Radiator and Coolant Flow System

Beginning with the radiator, there are materials which weigh approximately 50 mg m^{-2} that have been assumed in the design of solar satellites. Thus using an assumption that a radiator would weigh around 0.1 kg m^{-2} (including structural components) is reasonable. If the system dissipates 800 MW thermal with an operating temperature of 673 K to space (with a sink temperature of 4 K), the required area of the radiator can be estimated by the Stefan-Boltzmann Equation,

$$P = \epsilon\sigma A\left(T_{hot}^4 - T_{ambient}^4\right) \tag{5.51}$$

where, P is power radiated, ε is the emissivity (for an ideal material ε 1), $\sigma = 5.6703 \times 10^{-8}$ Wm^{-2} K^{-4}, $T_{hot}^4 = (673K)^4$, $T_{ambient}^4 = (4K)^4$.

The area of the ideal radiator is 58,500 m^2 with a mass of 5,850 kg.

As an example, consider a space based electrically driven laser. In order to estimate the total mass of the system each of the major components need to be considered ((1) Fuel Source, (2) Energy Convertor, (3) Energy Storage, (4) Laser System, (5) Radiator and Coolant Flow System).

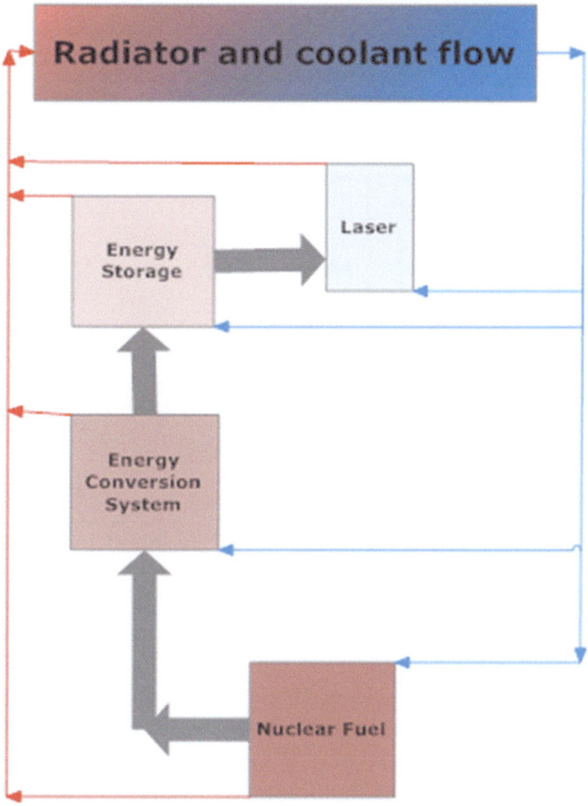

Fig. 5.37 Basic systems used in a space based laser system. In the case of an electrical pumped laser, the nuclear fuel would produce heat, the energy convertor would be a Stirling engine, the energy storage would be a capacitor bank & associated power circuits and the laser would be an independent system including optics, gas flow systems, tracking electronics, etc

- Fuel Source: The mass of the uranium fuel for a power producing reactor core has been discussed ($\gtrsim$100 kg).
- Energy Convertor: The means to convert the energy from the fuel to the energy form used to drive the laser is complex. With a nuclear reactor used as a heat source and a Stirling engine for the conversion of heat to electricity (with 30 % efficiency for an operating temperature of 800 K), the mass of the reactor structure and the Stirling engine would be about 25,000 kg.
- Energy Storage: The mass of the capacitors and power electronics needed to drive an electrically driven laser is about 1000 kg.
- Laser System: The mass of the laser and auxiliary systems is about 3000 kg.
- Radiator and Cooling System: The mass of the radiator and coolant is on the order of 2000 kg.

Thus the total mass of electrically driven space laser would be ~31,000 kg. In contrast a reflective aerosol core reactor with a beryllium moderator would have a total mass of about 12,350 kg.

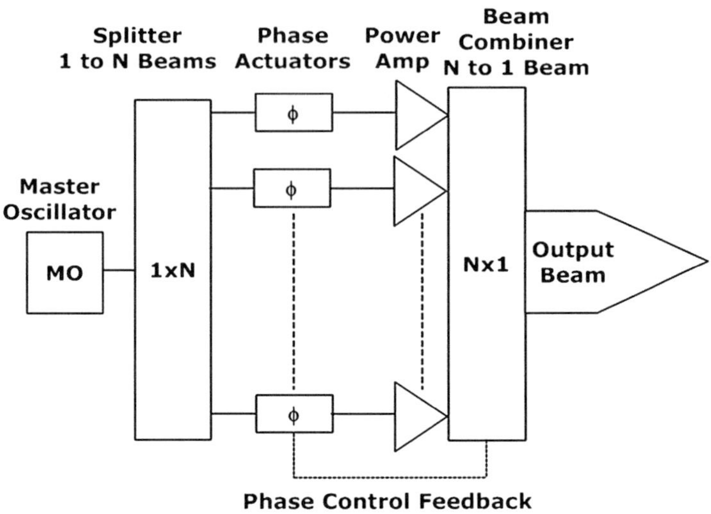

Fig. 5.38 A diagram of an actively phased locked coherent beam combination (CBC) master oscillator power amplifier (MOPA) array

Most of the reactor/nuclear-pumped laser system concepts have unit cells that consist of the fuel, the moderator and the laser medium. A core is typically made up of a large number of unit cells (136 for the reflective aerosol fueled reactor with a beryllium reactor described above). Each individual laser beam needs to be phased with the beams that make up the core. In the NASA UF_6 fueled laser (from 1979), there is a master oscillator and each of the 54 laser cells are single pass amplifiers. Significant progress has since been made in coherent beam combining. The critical issue in coherent beam combination is the identification of the optimum architectures and methods [61]. Fiber lasers have used coherent beam combination to produce 30–40 kW lasers which have been used to shoot down rockets and artillery shells [62]. One basic concept is the actively phased locked coherent beam combinations master oscillator power amplifier array (Fig. 5.38). The three major components for this array are: (1) A high gain/high power laser amplifier which maintains coherence properties of the master oscillator; (2) Optical systems which geometrically overlap the beams in the far field; (3) Active control system which co-phase the amplified output beams in a closed loop feedback system.

An actively phase locked CBCMOPA is a complicated system and scaling it using a reactor/nuclear-pumped laser with power levels >10 MW will be challenging. Specifically, the spatial temperature variation in the laser cell creates spatially dependent density differences in the laser medium. These density variations, $\frac{\Delta\rho}{\rho}$, create problems in phase locking the beams from each laser cavity [63].

Another method which has been used to combine multiple coherent broad-band pump beams into a single coherent Stokes wave is forward-stimulated Raman scattering [63–68]. The multiple pump waves should propagate in the direction of the Stokes wave propagation in order to maximize efficiency. This is a challenging condition.

Coherent beam combining has been a difficult problem to overcome.

5.7.4 Photolytic Nuclear-Pumped Laser Designs University of Missouri (1981–91)

As the prior sections discuss, the method of excitation is an important design factor which impacts everything from the system size, the type of laser that can be used to the method of coherent beam combining. For example, the use of nuclear-driven fluorescers combined with photolytically driven lasers has some important advantages.

Initially (1980–1985) the nuclear-pumped photolytically driven laser system research was focused on rare gas excimer fluorescers and gas photolytically driven lasers [18, 32, 33, 69, 70]. The fluorescer gas work was centered on the xenon and xenon bromide fluorescers and the xenon fluoride and iodine photolytic lasers [3, 22, 32, 33, 69, 71, 72].

This study of the aerosol core nuclear reactor driven flashlamp was based on the use of the DISNEL code (19 group diffusion code). The following materials were used:

- Fuel: UC, UO_2 , and PuO_2
- Moderator: Beryllium and Stainless Steel (fast reactor)
- Reflector: Beryllium and Stainless Steel (fast reactor)
- Fluorescer: Xenon
- Laser: Iodine and Argon
- Structure: Fused Silica and Stainless Steel

Results from this study indicate that a thermal reactor using plutonium oxide and beryllium will have the smallest critical parameters (for example a sphere with a 1-m radius core and a 0.2-m Be reflector had a $k_{eff} = 1.5022$). The uranium and beryllium reactor performed slightly poorer (a sphere with a 1-m radius core and a 0.2-m Be reflector had $k_{eff} = 1.299$). A study of potential fast reactor configurations demonstrated that Pu with a stainless steel reflector will, considering a spherical geometry, have a radius of about 15 m while uranium with a stainless steel reflector will have a radius of about 22 m.

In addition, the code provided a means to calculate the average neutron lifetime. With this value plus k_{eff} it was determined that the fast reactors would have pulse widths on the order of several hundred microseconds while the thermal reactors would have pulse widths on the order of milliseconds. It turns out that the plutonium fueled reactors had shorter lifetimes than the uranium fueled reactors. Clearly, with respect to size and pulse width, the plutonium thermal reactor is superior. The required laser power output will force the reactor to be much larger than a just

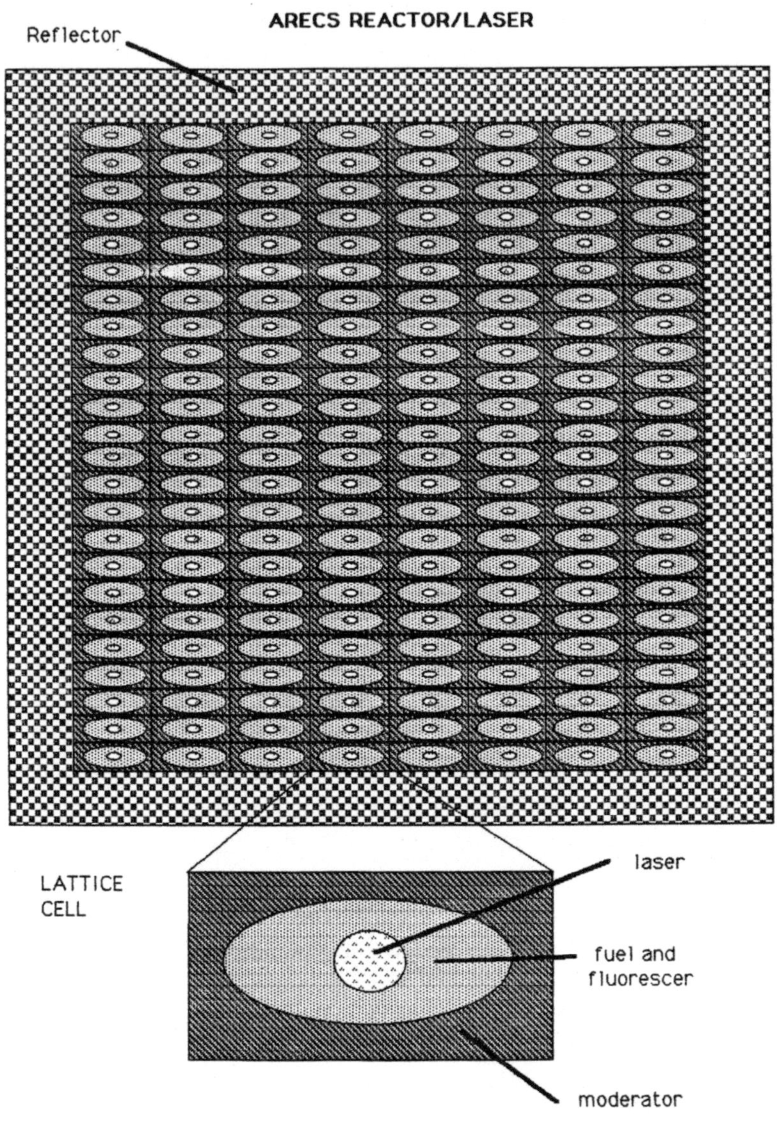

Fig. 5.39 An illustration of a photolytic nuclear-pumped iodine laser that is fueled with non-reflective (black) U-235 particles. The mass was estimated to be 15 metric tons and the volume 43 m^3 [23]

critical reactor (for example a just critical square right cylindrical reactor would be just critical for H = R = 108.4 cm). A 1-MJ laser pulse (1 ms FWHM pulse) is, hopefully, a conservative assumption for practical applications. The reactor volume can be calculated using unit cells (Fig. 5.39). The reactor volume was estimated to be around 43 m^3 using a black aerosol.

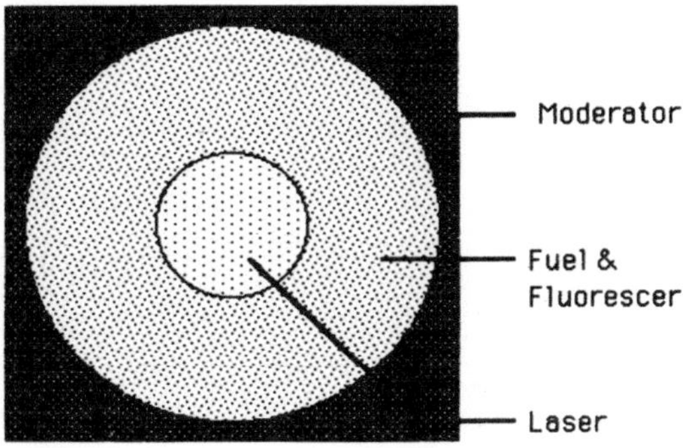

Fig. 5.40 A unit cell of a photolytic iodine nuclear-pumped laser fueled with reflective U-235 particles. The mass of the unit was estimated to be 1 metric ton with a volume of 0.6 m^3 [58]

A diffusion based optical transport model was used to calculate the sizing of a core fueled with reflective aerosol particles [58]. The size of the reactor is significantly reduced to about 0.6 m^3 per MJ of laser output with a mass of 1 Tonn (Fig. 5.40).

The transport of photons created in a weak plasma formed by fissioning aerosol mixed with a fluorescer gas is important for understanding how the scaling of a nuclear-driven flashlamp. The development of a transport code is very difficult [34]. A model based on diffusion theory was developed [35]. The fundamental aspect of the model is that the cross section of an aerosol particle can be approximated by assuming that the particle is spherical. The total cross section of a spherical particle is,

$$\sigma_t = \pi r_p^2 \tag{5.52}$$

where r_p is the radius of an aerosol particle.

The particles are in the micrometer range based on the transport range of fission fragments in the fuel material. The interaction of photons with a particle is modeled by the probability of a scattering event when a photon strikes a particle. This probability of scattering off of the particle is the reflection coefficient R_o. This reflection coefficient can contain information of the reflectivity of pellet wall as well as Rayleigh scattering (although the particle size distribution is such that the particles are generally larger than the wavelength of the photons). The scattering cross section is,

$$\sigma_s = R_0 \sigma_t \tag{5.53}$$

The absorption cross section is found by assuming that the photon either is scattered off the particle or absorbed by the particle. Thus the absorption cross section is,

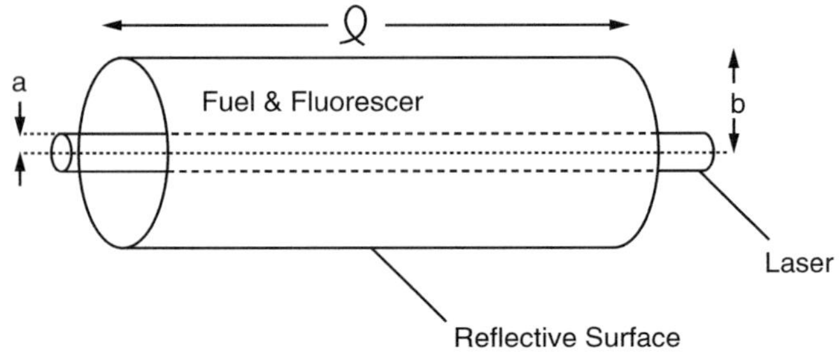

Fig. 5.41 A diagram of the cylindrical geometry chosen for the model

$$\sigma_a = (1 - R_0)\sigma_t \tag{5.54}$$

Three assumptions are made about the behavior of photons in an environment where particles are reflective in order to apply diffusion theory:

1. $\sigma_s > \sigma_a$
2. The intensity is a slowly varying function of position
3. Intensity does not vary with time

These assumptions closely resemble those of neutron diffusion theory used in reactor physics. The governing equation for diffusion theory is,

$$\nabla^2\Phi - {}^{\Phi}\!/_{L^2} = -{}^{S}\!/_{D} \tag{5.55}$$

where $L^2 = {}^{D}\!/_{\Sigma_a}$, $\Sigma_a = G\sigma_a$, G=aerosol density (cm^{-3}), $D = {}^{\Sigma_s}\!/_{(3\Sigma_t^2)}$ and S=photon source (photons cm^{-3} s^{-1}).

In order to solve Eq. (5.55), the geometry needs to be chosen. In this case reactors typically have a cylindrical geometry (but other geometries can be used). Unit cells are stacked in order to form the core and the laser is part of a unit cell as shown in Fig. 5.40. Thus the reactor/laser has many laser beamlets which need to be combined. In the example below, the geometry chosen was cylindrical and it was assumed that the core radius was much smaller than the core height so that the reactor physics was dominated by the dimension in the radial direction (Fig. 5.41). The core was also assumed to be homogeneous. With these assumptions Eq. (5.55) becomes,

$$\frac{1}{r}\frac{d}{dr}\left(r\frac{d\Phi(r)}{dr}\right) - {}^{\Phi(\mathrm{r})}\!/_{L^2} = -{}^{S}\!/_{D} \tag{5.56}$$

The boundary conditions come from the physics of the system. The laser is assumed to be a strong photon absorber so it would resemble a control rod in the analogous neutronics problem [38]. The boundary condition for an absorbing rod is,

$$\left[\frac{1}{\Phi(r)}\left(\frac{d\Phi(r)}{dr}\right)\right]_{r=a} = \frac{1}{d}$$

An alternate boundary condition assumed that no partial current traveling in the plus direction escapes the absorber.

$$J_+(a) = \frac{\Phi(r)}{4} - \frac{D}{2}\frac{d\Phi(r)}{dr}\bigg|_a = 0 \tag{5.57}$$

where a is the radius of the laser cell and d is the extrapolation distance. For a black absorber (a $> \lambda_{tr}$) $d \approx 0.71\lambda_{tr}$ and if the transport length is greater than the laser cell radius (a $< \lambda_{tr}$) then $d \approx 1.33\lambda_{tr}$.

The second boundary condition can be derived from the partial current at the fluorescer cell wall ($J_b^- = R_1 J_b^+ \, where J_\pm(b) = \frac{1}{4}\Phi(x) \mp \frac{D}{2}\frac{d\Phi(x)}{dx}\Big|_b$). The power balance (discussed below) of the cell can also be used as the second boundary. Both the partial current and power balance boundary condition gives the same results.

There are three potential loss mechanisms for a photon as it is transported in the unit cell: (1) It is absorbed by an aerosol particle, (2) It is absorbed by the wall, or (3) it is absorbed by the laser. The photon source strength per unit volume is a function of the physics of the fission fragment transporting out of the aerosol particles and into the gas. The source function has the form,

$$S = C_1 G r_p^3 e^{(-Ar_p)} \tag{5.58}$$

where C_1 and A are constants related to the fission rate and material properties. A is a function of the mean free path of a light fission fragment in the fuel (λ_l) and the mean free path of a heavy fission fragment in the fuel (λ_h). C_1 is a function of the power deposited in the fluorescer gas by the fission fragments exiting the fuel particle and the fluorescence production efficiency. Thus the term $e^{(-Ar_p)}$ represents the transport efficiency of the fission fragment energy out of the fuel particle (curve fit to Fig. 3.3 in Chap. 3 for ^{235}U). The term $C_1 G r_p^3$ (similar to Eq. 3.14 in Chap. 3) is the average power density deposited in the fluorescer gas from the aerosol fuel times the efficiency of the production of fluorescence from the gas (e.g., Table 3.3 in Chap. 3) divided by the energy per photon. This gives the photon produced per cubic centimeter per second.

A photon balance equation can be written realizing that the following must be true,

$$\begin{aligned}&\text{Photons Absorbed by Wall} + \text{Photons Absorbed by Aerosol}\\&\quad + \text{Photons by Absorbed Laser}\\&\quad = \text{S} \times \text{Volume}\end{aligned} \tag{5.59}$$

with,

$$Loss\ to\ Wall = (1 - R_1)\mathrm{J}_+(b)(2\pi b\ell) \tag{5.60}$$

where R_1 is the wall reflectivity, b is the radius of the cylindrical chamber, ℓ is the chamber length.

$$Loss\ to\ Aerosol = \int_a^b \Sigma_a \Phi(r)(2\pi r\ell)dr \tag{5.61}$$

$$Loss\ to\ Laser = \mathrm{J}_-(a)(2\pi a\ell) \tag{5.62}$$

Thus,

$$(1 - R_1)\mathrm{J}_+(b)(2\pi b\ell) + \int_a^b \Sigma_a \Phi(r)(2\pi r\ell)dr + \mathrm{J}_-(a)(2\pi a\ell) = S\left(b^2 - a^2\right)\pi\ell \tag{5.63}$$

The fraction of energy lost to the laser (or coupling efficiency η_L) is,

$$\eta_L = \frac{Photons\ Absorbed\ by\ Laser}{Total\ Photons\ Produced} = \frac{\mathrm{J}_-(a)(2\pi a\ell)}{S\left(b^2 - a^2\right)\pi\ell} = \frac{\mathrm{J}_-(a)(2a)}{S\left(b^2 - a^2\right)} \tag{5.64}$$

The fraction of energy absorbed by the aerosol particles (η_a) is,

$$\eta_c = \frac{Photons\ Absorbed\ by\ Aerosol}{Total\ Photons\ Produced} = \frac{\int_a^b \Sigma_a \Phi(r)(2\pi r\ell)dr}{S\left(b^2 - a^2\right)\pi\ell}$$

$$= \frac{2\Sigma_a \int_a^b \Phi(r)rdr}{S\left(b^2 - a^2\right)} \tag{5.65}$$

The fraction of energy absorbed by the wall (η_w) is,

$$\eta_w = \frac{Photons\ Absorbed\ by\ Wall}{Total\ Photons\ Produced} = \frac{(1 - R_1)\mathrm{J}_+(b)(2\pi b\ell)}{S\left(b^2 - a^2\right)\pi\ell}$$

$$= \frac{(1 - R_1)\mathrm{J}_+(b)(2b)}{S\left(b^2 - a^2\right)} \tag{5.66}$$

Solving equation 5.56, the photon intensity as a function of r is,

$$\Phi(r) = B_1 I_0\left(\frac{r}{L}\right) + B_2 K_0\left(\frac{r}{L}\right) + {}^{SL^2}\big/_{D} \tag{5.67}$$

The volume of a reactor/laser may be estimated based upon the properties of the fluorescer and the laser. The first is the laser efficiency (η_{laser}) which is defined as laser energy/energy absorbed by the laser medium and the second is the fluorescence efficiency ($\eta_{fluorescer}$) defined as fluorescence energy produced/energy absorbed by the fluorescer. Depending upon the fluorescer and the laser, these values may vary significantly. For photolytic lasers, η_{laser} is high provided that the absorption spectrum of the laser closely matches the fluorescence spectrum of the photon emitter. One laser which can achieve a high η_{laser} with an efficient fluorescer that overlaps the absorption spectrum of the medium is the I_2* excimer laser [17]. The laser efficiency can have a value of 30 %. Fortunately, a fluorescer which can match the absorption spectrum of the laser is the efficient Xe_2* excimer. This fluorescer has a fluorescence efficiency of 40–50 % [2]. With this information, the volume (V) need to produce 1 MJ of laser energy can be found from the following equation,

$$V = \frac{1 \times 10^6}{\left[P_{df}\left({}^{\pi\left(b^2-a^2\right)}\big/_{b^2}\right)\eta_L \eta_{fluorescer} \eta_{laser} \tau\right]} \tag{5.68}$$

where P_{dl} is the power density in the fluorescer W cm^{-3}, τ is the pulse width in seconds (usually about 1 ms) and x is the moderator to fuel volume ratio.

Equations (5.63), (5.64), (5.65), (5.66), (5.67) and (5.68) are used to scale the reflective aerosol core reactor/laser. The input parameters are:

- a the laser radius in cm
- b the fluorescer radius in cm
- r_p the aerosol particle diameter cm
- G the aerosol particle density (particles cm^{-3})
- R_0 the reflectivity coefficient of the aerosol particles
- R_1 the reflectivity of the wall
- Fuel (^{235}U, ^{233}U or ^{239}Pu)
- $\Sigma_t = G\pi r_p^2$
- $\Sigma_a = (1 - R_0)G\pi r_p^2$
- $\Sigma_s = R_0 G\pi r_p^2$
- G = aerosol density (cm^{-3})
- $D = {}^{\Sigma_s}\big/_{\left(3\Sigma_t^2\right)}$
- $L^2 = {}^{D}/_{\Sigma_a}$
- S = photon source W cm^{-3}

The parameters $\Phi(r)$, η_c , η_a , η_w and V are evaluated by choosing values for the above variables. The numbers shown below were picked by an iterative process and were based on being near optimum values,

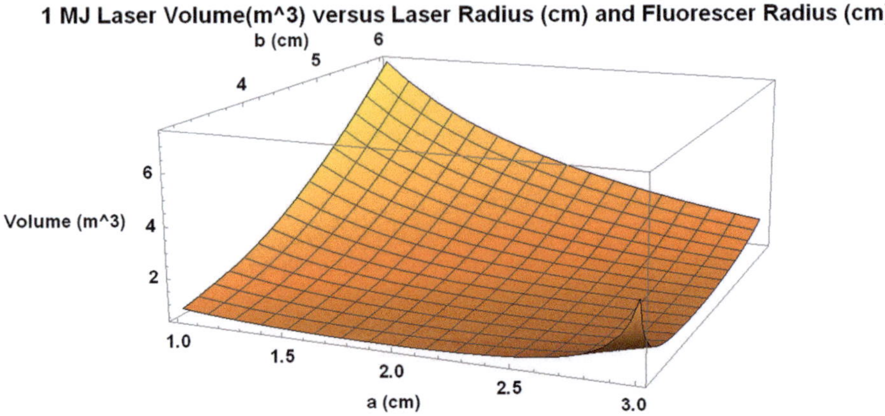

Fig. 5.42 Shown in the figure are the volume needed to produce 1 MJ of laser energy as a function of the laser radius (**a**) and fluorescence radius (**b**). This graph shows that there is an optimum for a laser radius of around 2.5 cm and a fluorescer cell radius of around 4 cm

- a = 2
- b = 4
- $r_p = 0.0005$
- G = 500,000
- $R_0 = 0.9$
- $R_1 = 0.9$
- Thermal neutron flux of 5×10^{16} cm^{-2} s^{-1}
- ^{235}U metal aerosol particles

Each parameter was varied and its effect on the laser volume was examined. The first parameter varied was laser radius and fluorescer radius (while holding the other parameters to the values shown above) and its impact on laser volume is shown in Fig. 5.42. The curve demonstrates an optimum laser radius near 2.5 cm and an optimum fluorescer radius near 4 cm. It should also be noted that the coupling efficiency continually improved as the laser radius was increased. It should be noted that as the laser radius increases, the volume of the nuclear driven fluorescer decreases thus resulting in an optimum volume for the laser. This effect can be more clearly seen in Fig. 5.43 which shows the laser energy emitted per unit cell as a function of laser radius and flourescer radius.

Figure 5.44 shows the photon absorption efficiency of the laser versus the laser cell radius (a) and the fluorescence cell radius (b) while holding the other parameters constant. As can be seen, the coupling efficiency increases as the radius difference between the fluorescer cell and laser cell decreases (b-a). This is to be expected because the solid angle for photon collection increases as (b-a) decreases.

Figure 5.45 shows the laser volume per MJ energy as a variation of the pellet radius (rp) and pellet density (G) while holding the other parameters constant. As can be seen, the laser volume decreases as a function of pellet radius. The coupling

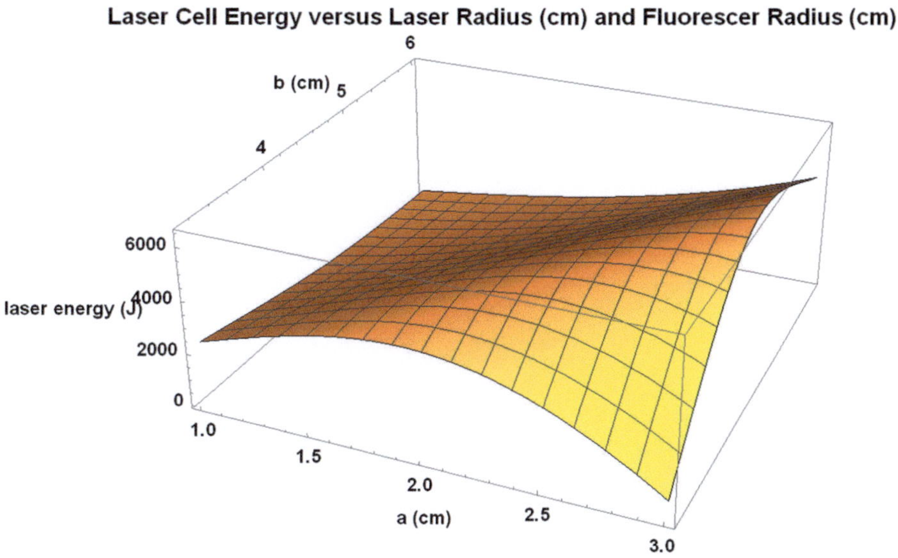

Fig. 5.43 Laser energy being emitted per unit cell as a function of laser cell radius (**a**) and fluorescence cell radius (**b**)

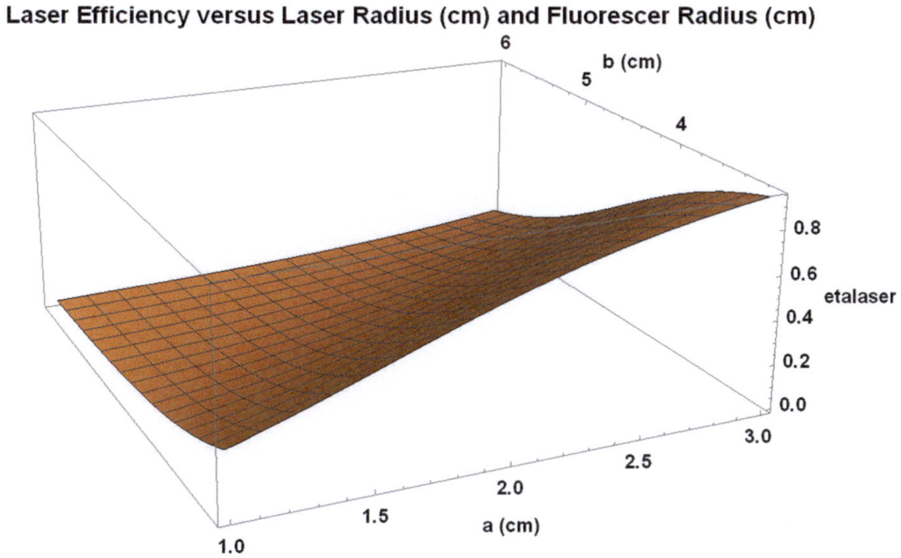

Fig. 5.44 Shown in the figure is the fraction of photons absorbed by the laser as a function of the fluorescer cell radius (**b**) and the laser cell radius (**a**)

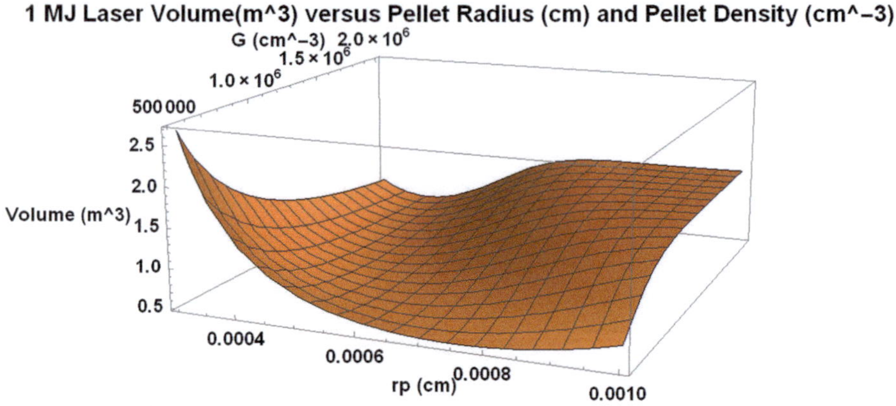

Fig. 5.45 Shown in the figure is the volume needed to produce 1 MJ of laser energy as a function of the pellet radius (**a**) and pellet density (**b**). This graph shows that there is a significant dependence on pellet radius and pellet density

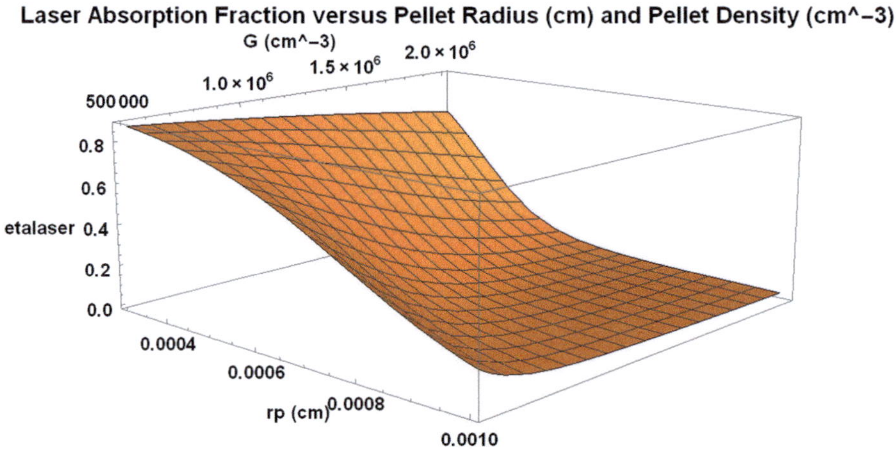

Fig. 5.46 Shown in the figure is the photon absorption efficiency in the laser as a function of pellet radius and pellet density. This graph shows a complicate relationship with pellet radius and pellet density

efficiency also drops slowly with the pellet radius. It should also be noted that as the pellet density increases, the laser volume decreases until an optimum value is reached then the laser volume increases. A pellet radius of 0.0005 cm and a pellet density of 1,000,000 cm-3 were chosen for the base case.

Figure 5.46 shows the photon absorption efficiency of the laser versus the laser cell radius (a) and the fluorescence cell radius (b) while holding the other parameters constant. As can be seen, the photon absorption efficiency decreases as the pellet density and pellet radius increases. The photon absorption efficiency of the laser plays an important role in the forming the contours of Fig. 5.45.

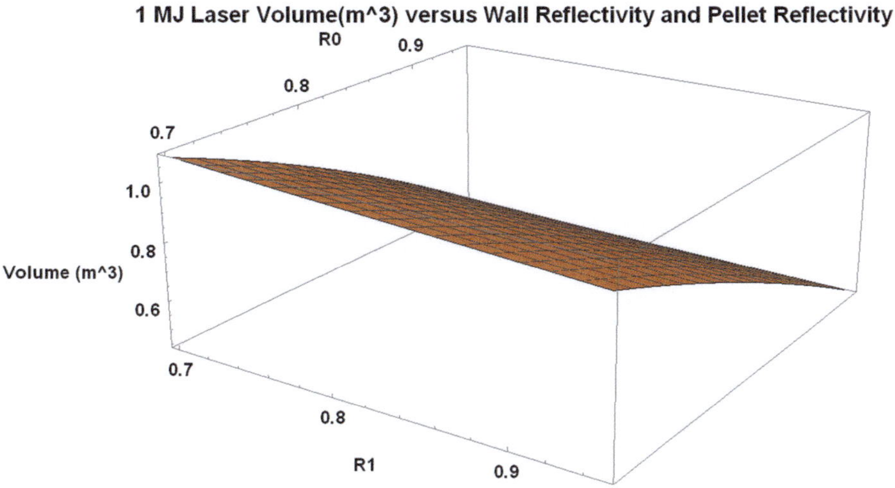

Fig. 5.47 Shown in the figure is the volume needed to produce 1 MJ of laser energy as a function of the pellet reflectivity (*R0*) and wall reflectivity (*R1*). This graph shows that there is a linear dependence on increasing pellet reflectivity

Figure 5.47 is the volume of for 1 MJ laser energy as a variation of the pellet reflectivity (R0) and wall reflectivity (R1) while holding the other parameters constant. As can be seen, the laser volume continues to decrease as both pellet and wall reflectivity increases. A pellet reflectivity of 0.9 for the base case was chosen based on the best available UV reflector (aluminum).

A reflective aerosol fueled reactor/photolytically pumped laser shows a significant reduction in the volume (5×10^5 cm^3 per MJ laser energy with a one ms pulse width and a flux of 5×10^{16} n cm^{-2} s^{-1}) and mass (1.0 Tonn per MJ laser energy) over a "black" aerosol fueled reactor. The intensity of the fluorescence on the surface of the photolytic laser is also significant as shown in Fig. 5.48. The maximum intensity is 21,500 W cm^2 (for G = 1,000,000) which is sufficient to drive most photolytic laser media. This study shows that there is a significant benefit in fueling a reactor with reflective aerosol particles.

From 1987 to 1991, the author began investigations on designs of solid-state photolytic laser media driven by self-critical reactor/ fluorescence sources in a project funded by the Department of Energy. [73, 74]. The motivation for this work was due to the low power density thresholds for some important solid-state lasers as seen in Table 5.16. Photolytic lasers can use the energy focus principle which enhances the power density absorbed by the laser medium that more than meets the power density threshold requirements of solid-state lasers [3]. Thus nuclear-pumped solid-state laser systems were deemed a very promising research endeavor [75–77]. The results from the first series of experiments which took place from September 1987 to December 1987 are described in Chap. 4. After these experiments were completed, comprehensive studies of improving flashlamp

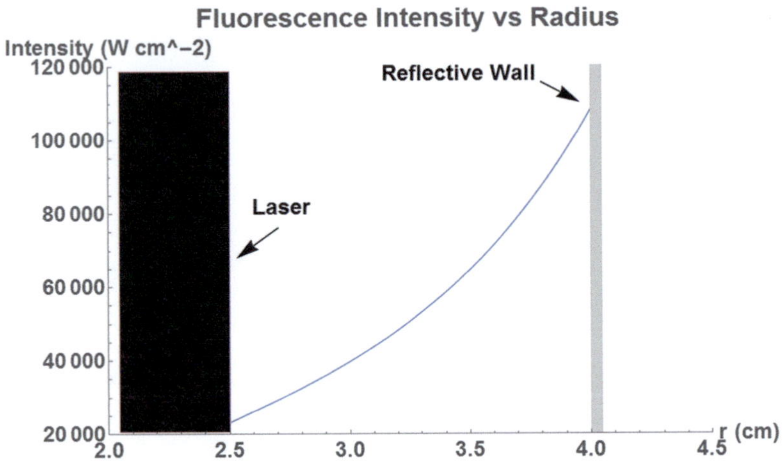

Fig. 5.48 The fluorescence intensity on the photolytic laser as a function of the fluorescence cell radius. The intensity on the laser cell is 21,500 W cm^{-2}, which is sufficient to drive most photolytic laser media

Table 5.16 The properties of solid-state laser crystals are shown in this table. The operational wavelength(s) (ω_1), stimulated emission cross section (σ_{se}), upper laser lifetime (τ_{ul}), thermal conductivity (k), threshold power density and current size limitations are listed

Medium	ω_1(μm)	σ_{se} (10^{-19} cm^2)	τ_{ul} (μs)	k (W m^{-1} K^{-1})	Threshold power density W cm^{-3}	Size (cm)
Nd:Cr: GSGG	1.06	1.3 at 300 K	222	6	115	9.6×19×0.5
Cr:GSGG	0.75–1.0	0.1 at 300 K	120	6	1149	9.6×19×0.5
Alexandrite	0.7–0.82	0.2 at 450 K	260	23	636	r=1. l=10
Ti:Al_2O_3	0.72–0.92	4.1 at 300 K	3.2	NA	3024	30.5×17×4.5
Emerald	0.73–0.81	0.2 at 300 K	200	NA	827	>0.7

efficiencies (that matched the absorption bands of solid-state lasers) and reactor system designs were undertaken by the author's students from 1987 to 1992 [78, 79].

The power density threshold of various photolytically driven gas lasers is shown in Table 5.17. Photolytically driven nuclear-pumped gas lasers can use the energy focus scheme, described in Chap. 4, in order to overcome the threshold power density requirements to drive a continuous wave laser with a steady-state reactor. Photolytic lasers also have some other advantages which can reduce the mass of large scale systems such as potentially higher system efficiencies. In addition, the gas flow can help with the thermal management [3, 22, 32, 69]. Gas lasers have also been studied using solar pumping [80] but the size and mass of these systems is much larger than that of nuclear-pumped systems.

Table 5.17 The properties of select gaseous photolytic lasers

Laser	Pump molecule	Pump wavelength (nm)	Laser wavelength (nm)	Threshold power density (W cm^{-3})
XeF*	XeF_2	160	480	15,000
I	C_3F_7I	282	1315	100
HgBr*	$HgBr_2$	193	502	15,000
HgCl	$HgCl_2$	<190	558	15,000

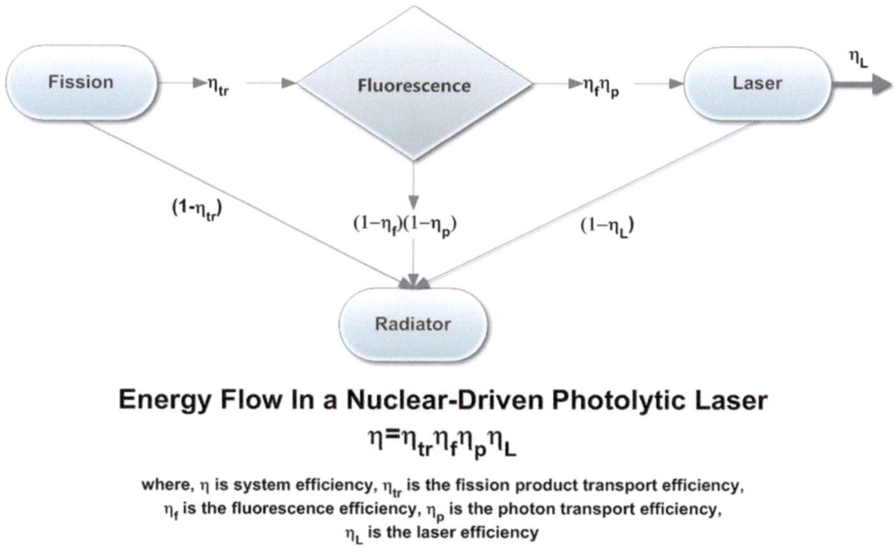

Fig. 5.49 The energy flow diagram for a nuclear-driven photolytic laser

An efficient coupling configuration for transporting light from a self-critical reactor/nuclear-driven fluorescer to a solid-state lasers is needed as shown in the energy flow diagram of Fig. 5.49 [81]. By applying a tapered reactor/nuclear-driven fluorescer cell design along with a large diameter-to-length ratio hollow lightpipe, the coupling efficiency of transporting photons from the nuclear-pumped fluorescer to the laser can be greater than 90 % [77]. Figure 5.50 shows a configuration using an aerosol core reactor/nuclear-driven fluorescer coupled to light pipes that transport the photons to a solid-state laser. The spectral matching of the photon source to the laser medium is very important for optically pumped lasers. By considering the wavelength of efficient fluorescers, such as the excimers, the laser crystal has to be chosen such that the absorption lines of laser crystal matches the wavelength of the fluorescer. The laser efficiency η_L is the product of the ratio of the pumping wavelength to the output wavelength of the laser and the extraction efficiency of

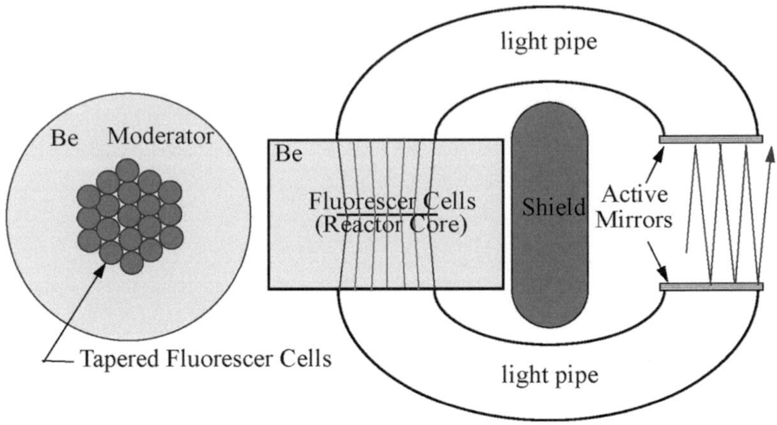

Fig. 5.50 Configuration of an aerosol reactor with coupling light pipes to active mirrors with an expected efficiency of 3 % and core volume of ~2.3 m^3 [28, 34]

Fig. 5.51 Mosaic of active mirror laser amplifiers [34]

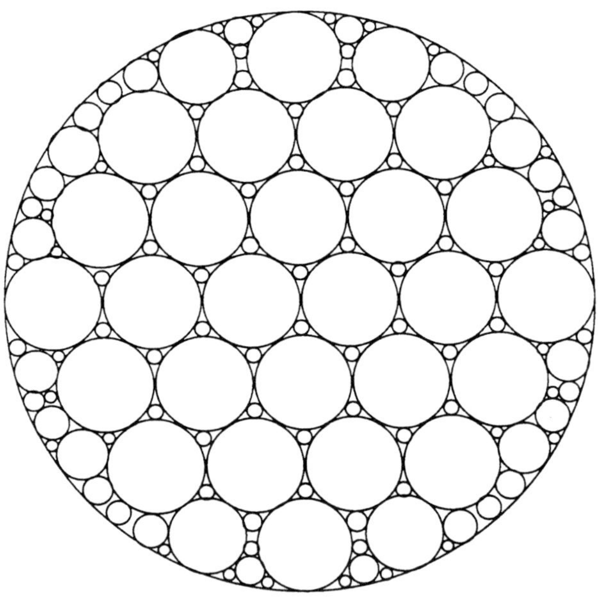

the laser cavity. The energy flow diagram shown in Fig. 5.49 determines how the system efficiency can be calculated for a compact nuclear-pumped solid-state laser.

As illustrated in Fig. 5.51, the light pipe [82] transports photons that are produced in the reactor to the mosaic amplifier of the Nd^{3+} doped crystal [76]. A conceptual large scale reactor/laser system utilizing the nuclear-driven fluorescer lamp to pump a remotely located solid-state laser concept is discussed. Total estimated efficiency is 3 %. The peak laser power is 12 MW, for 1-ms (FWHM)

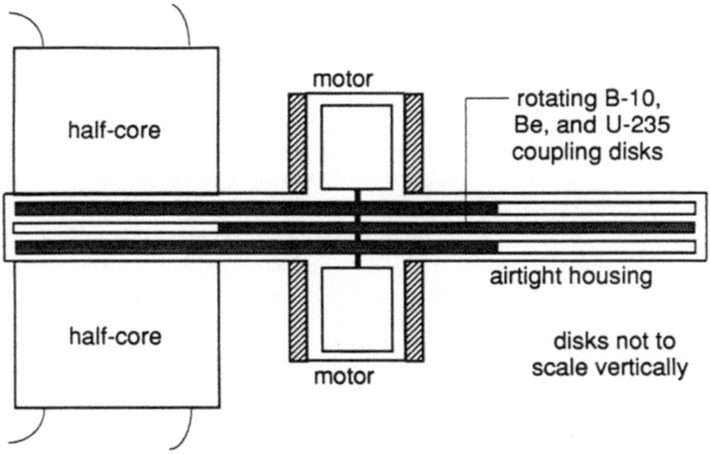

Fig. 5.52 1 ms, 1 Hz pulsed aerosol core nuclear reactor for pumping lasers. Two ^{235}U aerosol fluorescer cell half-cores are separated by three coupler disks, each composed of fuel and neutron absorber regions, rotating at different speeds. The peak power is 400 MW and the average power is 400 kW

pulses at a 1 Hz repetition rate, giving an average power output of 12 kW. The reactor power is 400 MW peak, 400 kW average. Active mirror solid-state laser amplifiers of Nd and Cr co-doped material, such as Nd,Cr:GSGG, are pumped by visible nuclear-driven alkali-metal excimer fluorescence. The fluorescence is transported around the nuclear radiation shield separating the fluorescer lamp/reactor from the laser by a large diameter-to-length-ratio hollow lightpipe. The reactor core is shown in Fig. 5.52 and consists of two aerosol fueled half reactor cores/nuclear-driven fluorescers separated by three ~15-m dia. thin coupling disks, made of neutron absorbing ^{10}B fibers and Be, each with a ^{11}B neutron non-absorbing region occupying one tenth of the circumference and rotating at 60 Hz, 600 Hz, and 6000 Hz, respectively, in a low pressure gas cooled housing.

Each reactor half-core is a 5° tapered, Ag reflector, large diameter to length ratio (d/L) conical fluorescer cell of length 1.2 m and an output end diameter of 1.2 m (Fig. 5.53). The fluorescer cell walls are made of two closely fitting Be partial shells, one fixed and the other rotating so that, between pulses, the holes in the two shells line up in order to allow the fuel/fluorescer aerosol mixture to flow, thus removing heat and fission products. The fluorescer cell walls are coated with Ag and over coated with ¼ λ diamond or sapphire to resist the alkali metals and abrasion. The thin output window is made of sapphire. With a d/L of ~1 and a 5° taper, the extraction efficiency of this cell should be limited mostly by the materials considerations, while that of the 0.3 d/L lightpipe should be >0.9.

The 52 full-size 20-cm diameter and numerous smaller active mirror amplifiers are grouped in two "mosaic" arrays in order to most efficiently utilize the pump fluorescence (Fig. 5.51). The smaller diameter amplifiers provide intermediate amplification of the output of the electrically pumped master oscillator so that a

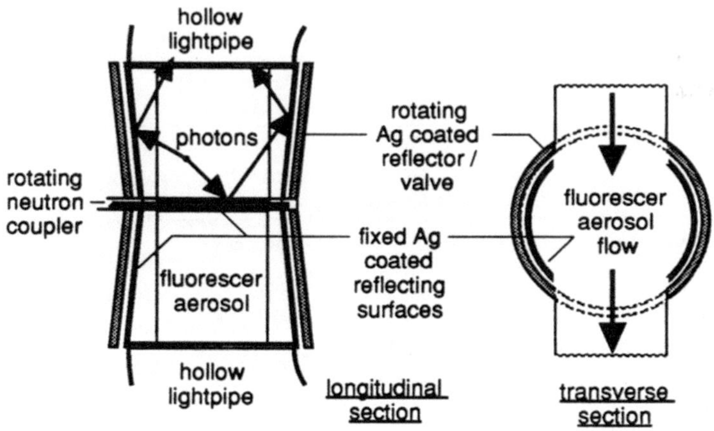

Fig. 5.53 A tapered large d/L lightpipe fluorescer cell is shown. A mirrored rotating valve allows for the flow of aerosol fuel/fluorescer/coolant mixture between pulses

Table 5.18 Parameters for a 1 ms, 1 Hz pulsed ^{235}U aerosol core nuclear reactor

Parameter	Value	Parameter	Value
Peak power	400 MW	Pulse length	1 ms
Average power	400 kW	Pulse rate	1 Hz
Power density	100 W cm^{-3}	Duty cycle	10^{-3}
Fuel loading	0.5 g cm^{-3}	Core volume	2.3 m^3
Peak flux	10^{17} n cm^{-2} s^{-1}		

Table 5.19 Potential efficiency of the solid-state laser pumped by a nuclear-driven fluorescer/reactor

Step	Efficiency range	Expected efficiency
Nuclear pumping	0.6	0.6
Alkali fluorescence	0.2–0.4	0.3
Cell extraction	0.7–0.9	0.7
Lightpipe transmission	0.85–0.95	0.9
Spectral matching	0.5–1	0.7
Active mirror absorption	0.8–1	0.8
$h\nu_{laser}/h\nu_{pump}$	0.4–0.7	0.6
Laser extraction	0.7–0.9	0.8
Total	0.008–0.12	0.03

highly saturating input is provided to the full aperture of each amplifier in order to enable maximum extraction of the stored energy. They also improve utilization of the pump fluorescence by filling the gaps between the larger diameter amplifiers.

The reactor parameters for this design are shown in Table 5.18. The potential efficiency for the solid-state laser pumped by a nuclear-driven fluorescer/reactor is shown in Table 5.19.

5.7.5 *Direct Drive Nuclear-Pumped Semiconductor Lasers (University of Missouri 2013)*

Semiconductor lasers (such as GaN) can be directly driven by ions from nuclear reactions. Fission fragments have a range of about ~10 μm in the laser medium. A design was developed for a high density reactor core made up of unit cells with a layer of 10 μm 93 % enriched ^{235}U, 1 μm diamond and 20 μm semiconductor micro-layers (Fig. 5.54). The cells are then stacked in a slab design to form the reactor. This stacked cell concept can achieve criticality in a volume of 0.1 m^3 and achieve a CW laser power level of >10 MW. The mass of the reactor and subsystems will be on the order of 2 metric tons.

In this design, the low-Z wide band gap material layer functions as the lasing medium (e.g., SiC, GaN and AlN). It does have some moderating properties. As mentioned previously, these layers must be composed of single-crystal material as the grain boundaries in polycrystalline material would present phasing problems for the laser.

The polycrystalline layers of diamond would function both as a high thermal conductivity material for heat removal and as a moderator. Diamond would be the ideal material for this purpose because of superior properties in radiation hardness, heat conduction and emissivity.

Highly enriched uranium (93 % U-235) would be optimum for the fissile layer of this design to maximize the flux to mass ratio of the core for the most compact assembly possible.

This design allows for a reactor small enough to be launched into orbit near an incoming asteroid and power output in the 10–100 MW range is sufficient to redirect asteroids as discussed. The payload mass for the laser is within the launch

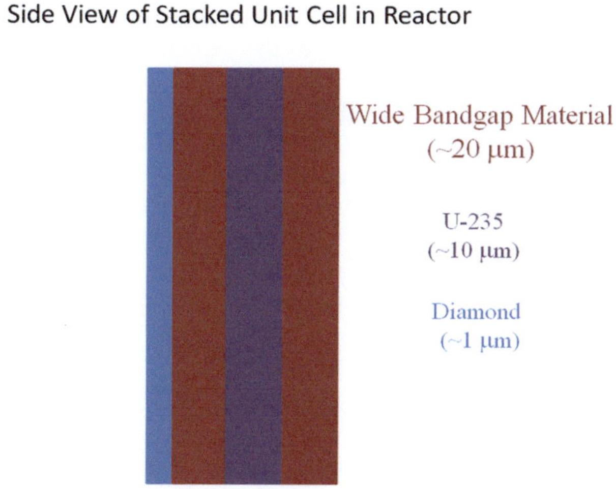

Fig. 5.54 Unit cell layout for solid-state nuclear-pumped laser/reactor

capability of current programs such as the Cassini–Huygens Spacecraft mission to Saturn with a mass of 5,574 kg.

5.7.6 *Thermally Pumped Carbon Dioxide Laser (University of Missouri 2014)*

As discussed in Chap. 1, the carbon dioxide laser can be thermally pumped. This is important because, as discussed in Chap. 2, nuclear fission can produce heat very efficiently as opposed to producing ionization which is limited to about 50 %. A thermally driven laser therefore would have some advantages in system efficiency. The problem is designing a reactor which can operate at temperatures of 1500 K and beyond. The design of a very high temperature reactor/laser was undertaken in 2014 [83]. The components of the reactor/thermally driven carbon dioxide laser are the high temperature core, the nitrogen flow channels and the laser/mixing cavity. The reactor fuels that were examined for the high temperature core were ZrC, UC and NbC (Table 5.20) based on their potential operating temperatures of 2000 K and beyond [84]. The second part of the design was the flow of nitrogen gas from the reactor core to the laser cavity. This part of the system must maintain the nitrogen temperature to limit the losses of the nitrogen vibrational state density. Finally the laser cavity mixes the hot N_2 with He and CO_2 gas mixtures and the mixture then undergoes adiabatic cooling through expansion.

In a thermally pumped CO_2 laser, a nitrogen gas flow is heated to high temperatures (e.g., 2000 K). At these temperatures, the nitrogen vibrational excitation states come into thermodynamic equilibrium with the gas. Thus a large population of the nitrogen gas resides in the vibrational states. The importance of this large population of nitrogen vibrational states is that the first vibrational state of nitrogen (N_2 ($\nu = 1$)) has an energy resonance (within 18 cm^{-1}) with the carbon dioxide upper laser level (CO_2 (001)). The mechanism for the transfer of energy from the excited nitrogen molecule to the carbon dioxide molecule is through collisional excitation (see Appendix 1 on the basic processes of plasma chemistry). The N_2 ($\nu = 1$) molecule undergoes a collision with a carbon dioxide molecule in the ground state (CO_2 (00 0)) to form the carbon dioxide upper energy level (CO_2 (001)). This is a resonance energy transfer mechanism. It is the dominant mechanism in the formation of the CO_2 (001) state and is the reason why the carbon dioxide laser has such a high efficiency [85, 86].

The core of the Ultra High Temperature Reactor (UHTR) is based on the Pebble Bed Reactor (PBR) combined with high temperature nuclear rocket fuel. Nitrogen gas, which is the gas coolant, flows through the core of the high-temperature nuclear reactor where it is heated.

The first step in the design is to find the density of the nitrogen vibrational levels after the gas comes into thermodynamic equilibrium with the core. Once a values for the nitrogen vibrational level density is found, then it is possible to adjust the

Table 5.20 Candidate materials for the high temperature reactor design comes from design studies of nuclear rockets [84]

Compound	Density g cm^{-3}	Melting point K	Linear expansion coefficient 10^{-6} deg^{-1}	Heat conduction W m K^{-1}	Elastic modulus GPa	Vickers harness GPa
Fuel materials						
UC	12.9	2500	10.4	19	220	9.0
UN	14.4	3074	9.3	18	265	8
ZrC+5%UC	6.9	3380	11.8	30	380	25
ZrC+5%UC +C	6.6	3250	11	32	350	20
ZrC+UC +NbC	7.6	5520	11	22	320	28
Construction materials						
ZrC	6.73	3690	8.6	30	390	27
ZrC+5%C	6.5	3180	5.5	52	230	18
NbC	7.8	3870	7.7	20–30	500	20
ZrC+50% NbC	7.3	3620	5.9	25	470	28
$ZrH_{1.9}$ εphase	5.6	2470	7.0	30	69	0.16
Pyrographic	1.7	4000	8.5	70	48	0.1

operating temperature of the reactor to determine how the operating temperature impacts the efficiency of the thermally pumped laser system. In a CO_2 transfer laser a collisional resonance exists between the several of the vibrational states of nitrogen (N_2 ($v = 1$–8)) and the ground state of carbon dioxide (CO_2 (00 0)) as shown in Fig. 5.55 [9]. Thus the first 8 vibrational levels of nitrogen are capable of populating the upper laser level of carbon dioxide (CO_2 (001)) [9]. Using the Maxwell Boltzmann distribution the probability that a nitrogen molecule, in thermodynamic equilibrium with the gas, resides in the $v = 1$ state is,

$$P(v = 1) = exp[-hc\omega_e(v + ½)/kT] \tag{5.69}$$

The fraction of the nitrogen molecules in vibrational states $v = 1$–8 is given by the sum of the probabilities P($v = 1$–8) at the energies, $hc\omega_e$ (v + ½) that occur, divided by the sum of the probabilities for all vibrational states, $v = 0$ to $v = \infty$ (Eq. (5.70)) [9]. The frequency, ω_e, from Fig. 5.55 is 2359.1 cm^{-1}, h is Plank's constant (6.62606957 × 10^{-34} m^2 kg / s), k is Boltzmann's constant (1.3806488 × 10^{-23} m^2 kg s^{-2} K^{-1}) and T is gas temperature in K.

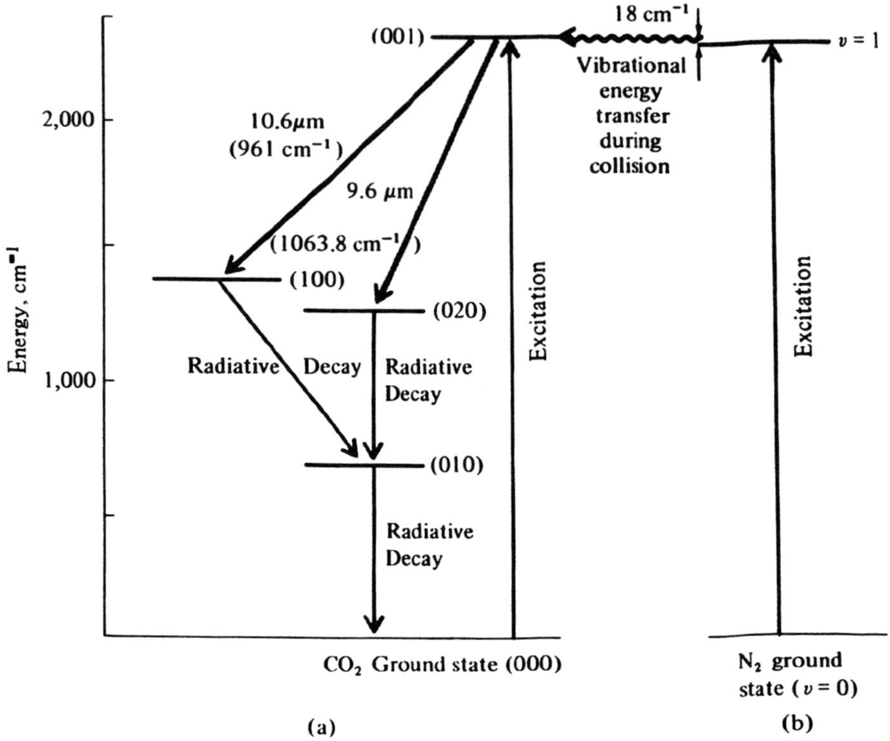

Fig. 5.55 Relevant energy levels for the carbon dioxide laser. There is a resonance between the $N_2(\nu = 1)$ state and the $CO_2(001)$ state (with a separation of 18 cm^{-1}). The minor separation is insignificant and at operational temperatures beyond 300 K, the kinetic energy of the molecule is sufficient to overcome this small differential energy [9, 87]

$$f = \frac{\sum_{1}^{8} exp[-hc\omega_e(v + ½)/kT]}{\sum_{0}^{\infty} exp[-hc\omega_e(v + ½)/kT]} \tag{5.70}$$

Equation (5.70) can be solved by evaluating the sums in the numerator and denominator. The result of this evaluation is shown in Table 5.21 at various temperatures.

A plot of the nitrogen molecules in vibrational states 1 through 8 divided by the total number of nitrogen states is shown in Fig. 5.56. This figure illustrates the potential operating range of Pebble Bed Reactor fueled with ZrC, UC, NbC and their alloys.

Clearly, higher temperatures lead to a greater fraction of nitrogen in the excited state, however this leads to difficult materials challenges. An appropriate

Table 5.21 Shown in the table is the ratio of the nitrogen molecules in the 1 through 8 vibrational states divided by the total number of nitrogen molecules in all energy states

T (K)	f Ratio of N_2 in states v=1 to v=8 to total N_2 states
1000	0.0334892
1500	0.103897
2000	0.183
2500	0.257012
3000	0.322293
3500	0.378756
4000	0.427306
4500	0.46899
5000	0.504756
5500	0.535407
6000	0.561613

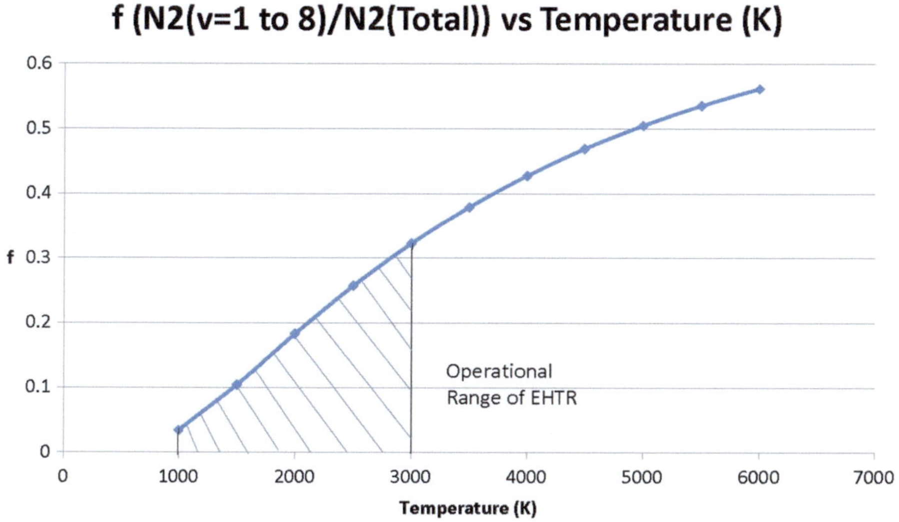

Fig. 5.56 Shown is the fraction of nitrogen in the 1 through 8 vibrational states as a function of temperature. The Potential operational range of the Ultra High Temperature Reactor (UHTR) is shown in the cross hatched area

temperature must be chosen to provide the workable fraction of nitrogen in the 1 through 8 vibrational states while minimizing the necessary temperatures. As an initial design point, a gas temperature of 2000 K is chosen. The percentage of N_2 in the excited vibrational states 1 through 8 is 18.3 % at 2000 K (Fig. 5.56).

Nuclear thermal rockets were designed by the United States Air Force (USAF) and by the Soviet Space Exploration Program that greatly exceeded an outlet temperature of 2000 K. These reactors have a different geometry than a Very High Temperature Reactor (VHTR) design, as they were engineered to provide maximum specific impulse per unit reactor mass, not to provide a steady stream of hot working fluid for a nuclear-pumped laser or a combined power cycle.

In the nuclear rocket programs the reactors were used to heat the working fluid (hydrogen) to a high temperature. The main obstacle in VHTR research was and continues to be in the materials selection. Based on a review of the literature, this study focuses on a reactor in a pebble bed configuration, inspired by the materials used in the American and Soviet NTR programs. The fuel design is analogous to the Tri-ISotopic (TRISO) fuel particles, but composed of different materials as potential fuel compositions for use at temperatures at or beyond 2000 K.

The core design was based on a USAF nuclear rocket design [88]. It consisted of 7 tubes of radius 1 m, placed in a graphite block in a hexagonal pattern 220 cm apart from each other. In each tube, 0.25 m of the radius taken out of the center and replaced with reactor grade graphite (NBG-18). The remaining part of the tube was filled with the pebbles (Fig. 5.57). Each pebble was 185 mm in diameter and each pebble contained approximately 1.5 million fuel particles as shown in Figs. 5.58 and 5.59.

The interior and exterior of the fuel bed channels has a thin skin of ZrC+5%C (the same material used as the clad) in order to prevent carbon dust from the moderator from being deposited into the channel and blown downstream into the laser cavity.

5.7.7 Reactor/Nuclear-Pumped Laser Designs in USSR and Russia (1985–2007)

The studies from the USSR and Russia have demonstrated very similar results to the studies performed in the USA. In 1985 a reactor composed of unit cells which consisted of (1) a neutron modifier, (2) a laser channel made from a metal tube coated on the inner surface with uranium fuel, (3) a laser gas channel and (4) a coolant (Fig. 5.60) [11]. The criticality studies used a multigroup diffusion model. The core parameters for reactor/laser using a He-Xe laser mixture is show in Table 5.22.

Approaches studied were reactor lasers with a transverse gas flow for heat removal outside of the core area (Fig. 5.61) and a reactor with heat storage placed in the core (Fig. 5.62). The parameters for these two concepts are shown in Table 5.23.

There were designs for nuclear flashlamp driven solid state lasers performed in the Former Soviet Union in 1996–2000 for the purpose of driving inertial confinement fusion reactors (a fission/fusion hybrid approach). The reactor used either uranium films or uranium hexafluoride. The laser energy was on the order of 10 MJ.

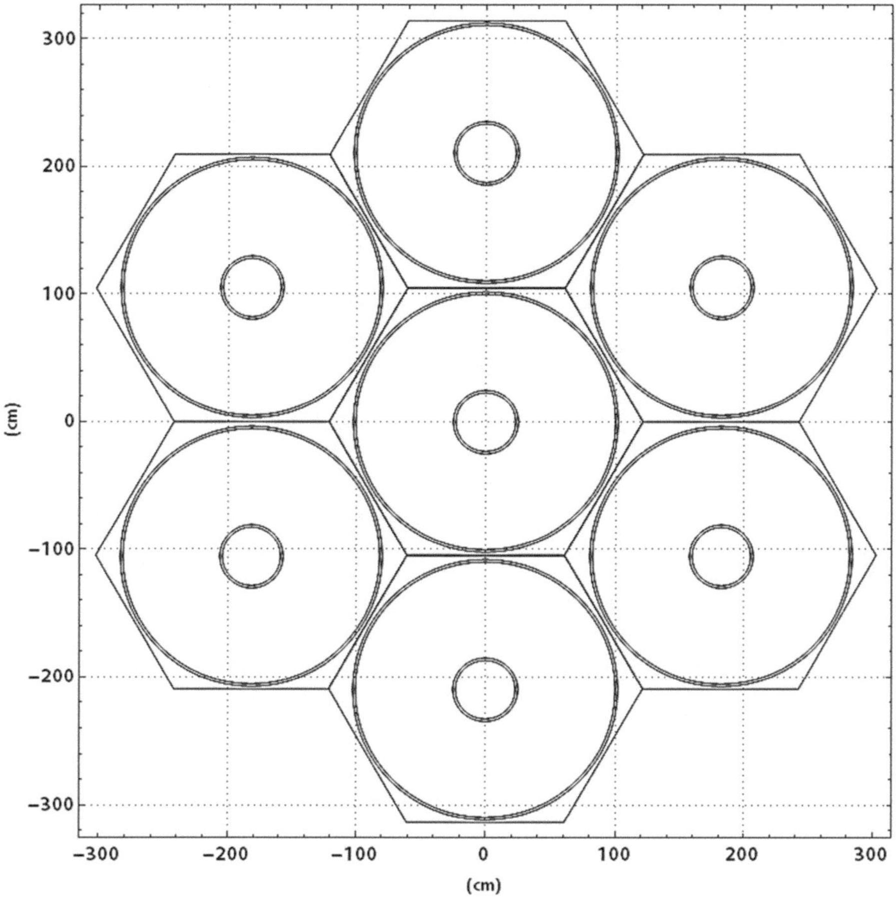

Fig. 5.57 The reactor layout is made up of 7 tubes of radius 1 m, placed in a graphite block in a hexagonal pattern 220 cm apart from each other. Each tube will have a smaller radius of 0.25 m taken out of the center and replaced with NBG-18. The remaining part of the tube will be filled with the pebbles [83]

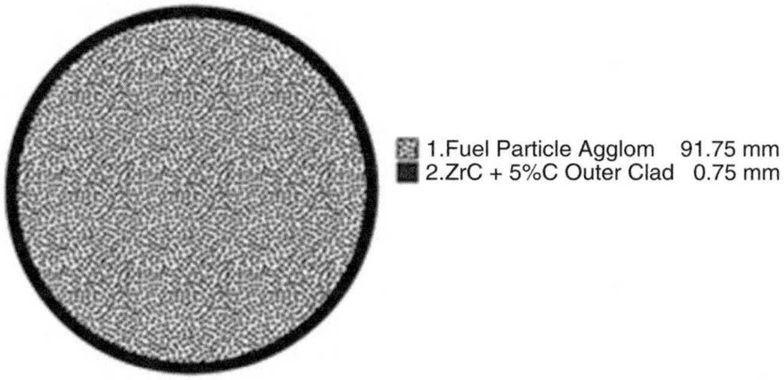

Fig. 5.58 Conceptual design of a fuel pebble for a UHTR [83]

Fig. 5.59 Conceptual design of a UHTR fuel particle [83]

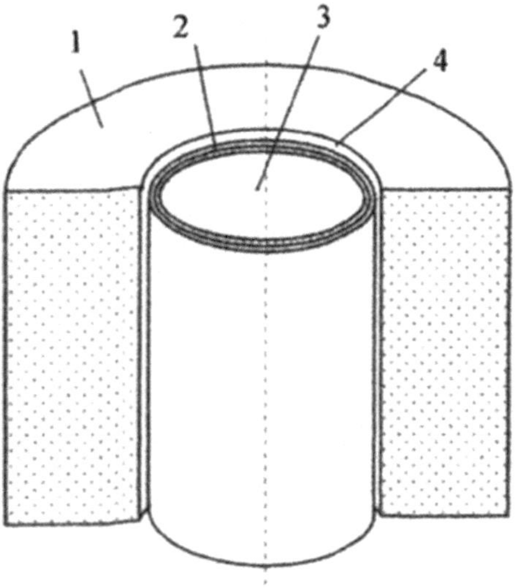

Fig. 5.60 A unit cell of a cylindrical laser module composed of (*1*) a neutron moderator, (*2*) a laser channel made from a metal tube coated on the inner surface with uranium fuel, (*3*) a laser gas channel and (*4*) a coolant channel [11] (With permission of Springer Science+Business Media)

Table 5.22 Characteristics of a thin film fueled reactor/laser using He-Xe mixtures [11]

Thermal power MW	150
Gas density cm^{-3}	3×10^{19}
Gas pressure atm	4
Power deposition in laser W cm^{-3}	3
Efficiency of energy transfer from fission fragment to laser %	16
Laser power efficiency %	1
Gas velocity m/s	150
Gas flow kg/	132
Power consumed by flowing gas through core MW	1.4
Gas temperature at core input K	1,080
Gas temperature at core output K	1,300
Maximum temperature of Uranium film K	1,657

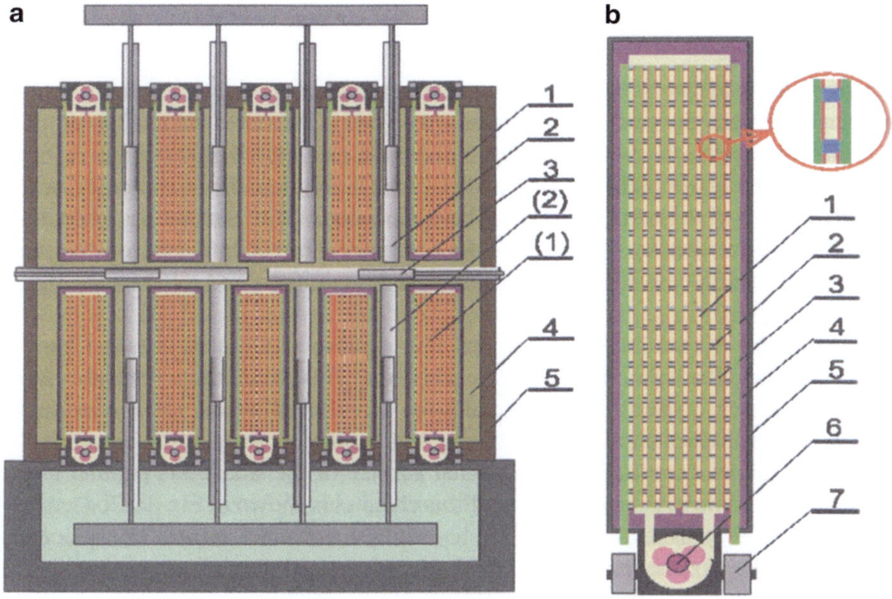

Fig. 5.61 Reactor/Laser core with transverse flow. (**a**) is the core with (*1*) core module; (*2*) control rods; (*3*) sensors; (*4*) reflector; (*5*) shielding. (**b**) is the reactor/laser module with (*1*) laser channel; (*2*) radiators; (*3*) moderator-coolant; (*4*) solid moderator; (*5*) enclosure; (*6*) gas flow system; (*7*) trolley for core motion [11] (With permission of Springer Science+Business Media)

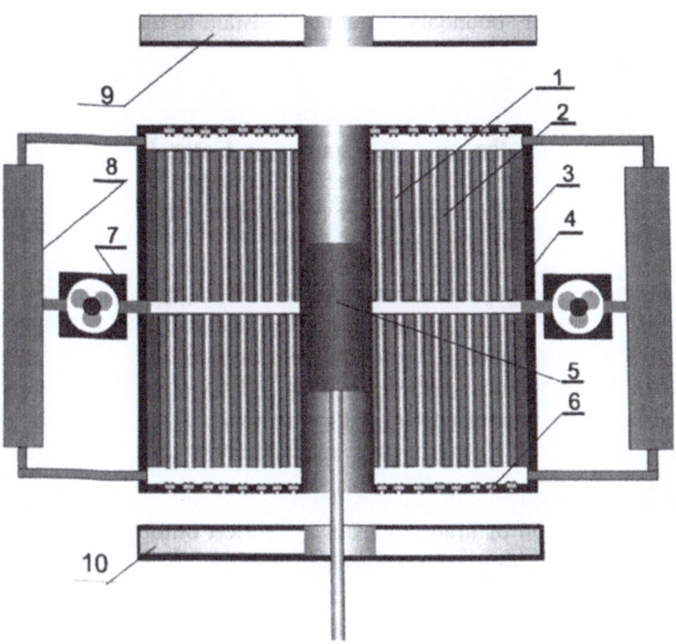

Fig. 5.62 The heat storage reactor/laser. The components are (*1*) the laser channel; (*2*) moderator; (*3*) reflector; (*4*) container; (*5*) control rods; (*6*) optics; (*7*) gas flow system; (*8*) heat exchangers; (*9,10*) optics [11] (With permission of Springer Science+Business Media)

Table 5.23 Parameters for the reactor/laser with gas flow and the reactor/laser with heat capacity [11]

	Reactor/Laser with gas flow	Reactor/Laser with heat capacity
Type of design	Rectangular, modular	Cylindrical, monoblock
Mode of operation	Steady-state (30 h)	Brief (100 s)
Overall dimensions, m	4.5 × 4.5 × 4.5	2.4 × 2.4
Mass of reactor Tonn	100	15
Number of channels	1,000	300
Uranium mass kg	70	15
Thermal power MW	800	65
Laser medium	He:(Ne,Ar):Xe	He:(Ne,Ar):Xe
Laser wavelength nm	1730, 2030, 2650	1730, 2030, 2650
Laser output power kW	1,000	200

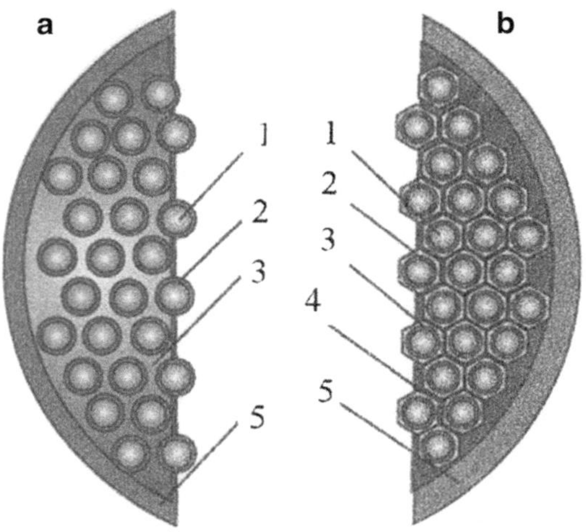

Fig. 5.63 A solid-state Nd doped laser pumped by a UF_6 fueled reactor. The components for a UF_6 fueled system illustrated by (**a**) are: (*1*) laser rods; (*2*) filters that absorb UV light; (*3*) fluorescer source; and (*5*) the module housing. The components for a uranium film fueled system illustrated by (**b**) are: (*1*) laser rods; (*2*) filters that absorb UV light; (*3*) fluorescer source; (*4*) the uranium thin films; and (*5*) the module housing [11] (With permission of Springer Science +Business Media)

The design utilized unit cells. The unit cells were placed in the reactor reflector (Fig. 5.63). Inside the 80 cm diameter module, there were about 100 Nd doped rods each having a diameter of 4.5 cm and a length of 1,200 cm. The space between rods which are arranged in a hexagonal shape is filled with UF_6-Xe at 2 atm. The rods are pumped by the optical emission from the plasma created by the gas fill (temperature being on the order of 10,000 K) which forms by the bombardment of a 3 ms neutron pulse from the coupled reactor (fluence of 3×10^{14} n cm^{-2}). The total energy from a module is 400 kJ with a 2.5 % efficiency.

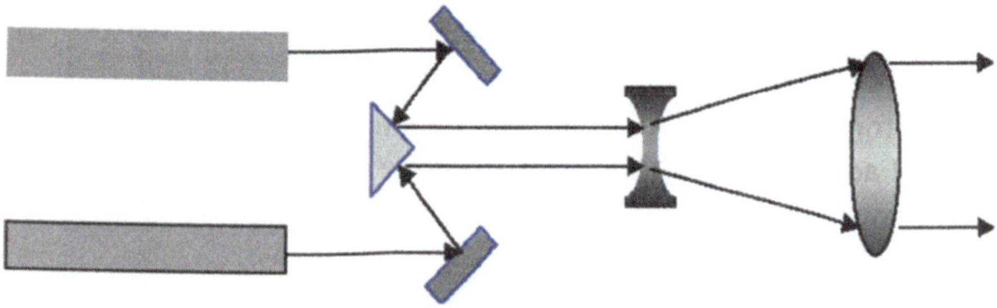

Fig. 5.64 Creating a mosaic radiation field from the laser unit cells with a telescoping system [11] (With permission of Springer Science+Business Media)

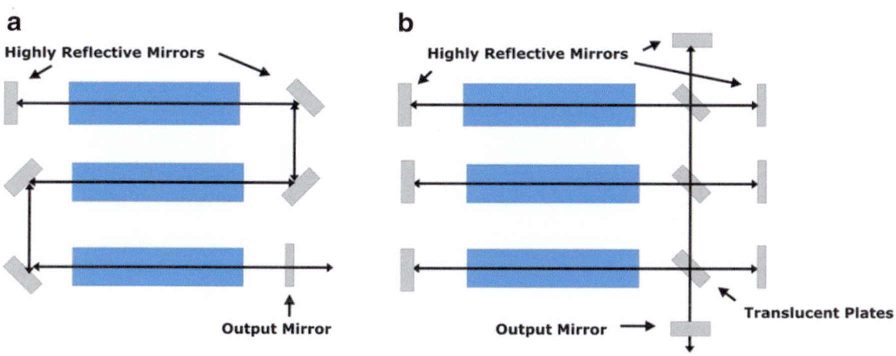

Fig. 5.65 An illustration of a serial mixing system is shown on side a. On side b, there is an illustration of a parallel mixing system. The components are high-reflectivity mirrors, output mirrors and translucent plates [11]

Work in the Former Soviet Union on beam combination was extensive. With a number of laser cells oscillating in the core, beam combining and coherent beam combining is also an important issue with the USSR and Russia reactor/laser core designs. The simplest approach is to add the radiation from each laser channel into a single incoherent mosaic beam using a periscope as shown in Fig. 5.64. The divergence of a single unit cell is 5×10^{-4} rad. After telescoping the combined beams the divergence is on the order of 5×10^{-5} rad.

If the unit cells are identical the laser beams can be combined by mixing the beams in a serial, parallel or mixed methodology. Figure 5.65 illustrated the concept of laser beam mixing. Side "a" shows a parallel method of mixing laser beams. The total length for the 10 channel parallel mix of cavities with a high Q-factor may be 30–50 m in length (limited to 10 channels because of losses in the semitransparent plates). Side "b" shows a mixed serial-parallel approach which can handle on the order of 100 beams. In this case beams are subdivided into groups each of which produces 100–200 kW. The combination of the groups of beams can

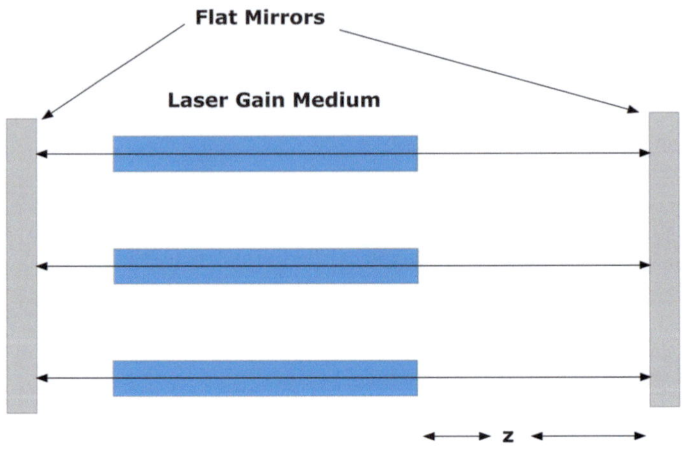

Fig. 5.66 Optical coupling of a many channel laser system to a flat mirror [11]

be combined outside of the reactor by an external optical system. When the reactor starts up, each unit cell will oscillate. The first laser to oscillate serves as the master oscillator. By using this master oscillator, it is possible to completely phase all lasers in the unit cell such that the radiation will have identical polarization, phase and frequency.

If the mirror is placed at the half way point of the periodic structure (Fig. 5.66), the image of the face plane of the channel is projected on the front flat mirror (provided the beams are in phase). The coupling of the radiation from the many channels is high because each spot contributes to the image. With the periodic arrangements of the unit cells in the reactor, there is a reproduction effect of the periodic structure of the monochromatic field at a specific distance. It has been shown that there is a distance z at which the mirror should be placed is equal to half of the Talbot distance which is defined as $Z_T = \beta a2/\lambda$ where a is the lattice pitch, λ is the laser wavelength and β is a numerical coefficient that is a function of the geometry.

Another method was discussed which uses a spatial intra-cavity filter is shown in Fig. 5.67. This configuration provides an in-phase mode which can be selected based on the design.

Another method that was discussed is the use of a nonlinear cell in the laser cavity. The proposed nonlinear cell was a Stimulated Brillouin Scattering (SBS) mirror (Fig. 5.68). At a power threshold the SBS mirror reflects the radiation back to the laser cavity. If a mixed beam is focused on the SBS cell, the reflected beams will be synchronized due to the redistribution of energy. To maintain stimulated Brillouin scattering, the total energy on the cell must exceed the threshold energy which is related to the threshold power by $E_{th} = W_{th} \times \tau$ (where τ is the pulse width). The threshold power is inversely proportional to the gain increment of the medium (g). The energy which is released from the nonlinear cell does develop other

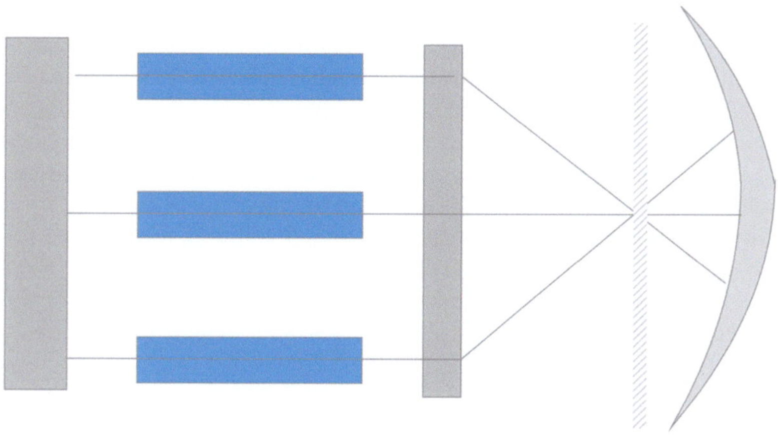

Fig. 5.67 A many channel laser system with a spatial intra-cavity filter [11]

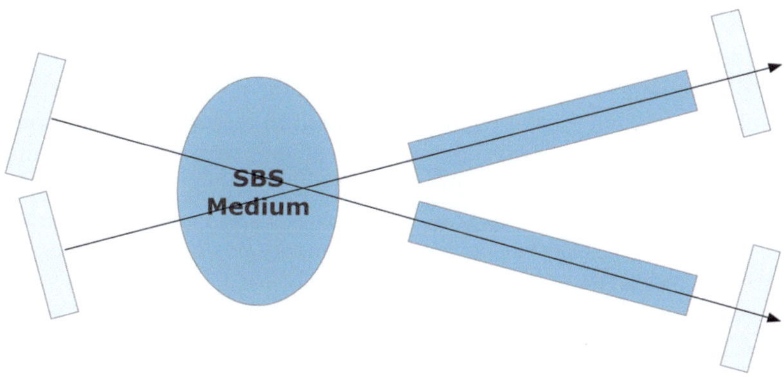

Fig. 5.68 A illustration of a cavity using a non-linear SBS medium as a mirror [11]

Table 5.24 Critical energies and gain increments for some SBS candidates [11]

Medium	g, cm/GW	E_{cr}, J	gE_{cr} cm × ns
TiCl3 (liquid)	14.2	1.1	16
CCl4 (liquid)	3.8	11	40
GLS-1 (neodymium glass)	1.2	20	24
Xe (gas at 50 atm)	90	5	480

processes which compete with SBS. Thus, there is a critical energy E_{cr} where the SBS effect sharply decreases. So, in choosing a SBS medium, the critical energy should exceed the threshold energy ($Ec_r > E_{th}$) or gE_{cr} (which only depends on material properties) should be as large as possible. Table 5.24 shows the characteristics of some potential SBS cells. From this table, it can be seen that high-purity xenon is preferable.

The effects of SBS over the range of 690–4200 nm was studied [11]. The potential of SBS mirrors was proven in that it was able to compensate for aberrations and increases in radial brightness, it was able to phase the radiation of a large number of laser cells (up to 1000) and it could operate at high frequency ($\geq$1 MHz). Experiments with a xenon SBS at 46 atm using lasers with wavelength of 1,315 nm showed W_{cr} of 3.2×10^5 W and E_{cr} of 5 J. The benefit of a non-linear mirror is that the beam divergence of a single unit cell laser would be about 0.5×10^{-3} rad. The use of linear combining (such as serial and parallel addition, telescoping, etc.) can improve the divergence of the combined beam to about 1×10^{-5} rad. The use of coherent beam combination methods lead to a beam divergence better than 1×10^{-6} rad. As will be discussed in Chap. 6, this improvement in beam divergence is important.

5.8 Summary

NPL technology was part of the U. S. Department of Defense research agenda from 1983 to 1990 as part of the Strategic Defense Initiative (SDI). When NPL research lost its priority, progress in the U. S. stagnated. In the meantime, researchers from the Former Soviet Union have continued research in NPLs. Other countries (e.g., China) have shown interest [89] and have developed working relationships with Russia to enhance the Chinese program [90]. The primary focus of both the Russian and Chinese program openly appears to be the 1.73 μm and 2.03 μm Xe laser because of the laser's 2 % efficiency and low pumping power density. This does not necessarily mean that there are not other efforts on classified NPL lasers systems. It simply means that this is only what Russian and Chinese scientists have openly published. Clearly it is desirable to have higher NPL efficiency thus advanced NPL concepts are necessary to achieve these goals. The concern expressed in the congressional report 851 [90] is that the technology has military applications in space defense and missile defense. This is a valid concern because of the scalability of NPL technology, specifically if the laser is directly coupled to the reactor fuel (like the design shown in Fig. 5.26). The case made in this book is that NPL/reactor technology can be constructed from unit cells which may be assembled to the desirable size to achieve the needed power level. Thus the requirements for ballistic missile defense may be achievable.

It is more interesting to consider non-military applications however. High power/high energy lasers have very important non-military applications. One example is for a device which can deflect asteroids [91]. The requirements for asteroid deflection are stringent, (>100 MW CW, capable of operating for years and the ability to beam energy on the asteroid at a high enough intensity to cause ablation of matter). NPL technology is promising for asteroid deflection because of its scalability to the very high power levels needed. As discussed, through the use of volume sources, energy focusing, photolytic pumping and semiconductor lasers, it may be feasible to achieve system sizes for asteroid deflections which can be launched with present rocket technology. The scalability and reduced mass of NPL technology is important for other non-military applications including space propulsion [92, 93], power transmission [94], long distance communication [11], photochemistry [11, 22], fog dispersion [11], illumination of large areas [11], heating of high melting point materials [11], and asteroid mining [91].

Problems

1. Using Fig. 5.1, estimate the system efficiency of an electrically pumped XeF laser (with a wall plug efficiency of 5 %). State your assumptions.
2. In Sect. 5.3, e-beam pumped semiconductor lasers are mentioned. How are e-beam pumped lasers similar to nuclear-pumped lasers?
3. How would you construct a thin membrane coated with a thin layer of ^{235}U (Fig. 5.6)?
4. Redo the example problem for a reactor/laser system illustrated in Fig. 5.15 for a bare homogeneous cylindrical reactor fueled with thin ^{235}U films. Explain the geometry of the unit cell and how it fits into the cylindrical reactor. What are the reactor dimensions and fuel loading?
5. Redo the example problem for a reactor/laser system illustrated in Fig. 5.15 for a bare homogeneous spherical reactor fueled with thin ^{235}U films. Explain the geometry of the unit cell and how it fits into the spherical reactor. What are the reactor dimensions and fuel loading?
6. Redo the example problem for a reactor/laser system illustrated in Fig. 5.15 for a bare homogeneous infinite cylindrical reactor fueled with ^{235}U aerosol particles. Explain the geometry of the unit cell and how it fits into the cylindrical reactor. What are the reactor dimensions and fuel loading?
7. Redo the example problem for a reactor/laser system illustrated in Fig. 5.15 for a bare homogeneous cylindrical reactor (with z and r variation) fueled with ^{235}U aerosol particles. Explain the geometry of the unit cell and how it fits into the cylindrical reactor. What are the reactor dimensions and fuel loading?
8. Redo the example problem for a reactor/laser system illustrated in Fig. 5.15 for a bare homogeneous infinite cylindrical reactor fueled with $^{235}UF_6$. Explain the geometry of the unit cell and how it fits into the cylindrical reactor. What are the reactor dimensions and fuel loading?
9. Redo the example problem for a reactor/laser system illustrated in Fig. 5.15 for a bare homogeneous infinite cylindrical reactor fueled with ^{235}U caged in a uranofullerene. Explain the geometry of the unit cell and how it fits into the cylindrical reactor. What are the reactor dimensions and fuel loading?
10. Develop a reflective aerosol core infinite slab reactor laser model (similar to the infinite cylindrical model in Eqs. (5.63), (5.64), (5.65), (5.66), (5.67), (5.68) and generate Figures similar to Figs. 5.42, 5.43, 5.44, 5.45, 5.46, and 5.47.
11. Redo the reflective aerosol core infinite cylinder reactor laser model (Eqs. (5.63), (5.64), (5.65), (5.66), (5.67), (5.68)) and regenerate Figs. 5.42, 5.43, 5.44, 5.45, 5.46, and 5.47.

12. Design a reactor/laser system choosing the pumping method, the fuel, the laser medium, the core design, etc. How does the laser system scale? What is the fuel load?
13. Suppose that the laser you designed in problem 10 was to be used in space. Design a radiator for heat removal.
14. Perform a bare homogeneous core analysis for a nuclear-pumped solid state laser using the unit cell shown in Fig. 5.53. Perform an optimization study of the moderator/fuel ratio. What is the fuel loading for an optimized reactor/laser?
15. Use Eq. (5.70) and regenerate Table 5.21.
16. What is the moderator to fuel ratio for the ultra high temperature reactor design shown in Sect. 5.7.6?
17. What are the major concepts for phase locking multiple laser beams? Contrast the concepts and discuss your preference for a large scale reactor/nuclear-pumped laser system.
18. Do a first cut design of a reactor/laser with a beam power of 1 GW. State all of your assumptions and provide all details of your work.

References

1. Prelas MA (1992) Solid state laser media driven by remote nuclear powered fluorescence, USA Patent 5,114,661, May 19, 1992
2. Prelas MA, Boody FP, Miley GH, Kunze JF (1988) Nuclear driven flashlamps. Laser Part Beams 6:25–62
3. Prelas MA, Jones GL (1982) Design studies of volume-pumped photolytic systems using a photon transport code. J Appl Phys 53:165–169
4. Schlie LA, Rathge RD (1996) Repetitively pulsed, 70 Joule photolytic iodine laser with excellent optical and long/reliable operation. (Phillips Laboratory Laser and Imaging Directorate, ed). Kirtland Air Force Base, Albuquerque (Report PL-TR-96-1046)
5. Boody FP, Prelas MA, Anderson JH, Nagalingam SJS, Miley GH (1978) Progress in nuclear-pumped lasers. In: Billman K (ed) Radiation energy conversion in space, vol 61. AIAA, New York, pp 379–410
6. Miley GH, Nagalingham SJS, Boody FP, Prelas MA (1978) Production of XeF(B) by nuclear-pumping. In: Vincent J (ed) International conference on lasers 78, Orlando, Florida. Corcoran, Society for Optical & Quantum Electronics, STS Press, MacLean
7. Miley GH, Boody FP, Nagalingham SJS, Prelas MA (1978) Production of XeF(B-X) by nuclear-pumping. Bull Am Phys Soc 24:117
8. Miley GH, McArthur DA, Deyoung RJ, Prelas MA (1989) Fission reactor pumped laser: history and prospects. In: Carlson JW, Behrens AD (eds) 50 years of nuclear power, National Academy of Science and NIST. American Nuclear Society, LaGrange Park, IL, pp 333–342
9. Verdeyen JT (2000) Laser electronics. Prentice Hall, Inc., Upper Saddle River
10. Matovich E (1968) In pursuit of a pulsed homogeneous nuclear laser. IEEE J Quantum Electron 4:379–379
11. Melnikov SP, Sizov AN, Sinyanskii AA, Miley GH (2015) Lasers with nuclear pumping. Springer, New York

12. Prelas MA (1989) Nuclear-driven solid-state lasers. In: Lasers '89; Proceedings of the international conference, New Orleans, LA, Dec. 3–8, 1989 (A91-41326 17–36). STS Press, McLean, 1990, pp 263–269
13. Ponomarenko VP, Filachev AM (2007) Infrared techniques and electro-optics in russia: a history 1946–2006. SPIE Press, Washington, ISBN 978081946355
14. Watermann ML, Prelas MA (2013) Integrated solid-state nuclear pumped laser/reactor design for asteroid redirection. Trans Am Nucl Soc 109:1531–1532, November 2013
15. Chung A, Prelas M (1984) Charged particle spectra from U-235 and B-10 micropellets and slab coatings. Laser Part Beams 2:201–211
16. Chung AK, Prelas MA (1984) The transport of heavy charged particles in a cylindrical nuclear-pumped plasma. Nucl Sci Eng 86:267–274
17. Platzmann RL (1961) Total ionization in gases by high energy particles: an appraisal of our understanding. Int J Appl Radiat Isot 10:116
18. Prelas MA Nuclear pumped photolytic energy focus 1979 to 1981. Nuclear Science and Engineering Institute, University of Missouri, Columbia. doi:10.13140/RG.2.1.4169.0728
19. Prelas M, Boody F (1982) Charged particle transport in uranium micropellets. In: 1982 IEEE International conference on plasma science, vol 82CH1770-7, Ottawa. doi:10.13140/RG.2.1.4967.4728
20. Lee MYJJ, Simones MMP, Kennedy JC, Us H, Makarewicz MPF, Neher DJA (2014) Thorium fuel cycle for a molten salt reactor: state of missouri feasibility study. In: ASEE annual conference, Indianappolis, IN. ASEE, Washington, DC, p 28. Available: http://www.asee.org/public/conferences/32/papers/10990/view
21. Mencin DJ, Prelas MA (1992) Gaseous like uranium reactors at low temperatures using C60 cages. In: Nuclear technologies for space exploration. vol. 2, NTSE '92. American Nuclear Society, La Grange Park, pp 403–433
22. Prelas MA, Loyalka SK (1981) A review of the utilization of energetic ions for the production of excited atomic and molecular states and chemical synthesis. Prog Nucl Energy 8:35–52
23. Martin AF (1964) Apparatus for producing controllable slow neutron chain reaction, USA Patent US3154473 (A), October 27, 1964
24. Conner JWP, Davis WE (1962) Use of nuclear fission in synthesizing organic compounds, Patents, US 3065159 A
25. Fellows AT (1966) Method and contact material for chemical conversion in presence of nuclear fission fragments. USA Patent 3,228,848
26. Coelseberg EA (1958) Investigation of a nuclear fuel making it possible to use the kinetic energy of fission products for chemical synthesis. In: Second international conference on the peaceful uses of atomic energy, vol 29. Addison-Wesley, Reading, MA, p 424
27. McArthur DA, Tollefsrud PB (1975) Observation of laser action in CO gas excited only by fission fragments. Appl Phys Lett 26:187
28. H. P. C. LTD (1957) Improvements in or relating to use of nuclear fission in synthesizing organic compounds. United Kingdom Patent GB 838361
29. Anderson A, Dominey D (1968) The radiolysis of carbon dioxide. Radiat Res Rev 1:269
30. Anderson RN, Selvadvray G, Goldstein MK (1981) Nuclear reactors capable of in-situ fuel processing. In: Alternative energy sources II. Hemisphere Publishing, New York, p 2339
31. Jalufka NW, DeYoung RJ, Hohl F, Williams MD (1976) Nuclear pumped ^{3}He-Ar laser excited by the ^{3}He(n, p)T reaction. Appl Phys Lett 29:189–190
32. Prelas MA, Boody FP, Zediker M (1984) A direct energy conversion technique based on an aerosol core reactor concept. In: IEEE international conference on plasma science, 84CH1958-8, IEEE, New York, NY, p 8
33. Prelas MA, Boody FP, Zediker MS (1985) An aerosol core nuclear reactor for space-based high energy/power nuclear-pumped lasers. In: El-Genk MS, Hoover M (eds) Space nuclear power systems. Orbit Book Company, Malabar

34. Prelas MA, Boody FP, Zediker MS (1984) Study of the basic transport properties (Charged particle transport, fluorescence transport & coupling efficiency) of the photon intermediate direct energy conversion technique, Research Gate. doi:10.13140/RG.2.1.4763.3121
35. Prelas M, Kunze J, Boody F (1986) A compact aerosol core reactor/laser fueled with reflective micropellets. In: Hora H, Miley G (eds) Laser interaction and related plasma phenomena. Springer, Boston, pp 143–154
36. Ghosh T, Prelas M (2009) Energy resources and systems, vol 1, Fundamentals and non-renewable resources. Springer, Dordrecht
37. Prelas MA, Akerman MA, Boody FP, Miley GH (1977) Direct nuclear-pumped 1.45 μm atomic laser in mixtures of He-CO and He-CO_2. Appl Phys Lett 31:679
38. Henry AF (1975) Nuclear-reactor analysis. MIT Press, Cambridge
39. DOE-HDBK-1019/1-93 (1993) Nuclear physics and reactor theory, Department of Energy, Washington DC. Available: http://www.steamtablesonline.com/pdf/Nuclear-Volume1.pdf
40. ThorEA_Wiki (2014) Thermal, epithermal and fast neutron spectra. Available: http://thorea.wikia.com/wiki/Thermal,_Epithermal_and_Fast_Neutron_Spectra. September 26, 2014
41. LANL (2014) MCNPX. Available: https://mcnpx.lanl.gov/, February 9, 2015
42. PolytechniqueMontreal (2014) Dragon code. Available: http://www.polymtl.ca/nucleaire/DRAGON/en/, February 9, 2015
43. ORNL (2015) Radiation transport group. Available: http://web.ornl.gov/sci/nsed/rnsd/rt/code.shtml
44. Lamarsh JR (1961) Introduction to nuclear reactor theory. Addison-Wesley, Reading
45. Duderstadt JJ, Hamilton LJ (1976) Nuclear reactor analysis. Wiley, New York
46. Kreyszig E (2000) Advanced engineering mathematics: maple computer guide. Wiley, New York
47. Kim TK, Yang WS, Taiwo TA, Khalil HS (2004) Whole-core depletion studies in support of fuel specification for the Next Generation Nuclear Plant (NGNP) core, Argonne National Laboratory Nuclear Energy Division (ed). U. S. Department of Energy, p 75. Available: http://www.ipd.anl.gov/anlpubs/2004/11/51497.pdf
48. Guoxiang G, Prelas MA, Kunze JF (1986) Studies of an aerosol core reactor/laser's critical properties. In: Hora H, Miley GH (eds) Laser interaction and related plasma phenomena. Springer, Boston, pp 603–611
49. McArthur DA, Schmidt TR, Tollefsrud PB, Walker JV (1975) Preliminary designs for large (−1 MJ) reactor-driven laser systems. In: IEEE International Conference Plasma Science. University of Michigan, Ann Arbor, p 75
50. Schmidt TR, McArthur DA (1976) Neutronics analysis for a subcritical nuclear laser driver excited by a fast pulse reactor, Department_of_Energy (ed). Sandia National Laboratory, Albuquerque. Available: https://www.ntis.gov/Search/Home/titleDetail/?abbr=SAND760139
51. McArthur D, Schmidt T, Philbin J, Tollefsrud P (1977) Concepts for the construction of large reactor-excited laser systems, Sandia National Laboratory, SAND76-0584. Available: http://www.osti.gov/scitech/biblio/5268082
52. Tollefsrud PB (1976) A high energy flowing nuclear laser. In: Report of work-shop on direct nuclear pumping of lasers. Naval Postgraduate School, Monterey. Available: http://ntrs.nasa.gov/archive/nasa/casi.ntrs.nasa.gov/19790014704.pdf
53. Rodgers RJ (1979) Initial conceptual design study of self-critical nuclear pumped laser systems, NASA (Ed). NASA Scientific and Technological Information Office, NASA, p 51. Available: http://ntrs.nasa.gov/archive/nasa/casi.ntrs.nasa.gov/19790014704.pdf
54. DeYoung RJ, Shiu YJ, Williams MD (1980) Fission-fragment nuclear laSing of Ar(He)-Xe. Appl Phys Lett 37:2175–2177
55. Gu G (1987) Analysis of high power/energy nuclear pumped laser/reactor concepts. PhD, Nuclear Engineering, University of Missouri
56. Gu G-X, Prelas MA, Kunze JF (1987) Space based nuclear-pumped laser/reactor concepts. In: Genk ME, Hoover M (eds) Transactions of the fourth symposium on space nuclear power systems. Sandia National Laboratories, Albuquerque, pp 143–147

57. Prelas M, Boody F, Zediker M (1984) A direct energy conversion technique based on an aerosol core reactor concept. In: IEEE Publication, 84CH1958-8, IEEE, New York, NY, p 8
58. Prelas MA, Boody FP, Kunze JF (1986) A compact aerosol core reactor/laser fueled with reflective micropellets. In: Hora H, Miley GH (eds) Lasers and related plasma phenomena, vol 7. Plenum Press, New York
59. Gu G, Kunze JF, Boody FP, Prelas MA (1988) Neutronics conceptual design of a UF_6-fueled gaseous laser system. In: Fifth symposium on space nuclear power systems 1988. Sandia National Laboratories, Albuquerque, p 24
60. Boody FP, Prelas MA (1992) Transient radiation-induced absorption in fused silica optical fibers, 450–950 nm. In: Proceedings of specialist conference on physics of nuclear induced plasmas and problems of nuclear-pumped lasers. Institute of Physics and Power Engineering, Obninsk. doi:10.13140/RG.2.1.4220.5923
61. Brignon A (2013) Coherent laser beam combining. Wiley, New York
62. Patel P (2014) Lockheed martin shows off high-power fiber laser weapon. IEEE Spectrum. Available: http://spectrum.ieee.org/tech-talk/aerospace/military/lockheed-martin-shows-off-highpower-fiber-laser-weapon, February 7, 2014
63. APS Study Group on Science and Technology of Directed Energy Weapons (1987) Report to the APS of the study group on science and technology of directed energy weapons. Rev Mod Phys 59:S1–S202
64. Eggleston J (1986) Steady-state coherent Raman beam combining with multiaxial mode lasers. IEEE J Quantum Electron 22:1942–1952
65. Flusche BM, Alley TG, Russell IH, Roh WB (2006) Multi-port beam combination and cleanup in large multimode fiber using stimulated Raman Scattering. Opt Express 14:11747–11755, November 27, 2006
66. Flusche BM (2006) Development of a multiple beam combiner using stimulated raman scattering in multimode fiber. MS, Department of Engineering Physics, Air Force Institute of Technology
67. Trainor DW, Smith MJ, Duzy C, Nicholson W, Appel C, Roberts T (1988) Xenon fluoride laser beam combining using stimulated raman scattering techniques. Proc SPIE 739:136–137
68. Tyson RK (2010) Principles of adaptive optics. CRC Press, Baco Raton
69. Boody FP, Prelas MA (1983) Photolytic dual-media nuclear pumping of excimer lasers. AIP Conf Proc 100:349
70. Prelas MA (1979) Nuclear pumping mechanisms in atomic carbon and in excimers. In: Nuclear-pumped lasers, NASA Langley Research Center, Hampton, Virginia. Available: http://ntrs.nasa.gov/search.jsp?R=19800005190
71. Prelas MA (1981) A potential UV fusion light bulb for energy conversion. Bull Am Phys Soc 26:1045
72. Prelas MA (1982) Nuclear powered space lasers: an evaluation of current technology, Nuclear-Pumped Laser Corporation, December 5, 1982. doi:10.13140/RG.2.1.1355.4405
73. Prelas MA (1987) Photolytic pumping of solid-state Nd_{3+} doped lasers using nuclear-driven flashlamps, Idaho National Engineering Laboratory, Experimental design and testing of a nuclear-driven flashlamp to pump NdYAG January 1987–December 1987. doi:10.13140/RG.2.1.1547.6329
74. Prelas MA (1991) Remote pumping of solid-state lasers with nuclear driven fluorescers, Department of Energy, Ed., DOE/ER/13029-T3, p. 1–35. Available: https://www.researchgate.net/publication/234355574_Remote_pumping_of_solid_state_lasers_with_nuclear_driven_fluorescers
75. Prelas MA (1990) Nuclear-driven solid-state lasers. In: Shay DG, Harris TM (eds) Proceeding of the international conference on lasers 89. STS Press, MacLean
76. Boody FP, Prelas MA (1992) Design of a large-scale nuclear-driven fluorescer pumped solid-state laser system. In: Specialist conference on physics of nuclear induced plasmas and problems of nuclear-pumped lasers, vol 1. Institute of Physics and Power Engineering, Obninsk. doi:10.13140/RG.2.1.2123.4401

77. Boody FP, Prelas MA (1991) Very high average power solid-state lasers pumped by remotely-located nuclear-driven fluorescers. In: Advanced solid-state lasers. Optical Society of America, Washington, DC, pp 192–199
78. Boody FP (1991) Nuclear-driven fluorescence pumped solid-state lasers PhD, Nuclear Engineering, University of Missouri, Columbia, MO, USA
79. Lin LTS (1994) Microwave and nuclear excitations of alkali metal vapors. PhD, Nuclear Engineering, University of Missouri
80. Taussig R et al (1979) Design investigation of solar powered lasers for space applications. Mathematical Sciences Northwest Incorporated, Bellevue WA. Available: http://www.dtic.mil/dtic/tr/fulltext/u2/a335155.pdf
81. Lin LTS, Prelas MA, He Z, Bahns JT, Stwalley WC, Miley GH et al (1995) Design of an ICF plant using a nuclear-driven solid-state laser. Laser Part Beams 13:95
82. Boody FP, Prelas MA (1992) Efficient light transport using large diameter-to-length ratio hollow lightpipes. In: Proceedings of specialist conference on physics of nuclear induced plasmas and problems of nuclear-pumped lasers, vol 1. Institute of Physics and Power Engineering, Obninsk. doi:10.13140/RG.2.1.3433.5207
83. Prelas M, White N, Wisniewski D, Boraas M, Kasiwattanawut H, Walton K, Correa S, De Castro J, De-La-Torre-Aguilar F, Knewtson T, Nelson S, Schutte J, Tchouaso MT, Watterman M (2014) Design of a space based very high temperature reactor thermally pumped carbon dioxide laser. Nuclear Science and Engineering Institute, University of Columbia, Columbia
84. Lanin AG, Fedik II (2011) Selecting and using materials for a nuclear rocket engine reactor. Phys Usp 54:305–318
85. Cheo PK (1979) CO_2 Lasers. In: Levine AK, DeMaria AJ (eds) Lasers, vol 3. Marcel Dekker, New York
86. Manes KR, Seguin HJ (1972) Analysis of the CO2 TEA laser. J Appl Phys 43:5073–5078
87. Yariv A (1976) Introduction to optical electronics. Holt, Rinehart and Winston, New York
88. Lawrence TJ (2005) Nuclear thermal rocket propulsion systems. DTIC Document. Available: http://www.usafa.edu/df/dfas/Papers/20042005/Nuclear%20Thermal%20Rocket%20Propulsion%20Systems%20-%20Lawrence.pdf
89. Yang C, Chen H, Zheng C, Zhao X, Han H (2002) The progress of nuclear pumped laser in CFBR-II reactor. Chin Opt Lett 1:292–293
90. Select-Committee-United-States-House-of-Representatives (1999) U.S. National Security and Military/Commercial concerns with the People's Republic of China. U.S. Government Printing Office, Report 105-851, Chapter 4 PRC Missile and Space Forces. Available: http://www.gpo.gov/fdsys/search/pagedetails.action?browsePath=105%2FHRPT%2F%5B800%3B%5D&granuleId=GPO-CRPT-105hrpt851-3&packageId=GPO-CRPT-105hrpt851&fromBrowse=true
91. Prelas MA, Watermann ML, Wisniewski DA, Neher JA, Weaver CL (2014) A review of nuclear pumped lasers and applications (Asteroid deflection). In: ASEE 121st annual conference and exposition. ASEE, Indianapolis, p 27. Available: http://www.asee.org/public/conferences/32/papers/10774/view
92. Forward RL (1995) Advanced propulsion systems. In: Humble RW, Henry GN, Larson WJ (eds) Space propulsion analysis and design. McGraw-Hill Co, New York, p 631
93. Bae YK (2012) Prospective of photon propulsion for interstellar flight. Phys Procedia 38:253–279
94. Summerer L, Purcell O (2009) Concepts for wireless energy transmission via laser, Advanced-Concepts-Team (ed), European Space Agency. Available: http://www.esa.int/gsp/ACT/doc/POW/ACT-RPR-NRG-2009-SPS-ICSOS-concepts-for-laser-WPT.pdf

Chapter 6
Applications

Abstract There are applications for high energy/high power laser beams that have motivated research on nuclear-pumped lasers. The applications are both long term and near term. This chapter examines a number of potential long term applications for high energy/high power laser beams. In addition, near term applications for nuclear-pumped laser technology are considered.

Keywords Asteroid deflection • ABM defense • Nuclear pumped lasers

6.1 Introduction

As discussed in previous chapters, it is feasible to build self-critical nuclear-pumped laser/reactors. However, the technology is not simple. There are many challenges remaining in order to bring the technology to fruition. The question is whether or not it is practical to do so and is there a reason to move forward. This question depends upon the application and associated problem(s):

- The strategic importance of the application to solve the problem;
- The requirements of the mission such as the laser power, the pulse length (short to CW), the required integrated energy produced over the lifetime of the laser, the power on target, etc.;
- And, are there any other devices that can do the job better and/or cheaper.

The reasons for moving forward in the development of nuclear-pumped laser technology are based on their capabilities and their properties. Nuclear-pumped lasers can produce sustained high power output over long periods of time. These lasers scale with volume rather than with surface (most other laser systems scale by surface area). Nuclear-pumped lasers use the tremendous energy storage capability of nuclear fuel (about 144,000,000 MJ/kg for Highly Enriched Uranium See Chap. 4). For some applications there may be no other option. One scenario that has widely been discussed in recent years is the possibility of an asteroid impact which could cause the collapse of society or even lead to a mass extinction event. There are very few options to deflect a large asteroid. The energy that is required to deflect a 2 km diameter asteroid is huge (4.9×10^{15} J). To put some perspective on this energy, one megaton of TNT (1×10^{6} t) is 4.184×10^{15} J. Lasers are a very

M. Prelas, *Nuclear-Pumped Lasers*, DOI 10.1007/978-3-319-19845-3_6

desirable option for deflecting asteroids [1]. However the laser would have to have a beam power of 315 MW which irradiates the asteroid's surface continuously for 180 days to be able to deflect a 2 km diameter asteroid. Nuclear-pumped lasers are the most viable option for scaling continuous wave space based lasers to these types of power levels.

In the following sections there are a number of areas in which high energy/high power lasers are needed. The viability of using nuclear-pumped lasers will be discussed for each of these areas.

6.2 Strategic Defense Initiative

The Strategic Defense Initiative (SDI) was initiated by President Ronald Reagan in 1983 with one of the main objectives being the development of space based laser platforms which were capable of generating a directed energy source to destroy the post boost phase vehicle (or reentry) vehicles of a ballistic missile when it enters space (see Chap. 1). SDI focused on two general technologies: kinetic energy weapons (e.g., rocket based interceptors) and directed energy weapons (e.g., lasers and particle beams). Nuclear-pumped lasers fell in the category of a directed energy weapon and were part of the SDI program due to the attributes described above. The available technologies for nuclear-powered space lasers have been discussed in Chaps. 4 and 5. In this section, the means of applying these technologies to the SDI mission are discussed. Primarily, nuclear energy is viewed as one of the most attractive energy sources for space because the fuel has a superior energy-to-mass ratio (E/M) [2]. This advantage can be seen by comparing a typical chemical fuel which produces a high chemical energy density per unit mass (e.g., $H_2 + (1/2) O_2 \rightarrow H_2O + 2.2$ eV) to a typical nuclear fuel which produces several hundred MeV per reaction ($^{235}U + n \rightarrow$ fission fragments $+ \nu n + 200$ MeV). The energy-to-mass ratios (E/M) for the two fuels are significantly different:

- E/M (hydrogen and oxygen) = 5,350,000 J/kg
- E/M (^{235}U) = 82,000,000,000,000 J/kg

Hence, for space applications where the cost of launching an object by orbital transport is very expensive per unit mass placed in orbit, large energy-to-mass ratios are needed for space based laser systems.

The SDI mission requirements are very demanding [3, 4]. The suggested criteria for the mission of defending against intercontinental ballistic missiles (ICBMs) were that the system needed to contend with 1,000 missiles in an 8-min period and it needed to produce a large energy burst per missile. The latter requirement is tied to the laser's brightness [4]. Brightness is measured in Watts/sr (or Joules/sr for a pulsed laser). A common measured quantity from a laser beam is its fluence (F) which is defined as the Joules cm^{-2} delivered to the target. If a laser has a power output of P (Watts), the fluence is then found by multiplying the laser power by its pulse length (Δt) and dividing this product by the spot size of the laser beam on the target (A).

$$F = {}^{P\Delta t}/_{A} \tag{6.1}$$

If the laser is in orbit and it is targeting the post boost phase vehicle or reentry vehicles, then there may be some transmission losses between the laser and target. These losses are taken into account by the factor L and Eq. 6.1 can be rewritten,

$$F = {}^{P\Delta t}/_{AL} \tag{6.2}$$

In making estimates of laser brightness, the pulse shape can be assumed to be rectangular and assume that L is equal to one [4]. The radius of the spot size on the target is related to the distance from the laser to the target (R). The divergence (θ) of a diffraction limited beam (or Gaussian beam) is related to the laser wavelength (λ) divided by the laser's aperture diameter (D) so,

$$A = {}^{\pi R^2 \lambda^2}\Big/_{D^2} \tag{6.3}$$

Substituting Eq. 6.3 into Eq. 6.1,

$$F = {}^{P\Delta t D^2}\Big/_{\pi R^2 \lambda^2} \tag{6.4}$$

Multiplying Eq. 6.4 by R^2 and dividing by Δt, the equation for brightness (B) is found,

$$B = {}^{FR^2}\Big/_{\Delta t} = {}^{PD^2}\Big/_{\pi \lambda^2} \tag{6.5}$$

It is interesting to note that in Eq. 6.5 that brightness is related to target parameters in the first term $({}^{FR^2}/_{\Delta t})$ and laser parameters in the second term $({}^{PD^2}/_{\pi\lambda^2})$.

In order to destroy any given missile, the target hardness factor (energy on target per unit area kJ cm^{-2}) must be considered. A soft target requires less energy and is defined as being in the range of 1 kJ cm^{-2}. A hardened target (one with reflective surfaces and insulators) requires a larger energy (~100 kJ cm^{-2}). Using a fluence of 10 kJ cm^{-2}, and assuming a distance from laser to target of 1000 km, the brightness time product ($B\Delta t = FR^2$) is 1×10^{20} J/sr. The brightness time product can then be applied to assess the required properties of the laser from Eq. 6.5.

$$B\Delta t = 1 \times 10^{20} = {}^{P\Delta t D^2}\Big/_{\pi \lambda^2} \tag{6.6}$$

If the laser was a steady-state nuclear-pumped carbon dioxide laser ($\lambda = 10,600\,\text{nm}$) with a beam diameter of 4 m and a dwell time on target of 1 s ($\Delta t = 1$), the required laser power needed to achieve the desired brightness-time product is,

$$P = 2.206 \times 10^9\,\text{Watts} \tag{6.7}$$

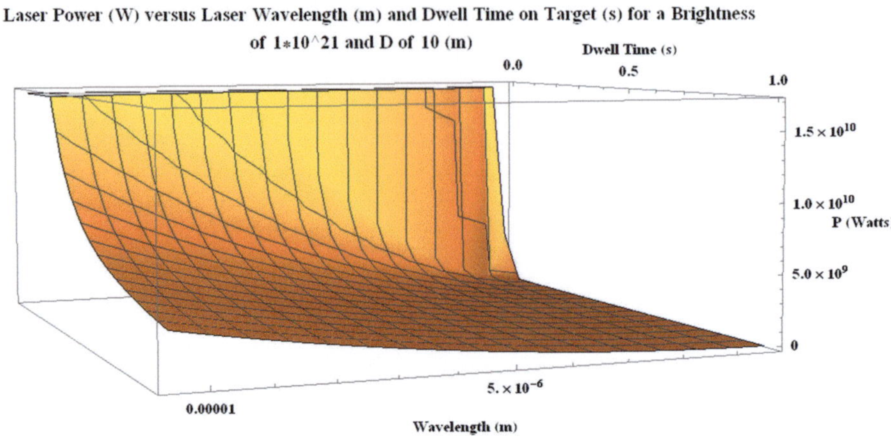

Fig. 6.1 The necessary laser power on a target that requires 10 kJ cm^{-2} as a function of laser wavelength and laser aperture diameter with a brightness of 1×10^{21} is shown

This means that the carbon dioxide laser would need a steady-state laser power of 2.2 GW.

On the other hand if the laser was a pulsed nuclear-pumped xenon fluoride laser ($\lambda = 351\,\text{nm}$) with a beam diameter of 4 m and a dwell time on target of 1 ms, the required laser power needed to achieve the desired brightness time product is,

$$P = 2.419 \times 10^9\,\text{Watts} \tag{6.8}$$

The beam energy for the pulsed xenon fluoride laser would have to be 2.419 MJ per post reentry vehicle.

The relationships of these parameters are shown in Figs. 6.1 and 6.2. These figures are plots of the space based laser power that is required to destroy a target (10 kJ cm^{-2}) as a function of laser wavelength, dwell time on target and a laser aperture diameter of 10 m for different brightness'. As can be seen, when the wavelength decreases so does the required power. This is due to the beam divergence being directly proportional to the laser wavelength. Also as the dwell time increases (to 1 s), the laser power decreases. This is due the beam divergence being inversely proportional to dwell time on the target.

A ground based laser would have a much higher power requirement due to the increased range (R) between the laser and target [4] (see Eq. 6.5). It would also have to contend with atmospheric distortion of the Gaussian beam.

Assuming a space based xenon fluoride laser that is approximately 10 % efficient driven by an electrical power generator with a conversion efficiency of 33 %, the total chemical energy needed to produce enough laser energy for a 1,000 ICBM mission is approximately 72,570 MJ. This number assumes that there is no excess

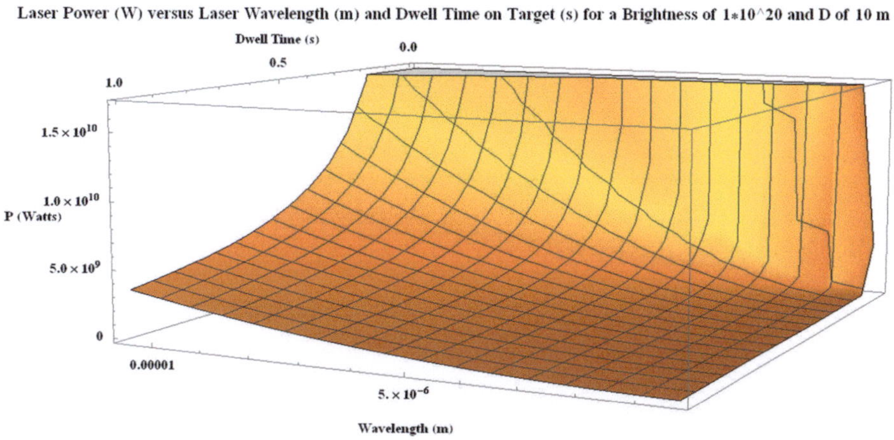

Fig. 6.2 The necessary laser power on a target that requires 10 kJ cm^{-2} as a function of laser wavelength and laser aperture diameter with a brightness of 1×10^{20} is shown

chemical energy stored, thus the chemical fuel will be completely exhausted once the mission is completed. A nuclear source on the other hand must maintain an excess inventory of fuel in order to maintain criticality (however due to the large amount of energy stored in nuclear fuel, the fuel mass would only have to about 10 kg). Thus in comparing the mass of the chemical and nuclear energy sources the mass advantage of nuclear fuel is clear:

- Chemicals-13,565 kg
- Nuclear- 10.0 kg

6.2.1 System Evaluation

There are two ways to utilize the energy in a nuclear source for powering space based lasers: an indirect method and a direct method. A description and preliminary evaluation of each type of system follows. The indirect method uses the energy source (fuel) to generate electricity so that the system needs to have auxiliary systems (e.g., electrical generator, power conditioning systems, transformers, capacitors, etc.), the optical cavity, optics, associated electronics and the cooling system. A directly-pumped system uses the ionizing radiation from nuclear reactions to excite the laser so that an electrical generator is not needed thus this type of scheme would have fewer auxiliary systems. However, the nuclear-pumped systems can fall into three categories: (1) an open cycle where the laser gas flows and is ejected (this requires the system to store some excess laser gas that is used during the mission); (2) a closed cycle where the laser gas is recycled (this type of system stores only the amount of gas needed to make the laser work and can operate

Table 6.1 Space based system and component mass estimates (in kg) for a chemical laser, an electrically pumped laser with a nuclear power electrical generator, a direct nuclear-pumped system with gas recycling and 1000 single shot nuclear-pumped laser modules [5]

Component	Chemical laser	Nuclear electrical plant and indirect laser	Direct nuclear-pumping with gas recycling	Single shot nuclear-pumped laser
Optical cavity and Electronics mass (kg)	1,000	1,300[a]	1,300[a]	520[b]
Fuel mass (kg)	1.50×10^5	20	10	10
Cooling system mass (kg)	0[c]	5.93×10^3 to 59.3×10^3	5.85×10^3 to 58.5×10^3	0
Auxiliary system mass (kg)	0	1.6×10^4	0	0
Total mass (kg) for 1000 Shots (80 MJ each)	1.53×10^5	2.43×10^4 to 7.76×10^4	0.817×10^4 to 6.0×10^4	5.3×10^5

[a]Laser gas is recycled
[b]Where there are 40 clusters of 25 single shot NPLs and each cluster shares optics and electronics
[c]Open cycle cooling

indefinitely limited only by the energy stored in the nuclear fuel); and (3) a single shot system where the laser is used once and then is designed to fail [5]. A chemical laser requires the least number of auxiliary systems since the energy is stored in the fuel and the laser excitation and optical extraction process takes place during the kinetics of combustion. This type of system needs an optical cavity and electronics. The mass of some of these space based system strategies was estimated in a 1982 study as seen in Table 6.1 [5].

6.2.2 *Polyus-Skif*

The feasibility of developing a continuous wave laser on an orbital platform using an electrical power generator was explored in the USSR with the Polyus-Skif orbital laser weapons platform. Polyus-Skif was designed to destroy potential United-States SDI satellites. The platform had a 1 MW carbon-dioxide laser [6]. Polyus comes from the Russian word for "the north pole" and Skif comes from the word Scythians (an ancient tribe of central Asian warriors). The carbon-dioxide laser module was originally developed and tested in 1984 as an aircraft based directed energy weapon designed to be used as a theater missile defense system. The Polyus-Skif was large and needed the Energia rocket to get it into orbit. The Energia rocket, which was used to launch the Buran space shuttle, was capable of lifting 95 t. The Polyus-Skif was 131 ft long and more than 13 ft in diameter. The mass of the Polyus-Skif was ~88,000 kg. It was made up of modular units. One module had rocket engines to place the Skif into its final orbit. There were solar panels which provided power for instruments. Another module carried the carbon dioxide tanks. Two turbo electrical power generators (to power the laser) were carried in another module. It included a rotating turret which pointed the laser beam. The Polyus-Skif was built long and thin which allowed it to fit on the side of the Energia attached to its central fuel tank.

The Polyus-Skif had its challenges. Tracking a long distance target and pointing a 1 MW beam so that it hits the target is very difficult when considering that the two large turbo generators have moving parts and that a high temperature gas exhaust from the generators and laser will induce a thrust on the platform. The Russians developed a sophisticated system which minimized the force of the expelled gas by sending it through deflectors. The device was also able to dampen the motions caused by exhaust gases, the moving parts of the turbo generators and the moving laser turret. On May 15, 1987 the Polyus-Skif demonstration platform was launched (Fig. 6.3). The launch was initially going very well. The Polyus-Skif actually separated from the Energia rocket successfully and was positioning itself to fire its rockets to put the platform into a higher orbit. A software glitch caused the Polyus-Skif to point in the wrong direction and fall back to earth. Had the Polyus-Skif demonstration project succeeded, the landscape of space based warfare would have changed completely. Polyus-Skif could have achieved the same type of eye opening effect on the United States that Sputnik had in October, 1957.

6.3 Laser Propulsion

The fundamental principle of laser propulsion is that a photon behaves like a particle in that it has momentum. This momentum can be found from,

$$p_{ph} = hv/c \tag{6.9}$$

Since a photon travels at the speed of light, the attraction of using photon propulsion is that it can, in theory, push the spacecraft to a velocity that approaches a significant fraction of the speed of light. The type of thrust that a photon propulsion system can generate is found by using Eq. 6.9 (a non-relativistic approach is assumed here, but it is understood that the relativistic problem needs to be solved as the spacecraft velocity approaches the speed of light and is treated elsewhere [8]).

The photon energy is $E_{ph} = hc/\lambda$ where $h = 6.626 \times 10 - 34$ J/s, and $c = 299792458$ m/s so,

$$E_{ph} = 6.626 \times 10^{-34} \times {2.9979 \times 10^{8}}\Big/{\lambda \times 10^{-9}} = {1.986 \times 10^{-16}}\Big/{\lambda} \text{ (Joules)} \tag{6.10}$$

Fig. 6.3 Polyus-Skif orbital laser platform on launch pad attached to the Energia rocket [7]

where λ is in nm

$$p_{ph} = \frac{1.986\times 10^{-16}}{\lambda\times 2.9979\times 10^{8}} = \frac{6.625\times 10^{-25}}{\lambda}\left(\mathrm{kg\,m\,s^{-1}}\right) \quad (6.11)$$

The simplest concept is to focus a beam from a laser fixed in space to the distant spaceship (Beamed Laser Propulsion or BPL) [9]. The beam hits a large reflective mirrored surface which is mounted on the spacecraft. This is known as a solar sail. The momentum transferred to the spacecraft by each photon will be 2 p_{ph} because the photon hits and then bounces off the mirror. If the photon beam energy is 100 MW and the photon wavelength is 10,600 nm,

$$E_{ph} = 1.874 \times 10^{-20}\mathrm{J} \quad (6.12)$$

$$p_{ph} = 6.251 \times 10^{-29} kg\,m\,s^{-1} \tag{6.13}$$

The number of photons in a 100 MW beam is,

$$N = 100 \times \frac{10^6}{1.874 \times 10^{-20}} = 5.336 \times 10^{27}\, {}^{\text{photons}}\!/_{\text{s}} \tag{6.14}$$

The total rate of momentum change (or force) from the photons bouncing off the photon sail is $2 \times \Delta P_{Total}$,

$$F = \frac{2\Delta P_{\text{Total}}}{\Delta t} = 1.067 \times 10^{28} \frac{\text{photons}}{s} \times 6.251 \times 10^{-29} \frac{\text{kgms}^{-1}}{s}$$
$$= 0.667\,\text{kg}\,\text{m}\,\text{s}^{-2} \tag{6.15}$$

This equation can be rewritten using the second law of motion,

$$F = ma = m\frac{dv}{dt} = 0.667 kg\,m\,s^{-2} \tag{6.16}$$

where, m is the mass of the space ship in kg.

The velocity of the space ship can be found by solving Eq. 6.16,

$$v(t) = ({}^{0.667}\!/_{m})t + \text{Constant} \tag{6.17}$$

The initial condition is $v(t = 0) = 0$ so the constant is equal to zero and Eq. 6.17 simplifies to,

$$v(t) = ({}^{0.667}\!/_{m})t \tag{6.18}$$

If the spaceship has a mass of 500,000 kg, the force is so small that even after 10 years its velocity only reaches 400 m/s using BPL (Fig. 6.4). BPL would need a much larger laser. Initial studies suggested a laser power of about one Terawatt.

The photons do not generate enough force to create significant thrust. However, if one realizes that photons can be recycled by using two mirrors set up as an oscillator. Recycling requires that one of the mirrors is anchored at a fixed point in space (like a moon, asteroid, planet). If the laser beam originates from the fixed mirror [10–12], the beam may be injected into an oscillator cavity formed by the fixed mirror and a mirror on the spacecraft. If the mirrors have the right radius of curvature the beam will bounce back and forth setting up an oscillation. This is called a Passive Resonant Optical Cavity (PROC). The initial momentum imparted to the space ship during the first bounce is equal to the momentum generated by the BPL. Each subsequent oscillation adds another push about equal to the momentum

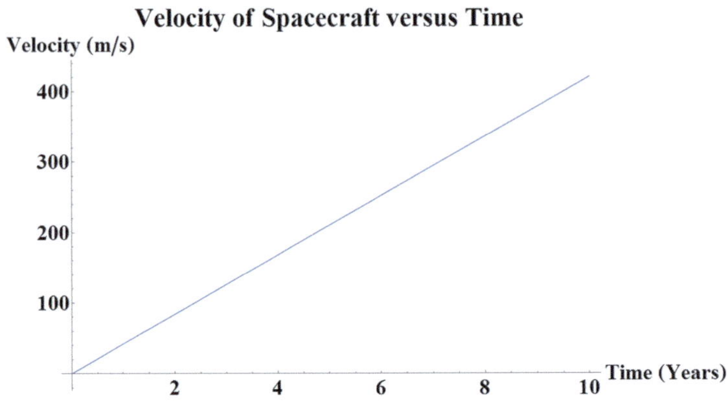

Fig. 6.4 The time evolution of velocity of a 500 t spacecraft using a 100 MW laser beam for photon sailing

generated by the first bounce (same as the BPL approach). Thus each time the beam bounces it adds to the momentum and a multiplication factor (M) on thrust is defined which is equal to the number of bounces in the oscillator. It is feasible to have a multiplication factor on the thrust generated by a BPL of 1000–1,000,000. Equation 6.18 can be multiplied by the number of bounces (M) to find the thrust of a PROC (it is understood that there will be photon losses during each oscillation and these losses can be accounted for in a revised theory),

$$v(t) = \left({}^{0.667}\!/_{m}\right)Mt \tag{6.19}$$

If there are 100,000 bounces, the velocity curve changes favorably as shown in Fig. 6.5. The minimum distance from the Earth to Mars is 54.6×10^9 m. Based on these numbers a trip to Mars could take less than a month.

Passive resonant optical cavities have been used in high-sensitivity optical detection systems such as cavity ring-down spectroscopy and have achieved bounce numbers of 20,000 [13]. The problem with this approach is that the laser beam needs to be of a single frequency and very narrow lasers tend to be inefficient in converting power into photons.

A way around the need for a single frequency laser for the PROC is an approach called the Photonic Laser Thruster (PLT) [8]. The PLT was conceived as a means to achieve high multiplication factors while overcoming the problems of injecting laser photons into a passive resonant optical cavity. The PLT forms a laser cavity between two space platforms (with mirrors) and the gain medium located between the mirrors (Fig. 6.6).

Experiments with a PLT demonstrated a multiplication factor of about 3000 and demonstrated that the cavity was stable against tilting, vibrations and motion of the mirrors. This result is promising and may represent a viable means of being able to

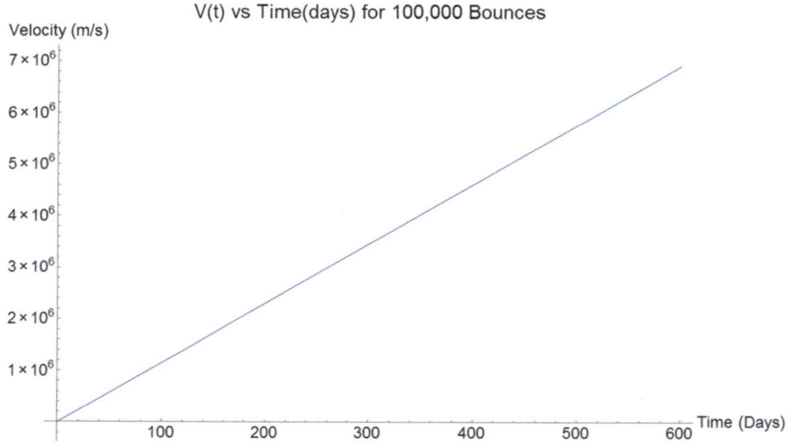

Fig. 6.5 The time evolution of velocity of a 500 t spacecraft using a 100 MW laser beam and an oscillating mirror system between the fixed laser source point and the spacecraft's mirror. The photons are recycled 100,000 times and no beam losses are assumed

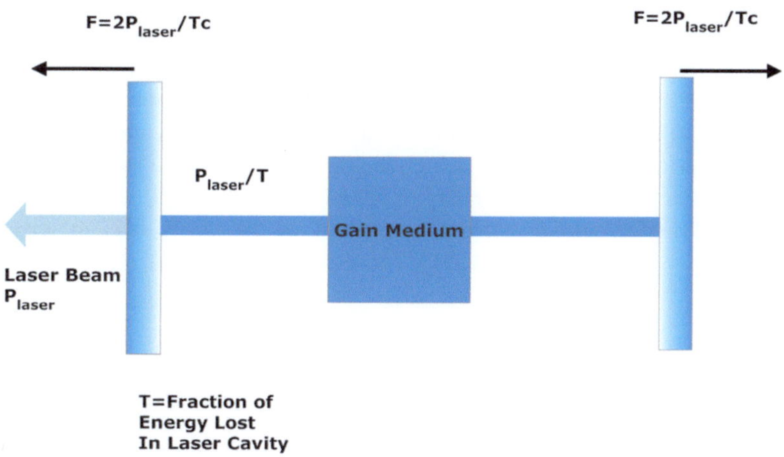

Fig. 6.6 Shown is a diagram of the PLT concept based on an active resonant optical cavity [8]

use the photon recycling strategy. A nuclear-pumped laser medium has the potential to serve as the gain cell. There are many issues such as tracking over large distances that need to be researched. Methods for reducing the demands on the technology have been proposed. A strategy was introduced to develop a photonic railway which would have cells located at strategic locations for interplanetary travel [8].

6.4 Asteroid Deflection

Asteroid deflection is an application that needs a very large amount of energy produced over the lifetime of the laser. The use of a nuclear-pumped laser in deflecting an asteroid is a mission which is suited to one of its strengths. The total fuel load of a reactor/laser system can be adjusted for required stored energy to deflect an asteroid. Take for example a large asteroid. Assume that the nuclear-pumped laser has an efficiency on the order of 1 % (considering all factors of energy transport from the fuel to the laser medium). As previously discussed the key to driving a laser is in the achievement of the minimum threshold power density. Wavelength is also a consideration because of the physics of diffraction limited optics which, as discussed in Sect. 6.1, governs the laser's aperture diameter and spot size on the target. To understand the significance of diffraction limited optics, consider the angular spread of the laser beam,

$$\Delta\theta \approx \lambda/d \tag{6.20}$$

where, $\Delta\theta$ is the angular spread, λ is wavelength in meters and d is the diameter of the optics in meters.

The angular spread of the beam determines the distance that the laser beam can be focused. As an example, assume that a xenon fluoride laser with a wavelength of 351 nm is tasked to deflect a 4000 m radius asteroid at a distance of 400 billion meters from earth. A diagram of the problem is shown in Fig. 6.7. Due to the angular spread of a laser, the beam is diagrammed from the laser's aperture to the asteroid (Fig. 6.7). The figure shows an earth based laser focused on a 4000 m radius asteroid in the asteroid belt which is 400 billion meters away. A laser beam must have enough intensity (and energy content per unit area) to ablate the surface of an asteroid. For a rocky asteroid, the required intensity is 10 MW/m^2 or greater that is necessary to evaporate stone [1]. The concept of applying a steady thrust on

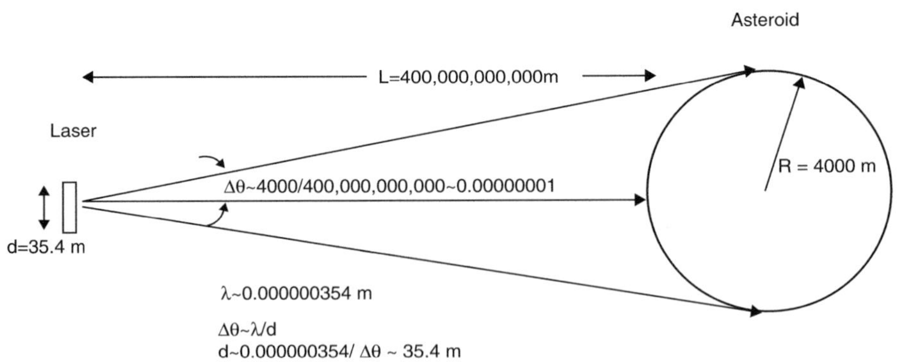

Fig. 6.7 Illustration of a laser in an earth orbit irradiating an asteroid

the asteroid has been discussed [1]. The change in velocity necessary to deflect an asteroid enough to miss the earth (a distance of one earth diameter) is:

$$v \approx 0.01 \frac{m}{s} \tag{6.21}$$

The resulting thrust which is imparted on an asteroid by steadily ablating the surface with a high intensity beam of photons is ~3.3×10^{-5} N W^{-1}. Using the model introduced by Melosh et. al. [1] for deflecting an asteroid using photons, an equation can be derived showing the relationship between laser power, asteroid diameter and irradiation time needed to deflect the orbit of an asteroid at least one earth diameter to prevent an earth strike. The force generated by the evaporation of mass is,

$$F = 0.5 \frac{dm}{dt} v_e \tag{6.22}$$

where $\frac{dm}{dt}$ is the mass ejection rate and v_e is the velocity of the ejected material.

The heat of evaporation for a silicate asteroid is 15 MJ/kg and the velocity of the atoms being ablated (v_e) is about 1000 m/s. Setting the mass ejection rate equal to the laser power (L_p, Watts) divided by the heat of evaporation, Eq. 6.22 can be rewritten,

$$m_a a_a \approx 0.5 \frac{L_P(W)}{15 \times 10^6 (J/kg)} 1000 \left(\frac{m}{s}\right) \tag{6.23}$$

where m_a is the mass of the asteroid in kg, and a_a is the acceleration of the asteroid.

The mass of a silicate asteroid is its density (3900 kg/m^3) times its volume ($\frac{4}{3}\pi \frac{(d_a)^3}{8}$), where d_a is the asteroid diameter in meters. Equation 6.23 can now be solved for L_p,

$$L_p \approx 3.333 \times 10^{-5} m_a a_a \tag{6.24}$$

Using $m_a = 3900 \left(\frac{kg}{m^3}\right) \frac{4}{3}\pi \frac{(d_a)^3}{8}$, $a_a = \frac{\Delta v}{\Delta t}$ and calling Δv the velocity change required to deflect an asteroid one earth radius (a velocity change of about 0.01 m/s) [1], an equation for laser power as a function of asteroid diameter and irradiation time, t (where Δt =irradiation time t) can be found,

$$L_p = 6.125 \times 10^5 \frac{(d_a)^3}{t} \tag{6.25}$$

where d_a is asteroid diameter in m, t is irradiation time in s; L_p is laser power in watts.

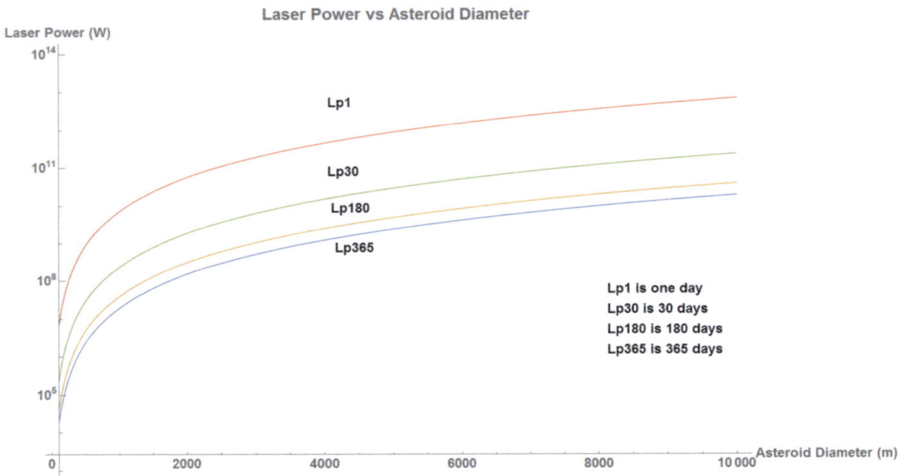

Fig. 6.8 Laser power vs. asteroid diameter for various irradiation times

The laser power required to deflect asteroids of various diameters is shown in Fig. 6.8 for various irradiation times. With smaller asteroids (~500 m), the laser power and irradiation times are moderate (30 days and < 51 MW). However, for very large asteroids (~10,000 m) the power and irradiation time is significant (365 days and > 17 GW).

Something in common for all high power/energy laser applications is a requirement on the beam intensity on the target (e.g., asteroids needing > 10 MW/m^2 to evaporate the surface). An analysis similar to that of asteroid deflection was done for SDI in Sect. 6.1. There are commonalities in these applications. SDI applications, like asteroid deflection, have certain beam intensity and irradiation time requirements. In both applications, the energy per unit area (J/m^2) is called the fluence (F). Each mission has a fluence and total energy requirement on target in order to achieve a goal; e.g., deflection of an asteroid or destruction of a ballistic missile. The fluence is calculated from the laser power, L_p (W), the irradiation time, t (s), the area of the spot size on target, A (m^2), and the laser beam transmission efficiency from the laser aperture to target (where beam energy can be lost transmitting through air for example), T (see Eq. 6.26) [14].

$$F = \frac{T L_p t}{A} \tag{6.26}$$

The spot size is a function of the far field divergence angle, $\Delta\theta$, the distance to the target, R, the laser wavelength, λ, and the diameter of the laser optics, d (see Eq. 6.27).

$$\Delta\theta \cong \frac{\lambda}{d} \tag{6.27}$$

The spot size is,

$$A = \frac{\pi R^2 \lambda^2}{d^2} \tag{6.28}$$

So assuming T=1,

$$F = \frac{L_p t d^2}{\pi R^2 \lambda^2} \tag{6.29}$$

A specific example can be examined to understand the needs of the mission and the factors which are important in scaling the laser system. As previously discussed the laser power must be delivered to the asteroid in a beam with an intensity of at least 10 MW/m^2 in order to evaporate the surface and create thrust. For example, assume that the asteroid is 1.4 km in diameter. Looking at Fig. 6.8 for a 180 day irradiation, the laser power should be at least 108 MW continuous wave. Given these assumptions, the distance of the laser system from the asteroid can be determined. This distance can be calculated by dividing the power of the laser by its spot size (A) and setting that quantity equal to 10 MW/m^2 (Table 6.2).

$$10\frac{\text{MW}}{\text{m}^2} = 108\,\text{MW}/\text{A} \tag{6.30}$$

Table 6.2 The total energy needed to deflect an asteroid of a given diameter and the minimum mass of U-235 or hydrogen needed to store that amount of energy

Laser power (W)	Asteroid diameter (km)	Lp*t (J) (t = 180 day)	Mass U235 (kg)	Mass H_2 (kg)
3.94E+04	0.1	6.13E+11	4.25E-03	4.28E+03
4.92E+06	0.5	7.66E+13	5.32E-01	5.35E+05
3.94E+07	1	6.13E+14	4.25E+00	4.28E+06
3.15E+08	2	4.90E+15	3.40E+01	3.43E+07
1.06E+09	3	1.65E+16	1.15E+02	1.16E+08
2.52E+09	4	3.92E+16	2.72E+02	2.74E+08
4.92E+09	5	7.66E+16	5.32E+02	5.35E+08
8.51E+09	6	1.32E+17	9.19E+02	9.25E+08
1.35E+10	7	2.10E+17	1.46E+03	1.47E+09
2.02E+10	8	3.14E+17	2.18E+03	2.19E+09
2.87E+10	9	4.47E+17	3.10E+03	3.12E+09
3.94E+10	10	6.13E+17	4.25E+03	4.28E+09

Equation 6.28 for "A" is substituted into Eq. 6.30 and the resulting equation is rearranged to find the following relationship for R,

$$R = \left(10.8\,d^2/\pi\lambda^2\right)^{\frac{1}{2}} \tag{6.31}$$

The importance of wavelength can be demonstrated through diffraction limited optics by seeing how two different lasers compare: a CO_2 laser with a wavelength of 10,600 nm and a XeF laser with a wavelength of 351 nm. The required distance between the laser (R_{CO_2} for CO_2 and R_{XeF} for XeF) and the asteroid is calculated using Eq. 6.31 and the results are shown in Eq. 6.32 for laser aperture diameters of 1 m and 10 m,

$$R_{CO_2} = \begin{cases} 175\,km\ for\ d = 1\,m \\ 553\,km\ for\ d = 10\,m \end{cases} \tag{6.32}$$

$$R_{XeF} = \begin{cases} 167{,}000\,km\ for\ d = 1\,m \\ 528{,}000\,km\ for\ d = 10\,m \end{cases}$$

Thus, a CO_2 laser must be much closer to the asteroid than the XeF laser in order to meet the 10 MW/m^2 criterion needed to ablate the surface.

In the example of the deflection of a 1.4 km diameter asteroid, the product of laser power and irradiation time, L_pt, must be greater than or equal to 1.68×10^{15} J and is related to the total momentum change generated by the evaporating mass from the asteroid surface. The laser must operate continuously for 180 days at a power level of 108 MW in order to accumulate enough energy to evaporate enough mass from the asteroid in order to deflect it. In contrast, for a ballistic missile defense system, the L_pt product must be on the order of 10^5 J (for a hardened target) to shoot down a ballistic missile [14]. The 1×10^5 J on target for ballistic missile defense is feasible with a very large pulsed laser. However, the 1.68×10^{15} J needed for asteroid deflection can only be delivered by a very large CW laser over long irradiation times. The challenge of building a high power laser for the ballistic missile defense mission or the high energy CW laser needed for asteroid deflection both fit very well with the strengths of NPL technology.

6.5 Space Based Mining

There are three classifications for asteroids that need to be discussed. Each type of asteroid has its own resource value and laser power needs for mining. These asteroid types are:

- "C-type" are dark in color and carbon rich. These asteroids have an abundance of water that is bound up as hydrated clay minerals. There is organic material in the form of organic carbon, phosphorous and other key elements. These chemicals are useful for fertilizers. These are similar to carbonaceous chondrite meteorites.

- S-type asteroids contain large quantities of metal (e.g. iron, nickel and cobalt) and traces of minable gold, platinum and rhodium. It has been estimated that a 10 m asteroid of this type contains about 650,000 kg of metal, with about 50 kg of platinum and gold. These are similar to LL Chondrite meteorites.
- M-type asteroids are rare but are believed to contain ten times more metal than S-type asteroids including platinum group metals. These are similar to iron meteorites.

It has been estimated that key resources (e.g., phosphorus, antimony, zinc, tin, silver, lead, indium, gold, copper, etc.) needed for industry and food production could be exhausted on earth in 50 years [15]. Space based mining of asteroids might be the only viable means of obtaining these resources in the future. Asteroid mining has some similarities to ablative asteroid deflection. The photon energy is absorbed by the rocky surface and material is ablated. From Sect. 6.3, the heat of vaporization for rock is about 15 MJ/kg. If the object which is being mined is an asteroid, the laser can be focused on the surface such that the average beam intensity is at least 10 MW/m^2 in order to cause evaporation of the surface. So in using a 10 GW laser beam as discussed in the prior section, a surface area of 1000 m^2 can be ablated. The mass ejection rate is then,

$$\dot{m} = {10\times10^9 \frac{J}{s}} \Big/ {15\times10^6 \frac{J}{kg}} = 666.7 \left(kg\, s^{-1}\right) \tag{6.33}$$

The rate of ejected material would be on the order of 2,400 t/h. The average velocity of the ejected molecules and atoms is about 1000 m/s (as discussed in Sect. 6.3). This material can be collected by large catchers and periodically transported back to earth. The value of the material is dependent on the content of the asteroid that is mined. It has been estimated that the average value of \$14,990/t [16] (\$14,849 Fe & Ni, \$69.26 Pt, \$28.68 Ir, \$17.74 Pd, \$12.17 Au, \$8.53 Rh, \$4.27 Ru and \$1.16 Os). The mining of an S type asteroid could be worth 157 billion dollars per year in 2012 dollars. Asteroid mining could be a very important resource for future generations.

6.6 Power Beaming

Lasers have been proposed in power beaming applications where the energy stored in the laser photons are converted to an energy form (e.g., electricity or propulsion). The basic concept of the conversion of electromagnetic radiation as a means of transmitting power dates back Nikola Tesla's invention of wireless power transfer [17, 18]. The concept is illustrated in Fig. 6.9. In this figure there is a power source, a transmitter (which generates the electromagnetic wave), an antenna (which projects the electromagnetic wave into space), a receiving antenna (which absorbs the electromagnetic wave power), a receiver which converts the wave into electrical energy and a load which uses the electric energy. Tesla used incoherent radiofrequency waves to transmit the power (focusing incoherent waves is analogous to focusing a beam from a flashlight). The fundamental principles used by

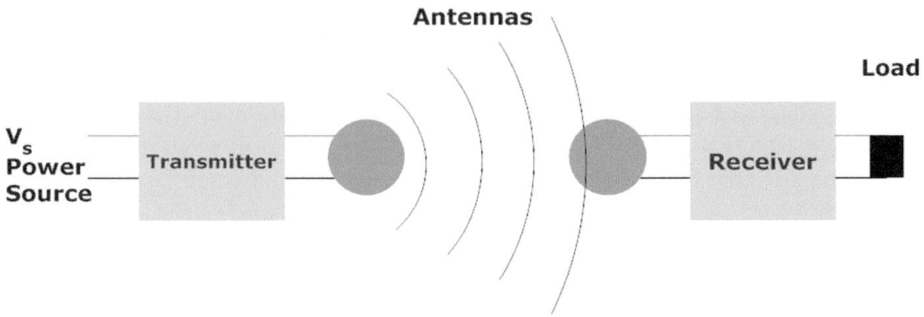

Fig. 6.9 General diagram of a wireless power beaming system

Tesla are true for laser beam power transmission. The transmitter is the laser (which is powered by a source like nuclear energy), the coherent electromagnetic wave is projected into space by optics which focus the beam, and the beam energy is absorbed by a converter (which converts photons to heat or electricity) and a load which uses the power.

An example system design used a 50–70 % efficient CO gas dynamic laser to beam power to the moon [19, 20]. There are a number of potential energy conversion methods for photons (e.g., photovoltaic cells, optical diodes, Thermo-Electronic Laser Energy Converter (TELEC), thermoelectric convertors and thermodynamic systems) [21, 22].

Earth based power beaming applications have also been discussed. Space based lasers can be used to heat gases to produce thrust in aircraft [23–25]. The laser beam can be focused onto a receiver of an aircraft which then induces gas heating. The heated gases produce thrust which is used to propel the aircraft (Fig. 6.10).

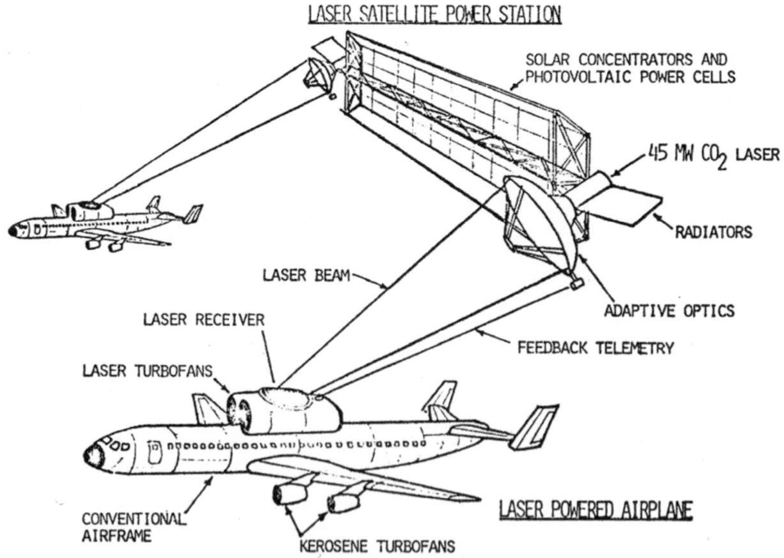

Fig. 6.10 Diagram of a space-based laser providing the energy for a flight transportation system [23]

6.7 Near Term Applications of Nuclear-Pumped Laser Technology

The nuclear-pumped laser applications described above are forward looking. Technology developed in the quest of nuclear-pumped lasers can have near term applications. This section discussed potential near term applications which can be derived from the technology.

An example of a near term example depends on the capability of measuring low levels of line spectra from gases excited by ionizing radiation. Spent nuclear fuel is stored in Dry Storage Containers (DCS). Once spent fuel is loaded in the DCS, it is filled with helium gas and sealed [26, 27]. The spent nuclear fuel emits a large amount of gamma rays (ionizing radiation as described in Chap. 2) which creates ionization and excitation in the helium gas. As discussed in Chap. 1, ionized and excited gas emits a spectra that is characteristic of the atoms and molecules in that gas. The intensity of the spectra is dependent upon the power density absorbed in the gas. For a DCS, the power density deposited in the gas is shown in Table 6.3. As shown in the table, the gamma radiation decays with time and the power density also decreases with time. The power density is dependent upon the gas pressure and this is indicated by showing results for helium gas pressures of 1, 2 and 5 atm.

As long as the cask is stable, the spectra will come from the atoms which make up the background gas (He, some oxygen and nitrogen which come off the walls and other impurities common in research grade helium gas). The major concerns with a DCS storage system is corrosion which comes from improper drying of the fuel after loading (e.g., excess water vapor in the DCS) and fuel elements cracking and leaking. If either of these events occurs, then a system designed to optically monitor the spectral emission of the background gas will show irregularities in the line spectra from hydrogen (from water vapor), enhanced oxygen levels (from water vapor), xenon (from a fuel breach), krypton (from a fuel breach), cesium (from a fuel breach) and iodine (from a fuel breach). The importance of monitoring the content of the DCS is that it may be required to store spent fuel for a very long time until the issue of spent nuclear fuel storage in a waste repository is settled. This

Table 6.3 Power density deposited in the background gas over the potential storage cask lifetime with fill pressures of 1, 2 and 5 atm

Time (year)	Total photon/ seconds	Power density (W/cm^3), 1 atm	Power density (W/cm^3), 2 atm	Power density (W/cm^3) 5 atm
0	8.33E+18	8.04E-05	1.61E-04	4.01E-04
5	1.68E+17	1.62E-06	3.24E-06	8.09E-06
10	1.00E+17	9.69E-07	1.94E-06	4.84E-06
30	5.46E+16	5.27E-07	1.05E-06	2.63E-06
50	3.42E+16	3.30E-07	6.59E-07	1.65E-06
100	1.13E+16	1.09E-07	2.19E-07	5.46E-07

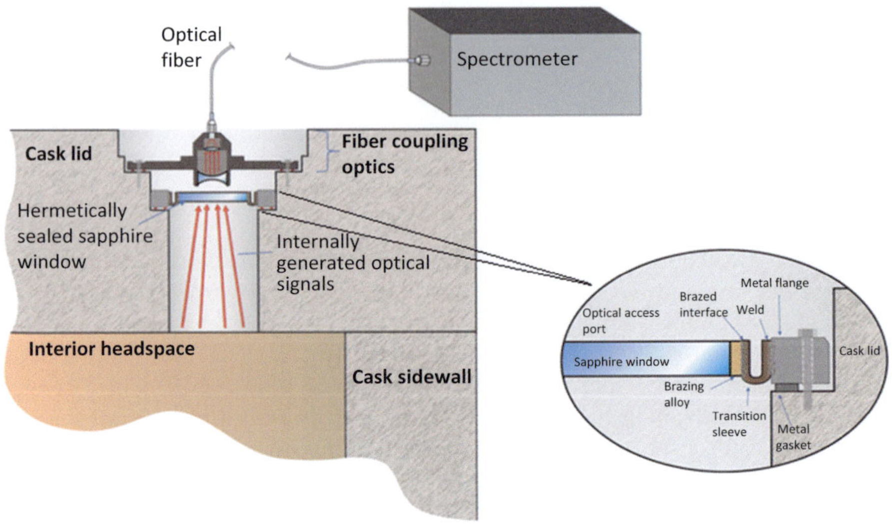

Fig. 6.11 A possible configuration for an optical monitoring system on a DCS unit

issue is controversial and the DCS may be forced to serve either as a long term or as a permanent spent fuel storage system. Thus a passive means of monitoring the contents of the DCS will be a critical diagnostic capability. The Optical Monitoring System (which will be referred to as the OMS) is self-powered and is passive thus making it ideal for this mission.

The OMS is illustrated by Fig. 6.11 which shows one possible means of interfacing an OMS to a DCS unit. Studies of the sensitivity of the OMS were performed and it was found that through the use of photon counting methods and efficient optical collection systems that the device can detect impurities in the DCS on the order of 0.01 parts per billion [26].

Another potential near term application is based on the nuclear-driven flashlamp. Incoherent nuclear-driven fluorescence itself can be a useful end product. One example of how useful nuclear-driven fluorescence may be is in the growth of organisms that use photosynthesis. It is possible to build large reactors with low power densities that are capable of producing fluorescence that matches the absorption spectra of chlorophyll (Fig. 6.12). In addition, the reactors byproduct of heat can be used to warm the plants.

A bionuclear plant (or high tech marsh) consisting of a one GW thermal plant that would generate nuclear driven fluorescence as a photon source for cultivating a single cell seaweed (chlorella) was proposed [28]. Chlorella has the ability to double its mass in one day under ideal conditions (temperatures of about 25 C and tailored light exposure of about 0.05 W cm^{-2}) and feed (water, carbon dioxide and mineral salt). A homogeneous reactor fueled with uranium-235 and thorium-232 was used in the base study. Each 1 GWth reactor was shown to be capable of producing about 4×10^{16} J/year of organic fuel.

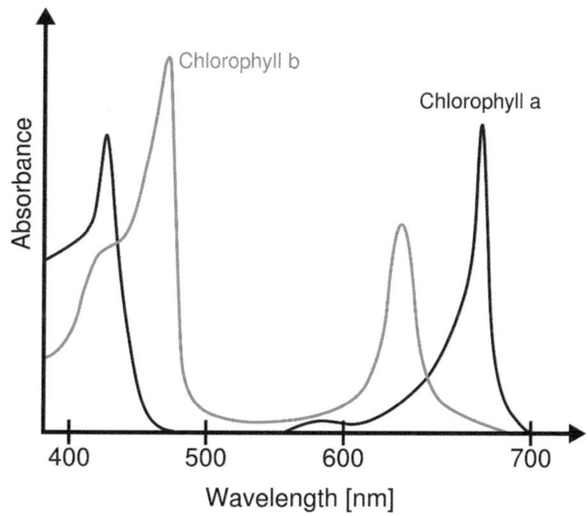

Fig. 6.12 Absorption spectrum of chlorophyll

6.8 Summary

Nuclear-pumped laser technology is one of the few potential laser systems that can scale to very high power/energy levels. Reactor/lasers scale with volume thus allowing the systems to be more compact than conventional laser technologies which scale by surface area. Additionally, nuclear fuel stores significantly more energy than chemical fuels. The reduction in required fuel mass makes reactor/laser technology idea for space applications. Several far term applications which need all of the advantages offered by nuclear-pumped laser technology were discussed.

Near term benefits of technologies developed as part of the research into nuclear-pumped lasers are also feasible. One technology for the monitoring of spent nuclear fuel, the Optical Monitoring System, was examined. In summary there are important advantages of nuclear-pumped laser systems which can be exploited where other systems have limited viability.

Problems

1. A reactor/laser system is designed to beam 100 MW carbon dioxide laser from an earth orbit to a moon base. At the moon base there is a thermionic energy converter (using an advanced photovoltaic technology matched to the 10,600 nm photons of the laser) that converts 80 % of the laser power to electricity. The reactor has a fuel loading of 20 kg of ^{235}U. The efficiency of the carbon dioxide reactor/laser is 40 %. When the ^{235}U fuel loading drops to 11 kg, the useful life of the reactor/laser system is completed. What is the total energy beamed to the moon base from the reactor/laser during its lifetime? What is the service lifetime of the reactor/laser? If the base obtained its power from a 60 % efficient fuel cell, what is the total mass of hydrogen and oxygen needed to match the energy that was beamed to the base from the reactor/laser over its lifetime?

2. Why are short wavelength lasers desirable when beaming power to a distant object?
3. If a reentry vehicle is hardened such that 100 kJ cm^{-2} is required to destroy it, what is the required brightness needed from a space based laser antiballistic missile system?
4. For the brightness found in problem 3, discuss the necessary parameters needed for this mission with a xenon fluoride laser, an iodine laser and a carbon dioxide laser.
5. Compare and contrast a carbon dioxide and xenon fluoride based nuclear-pumped laser system for asteroid deflection.
6. Discuss the laser parameters needed to deflect a 3 km, 6 km and 10 km diameter asteroids. Do you think lasers of this scale are possible?
7. How would one go about building a bionuclear reactor which produces biomass? What are the reactor's specification, how does it interface to the single cell organisms, and what are the major safety issues?
8. Is a space based SDI laser feasible with present technology? Provide details in your analysis.
9. Is a laser propulsion system for interplanetary travel feasible with present technology? Provide details in your analysis.
10. Is a laser based asteroid deflections system feasible with present technology? Provide details in your analysis.
11. What other applications can you think of for high-power/high energy reactor/nuclear-pumped laser systems?
12. Discuss the technical challenges of asteroid mining.

References

1. Melosh HJ, Nemchinov IV, Zetzer YI (1994) Non-nuclear strategies for deflecting comets and asteroids. In: Gehrels T (ed) Hazards due to comets and Asteroids. University of Arizona Press, Tucson, pp 1111–1132
2. Klass PJ (1982) House unit alters laser goals. Aviation Week & Space Technology, 26 April 1982
3. Tsipis K (1981) Laser weapons. Sci Am 245:51–57
4. APS Study Group on Science and Technology of Directed Energy Weapons (1987) Report to the APS of the study group on science and technology of directed energy weapons. In: Review of modern physics, vol 59., pp S1–S202
5. Prelas MA (1982) Nuclear powered space lasers: an evaluation of current technology, Nuclear-Pumped Laser Corporation, December 5, 1982. doi:10.13140/RG.2.1.1355.4405
6. Day DA, Kennedy III RG (2010) Soviet star wars: the launch that saved the world from orbiting laser battle stations. Air & Space Magazine. Available: http://www.airspacemag.com/space/soviet-star-wars-8758185/?no-ist. January 2010
7. Buran-Energia (2015). Archives for "Polyus". Available: http://www.buran-energia.com/blog/category/polyus/. February 28 2015
8. Bae YK (2012) Prospective of photon propulsion for interstellar flight. Phys Procedia 38:253–279
9. Forward RL (1995) Advanced propulsion systems. In: Humble RW, Henry GN, Larson WJ (eds) Space propulsion analysis and design. McGraw-Hill Co, New York, p 631

10. Meyer TR, Mckay CP, Mckenna PM (1987) The laser elevator: momentum transfer using an optical resonator. In: 38th IAF conference, Brighton, UK, p. 6. Available: http://ntrs.nasa.gov/search.jsp?R=19880028776
11. Meyer TR, Pryor WR, McKay CP, McKenna PM (2002) Laser elevator: momentum transfer using an optical resonator. J Spacecr Rocket 39:258–266, March–April 2002
12. Simmons JFL, McInnes CR (1993) Was Marx right? Or how efficient are laser driven interstellar spacecraft? Am J Phys 61:205–207
13. Romanini D, Kachanov AA, Stoeckel F (1997) Cavity ringdown spectroscopy: broad band absolute absorption measurements. Chem Phys Lett 270:546–550, May 30 1997
14. Bloembergen N, Patel CKN, Avizonis P, Clem RG, Hertzberg A, Johnson TH et al (1987) Report to the American Physical Society of the study group on science and technology of directed energy weapons. Rev Mod Phys 59:S1–S201
15. Cohen D (2007) Earth's natural wealth: an audit. New Sci, May 23, 2007
16. Fiegl A (2012) Asteroid metals. National Geographic. Available: http://news.nationalgeographic.com/news/2013/01/130122-interactive-asteroid-mining-metal-abundance/?rptregcta=reg_free_np&rptregcampaign=2015012_invitation_ro_all#
17. Tesla N (1905) Art of transmitting electrical energy through the natural mediums, Google Patents US787412 A
18. Tesla N (1914) Apparatus for transmitting electrical energy, Google Patents US1119732
19. Coneybear JF (1978) Use of lasers for the transmission of power. In: Billmann K (ed) Radiation energy conversion in space, vol 61. AIAA, New York, pp 279–310
20. Rather JD (1978) New candidate lasers for power beaming and discussion of their applications. In: Billman K (ed) Radiation energy conversion in space, vol 61. AIAA, New York, pp 313–332
21. Britt E (1978) The TELEC-A plasma type of direct energy conversion. In: Billman K (ed) Radiation energy conversion in space. AIAA, New York, pp 421–449
22. Taussig R, Rose P, Zumdieck J, Hertzberg A (1978) Energy exchanger technology applied to laser heated engines. In: Billman K (ed) Radiation energy conversion in space, vol 61, 61st edn. AIAA, New York, pp 465–478
23. Hertzberg A, Sun K (1978) Laser aircraft propulsion. In: Billman K (ed) Radiation energy conversion in space, vol 61. AIAA, New York, pp 243–263
24. Douglas-Hamilton DH, Kantrowitz AR, Reilly DA (1978) Laser assisted propulsion research. In: Billman K (ed) Radiation energy conversion in space. AIAA, New York, pp 271–278
25. Jones WS (1978) Laser-powered aircraft and rocket systems with laser energy relay units. In: Billman K (ed) Radiation energy conversion in space, vol 61. AIAA, New York, pp 264–269
26. Prelas MA (2010) In situ optical monitoring device using line emission from the interaction of radiation from fresh and used nuclear fuel in dry storage casks or gas cooled reactors or Molten core reactors to monitor leakage from fuel rods or moisture content, University of Missouri, 2010. Available: https://tech.missouri.edu/UREF107236613242999/TF_Technology_Input/K6AG3801105742248878/A11113244140
27. Meyer R, Casella A, Suter J, Quao H, Prelas M (2014) Passive optical impurity monitoring of dry storage containers for spent fuel. Trans Am Nucl Soc 110:169–172, June 2014
28. Voinov AM, Dovbysh LE, Antsyshkina ND, Rudnev AV, Marshalkin VE, Vorontsov SV et al (1996) Bionuclear technology for organic fuel production in uranium-thorium cycle. RFNC-VNIIEF/University of Missouri, Columbia

Appendices

A1 Neutron Cross Sections from Neutron Resonance Parameters and Thermal Cross Sections [1–3]

Neutron interaction cross sections are energy dependent. Cross sections are typically dependent on the inverse velocity of the neutron thus the largest interaction cross sections occur at thermal energies. Table A1.1 lists the neutron interaction cross sections for a variety of conditions:

- σ – cross section measured in a Maxwellian neutron flux
- $\sigma(0)$ – cross section measured at 2200 m/s
- $\sigma(s)$ – "free" neutron nuclear scattering cross section
- $\sigma(r)$ – cross section measured with reactor neutrons
- $\sigma(a)$ – cross section for alpha particle emission
- $\sigma(p)$ – cross section for proton emission
- $\sigma(F)$ – cross section for neutron induced fission
- $\sigma(A)$ – neutron absorption cross section
- $\sigma(T)$ – total neutron cross section
- $\sigma(f)$ – cross section for neutron induced fission (reactor spectrum)
- $\sigma(R)$ – cross section for alpha particle emission (reactor spectrum)
- $\sigma(g)$ – cross section for neutron induced fission (2200 m/s)
- $\sigma(S)$ – neutron absorption cross section (reactor spectrum)
- $\sigma(P)$ – cross section from proton emission (2200 m/s)
- $\sigma(\alpha)$ – cross section from alpha particle emission (2200 m/s)

Also Table A1.1 lists the resonance integral for each element. Resonances occur in the epithermal energy range (because most thermal reactors have an epithermal flux that varies with neutron energy (E) as approximately 1/E. If the microscopic

M. Prelas, *Nuclear-Pumped Lasers*, DOI 10.1007/978-3-319-19845-3

absorption cross section ($\sigma_a(E)$) is used and assuming no self-shielding, the epithermal resonance integral is,

$$RI = \int_{E_0}^{\infty} \sigma_a(E) \frac{dE}{E} \qquad \text{(A1)}$$

where E_0 is normally taken to be five times kT which characterizes the Maxwellian thermal flux distribution. For a temperature of 20 C, kT is 0.0253 eV.

Table A1.1 Neutron cross sections from neutron resonance parameters and thermal cross sections in the elements

A	Element	Z	Type of cross section	Cross section barns	Uncertainty	RI	
						Resonance integral	Uncertainty
1	H	1	σ(0)	3.33E−01	7.00E−04	0.00E+00	0.00E+00
1	H	1	σ(s)	2.05E+01	1.40E−02	0.00E+00	0.00E+00
2	H	1	σ	5.19E−04	7.00E−06	0.00E+00	0.00E+00
2	H	1	σ(s)	3.39E+00	1.20E−02	0.00E+00	0.00E+00
3	H	1	σ	6.00E−06	0.00E+00	0.00E+00	0.00E+00
3	H	1	σ(s)	1.70E+00	3.00E−02	0.00E+00	0.00E+00
3	He	2	σ	3.10E−05	9.00E−06	0.00E+00	0.00E+00
3	He	2	σ(s)	3.10E+00	1.30E−01	0.00E+00	0.00E+00
3	He	2	σ(P)	5.33E+03	7.00E+00	0	0
4	He	2	σ(s)	7.60E−01	1.00E−02	0.00E+00	0.00E+00
6	Li	3	σ(s)	7.50E−01	2.00E−02	0.00E+00	0.00E+00
6	Li	3	σ	3.85E−02	3.00E−03	0.00E+00	0.00E+00
6	Li	3	σ(a)	9.40E+02	4.00E+00	0.00E+00	0.00E+00
7	Li	3	σ(s)	9.70E−01	4.00E−02	0.00E+00	0.00E+00
7	Li	3	σ	4.54E−02	3.00E−03	0.00E+00	0.00E+00
7	Be	4	σ(p)	4.80E+04	9.00E+03	0	0
7	Be	4	σ(a)	1.00E−01	0.00E+00	0	0
9	Be	4	σ(s)	6.15E+00	5.00E−03	0.00E+00	0.00E+00
9	Be	4	σ	7.60E−03	8.00E−04	4.00E−03	4.00E−04
10	Be	4	σ	1.00E−03	0.00E+00	0.00E+00	0.00E+00
10	B	5	σ	5.00E−01	2.00E−01	0.00E+00	0.00E+00
10	B	5	σ(A)	3.84E+03	9.00E+00	1.72E+03	5.00E+00
10	B	5	σ(s)	2.23E+00	6.00E−02	0.00E+00	0.00E+00
11	B	5	σ(s)	4.84E+00	4.00E−02	0.00E+00	0.00E+00

(continued)

Table A1.1 (continued)

A	Element	Z	Type of cross section	Cross section barns	Uncertainty	RI Resonance integral	Uncertainty
11	B	5	σ	5.50E−03	3.30E−03	0.00E+00	0.00E+00
12	C	6	σ(s)	4.75E+00	2.00E−03	0.00E+00	0.00E+00
12	C	6	σ	3.53E−03	7.00E−05	1.57E−03	5.00E−05
13	C	6	σ(s)	4.19E+00	1.20E−01	0.00E+00	0.00E+00
13	C	6	σ	1.37E−03	4.00E−05	1.70E−03	2.00E−04
14	C	6	σ	1.00E−06	0.00E+00	0.00E+00	0.00E+00
14	N	7	σ	7.50E−02	7.50E−03	3.40E−02	0.00E+00
14	N	7	σ(s)	1.01E+01	1.20E−01	0.00E+00	0.00E+00
14	N	7	σ(p)	1.83E+00	3.00E−02	0	0
15	N	7	σ(s)	4.59E+00	5.00E−02	0.00E+00	0.00E+00
15	N	7	σ	2.40E−05	8.00E−06	1.10E−04	0.00E+00
16	O	8	σ(s)	3.76E+00	6.00E−03	0.00E+00	0.00E+00
16	O	8	σ	1.90E−04	1.90E−05	3.60E−04	0.00E+00
17	O	8	σ(s)	3.61E+00	6.00E−02	0.00E+00	0.00E+00
17	O	8	σ	5.38E−04	6.50E−05	3.90E−04	0.00E+00
17	O	8	σ(a)	2.35E−01	1.00E−02	0	0
18	O	8	σ(to 0)	1.60E−04	1.00E−05	0.00E+00	0.00E+00
19	F	9	σ	9.60E−03	5.00E−04	2.10E−02	3.00E−03
19	F	9	σ(s)	3.64E+00	1.00E−02	0.00E+00	0.00E+00
20	Ne	10	σ	3.70E−02	4.00E−03	1.75E−02	0.00E+00
20	Ne	10	σ(s)	2.47E+00	3.00E−02	0.00E+00	0.00E+00
21	Ne	10	σ(s)	5.10E+00	3.00E−01	0.00E+00	0.00E+00
21	Ne	10	σ	6.66E−01	1.10E−01	2.96E−01	0.00E+00
21	Ne	10	σ(a)	1.50E+00	0.00E+00	0	0
22	Ne	10	σ(s)	1.71E+00	9.00E−03	0.00E+00	0.00E+00

22	Ne	10	σ	4.55E−02	6.00E−03	2.30E−02	0.00E+00
22	Na	11	σ	2.90E+04	1.00E+03	2.00E+05	5.00E+04
23	Na	11	σ(s)	3.03E+00	2.00E−02	0.00E+00	0.00E+00
23	Na	11	σ(to 472.2)	4.00E−01	3.00E−02	0.00E+00	0.00E+00
23	Na	11	σ(0)	5.30E−01	5.00E−03	3.11E−01	1.00E−02
24	Mg	12	σ(s)	3.74E+00	4.00E−02	0.00E+00	0.00E+00
24	Mg	12	σ(0)	5.10E−02	5.00E−03	3.20E−02	4.00E−03
25	Mg	12	σ	1.90E−01	3.00E−02	9.80E−02	1.50E−02
26	Mg	12	σ(0)	3.82E−02	8.00E−04	2.60E−02	2.00E−03
26	Mg	12	σ(s)	2.83E+00	1.70E−01	0.00E+00	0.00E+00
27	Mg	12	σ(to 0)	7.00E−02	2.00E−02	0.00E+00	0.00E+00
27	Al	13	σ(0)	2.31E−01	3.00E−03	1.70E−01	7.00E−02
27	Al	13	σ(s)	1.41E+00	1.00E−03	0.00E+00	0.00E+00
28	Si	14	σ(s)	1.99E+00	6.00E−03	0.00E+00	0.00E+00
28	Si	14	σ	1.77E−01	5.00E−03	1.10E−01	1.50E−02
29	Si	14	σ	1.01E−01	1.40E−02	7.70E−02	1.50E−02
30	Si	14	σ(0)	1.07E−01	2.00E−03	7.10E−01	6.00E−02
30	Si	14	σ(s)	2.49E+00	4.00E−02	0.00E+00	0.00E+00
31	Si	14	σ(to 0)	1.80E−01	4.00E−02	0.00E+00	0.00E+00
31	P	15	σ(s)	3.13E+00	1.00E−02	0.00E+00	0.00E+00
31	P	15	σ(0)	1.72E−01	6.00E−03	8.50E−02	1.00E−02
32	S	16	σ(s)	9.43E−01	2.10E−03	0.00E+00	0.00E+00
32	S	16	σ	5.30E−01	4.00E−02	8.00E−02	0.00E+00
32	S	16	σ(a)	7.00E−03	4.00E−03	0	0
33	S	16	σ	3.50E−01	4.00E−02	9.70E−02	0.00E+00
33	S	16	σ(s)	2.80E+00	7.00E−01	0.00E+00	0.00E+00
33	S	16	σ(p)	2.00E−03	1.00E−03	0	0

(continued)

Table A1.1 (continued)

A	Element	Z	Type of cross section	Cross section barns	Uncertainty	RI	
						Resonance integral	Uncertainty
33	S	16	σ(a)	1.90E−01	8.00E−02	0	0
34	S	16	σ	2.27E−01	5.00E−03	1.50E−03	0.00E+00
35	Cl	17	σ(s)	2.06E+01	3.00E−01	0.00E+00	0.00E+00
35	Cl	17	σ	4.36E+01	4.00E−01	1.80E+01	2.00E+00
35	Cl	17	σ(p)	4.89E−01	1.40E−02	0	0
35	Cl	17	σ(a)	8.00E−05	4.00E−05	0	0
36	S	16	σ	1.50E−01	3.00E−02	1.70E−01	4.00E−02
36	Cl	17	σ	1.00E+01	0.00E+00	0.00E+00	0.00E+00
36	Ar	18	σ	5.20E+00	5.00E−01	0.00E+00	0.00E+00
36	Ar	18	σ(s)	7.37E+01	4.00E−01	0.00E+00	0.00E+00
36	Ar	18	σ(a)	5.50E−03	1.00E−04	0	0
37	Cl	17	σ(to 0+671.4)	4.33E−01	6.00E−03	3.00E−01	4.00E−02
37	Cl	17	σ(s)	1.15E+00	5.00E−02	0.00E+00	0.00E+00
37	Cl	17	σ(to 671.4)	4.70E−02	1.00E−02	0.00E+00	0.00E+00
37	Ar	18	σ(p)	6.90E+01	1.40E+01	0	0
37	Ar	18	σ(a)	1.97E+03	3.30E+02	0	0
38	Ar	18	σ(s)	1.50E+00	1.50E+00	0.00E+00	0.00E+00
38	Ar	18	σ	8.00E−01	2.00E−01	0.00E+00	0.00E+00
39	Ar	18	σ	6.00E+02	3.00E+02	0.00E+00	0.00E+00
39	K	19	σ	2.10E+00	2.00E−01	1.10E+00	1.00E−01
39	K	19	σ(a)	4.30E−03	5.00E−04	0	0
40	Ar	18	σ	6.60E−01	1.00E−02	4.10E−01	3.00E−02
40	K	19	σ	3.00E+01	8.00E+00	1.30E+01	4.00E+00
40	K	19	σ(p)	4.40E+00	3.00E−01	0	0
40	K	19	σ(a)	3.90E−01	3.00E−02	0	0

40	Ca	20	σ(s)	3.01E+00	8.00E−02	0.00E+00	0.00E+00
40	Ca	20	σ(0)	4.10E−01	2.00E−02	2.20E−01	2.00E−02
40	Ca	20	σ(a)	2.50E−03	1.10E−03	0	0
41	Ar	18	σ	5.00E−01	1.00E−01	0.00E+00	0.00E+00
41	K	19	σ	1.46E+00	3.00E−02	1.42E+00	6.00E−02
41	Ca	20	σ	4.00E+00	0.00E+00	0.00E+00	0.00E+00
42	Ca	20	σ	6.80E−01	7.00E−02	3.90E−01	4.00E−02
42	Ca	20	σ(s)	1.20E+00	2.00E−01	0.00E+00	0.00E+00
43	Ca	20	σ	6.20E+00	6.00E−01	3.93E+00	1.50E−01
44	Ca	20	σ(0)	8.80E−01	5.00E−02	5.60E−01	1.00E−02
45	Ca	20	σ	1.50E+01	0.00E+00	0.00E+00	0.00E+00
45	Sc	21	σ(s)	2.24E+01	4.00E−01	0.00E+00	0.00E+00
45	Sc	21	σ(to 142.5)	9.80E+00	1.10E+00	5.40E+00	6.00E−01
45	Sc	21	σ(to 0)	1.74E+01	1.10E+00	0.00E+00	0.00E+00
45	Sc	21	σ(0)	2.72E+01	2.00E−01	0	0
46	Ca	20	σ	7.40E−01	7.00E−02	9.60E−01	1.00E−01
46	Sc	21	σ(r)	8.00E+00	1.00E+00	0.00E+00	0.00E+00
46	Ti	22	σ	5.90E−01	1.80E−01	3.00E−01	9.00E−02
46	Ti	22	σ(s)	2.78E+00	2.40E−01	0.00E+00	0.00E+00
47	Ti	22	σ	1.70E+00	2.00E−01	1.50E+00	2.00E−01
47	Ti	22	σ(s)	3.10E+00	2.00E−01	0.00E+00	0.00E+00
48	Ca	20	σ	1.09E+00	1.40E−01	8.90E−01	1.80E−01
48	Ti	22	σ	7.84E+00	2.50E−01	3.90E+00	2.00E−01
48	Ti	22	σ(s)	4.10E+00	2.00E−01	0.00E+00	0.00E+00
49	Ti	22	σ	2.20E+00	3.00E−01	1.20E+00	2.00E−01
49	Ti	22	σ(s)	7.00E−01	3.00E−01	0.00E+00	0.00E+00
50	Ti	22	σ(0)	1.79E−01	3.00E−03	1.18E−01	1.10E−02

(continued)

Table A1.1 (continued)

A	Element	Z	Type of cross section	Cross section barns	Uncertainty	RI Resonance integral	Uncertainty
50	Ti	22	σ(s)	3.70E+00	3.00E−01	0.00E+00	0.00E+00
50	V	23	σ(r)	6.00E+01	4.00E+01	4.30E+01	1.50E+01
50	V	23	σ(s)	7.50E+00	1.00E+00	0.00E+00	0.00E+00
50	Cr	24	σ	1.59E+01	2.00E−01	7.80E+00	4.00E−01
51	V	23	σ(0)	4.90E+00	1.00E−01	2.70E+00	1.00E−01
51	V	23	σ(s)	4.80E+00	2.00E−01	0.00E+00	0.00E+00
52	Cr	24	σ(s)	2.96E+00	2.00E−02	0.00E+00	0.00E+00
52	Cr	24	σ	7.60E−01	6.00E−02	4.80E−01	0.00E+00
53	Cr	24	σ(s)	7.78E+00	2.00E−01	0.00E+00	0.00E+00
53	Cr	24	σ	1.82E+01	1.50E+00	8.90E+00	0.00E+00
53	Mn	25	σ(r)	7.00E+01	1.00E+01	3.00E+01	5.00E+00
54	Cr	24	σ(s)	2.54E+00	1.00E−01	0.00E+00	0.00E+00
54	Cr	24	σ	3.60E−01	4.00E−02	1.80E−01	4.00E−02
54	Mn	25	σ(r)	3.80E+01	0.00E+00	1.70E+01	0.00E+00
54	Fe	26	σ(s)	2.17E+00	1.00E−01	0.00E+00	0.00E+00
54	Fe	26	σ	2.25E+00	1.80E−01	1.20E+00	2.00E−01
55	Mn	25	σ(0)	1.33E+01	2.00E−01	1.40E+01	3.00E−01
55	Mn	25	σ(s)	2.20E+00	2.00E−01	0.00E+00	0.00E+00
56	Fe	26	σ(s)	1.25E+01	4.90E−01	0.00E+00	0.00E+00
56	Fe	26	σ	2.59E+00	1.40E−01	1.40E+00	2.00E−01
57	Fe	26	σ	2.48E+00	3.00E−01	1.60E+00	2.00E−01
58	Fe	26	σ(0)	1.28E+00	5.00E−02	1.70E+00	1.00E−01
58	Co	27	σ(for 0)	1.90E+03	2.00E+02	7.00E+03	1.00E+03
58	Co	27	σ(for 24.9)	1.40E+05	1.00E+04	0.00E+00	0.00E+00
58	Ni	28	σ(s)	2.53E+01	4.00E−01	0.00E+00	0.00E+00

58	Ni	28	σ	4.60E+00	3.00E−01	2.20E+00	2.00E−01
58	Ni	28	σ(a)(to 0)	3.00E−05	0.00E+00	0	0
59	Co	27	σ(s)	6.00E+00	6.00E−02	0.00E+00	0.00E+00
59	Co	27	σ(0)(to 58.6)	2.04E+01	8.00E−01	3.90E+01	2.00E+00
59	Co	27	σ(0)(to 0 + 58.6)	3.72E+01	6.00E−02	7.40E+01	2.00E+00
59	Ni	28	σ(a)	1.23E+01	6.00E−01	0	0
59	Ni	28	σ(p)	2.00E+00	5.00E−01	0	0
59	Ni	28	σ(p)(to 0)	1.34E+00	1.80E−01	0	0
59	Ni	28	σ(A)	9.20E+01	4.00E+00	1.38E+02	8.00E+00
59	Ni	28	σ(0)	7.77E+01	4.10E+00	0.00E+00	0.00E+00
60	Co	27	σ(for 0)	2.00E+00	2.00E−01	4.30E+00	9.00E−01
60	Co	27	σ(for 58.6)	5.80E+01	8.00E+00	2.30E+02	5.00E+01
60	Ni	28	σ	2.90E+00	2.00E−01	1.50E+00	2.00E−01
60	Ni	28	σ(s)	9.80E−01	7.00E−02	0.00E+00	0.00E+00
61	Ni	28	σ(s)	9.00E+00	1.00E+00	0.00E+00	0.00E+00
61	Ni	28	σ	2.50E+00	8.00E−01	1.50E+00	4.00E−01
61	Ni	28	σ(a)(to 0)	3.00E−05	0.00E+00		
62	Ni	28	σ	1.45E+01	3.00E−01	6.60E+00	2.00E−01
62	Ni	28	σ(s)	9.10E+00	4.00E−01	0.00E+00	0.00E+00
63	Ni	28	σ	2.44E+01	3.00E+00	0.00E+00	0.00E+00
63	Cu	29	σ(s)	5.10E+00	2.00E−01	0.00E+00	0.00E+00
63	Cu	29	σ(0)	4.50E+00	2.00E−02	0.00E+00	0.00E+00
64	Ni	28	σ(s)	1.40E−03	3.00E−03	0.00E+00	0.00E+00
64	Ni	28	σ(0)	1.52E+00	3.00E−02	9.80E−01	1.50E−01
64	Cu	29	σ(r)	6.00E+03	0.00E+00	0.00E+00	0.00E+00
64	Zn	30	σ	7.60E−01	2.00E−02	1.45E+00	6.00E−02
64	Zn	30	σ(s)	3.90E+00	3.00E−01	0.00E+00	0.00E+00

(continued)

Table A1.1 (continued)

A	Element	Z	Type of cross section	Cross section barns	Uncertainty	RI Resonance integral	Uncertainty
65	Ni	28	σ(r)	2.24E+01	2.00E+00	0.00E+00	0.00E+00
65	Cu	29	σ(0)	2.17E+00	3.00E−02	0.00E+00	0.00E+00
65	Cu	29	σ(s)	1.41E+01	5.00E−01	0.00E+00	0.00E+00
65	Zn	30	σ(R)	2.50E+02	1.50E+02	0	0
66	Cu	29	σ(r)	1.35E+02	1.00E+01	0.00E+00	0.00E+00
66	Zn	30	σ(s)	4.90E+00	3.00E−01	0.00E+00	0.00E+00
66	Zn	30	σ	8.50E−01	2.00E−01	1.77E+00	0.00E+00
66	Zn	30	σ(a)	2.00E−05	0.00E+00	0	0
67	Zn	30	σ	6.80E+00	8.00E−01	2.52E+01	0.00E+00
67	Zn	30	σ(a)	6.00E−06	4.00E−06	0	0
68	Zn	30	σ(to 438.6)	7.20E−02	4.00E−03	0.00E+00	0.00E+00
68	Zn	30	σ(to 0)	1.00E+00	1.00E−01	0.00E+00	0.00E+00
68	Zn	30	σ(s)	5.40E+00	3.00E−01	0.00E+00	0.00E+00
68	Zn	30	σ(a)	2.00E−05	0.00E+00	0	0
69	Ga	31	σ(0)(to 0)	1.68E+00	7.00E−02	1.56E+01	1.50E+00
70	Zn	30	σ(to 157.7)	8.70E−03	5.00E−04	0.00E+00	0.00E+00
70	Zn	30	σ(to 0)	8.30E−02	5.00E−03	0.00E+00	0.00E+00
70	Ge	32	σ(to 0)	3.15E+00	1.60E−01	0.00E+00	0.00E+00
70	Ge	32	σ(to 198.4)	2.80E−01	7.00E−02	0.00E+00	0.00E+00
70	Ge	32	σ(s)	8.80E+00	8.00E−01	0.00E+00	0.00E+00
71	Ga	31	σ(0)(to 119.7)	1.50E−01	5.00E−02	0.00E+00	0.00E+00
71	Ga	31	σ(0)(to 0 + 119.7)	4.71E+00	2.30E−01	3.12E+01	1.90E+00
72	Ge	32	σ(s)	7.50E+00	7.00E−01	0.00E+00	0.00E+00
72	Ge	32	σ(r)	9.80E−01	9.00E−02	7.60E−01	0.00E+00
73	Ge	32	σ(r)	1.50E+01	2.00E+00	6.37E+01	0.00E+00

74	Ge	32	σ(s)	6.10E+00	4.00E−01	0.00E+00	0.00E+00
74	Ge	32	σ(to 139.7)	1.70E−01	3.00E−02	4.10E−01	7.00E−02
74	Ge	32	σ(to 0)	3.40E−01	8.00E−02	0.00E+00	0.00E+00
74	Se	34	σ	5.18E+01	1.20E+00	5.20E+02	9.00E+01
75	As	33	σ(0)	4.50E+00	1.00E−01	6.10E+01	4.00E+00
75	As	33	σ(s)	5.43E+00	3.00E−02	0.00E+00	0.00E+00
76	Ge	32	σ(to 0)	6.00E−02	1.00E−02	8.00E−01	2.00E−01
76	Ge	32	σ(to 159.7)	1.00E−01	1.00E−02	1.20E+00	2.00E−01
76	Se	34	σ(s)	1.84E+01	3.00E−01	0.00E+00	0.00E+00
76	Se	34	σ(r)(to 161.8)	2.20E+01	1.00E+00	1.70E+01	2.00E+00
76	Se	34	σ(to 0 + 161.8)	8.50E+01	7.00E+00	4.03E+01	0.00E+00
76	Br	35	σ(p)	2.24E+02	4.20E+01	0	0
77	Se	34	σ(s)	8.43E+00	1.60E−01	0.00E+00	0.00E+00
77	Se	34	σ(r)	4.20E+01	4.00E+00	3.01E+01	0.00E+00
78	Se	34	σ(to 95.7)	3.80E−01	2.00E−02	3.70E+00	6.00E−01
78	Se	34	σ(to 0)	5.00E−02	1.00E−02	0.00E+00	0.00E+00
78	Se	34	σ(s)	8.40E+00	2.00E−02	0.00E+00	0.00E+00
78	Kr	36	σ(to 0 + 130.0)	6.20E+00	9.00E−01	1.95E+01	2.00E+00
78	Kr	36	σ(to 130.0)	1.70E−01	2.00E−02	0.00E+00	0.00E+00
79	Br	35	σ(0)(to 86.3)	2.40E+00	6.00E−01	3.20E+01	9.00E+00
79	Br	35	σ(0)(to 0)	8.60E+00	4.00E−01	9.50E+01	1.10E+01
80	Se	34	σ(to 0)	5.30E−01	4.00E−02	0.00E+00	0.00E+00
80	Se	34	σ(to 103.0)	8.00E−02	1.00E−02	3.40E−01	9.00E−02
80	Se	34	σ(s)	6.95E+00	6.00E−02	0.00E+00	0.00E+00
80	Kr	36	σ(to 190.5)	4.55E+00	6.50E−01	0.00E+00	0.00E+00
81	Br	35	σ(0)(to 0)	2.60E−01	0.00E+00	0.00E+00	0.00E+00
81	Br	35	σ(0)(to 46)	2.43E+00	4.00E−01	0.00E+00	0.00E+00

(continued)

Table A1.1 (continued)

A	Element	Z	Type of cross section	Cross section barns	Uncertainty	RI	
						Resonance integral	Uncertainty
82	Se	34	σ(to 228.5)	3.90E−02	3.00E−03	3.90E−02	4.00E−03
82	Se	34	σ(to 0)	5.20E−03	4.00E−04	0.00E+00	0.00E+00
82	Se	34	σ(s)	5.00E+00	2.00E−01	0.00E+00	0.00E+00
82	Kr	36	σ(to 41.5)	1.40E+01	2.50E+00	0.00E+00	0.00E+00
82	Kr	36	σ(to 0 + 41.5)	2.80E+01	2.00E+01	1.30E+02	1.30E+01
83	Kr	36	σ	1.80E+02	3.00E+01	1.83E+02	2.50E+01
84	Kr	36	σ(to 0)	4.20E−02	4.00E−03	0.00E+00	0.00E+00
84	Kr	36	σ(to 304.9)	9.00E−02	1.30E−02	0.00E+00	0.00E+00
84	Rb	37	σ(p)	1.20E+01	2.00E+00	0	0
84	Sr	38	σ(to 0)	3.50E−01	7.00E−02	6.72E+00	1.30E+00
84	Sr	38	σ(to 238.7)	6.00E−01	6.00E−02	4.59E+00	1.50E−01
85	Kr	36	σ	1.66E+00	2.00E−01	1.80E+00	1.00E+00
85	Rb	37	σ(to 0)	4.27E−01	1.10E−02	0.00E+00	0.00E+00
85	Rb	37	σ(to 556.0)	5.30E−02	5.00E−03	0.00E+00	0.00E+00
86	Kr	36	σ(to 0)	3.00E−03	2.00E−03	0.00E+00	0.00E+00
86	Sr	38	σ(to 388.5)	8.40E−01	6.00E−02	4.79E+00	2.40E−01
86	Sr	38	σ(to 0)	2.00E−01	3.00E−02	0.00E+00	0.00E+00
87	Kr	36	σ(to 0)	1.26E+04	0.00E+00	0.00E+00	0.00E+00
87	Rb	37	σ (to 0)	1.20E−01	3.00E−02	1.90E+00	2.00E−01
87	Sr	38	σ	1.60E+01	3.00E+00	1.18E+02	3.00E+01
88	Rb	37	σ(r)	1.00E+00	3.00E−01	0.00E+00	0.00E+00
88	Sr	38	σ	5.80E−02	4.00E−04	6.50E−02	0.00E+00
89	Sr	38	σ	4.20E−01	4.00E−02	0.00E+00	0.00E+00
89	Y	39	σ(s)	7.67E+00	6.00E−02	0.00E+00	0.00E+00
89	Y	39	σ(to 682.0)	1.00E−03	2.00E−04	0.00E+00	0.00E+00

89	Y	39	σ(0)	1.28E+00	2.00E−02	1.00E+00	1.00E−01
90	Sr	38	σ	9.00E−01	5.00E−01	0.00E+00	0.00E+00
90	Y	39	σ(r)	6.50E+00	0.00E+00	0.00E+00	0.00E+00
90	Zr	40	σ	1.10E−02	5.00E−03	1.40E−01	0.00E+00
90	Zr	40	σ(s)	5.30E+00	3.00E−01	0.00E+00	0.00E+00
91	Y	39	σ(r)	1.40E+00	3.00E−01	0.00E+00	0.00E+00
91	Zr	40	σ	1.24E+00	2.50E−01	5.20E+00	7.00E−01
91	Zr	40	σ(s)	1.07E+01	6.00E−01	0.00E+00	0.00E+00
92	Zr	40	σ	2.20E−01	6.00E−02	6.30E−01	0.00E+00
92	Mo	42	σ	1.90E−02	0.00E+00	8.10E−01	0.00E+00
93	Zr	40	σ	2.70E+00	1.40E+00	0.00E+00	0.00E+00
93	Nb	41	σ(0)(to 0 + 40.9)	1.15E+00	5.00E−02	8.50E+00	5.00E−01
93	Nb	41	σ(s)	6.37E+00	7.00E−02	0.00E+00	0.00E+00
94	Zr	40	σ(s)	6.10E+00	4.00E−01	0.00E+00	0.00E+00
94	Zr	40	σ	4.99E−02	2.40E−03	2.30E−01	1.00E−02
94	Nb	41	σ(0)(to 0)	1.49E+01	1.00E+00	0.00E+00	0.00E+00
94	Nb	41	σ(to 235.7)	6.00E−01	1.00E−01	0.00E+00	0.00E+00
94	Mo	42	σ	1.50E−02	0.00E+00	8.20E−01	0.00E+00
95	Nb	41	σ(r)	7.00E+00	0.00E+00	0.00E+00	0.00E+00
95	Mo	42	σ	1.40E+01	5.00E−01	1.09E+02	5.00E+00
96	Zr	40	σ(s)	6.60E+00	4.00E−01	0.00E+00	0.00E+00
96	Zr	40	σ	2.29E−02	1.00E−03	5.30E+00	3.00E−01
96	Mo	42	σ	5.00E−01	2.00E−01	1.70E+01	3.00E+00
96	Ru	44	σ(0)	2.90E−01	2.00E−02	7.34E+00	8.00E−02
97	Mo	42	σ	2.10E+00	5.00E−01	1.40E+01	3.00E+00
98	Mo	42	σ(0)	1.30E−01	6.00E−03	6.90E+00	3.00E−01
98	Tc	43	σ(r)(to 142.7)	9.30E−01	2.00E−01	0.00E+00	0.00E+00

(continued)

Table A1.1 (continued)

A	Element	Z	Type of cross section	Cross section barns	Uncertainty	RI	
						Resonance integral	Uncertainty
98	Ru	44	σ(r)	8.00E+00	0.00E+00	0.00E+00	0.00E+00
99	Tc	43	σ(0)	2.00E+01	1.00E+00	0.00E+00	0.00E+00
99	Ru	44	σ(0)	7.10E+00	1.00E+00	1.60E+02	2.00E+01
100	Mo	42	s	1.99E−01	3.00E−03	3.75E+00	1.50E−01
100	Ru	44	σ(0)	5.00E+00	6.00E−01	1.12E+01	1.10E+00
101	Ru	44	σ(0)	3.40E+00	9.00E−01	1.00E+02	2.00E+01
102	Ru	44	σ(0)	1.21E+00	7.00E−02	4.20E+00	1.00E−01
102	Pd	46	σ(r)	3.40E+00	3.00E−01	1.00E+01	2.00E+00
103	Rh	45	σ(0)(to 0 + 129.0)	1.45E+02	2.00E+00	1.10E+03	5.00E+01
103	Rh	45	σ(0)(to 129.0)	1.00E+01	1.00E+00	7.50E+01	5.00E+00
104	Ru	44	σ(0)	3.20E−01	2.00E−02	4.30E+00	1.00E−01
104	Rh	45	σ(r)(for 0)	4.00E+01	3.00E+01	0.00E+00	0.00E+00
104	Rh	45	σ(r)(for 129.0)	8.00E+02	1.00E+02	0.00E+00	0.00E+00
104	Pd	46	σ	6.00E−01	3.00E−01	1.60E+01	2.00E+00
105	Ru	44	σ(0)	3.90E−01	6.00E−02	0.00E+00	0.00E+00
105	Rh	45	σ(r)(to 0)	1.10E+04	3.00E+03	0.00E+00	0.00E+00
105	Rh	45	σ(r)(to 137)	5.00E+03	1.00E+03	0.00E+00	0.00E+00
105	Pd	46	σ(s)	5.00E+00	6.00E−01	0.00E+00	0.00E+00
105	Pd	46	σ	2.00E+01	3.00E+00	6.22E+01	0.00E+00
106	Ru	44	σ(0)	1.46E−01	4.50E−02	2.00E+00	6.00E−01
106	Pd	46	σ(0)(to 0)	2.92E−01	2.90E−02	0.00E+00	0.00E+00
106	Pd	46	σ(s)	5.10E+00	6.00E−01	0.00E+00	0.00E+00
106	Pd	46	σ(0)(to 241.9)	1.30E−02	2.00E−03	0.00E+00	0.00E+00
106	Cd	48	σ(r)	1.00E+00	0.00E+00	4.00E+00	0.00E+00
107	Pd	46	σ	1.80E+00	2.00E−01	8.66E+01	0.00E+00

107	Ag	47	σ(0)(to 109.4)	3.30E−01	8.00E−02	1.20E+00	2.00E−01
107	Ag	47	σ(s)(to 0)	7.44E+00	9.00E−02	0.00E+00	0.00E+00
107	Ag	47	σ(0)(to 0 + 109.4)	3.76E+01	1.20E+00	1.00E+02	5.00E+00
108	Pd	46	σ(to 0)	8.30E+00	5.00E−01	0.00E+00	0.00E+00
108	Pd	46	σ(to 188.9)	1.83E−01	3.30E−02	0.00E+00	0.00E+00
108	Cd	48	σ(r)	1.10E+00	3.00E−01	1.10E+01	0.00E+00
109	Ag	47	σ(0)(to 117.6)	4.70E+00	2.00E−01	7.23E+01	4.00E+00
109	Ag	47	σ(s)	2.55E+00	6.00E−02	0.00E+00	0.00E+00
109	Ag	47	σ(0)(to 0)	8.63E+01	3.00E+00	0.00E+00	0.00E+00
109	Cd	48	σ(r)	7.00E+02	1.00E+02	0.00E+00	0.00E+00
109	Cd	48	σ(a)	5.00E−02	0.00E+00	0	0
110	Pd	46	σ(r)(to 172.2)	3.70E−02	6.00E−03	7.00E−01	2.00E−01
110	Pd	46	σ(r)(to 0)	1.90E−01	3.00E−02	2.40E+00	3.00E−01
110	Ag	47	σ(0)(for 117.6)	8.20E+01	1.10E+01	0.00E+00	0.00E+00
110	Cd	48	σ(r)(to 396.2)	1.40E−01	5.00E−02	3.00E+00	1.00E+00
110	Cd	48	σ(r)(to 0+396.2)	1.10E+01	1.00E+00	3.70E+01	0.00E+00
111	Ag	47	σ(r)	3.00E+00	2.00E+00	1.05E+02	2.00E+01
111	Cd	48	σ(s)	5.00E+00	1.00E+00	0.00E+00	0.00E+00
111	Cd	48	σ(r)	2.40E+01	3.00E+00	5.00E+01	0.00E+00
112	Cd	48	σ(r)	2.20E+00	5.00E−01	1.20E+01	0.00E+00
112	Sn	50	σ(to 77.4)	3.00E−01	4.00E−02	0.00E+00	0.00E+00
112	Sn	50	σ(to 0)	7.10E−01	1.00E−01	0.00E+00	0.00E+00
113	Cd	48	σ(0)	2.06E+04	4.00E+02	3.90E+02	0.00E+00
113	In	49	σ(s)	3.75E+00	7.00E−02	0.00E+00	0.00E+00
113	In	49	σ(0)(to 0)	3.90E+00	4.00E−01	0.00E+00	0.00E+00
113	In	49	σ(0)(to 190.3)	8.10E+00	8.00E−01	0.00E+00	0.00E+00
113	In	49	σ(0)(to 502.0)	3.10E+00	7.00E−01	0.00E+00	0.00E+00

(continued)

Table A1.1 (continued)

A	Element	Z	Type of cross section	Cross section barns	Uncertainty	RI Resonance integral	Uncertainty
114	Cd	48	σ(s)	6.00E+00	1.00E+00	0.00E+00	0.00E+00
114	Cd	48	σ(to 181.0)	3.60E−02	7.00E−03	0.00E+00	0.00E+00
114	Cd	48	σ(to 0)	3.00E−01	2.00E−02	0.00E+00	0.00E+00
114	Sn	50	σ	1.15E−01	3.00E−02	5.10E+00	1.50E+00
114	Sn	50	σ(s)	4.60E+00	5.00E−01	0.00E+00	0.00E+00
115	In	49	σ(s)	2.60E+00	1.00E−01	0.00E+00	0.00E+00
115	In	49	σ(0)(to 127.3)	1.62E+02	7.00E−01	0.00E+00	0.00E+00
115	In	49	σ(0)(to 289.7)	8.10E+01	8.00E+00	0.00E+00	0.00E+00
115	In	49	σ(0)(to 0)	4.00E+01	2.00E+00	6.50E+02	3.00E+01
115	Sn	50	σ(r)	3.00E+01	7.00E+00	2.90E+01	6.00E+00
116	Cd	48	σ(to 136.4)	2.50E−02	1.00E−02	0.00E+00	0.00E+00
116	Cd	48	σ(s)(to 0)	6.00E+00	1.00E+00	0.00E+00	0.00E+00
116	Cd	48	σ(to 0)	5.00E−02	8.00E−03	0.00E+00	0.00E+00
116	Sn	50	σ(r)(to 314.6)	6.00E−03	2.00E−03	4.90E−01	1.60E−01
116	Sn	50	σ(r)	1.40E−01	3.00E−02	1.13E+01	2.00E+00
117	Sn	50	σ(r)	2.30E+00	5.00E−01	1.60E+01	5.00E+00
118	Sn	50	σ(s)	4.26E+00	1.50E−01	0.00E+00	0.00E+00
118	Sn	50	σ(to 89.5)	1.00E−02	6.00E−03	0.00E+00	0.00E+00
118	Sn	50	σ	2.20E−01	5.00E−02	4.70E+00	5.00E−01
119	Sn	50	σ(r)	2.20E+00	5.00E−01	2.90E+00	5.00E−01
120	Sn	50	σ(to 0)	1.40E−01	3.00E−02	1.20E+00	3.00E−01
120	Sn	50	σ(to 6.3)	1.00E−03	1.00E−03	0.00E+00	0.00E+00
120	Sn	50	σ(s)	5.17E+00	1.60E−01	0.00E+00	0.00E+00
120	Te	52	σ(r)(to 0)	2.00E+00	3.00E−01	0.00E+00	0.00E+00
120	Te	52	σ(r)(to 294.0)	3.40E−01	6.00E−02	0.00E+00	0.00E+00

121	Sb	51	σ(0)(to 163.6)	6.00E−02	1.00E−02	1.30E+01	1.00E+00
121	Sb	51	σ(0)(to 0+163.6)	5.90E+00	2.00E−01	2.00E+02	2.00E+01
122	Sn	50	σ(to 24.6)	1.80E−01	2.00E−02	0.00E+00	0.00E+00
122	Sn	50	σ(to 0)	1.00E−03	1.00E−03	0.00E+00	0.00E+00
122	Te	52	σ(to 247.5)	1.10E+00	5.00E−01	0.00E+00	0.00E+00
122	Te	52	σ	3.40E+00	5.00E−01	8.00E+01	1.00E+01
123	Sb	51	σ(to 0)	4.10E+00	1.00E−01	0.00E+00	0.00E+00
123	Sb	51	σ(0)(to 36.8)	1.90E−02	1.00E−02	0.00E+00	0.00E+00
123	Sb	51	σ(0)(to 10.9)	3.70E−02	1.00E−02	0.00E+00	0.00E+00
123	Te	52	σ	4.18E+02	3.00E+01	5.63E+03	3.25E+02
123	Te	52	σ(a)	4.60E−05	6.00E−06		
124	Sn	50	σ(to 27.5)	1.30E−01	5.00E−03	8.00E+00	2.00E−01
124	Sn	50	σ(to 0)	4.00E−03	2.00E−03	0.00E+00	0.00E+00
124	Sn	50	σ(s)	4.41E+00	3.00E−01	0.00E+00	0.00E+00
124	Sb	51	σ(r)	1.74E+01	2.80E+00	0.00E+00	0.00E+00
124	Te	52	σ	6.80E+00	1.30E+00	5.20E+00	7.00E−01
124	Te	52	σ(s)	3.80E+00	4.00E−01	0.00E+00	0.00E+00
124	Te	52	σ(to 144.8)	4.00E−02	2.50E−02	0.00E+00	0.00E+00
124	Xe	54	σ(0)(to 252.8)	2.80E+01	5.00E+00	6.00E+02	1.00E+02
124	Xe	54	σ(0)(to 0 + 252.8)	1.65E+02	2.00E+01	3.60E+03	7.00E+02
125	Te	52	σ	1.55E+00	1.60E−01	2.10E+01	4.00E+00
125	I	53	σ	8.94E+02	9.00E+01	0.00E+00	0.00E+00
125	Xe	54	σ(a)	3.00E−02	0.00E+00	0	0
126	Te	52	σ(to 0)	9.00E−01	1.50E−01	0.00E+00	0.00E+00
126	Te	52	σ(to 88.3)	1.35E−01	2.30E−02	0.00E+00	0.00E+00
126	I	53	σ(r)	5.96E+03	0.00E+00	4.06E+04	0.00E+00
126	Xe	54	σ(0)(to 0 + 297.1)	3.50E+00	8.00E−01	6.00E+01	1.00E+01

(continued)

Table A1.1 (continued)

A	Element	Z	Type of cross section	Cross section barns	Uncertainty	RI Resonance integral	Uncertainty
126	Xe	54	σ(0)(to 297.1)	4.50E−01	1.30E−01	8.00E+00	2.00E+00
127	Te	52	σ(for 88.3)	3.38E+03	5.10E+02	1.14E+03	1.70E+02
127	I	53	σ(0)	6.20E+00	2.00E−01	1.47E+02	6.00E+00
127	Xe	54	σ(a)	1.00E−02	0.00E+00	0	0
128	Te	52	σ(to 105.5)	1.50E−02	1.00E−03	7.70E−02	5.00E−03
128	Te	52	σ(to 0)	2.00E−01	8.00E−03	1.58E+00	6.00E−02
128	Xe	54	σ(r)(to 236.1)	4.80E−01	1.00E−01	3.80E+01	1.00E+01
128	Xe	54	σ(r)(to 0 + 236.1)	8.00E+00	0.00E+00	0.00E+00	0.00E+00
129	I	53	σ(0)(to 40.0)	1.80E+01	2.00E+00	0.00E+00	0.00E+00
129	I	53	σ(0)(to 0)	9.00E+00	1.00E+00	0.00E+00	0.00E+00
129	Xe	54	σ(0)	2.10E+01	5.00E+00	2.50E+02	5.00E+01
130	Te	52	σ(to 0)	2.70E−01	6.00E−02	0.00E+00	0.00E+00
130	Te	52	σ(to 182.2)	2.00E−02	1.00E−02	0.00E+00	0.00E+00
130	I	53	σ(r)	1.80E+01	3.00E+00	0.00E+00	0.00E+00
130	Xe	54	σ(0)(to 163.9)	4.50E−01	1.00E−01	1.60E+01	4.00E+00
130	Xe	54	σ(r)(to 0+163.9)	2.60E+01	0.00E+00	0.00E+00	0.00E+00
130	Ba	56	σ(to 0)	8.80E+00	9.00E−01	0.00E+00	0.00E+00
130	Ba	56	σ(to 187.1)	2.50E+00	3.00E−01	0.00E+00	0.00E+00
131	I	53	σ(0)(to 0)	8.00E+01	5.00E+01	0.00E+00	0.00E+00
131	Xe	54	σ(0)	8.50E+01	1.00E+01	9.00E+02	1.00E+02
132	Xe	54	σ(0)(to 0+233.2)	4.50E−01	6.00E−02	4.60E+00	6.00E−01
132	Xe	54	σ(0)(to 233.2)	5.00E−02	1.00E−02	9.00E−01	2.00E−01
132	Cs	55	σ(a)	1.50E−01	0.00E+00	0	0
132	Ba	56	σ(r)(to 288.2)	5.00E−01	0.00E+00	2.80E+00	0.00E+00
132	Ba	56	σ(r)(to 0)	6.50E+00	8.00E−01	0.00E+00	0.00E+00

133	Xe	54	σ(r)	1.90E+02	9.00E+01	0.00E+00	0.00E+00
133	Cs	55	σ(0)(to 0 + 138.7)	2.90E+01	1.50E+00	4.37E+02	2.60E+01
133	Cs	55	σ(0)(to 138.7)	2.50E+00	2.00E−01	0.00E+00	0.00E+00
134	Xe	54	σ(0)(to 526.6)	3.00E−03	3.00E−04	0.00E+00	0.00E+00
134	Xe	54	σ(0)(to 0+526.6)	2.65E−01	2.00E−02	0.00E+00	0.00E+00
134	Cs	55	σ(r)	1.40E+02	1.20E+01	5.42E+01	0.00E+00
134	Ba	56	σ(r)(to 268.2)	1.58E−01	2.40E−02	2.39E+01	3.80E+00
134	Ba	56	σ(r)	2.00E+00	1.60E+00	2.30E+01	5.00E+00
135	Xe	54	σ(0)	2.65E+06	1.10E+05	7.60E+03	5.00E+02
135	Cs	55	σ(to 0)	8.70E+00	5.00E−01	0.00E+00	0.00E+00
135	Ba	56	σ(to 2030.5)	1.39E−02	7.00E−04	4.65E−01	7.00E−02
135	Ba	56	σ	5.80E+00	9.00E−01	1.31E+02	1.00E+01
136	Xe	54	σ(0)	2.60E−01	2.00E−02	7.40E−01	2.10E−01
136	Ba	56	σ(to 661.7)	1.00E−02	1.00E−03	0.00E+00	0.00E+00
136	Ba	56	σ	4.00E−01	4.00E−01	1.60E+00	3.00E−01
136	Ce	58	σ(to 254.3)	9.50E−01	2.50E−01	0.00E+00	0.00E+00
136	Ce	58	σ(to 0)	6.30E+00	1.50E+00	0.00E+00	0.00E+00
137	Cs	55	σ(r)(to 0)	1.10E−01	3.30E−02	0.00E+00	0.00E+00
137	Ba	56	σ	5.10E+00	4.00E−01	4.30E+00	1.00E+00
138	Ba	56	σ	3.60E−01	3.60E−02	0.00E+00	0.00E+00
138	La	57	σ(0)	5.72E+01	5.70E+00	4.09E+02	9.00E+01
138	Ce	58	σ(to 754.2)	1.50E−02	5.00E−03	0.00E+00	0.00E+00
138	Ce	58	σ(to 0)	1.10E+00	3.00E−01	0.00E+00	0.00E+00
139	Ba	56	σ(r)	6.20E+00	1.60E+00	0.00E+00	0.00E+00
139	La	57	σ(0)	8.93E+00	4.00E−02	1.18E+01	8.00E−01
139	La	57	σ(s)	1.01E+01	2.20E−01	0.00E+00	0.00E+00
139	Ce	58	σ	5.00E+02	0.00E+00	0.00E+00	0.00E+00

(continued)

Table A1.1 (continued)

A	Element	Z	Type of cross section	Cross section barns	Uncertainty	RI Resonance integral	Uncertainty
140	Ba	56	σ(0)	1.60E+00	3.00E−01	1.36E+01	1.40E+00
140	La	57	σ	2.70E+00	3.00E−01	6.90E+01	4.00E+00
140	Ce	58	σ(s)	2.83E+00	1.10E−01	0.00E+00	0.00E+00
140	Ce	58	σ	5.70E−01	4.00E−02	4.70E−01	5.00E−02
141	Ce	58	σ	2.90E+01	3.00E+00	0.00E+00	0.00E+00
141	Pr	59	σ(s)	2.54E+00	6.00E−02	0.00E+00	0.00E+00
141	Pr	59	σ(0)(to 0+3.7)	1.15E+01	3.00E−01	1.74E+01	2.00E+00
141	Pr	59	σ(0)(to 3.7)	3.90E+00	3.00E−01	0.00E+00	0.00E+00
142	Ce	58	σ(r)	9.50E−01	5.00E−02	1.15E+00	5.00E−02
142	Pr	59	σ(r)	2.00E+01	3.00E+00	0.00E+00	0.00E+00
142	Nd	60	σ	1.87E+01	7.00E−01	3.40E+01	1.10E+01
142	Nd	60	σ(s)	7.70E+00	5.00E−01	0.00E+00	0.00E+00
143	Ce	58	σ	6.00E+00	7.00E−01	0.00E+00	0.00E+00
143	Pr	59	σ(0)	9.00E+01	1.00E+01	1.90E+02	2.50E+01
143	Nd	60	σ(s)	8.00E+01	2.00E+00	0.00E+00	0.00E+00
143	Nd	60	σ(0)	3.25E+02	1.00E+01	1.28E+02	3.00E+01
143	Nd	60	σ(a)	1.74E−02	1.60E−03	0	0
144	Ce	58	σ	1.00E+00	1.00E−01	2.60E+00	3.00E−01
144	Nd	60	σ(s)	1.00E+00	2.00E−01	0.00E+00	0.00E+00
144	Nd	60	σ	3.60E+00	3.00E−01	3.90E+00	5.00E−01
144	Sm	62	σ	7.00E−01	3.00E−01	0.00E+00	0.00E+00
145	Nd	60	σ(0)	4.20E+01	2.00E+00	2.40E+02	3.50E+01
145	Nd	60	σ(a)	1.00E−04	0.00E+00	0.00E+00	0.00E+00
145	Sm	62	σ	2.20E+02	0.00E+00	0.00E+00	0.00E+00
146	Nd	60	σ(0)	1.40E+00	1.00E−01	3.20E+00	5.00E−01

146	Nd	60	σ(s)	9.60E+00	4.00E−01	0.00E+00	0.00E+00
146	Pm	61	σ(r)	8.40E+03	1.68E+03	0.00E+00	0.00E+00
147	Nd	60	σ	4.40E+02	1.50E+02	4.05E+02	0.00E+00
147	Pm	61	σ(0)(to 0)	9.60E+01	1.80E+00	2.06E+03	1.00E+02
147	Pm	61	σ(0)(to 137.9)	7.24E+01	3.00E+00	7.90E+02	1.00E+02
147	Pm	61	σ(s)	2.10E+01	1.50E+00	0.00E+00	0.00E+00
147	Sm	62	σ(0)	5.70E+01	3.00E+00	0.00E+00	0.00E+00
147	Sm	62	σ(α)(to 0)	1.31E−04	2.00E−05	0.00E+00	0.00E+00
147	Sm	62	σ(α)(to 696.5)	3.70E−04	5.00E−05	0.00E+00	0.00E+00
148	Nd	60	σ(s)	4.00E+00	5.00E−01	0.00E+00	0.00E+00
148	Nd	60	σ(0)	2.50E+00	2.00E−01	1.40E+01	1.00E+00
148	Pm	61	σ(r)(for 0)	2.00E+03	1.00E+03	0.00E+00	0.00E+00
148	Pm	61	σ(0)(for 137.9)	1.06E+04	0.00E+00	3.60E+03	2.40E+03
148	Sm	62	σ	2.40E+00	6.00E−01	2.70E+01	1.40E+01
149	Pm	61	σ	1.40E+03	3.00E+02	0.00E+00	0.00E+00
149	Sm	62	σ(0)	4.01E+04	6.00E+02	3.39E+03	0.00E+00
149	Sm	62	σ(α)(to 0)	5.50E−03	4.00E−04	0	0
149	Sm	62	σ(α)(to 453.9)	2.35E−02	0.00E+00	0	0
149	Sm	62	σ(s)	1.97E+02	0.00E+00	0.00E+00	0.00E+00
150	Nd	60	σ(s)	3.50E+00	5.00E−01	0.00E+00	0.00E+00
150	Nd	60	σ(0)	1.20E+00	2.00E−01	1.40E+01	2.00E+00
150	Sm	62	σ(0)	1.04E+02	4.00E+00	3.58E+02	5.00E+01
151	Pm	61	σ	7.00E+02	0.00E+00	0.00E+00	0.00E+00
151	Sm	62	σ(0)	1.52E+04	3.00E+02	3.52E+03	1.60E+02
151	Eu	63	σ(0)(to 0)	5.90E+03	2.00E+02	1.51E+03	3.30E+02
151	Eu	63	σ(r)(to 147.8)	4.00E+00	2.00E+00	0.00E+00	0.00E+00
151	Eu	63	σ(0)(to 45.6)	3.30E+03	2.00E+02	1.79E+03	1.40E+02

(continued)

Table A1.1 (continued)

A	Element	Z	Type of cross section	Cross section barns	Uncertainty	RI Resonance integral	Uncertainty
151	Eu	63	σ(a)	8.90E−06	1.80E−06	0	0
151	Eu	63	σ(s)	6.30E+00	4.00E−01	0.00E+00	0.00E+00
152	Sm	62	σ(0)	2.06E+02	6.00E+00	2.97E+03	1.00E+02
152	Sm	62	σ(s)	3.00E+00	2.00E−01	0.00E+00	0.00E+00
152	Eu	63	σ(0)	1.28E+04	6.00E+02	1.58E+03	0.00E+00
152	Gd	64	σ	7.35E+02	2.00E+01	2.02E+03	1.60E+02
152	Gd	64	σ(a)	7.00E−03	0.00E+00	0	0
153	Sm	62	σ(0)	4.20E+02	1.80E+02	0.00E+00	0.00E+00
153	Eu	63	σ(s)	9.70E+00	7.00E−01	0.00E+00	0.00E+00
153	Eu	63	σ(0)	3.12E+02	7.00E+00	1.42E+03	1.00E+02
153	Eu	63	σ(a)	1.00E−06	0.00E+00	0	0
153	Gd	64	σ(r)	3.60E+04	4.00E+03	0.00E+00	0.00E+00
153	Gd	64	σ(a)(to 0)	2.10E−02	0.00E+00	0.00E+00	0.00E+00
153	Gd	64	σ(a)(to 334.0)	8.00E−03	0.00E+00	0.00E+00	0.00E+00
153	Gd	64	σ(a)(to 1071.4)	4.00E−03	0.00E+00	0.00E+00	0.00E+00
154	Sm	62	σ(s)	1.10E+01	3.00E+00	0.00E+00	0.00E+00
154	Sm	62	σ	8.40E+00	5.00E−01	3.20E+01	6.00E+00
154	Eu	63	σ(0)	1.34E+03	1.30E+02	8.02E+02	0.00E+00
154	Gd	64	σ(0)	8.50E+01	1.20E+01	3.35E+02	5.00E+01
154	Gd	64	σ(to 121.5)	4.90E−02	1.50E−02	0.00E+00	0.00E+00
155	Eu	63	σ	3.95E+03	1.25E+02	2.32E+04	3.00E+02
155	Gd	64	σ(s)	6.00E+01	0.00E+00	0.00E+00	0.00E+00
155	Gd	64	σ(0)	6.09E+04	5.00E+02	1.45E+03	1.00E+02
155	Gd	64	σ(a)(to 0)	1.00E−06	0.00E+00	0	0
155	Gd	64	σ(a)(to 121.8)	6.70E−05	9.00E−06	0	0

155	Gd	64	σ(a)(to 366.5)	1.50E−05	2.00E−06	0	0
156	Gd	64	σ	1.50E+00	1.20E+00	1.04E+02	1.50E+01
156	Dy	66	σ(0)	3.30E+01	3.00E+00	8.84E+02	8.00E+01
157	Gd	64	σ(s)	1.00E+03	0.00E+00	0.00E+00	0.00E+00
157	Gd	64	σ(0)	2.54E+05	8.15E+02	7.00E+02	2.00E+01
157	Gd	64	σ(a)(to 0)	5.00E−04	0.00E+00	0	0
157	Gd	64	σ(a)(to 82.0)	5.00E−05	0.00E+00	0	0
158	Gd	64	σ(0)	2.20E+00	2.00E−01	7.30E+01	7.00E+00
158	Dy	66	σ(0)	4.30E+01	6.00E+00	1.20E+02	1.00E+01
159	Tb	65	σ(s)	6.92E+00	6.00E−02	0.00E+00	0.00E+00
159	Tb	65	σ(0)	2.34E+01	4.00E−01	4.18E+02	2.00E+01
160	Gd	64	σ	7.70E−01	2.00E−02	7.20E+00	1.00E+00
160	Tb	65	σ(r)	5.25E+02	1.00E+02	0.00E+00	0.00E+00
160	Dy	66	σ(s)	5.60E+00	7.00E−01	0.00E+00	0.00E+00
160	Dy	66	σ(0)	5.60E+01	5.00E+00	1.16E+03	1.30E+02
161	Gd	64	σ(r)	3.10E+04	1.20E+04	0.00E+00	0.00E+00
161	Dy	66	σ(s)	1.65E+01	1.50E+00	0.00E+00	0.00E+00
161	Dy	66	σ(0)	6.00E+02	2.50E+01	1.20E+03	1.00E+02
161	Dy	66	σ(a)	3.00E−05	0.00E+00	0	0
162	Dy	66	σ(s)	2.80E−01	2.00E−01	0.00E+00	0.00E+00
162	Dy	66	σ(0)	1.94E+02	1.00E+01	2.76E+03	2.70E+02
162	Er	68	σ	1.90E+01	2.00E+00	4.80E+02	5.00E+01
162	Er	68	σ(a)	1.10E−02	0.00E+00	0	0
163	Dy	66	σ(s)	3.20E+00	5.00E−01	0.00E+00	0.00E+00
163	Dy	66	σ(0)	1.24E+02	7.00E+00	1.47E+03	1.00E+02
163	Dy	66	σ(a)	2.00E−05	0.00E+00	0	0
164	Dy	66	σ(0)(to 0)	1.04E+03	1.40E+02	0.00E+00	0.00E+00

(continued)

Table A1.1 (continued)

A	Element	Z	Type of cross section	Cross section barns	Uncertainty	RI	
						Resonance integral	Uncertainty
164	Dy	66	σ(0)(to 108.2)	1.61E+03	2.40E+02	0.00E+00	0.00E+00
164	Dy	66	σ(s)	3.19E+02	1.00E+01	0.00E+00	0.00E+00
164	Er	68	σ	1.30E+01	2.00E+00	1.05E+02	1.00E+01
164	Er	68	σ(a)	1.20E−03	0.00E+00	0	0
165	Dy	66	σ(for 0)	3.60E+03	3.00E+02	2.20E+04	3.00E+03
165	Dy	66	σ(r)(for 108.2)	2.00E+03	6.00E+02	0.00E+00	0.00E+00
165	Ho	67	σ(s)	8.65E+00	1.40E−01	0.00E+00	0.00E+00
165	Ho	67	σ(0)(to 0)	6.12E+01	1.10E+00	6.50E+02	2.20E+01
165	Ho	67	σ(to 6.0)	3.50E+00	4.00E−01	2.00E+01	9.00E+00
165	Ho	67	σ(a)	2.00E−05	0.00E+00	0	0
166	Er	68	σ(s)	1.42E+01	2.10E+00	0.00E+00	0.00E+00
166	Er	68	σ(to 207.8)	1.50E+01	2.00E+00	0.00E+00	0.00E+00
166	Er	68	σ(to 0)	4.60E+00	2.50E+00	0.00E+00	0.00E+00
166	Er	68	σ(a)	7.00E−05	0.00E+00	0.00E+00	0.00E+00
167	Er	68	σ(s)	1.60E+00	9.00E−01	0.00E+00	0.00E+00
167	Er	68	σ(0)	6.59E+02	1.60E+01	2.97E+03	7.00E+01
167	Er	68	σ(a)	7.00E−05	0.00E+00	0	0
168	Er	68	σ(s)	8.30E+00	7.00E−01	0.00E+00	0.00E+00
168	Er	68	σ(0)	2.74E+00	8.00E−02	3.70E+01	5.00E+00
168	Er	68	σ(a)	9.00E−05	0.00E+00	0	0
168	Yb	70	σ(0)	2.30E+03	1.70E+02	2.13E+04	1.00E+03
168	Yb	70	σ(a)	1.00E−04	0.00E+00	0	0
169	Tm	69	σ(0)	1.05E+02	2.00E+00	1.72E+03	3.00E+01
169	Tm	69	σ(to 183.2)	8.20E+00	1.70E+00	0.00E+00	0.00E+00
169	Tm	69	σ(a)	1.00E−05	0.00E+00	0.00E+00	0.00E+00

169	Tm	69	σ(s)	6.30E+00	1.00E−01	0.00E+00	0.00E+00
169	Yb	70	σ	3.60E+03	3.00E+02	5.20E+03	5.00E+02
170	Er	68	σ	5.80E+00	3.00E−01	4.50E+01	3.00E+00
170	Er	68	σ(s)	1.10E+01	1.90E+00	0.00E+00	0.00E+00
170	Tm	69	σ	9.20E+01	4.00E+00	4.60E+02	5.00E+01
170	Yb	70	σ(s)	5.85E+00	1.70E−01	0.00E+00	0.00E+00
170	Yb	70	σ(0)	1.14E+01	1.00E+00	3.20E+02	3.00E+01
170	Yb	70	σ(a)	1.00E−05	0.00E+00	0	0
171	Er	68	σ	2.80E+02	3.00E+01	0.00E+00	0.00E+00
171	Tm	69	σ	1.60E+02	1.00E+02	1.18E+02	6.00E+00
171	Yb	70	σ(0)	4.86E+01	2.50E+00	3.15E+02	3.00E+01
171	Yb	70	σ(a)	1.50E−06	0.00E+00	0	0
172	Yb	70	σ(0)	8.00E−01	4.00E−01	2.50E+01	3.00E+00
172	Yb	70	σ(a)	1.00E−06	0.00E+00	0	0
172	Yb	70	σ(s)	1.13E+01	2.00E−01	0.00E+00	0.00E+00
173	Yb	70	σ	1.71E+01	1.30E+00	3.80E+02	3.00E+01
173	Yb	70	σ(a)	1.00E−06	0.00E+00	0	0
174	Yb	70	σ(s)	4.63E+01	5.00E−01	0.00E+00	0.00E+00
174	Yb	70	σ(0)	6.94E+01	5.00E+00	2.70E+01	3.00E+00
174	Hf	72	σ(s)	1.50E+01	3.00E+00	0.00E+00	0.00E+00
174	Hf	72	σ(0)	5.61E+02	3.50E+01	4.36E+02	3.50E+01
175	Lu	71	σ(0)(to 123.0)	1.62E+01	5.00E−01	5.50E+02	3.00E+01
175	Lu	71	σ(s)	7.10E+00	4.00E−01	0.00E+00	0.00E+00
175	Lu	71	σ(0)(to 0)	6.90E+00	1.30E+00	6.10E+02	5.00E+01
175	Lu	71	σ(a)	6.00E−05	0.00E+00	0.00E+00	0.00E+00
176	Yb	70	σ(0)	2.85E+00	5.00E−02	6.30E+00	6.00E−01
176	Yb	70	σ(a)	1.00E−06	0.00E+00	0	0

(continued)

Table A1.1 (continued)

A	Element	Z	Type of cross section	Cross section barns	Uncertainty	RI Resonance integral	Uncertainty
176	Yb	70	σ(s)	9.66E+00	2.20E−01	0.00E+00	0.00E+00
176	Lu	71	σ(0)	2.09E+03	7.00E+01	0.00E+00	0.00E+00
176	Lu	71	σ(0)(to 970.1)	2.80E+00	7.00E−01	0.00E+00	0.00E+00
176	Lu	71	σ(to 150.4)	3.17E+02	5.80E+01	0.00E+00	0.00E+00
176	Lu	71	σ(a)	2.00E−03	0.00E+00	0.00E+00	0.00E+00
176	Lu	71	σ(s)	3.10E+00	8.00E−01	0.00E+00	0.00E+00
176	Hf	72	σ(0)	2.35E+01	3.10E+00	8.80E+02	4.00E+01
176	Hf	72	σ(s)	5.60E+00	3.00E−01	0.00E+00	0.00E+00
177	Hf	72	σ(0)	3.73E+02	1.00E+01	7.17E+03	2.00E+02
177	Hf	72	σ(to 2446.0)	2.00E−07	1.00E−07	0.00E+00	0.00E+00
177	Hf	72	σ(to 1147.4)	9.60E−01	5.00E−02	0.00E+00	0.00E+00
177	Hf	72	σ(s)	2.00E−01	2.00E−01	0.00E+00	0.00E+00
178	Hf	72	σ(0)	8.40E+01	4.00E+00	1.95E+03	1.20E+02
178	Hf	72	σ(s)	4.50E+00	2.00E−01	0.00E+00	0.00E+00
178	Hf	72	σ(0)(to 375.0)	5.30E+01	6.00E+00	0.00E+00	0.00E+00
179	Hf	72	σ(0)(to 1141.5)	4.45E−01	3.00E−03	6.90E+00	6.00E−01
179	Hf	72	σ(0)	4.10E+01	3.00E+00	6.30E+02	3.00E+01
179	Hf	72	σ(s)	7.00E+00	3.00E−01	0.00E+00	0.00E+00
180	Hf	72	σ(s)	2.20E+01	1.00E+00	0.00E+00	0.00E+00
180	Hf	72	σ(0)	1.30E+01	7.00E−02	3.50E+01	1.00E+00
180	Hf	72	σ(a)	1.00E−06	0.00E+00	0	0
180	Ta	73	σ(0)	5.63E+02	6.00E+01	1.35E+03	1.00E+02
180	W	74	σ	3.00E+01	1.20E+02	2.14E+02	3.00E+01
181	Hf	72	σ(r)	4.00E+01	3.00E+01	0.00E+00	0.00E+00
181	Ta	73	σ(0)(to 0)	2.05E+01	5.00E−01	6.60E+02	2.30E+01

181	Ta	73	σ(to 519.6)	1.10E−02	2.00E−03	4.15E−01	1.10E−01
181	Ta	73	σ(s)	6.12E+00	1.50E−01	0.00E+00	0.00E+00
182	Ta	73	σ(0)	8.70E+03	5.00E+02	8.62E+02	9.00E+01
182	W	74	σ(0)	2.07E+01	5.00E−01	6.04E+02	9.00E+01
182	W	74	σ(s)	8.84E+00	3.00E−01	0.00E+00	0.00E+00
183	W	74	σ(s)	2.40E+00	6.00E−01	0.00E+00	0.00E+00
183	W	74	σ(0)	1.01E+01	3.00E−01	3.37E+02	5.00E+01
184	W	74	σ(to 197.4)	2.00E−03	1.00E−03	0.00E+00	0.00E+00
184	W	74	σ(s)	7.35E+00	1.70E−01	0.00E+00	0.00E+00
184	W	74	σ(0)	1.70E+00	1.00E−01	1.47E+01	1.50E+00
184	Re	75	σ(r)	8.90E+03	0.00E+00	0.00E+00	0.00E+00
184	Os	76	σ	3.00E+03	1.50E+02	6.01E+02	5.10E+01
184	Os	76	σ(a)	1.00E−02	0.00E+00	0	0
185	Re	75	σ(0)	1.12E+02	2.00E+00	1.72E+03	5.00E+01
185	Re	75	σ(s)	1.06E+01	6.00E−01	0.00E+00	0.00E+00
186	W	74	σ(s)	1.47E−01	1.20E−02	0.00E+00	0.00E+00
186	W	74	σ(0)	3.79E+01	6.00E−01	4.85E+02	1.50E+01
186	Os	76	σ(0)	8.00E+01	1.30E+01	2.80E+02	3.00E+01
186	Os	76	σ(s)	1.80E+01	5.00E+00	0.00E+00	0.00E+00
187	W	74	σ(r)	6.40E+01	1.00E+01	2.76E+03	5.50E+02
187	Re	75	σ(s)	1.18E+01	4.00E−01	0.00E+00	0.00E+00
187	Re	75	σ(0)(to 172.1)	2.80E+00	1.00E−01	0.00E+00	0.00E+00
187	Re	75	σ(0)	7.64E+01	1.00E+00	3.00E+02	2.00E+01
187	Os	76	σ(s)	7.50E+00	6.00E+00	0.00E+00	0.00E+00
187	Os	76	σ(0)	3.20E+02	1.00E+01	5.00E+02	7.00E+01
187	Os	76	σ(a)	1.00E−04	0.00E+00	0	0
188	Re	75	σ(r)	2.00E+00	0.00E+00	0.00E+00	0.00E+00

(continued)

Table A1.1 (continued)

A	Element	Z	Type of cross section	Cross section barns	Uncertainty	RI	
						Resonance integral	Uncertainty
188	Os	76	σ(0)	4.70E+00	5.00E−01	1.52E+02	2.00E+01
188	Os	76	σ(a)	3.00E−05	0.00E+00	0	0
188	Os	76	σ(s)	7.70E+00	6.00E−01	0.00E+00	0.00E+00
189	Os	76	σ(s)	1.55E+01	1.00E+00	0.00E+00	0.00E+00
189	Os	76	σ(0)	2.50E+01	4.00E+00	6.74E+02	7.00E+01
189	Os	76	σ(to 1705.4)	2.60E−04	3.00E−05	1.30E−02	1.00E−03
190	Os	76	σ(s)	1.63E+01	9.00E−01	0.00E+00	0.00E+00
190	Os	76	σ(0)(to 74.4)	9.20E+00	7.00E−01	2.21E+01	1.70E+00
190	Os	76	σ(0)(to 0)	3.90E+00	6.00E−01	0.00E+00	0.00E+00
190	Pt	78	σ(0)	1.52E+02	4.00E+00	7.20E+01	1.00E+01
190	Pt	78	σ(a)	8.00E−03	0.00E+00	0	0
191	Os	76	σ	3.83E+02	5.40E+01	0.00E+00	0.00E+00
191	Ir	77	σ(s)	1.50E+01	0.00E+00	0.00E+00	0.00E+00
191	Ir	77	σ(to 155.2)	1.60E−01	7.00E−02	0.00E+00	0.00E+00
191	Ir	77	σ(0)(to 56.7)	6.45E+02	3.20E+01	0.00E+00	0.00E+00
191	Ir	77	σ(0)(to 0)	3.09E+02	3.00E+01	0.00E+00	0.00E+00
192	Os	76	σ(0)	2.00E+00	1.00E−01	4.60E+00	2.00E−01
192	Ir	77	σ(0)	1.42E+03	2.00E+02	3.24E+03	0.00E+00
192	Pt	78	σ(to 149.8)	2.20E+00	8.00E−01	0.00E+00	0.00E+00
192	Pt	78	σ(s)	1.23E+01	1.20E+00	0.00E+00	0.00E+00
192	Pt	78	σ(0)	1.00E+01	2.50E+00	1.15E+02	2.00E+01
192	Pt	78	σ(a)	2.00E−04	0.00E+00	0	0
193	Os	76	σ	3.80E+01	1.00E+01	0.00E+00	0.00E+00
193	Ir	77	σ(to 112+yms)	5.80E+00	2.00E+00	0.00E+00	0.00E+00
193	Ir	77	σ(0)	1.11E+02	5.00E+00	1.35E+03	1.00E+02

194	Pt	78	σ(s)	1.40E+01	2.00E−01	0.00E+00	0.00E+00
194	Pt	78	σ(to 259.3)	9.80E−02	1.10E−02	3.10E+00	1.00E−01
194	Pt	78	σ(0)	1.44E+00	1.90E−01	4.00E+00	1.00E+00
194	Pt	78	σ(a)	5.00E−06	0.00E+00	0	0
195	Pt	78	σ(s)	1.01E+01	2.00E−01	0.00E+00	0.00E+00
195	Pt	78	σ(0)	2.75E+01	1.20E+00	3.65E+02	5.00E+01
195	Pt	78	σ(a)	5.00E−06	0.00E+00	0	0
196	Pt	78	σ(0)	7.20E−01	4.00E−02	5.10E+00	3.00E−01
196	Pt	78	σ(0)(to 399.6)	4.40E−02	4.00E−03	3.50E−01	7.00E−02
196	Pt	78	σ(s)	1.23E+01	2.00E−01	0.00E+00	0.00E+00
196	Hg	80	σ	3.08E+03	1.80E+02	4.72E+02	1.50E+01
196	Hg	80	σ(to 298.9)	1.09E+02	6.00E+00	5.89E+01	2.40E+00
197	Au	79	σ(0)	9.87E+01	9.00E−02	1.55E+03	2.80E+01
197	Au	79	σ(s)	7.84E+00	1.30E−01	0.00E+00	0.00E+00
198	Pt	78	σ(0)	3.66E+00	1.90E−01	5.40E+01	4.00E+00
198	Pt	78	σ(s)	7.70E+00	2.00E−01	0.00E+00	0.00E+00
198	Pt	78	σ(to 424)	3.50E−01	4.00E−02	6.00E+00	7.00E−01
198	Au	79	σ	2.51E+04	3.70E+02	0.00E+00	0.00E+00
198	Hg	80	σ	2.00E+00	3.00E−01	7.10E+01	2.00E+00
198	Hg	80	σ(to 532.5)	1.80E−02	4.00E−03	1.80E+00	3.00E−01
199	Pt	78	σ	1.50E+01	1.00E+01	0.00E+00	0.00E+00
199	Au	79	σ(r)	3.00E+01	1.50E+01	0.00E+00	0.00E+00
199	Hg	80	σ(s)	6.57E+01	2.00E+00	0.00E+00	0.00E+00
199	Hg	80	σ(0)	2.15E+03	4.80E+01	4.35E+02	2.00E+01
200	Hg	80	σ	6.00E+01	0.00E+00	2.10E+00	0.00E+00
201	Hg	80	σ(0)	7.80E+00	2.00E+00	3.00E+01	3.00E+00
202	Hg	80	σ(0)	4.89E+00	5.00E−02	4.20E+00	2.00E−01

(continued)

Table A1.1 (continued)

A	Element	Z	Type of cross section	Cross section barns	Uncertainty	RI Resonance integral	Uncertainty
203	Tl	81	σ(0)	1.14E+01	2.00E−01	4.30E+01	2.00E+00
203	Tl	81	σ(a)	2.85E−02	0.00E+00	0	0
203	Tl	81	σ(s)	6.29E+00	2.90E−01	0.00E+00	0.00E+00
204	Hg	80	σ	4.30E−01	1.00E−01	8.50E−01	2.00E−01
204	Tl	81	σ	2.16E+01	2.00E+00	9.40E+01	1.90E+01
204	Pb	82	σ(0)	6.61E−01	7.00E−02	2.00E+00	2.00E−01
205	Tl	81	σ	1.04E−01	1.70E−02	6.20E−01	1.50E−01
205	Tl	81	σ(s)	1.16E+01	6.80E−01	0.00E+00	0.00E+00
206	Pb	82	σ(to 1633.4)	6.30E−03	1.30E−03	9.70E−02	1.40E−02
206	Pb	82	σ(0)	3.06E−02	8.00E−04	0	0
207	Pb	82	σ(0)	7.12E−01	1.00E−02	3.90E−01	1.00E−02
208	Pb	82	σ(0)	4.90E−04	3.00E−05	2.00E−03	2.00E−04
209	Bi	83	σ(s)	9.30E+00	3.00E−03	0.00E+00	0.00E+00
209	Bi	83	σ(0)(to 271.3)	9.60E−03	8.00E−04	0.00E+00	0.00E+00
209	Bi	83	σ(0)(to 0)	2.42E−02	4.00E−04	0.00E+00	0.00E+00
209	Bi	83	σ(a)	1.50E−04	2.00E−05	0.00E+00	0.00E+00
210	Pb	82	σ	5.00E−01	5.00E−01	0.00E+00	0.00E+00
210	Bi	83	σ(to m271.3)	5.40E+00	2 5.0000E−0	3 2.0000E−0	1 3.0000E−02
210	Po	84	σ(to 0)	3.00E−02	0.00E+00	0.00E+00	0.00E+00
210	Po	84	σ(to 1462)	5.00E−04	0.00E+00	0.00E+00	0.00E+00
210	Po	84	σ(a)	2.00E−03	0.00E+00	0.00E+00	0.00E+00
220	Rn	86	σ(p)	2.00E−01	0.00E+00	0.00E+00	0.00E+00
222	Rn	86	s	7.20E−01	7.00E−02	0.00E+00	0.00E+00
223	Ra	88	s	1.30E+02	2.00E+01	0.00E+00	0.00E+00
223	Ra	88	σ(F)	7.00E−01	3.00E−01	0.00E+00	0.00E+00

224	Ra	88	σ	1.20E+01	5.00E−01	0.00E+00	0.00E+00
226	Ra	88	σ(0)	1.28E+01	1.50E+00	2.80E+02	0.00E+00
226	Ra	88	σ(F)	5.00E−05	0.00E+00	0.00E+00	0.00E+00
227	Ac	89	σ	8.90E+02	3.00E+01	1.66E+03	3.00E+01
227	Ac	89	σ(F)	2.90E−04	0.00E+00	0.00E+00	0.00E+00
227	Th	90	σ(F)	2.02E+02	1.30E+01	0.00E+00	0.00E+00
228	Ra	88	σ(F)	2.00E+00	0.00E+00	0.00E+00	0.00E+00
228	Ra	88	σ	3.60E+01	5.00E+00	0.00E+00	0.00E+00
228	Th	90	σ(F)	3.00E−01	0.00E+00	0.00E+00	0.00E+00
228	Th	90	σ	1.23E+02	1.50E+01	1.01E+03	0.00E+00
229	Th	90	σ(F)	3.08E+01	1.50E+00	4.66E+02	7.50E+01
229	Th	90	σ(A)(to calc)	9.20E+01	6.00E+00	0.00E+00	0.00E+00
229	Th	90	σ(to calc)	6.10E+01	6.00E+00	1.00E+03	1.75E+02
230	Th	90	σ	2.30E+01	3.00E−01	1.01E+03	3.50E+01
230	Th	90	σ(F)	1.20E−03	0.00E+00	0.00E+00	0.00E+00
230	Pa	91	σ(F)	1.50E+03	2.50E+02	0.00E+00	0.00E+00
230	U	92	σ(F)	2.50E+01	1.00E+01	0.00E+00	0.00E+00
231	Pa	91	σ(0)	2.01E+02	2.30E+00	5.25E+02	6.00E+01
231	Pa	91	σ(F)	2.00E−02	1.00E−03	0.00E+00	0.00E+00
231	U	92	σ(F)	4.00E+02	3.00E+02	0.00E+00	0.00E+00
232	Th	90	σ(0)	7.37E+00	6.00E−02	8.50E+01	3.00E+00
232	Th	90	σ(F)	2.50E−06	0.00E+00	0.00E+00	0.00E+00
232	Pa	91	σ(f)	7.00E+02	1.00E+02	0.00E+00	0.00E+00
232	Pa	91	σ	4.64E+02	9.50E+01	3.00E+02	7.00E+01
232	U	92	σ(s)	1.33E+01	2.20E+00	0.00E+00	0.00E+00
232	U	92	σ(g)	7.68E+01	4.80E+00	3.50E+02	3.00E+01
232	U	92	σ(A)	1.52E+02	4.60E+00	0.00E+00	0.00E+00

(continued)

Table A1.1 (continued)

A	Element	Z	Type of cross section	Cross section barns	Uncertainty	RI Resonance integral	Uncertainty
232	U	92	σ(0)	7.49E+01	1.60E+00	2.80E+02	1.50E+01
233	Th	90	σ	1.50E+03	1.00E+02	4.00E+02	1.00E+02
233	Th	90	σ(F)	1.50E+01	2.00E+00	0.00E+00	0.00E+00
233	Pa	91	σ(to 0)	1.94E+01	3.00E+00	4.32E+02	7.00E+01
233	Pa	91	σ(F)	1.00E−01	0.00E+00	3.00E+00	0.00E+00
233	Pa	91	σ(to ~75)	2.01E+01	3.20E+00	4.38E+02	7.00E+01
233	U	92	σ(g)	5.29E+02	1.20E+00	7.60E+02	1.70E+01
233	U	92	σ(s)	1.28E+01	3.00E−01	0.00E+00	0.00E+00
233	U	92	σ(0)	4.55E+01	7.00E−01	1.37E+02	6.00E+00
233	U	92	σ(A)	5.75E+02	1.00E+00	1.37E+02	0
234	Th	90	σ	1.80E+00	5.00E−01	0.00E+00	0.00E+00
234	Th	90	σ(F)	1.00E−02	0.00E+00	0.00E+00	0.00E+00
234	Pa	91	σ(F)(to 0)	5.00E+03	0.00E+00	0.00E+00	0.00E+00
234	Pa	91	σ(F)(to −~75)	5.00E+02	0.00E+00	0.00E+00	0.00E+00
234	U	92	σ(s)	1.96E+01	1.00E+00	0.00E+00	0.00E+00
234	U	92	σ(F)	6.50E−01	0.00E+00	6.50E+00	0.00E+00
234	U	92	σ(0)	9.98E+01	1.30E+00	6.60E+02	7.00E+01
234	Np	93	σ(F)	9.00E+02	3.00E+02	0.00E+00	0.00E+00
235	U	92	σ(0)	9.83E+01	8.00E−01	1.44E+02	6.00E+00
235	U	92	σ(s)	1.43E+01	5.00E−01	0.00E+00	0.00E+00
235	U	92	σ(A)	6.81E+02	1.10E+00	0.00E+00	0.00E+00
235	U	92	σ(F)	5.83E+02	1.10E+00	2.75E+02	5.00E+00
235	Np	93	σ(to 22.5h)	1.50E+02	2.00E+00	0.00E+00	0.00E+00
236	U	92	σ(F)	7.00E−02	0.00E+00	7.80E+00	1.60E+00
236	U	92	σ(0)	5.11E+00	2.10E−01	3.60E+02	1.50E+01

236	Np	93	σ(F)	2.50E+03	1.50E+02	0.00E+00	0.00E+00
236	Pu	94	σ(F)	1.70E+02	3.50E+01	0.00E+00	0.00E+00
237	U	92	σ(F)	3.50E−01	0.00E+00	0.00E+00	0.00E+00
237	U	92	σ	4.43E+02	1.67E+02	1.20E+03	2.00E+02
237	Np	93	σ(0)	1.76E+02	2.90E+00	6.40E+02	5.00E+01
237	Np	93	σ(g)	2.15E−02	2.40E−03	6.90E+00	1.00E+00
237	Pu	94	σ(F)	2.46E+03	2.95E+02	0.00E+00	0.00E+00
238	U	92	σ(0)	2.68E+00	1.90E−02	2.77E+02	3.00E+00
238	U	92	σ(F)	4.00E−06	0.00E+00	1.54E−03	1.50E−04
238	U	92	σ(a)	1.30E−06	6.00E−07	0.00E+00	0.00E+00
238	U	92	σ(s)	9.38E+00	9.00E−02	0.00E+00	0.00E+00
238	Np	93	σ(F)	2.09E+03	3.00E+01	8.83E+02	7.00E+01
238	Pu	94	σ(s)	2.50E+01	1.80E+00	0.00E+00	0.00E+00
238	Pu	94	σ(g)	1.79E+01	4.00E−01	3.30E+01	5.00E+00
238	Pu	94	σ(0)	5.40E+02	7.00E+00	1.62E+02	1.50E+01
239	U	92	σ	2.20E+01	5.00E+00	0.00E+00	0.00E+00
239	U	92	σ(F)	1.40E+01	3.00E+00	0.00E+00	0.00E+00
239	Np	93	σ(r)(to <100)	3.20E+01	6.00E+00	0.00E+00	0.00E+00
239	Np	93	σ(r)(to 0)	3.60E+01	8.00E+00	0.00E+00	0.00E+00
239	Pu	94	σ(s)	7.60E+00	6.00E−01	0.00E+00	0.00E+00
239	Pu	94	σ(g)	7.48E+02	2.00E+00	3.01E+02	1.00E+01
239	Pu	94	σ(0)	2.69E+02	2.90E+00	2.00E+02	2.00E+01
240	Pu	94	σ(0)	2.90E+02	1.40E+00	8.10E+03	2.00E+02
240	Pu	94	σ(F)	5.60E−02	3.00E−02	8.80E+00	0.00E+00
240	Pu	94	σ(s)	1.64E+00	9.00E−02	0.00E+00	0.00E+00
241	Pu	94	σ(s)	9.00E+00	1.00E+00	0.00E+00	0.00E+00
241	Pu	94	σ(g)	1.01E+03	6.20E+00	5.70E+02	1.50E+01

(continued)

Table A1.1 (continued)

A	Element	Z	Type of cross section	Cross section barns	Uncertainty	RI Resonance integral	Uncertainty
241	Pu	94	σ(F)	1.01E+03	6.20E+00	5.70E+02	1.50E+01
241	Pu	94	σ(O)	1.37E+03	7.70E+00	0	0
241	Pu	94	σ(0)	3.58E+02	5.10E+00	1.62E+02	8.00E+00
241	Am	95	σ(g)	3.20E+00	9.00E−02	1.44E+01	1.00E+00
241	Am	95	σ(0)(to 0)	5.33E+02	1.30E+01	1.23E+03	1.00E+02
241	Am	95	σ(0)(to 48.6)	5.40E+01	5.00E+00	1.95E+02	2.00E+01
242	Pu	94	σ(0)	1.85E+01	5.00E−01	1.12E+03	4.00E+01
242	Pu	94	σ(s)	8.44E+00	2.10E−01	0.00E+00	0.00E+00
242	Pu	94	σ(F)	2.00E−01	0.00E+00	5.00E+00	2.00E+00
242	Am	95	σ(F)(for 0)	2.10E+03	2.00E+02	0.00E+00	0.00E+00
242	Am	95	σ(for 48.6)	2.00E+03	6.00E+02	0.00E+00	0.00E+00
242	Am	95	σ(F)(for 48.6)	6.95E+03	2.80E+02	1.80E+03	6.50E+01
242	Cm	96	σ(F)	5.00E+00	0.00E+00	0.00E+00	0.00E+00
242	Cm	96	σ	1.60E+01	5.00E+00	1.10E+02	2.00E+01
243	Pu	94	σ	8.70E+01	1.00E+01	2.70E+02	3.50E+01
243	Pu	94	σ(F)	1.96E+02	1.60E+01	0	0
243	Am	95	σ(0)	7.51E+01	1.80E+00	1.82E+03	7.00E+01
243	Am	95	σ(F)	1.98E−01	4.30E−03	9.00E+00	1.00E+00
243	Am	95	σ(0)(to 0)	3.80E+00	4.00E−01	9.40E+01	9.00E+00
243	Cm	96	σ(A)	7.47E+02	2.30E+01	0.00E+00	0.00E+00
243	Cm	96	σ(g)	6.17E+02	2.00E+01	1.57E+03	1.00E+02
243	Cm	96	σ(0)	1.30E+02	1.00E+01	2.15E+02	2.00E+01
244	Pu	94	σ	1.70E+00	1.00E−01	4.06E+01	2.90E+00
244	Am	95	σ(F)(for 0)	2.30E+03	3.00E+02	0.00E+00	0.00E+00
244	Am	95	σ(F)(for 70)	1.60E+03	3.00E+02	0.00E+00	0.00E+00

244	Cm	96	σ(s)	1.16E+01	7.00E−01	0.00E+00	0.00E+00
244	Cm	96	σ(g)	1.04E+00	2.00E−01	1.25E+01	2.50E+00
244	Cm	96	σ(0)	1.52E+01	1.20E+00	6.50E+02	3.00E+01
245	Pu	94	σ	1.50E+02	3.00E+01	2.20E+02	4.00E+01
245	Cm	96	σ(0)	3.69E+02	1.70E+01	1.01E+02	8.00E+00
245	Cm	96	σ(A)	2.51E+03	6.00E+01	0.00E+00	0.00E+00
245	Cm	96	σ(g)	2.15E+03	5.80E+01	8.40E+02	4.00E+01
246	Cm	96	σ(s)	1.11E+01	2.00E−01	0.00E+00	0.00E+00
246	Cm	96	σ(0)	1.22E+00	1.60E−01	1.21E+02	7.00E+00
246	Cm	96	σ(g)	1.40E−01	5.00E−02	1.02E+01	4.00E−01
247	Cm	96	σ(0)	5.70E+01	1.00E+01	5.30E+02	3.00E+01
247	Cm	96	σ(F)	8.19E+01	4.40E+00	7.60E+02	5.00E+01
248	Cm	96	σ(F)	3.70E−01	5.00E−02	1.50E+01	0.00E+00
248	Cm	96	σ(0)	2.63E+00	2.60E−01	2.70E+02	1.50E+01
249	Cm	96	σ	1.60E+00	8.00E−01	0.00E+00	0.00E+00
249	Bk	97	σ(0)	7.46E+02	4.00E+01	1.10E+03	1.00E+02
249	Cf	98	σ(A)	2.14E+03	3.90E+01	0.00E+00	0.00E+00
249	Cf	98	σ	4.97E+02	2.10E+01	7.65E+02	3.50E+01
249	Cf	98	σ(g)	1.64E+03	3.30E+01	2.38E+03	8.50E+01
250	Bk	97	σ(F)	9.60E+02	1.50E+02	0.00E+00	0.00E+00
250	Bk	97	σ	3.50E+02	0.00E+00	0.00E+00	0.00E+00
250	Cf	98	σ	2.03E+03	2.00E+02	1.16E+04	5.00E+02
251	Cf	98	σ	2.85E+03	1.50E+02	1.60E+03	3.00E+01
251	Cf	98	σ(A)	7.75E+03	2.90E+02	0.00E+00	0.00E+00
251	Cf	98	σ(F)	4.90E+03	2.50E+02	5.90E+03	1.00E+02

(continued)

Table A1.1 (continued)

A	Element	Z	Type of cross section	Cross section barns	Uncertainty	RI Resonance integral	Uncertainty
252	Cf	98	σ(F)	3.20E+01	4.00E+00	1.10E+02	3.00E+01
252	Cf	98	σ	2.04E+01	1.50E+00	4.35E+01	3.00E+00
253	Cf	98	σ(F)	1.30E+03	2.40E+02	2.00E+03	4.00E+02
253	Cf	98	σ	1.76E+01	1.80E+00	1.30E+01	3.00E+00
253	Es	99	σ(to 0)	5.80E+00	7.00E−01	1.14E+02	7.00E+00
253	Es	99	σ(to 78)	1.78E+02	1.50E+01	3.75E+03	2.00E+02
254	Cf	98	σ	4.50E+00	1.50E+00	0.00E+00	0.00E+00
254	Es	99	σ(A)	1.99E+03	1.60E+02	0.00E+00	0.00E+00
254	Es	99	σ(F)(for 0)	1.97E+03	1.60E+02	1.20E+03	2.50E+02
254	Es	99	σ(F)(for 78)	1.83E+03	8.00E+01	0.00E+00	0.00E+00
254	Es	99	σ(to 0)	2.83E+01	2.50E+00	1.82E+01	1.50E+00
254	Fm	100	σ	7.60E+01	0.00E+00	0.00E+00	0.00E+00
254	Fm	100	σ(S)	1.40E+03	0.00E+00	0.00E+00	0.00E+00
255	Es	99	σ(r)	5.50E+01	1.00E+01	0.00E+00	0.00E+00
255	Fm	100	σ(F)	3.36E+03	1.70E+02	0.00E+00	0.00E+00
255	Fm	100	σ(A)	3.39E+03	1.70E+02	0.00E+00	0.00E+00
255	Fm	100	σ	2.60E+01	3.00E+00	0.00E+00	0.00E+00
256	Fm	100	σ	4.50E+01	0.00E+00	0.00E+00	0.00E+00
257	Fm	100	σ(F)	2.95E+03	1.60E+02	0.00E+00	0.00E+00

References

1. Fisher HOM (1989). In: Los_Alamos_National_Laboratory (ed) A nuclear cross section data handbook, vol UC-705. University of California
2. Baumann NP (1963). In: S. R. Laboratory (ed) Resonance integrals and self-shielding factors for detector foils. Office of Technical Services, U. S. Department of Commerce
3. Mughabghab SF, Holden NE (1981) Neutron resonance parameters and thermal cross sections. Academic Press, New York

A2 Plasma Basics

Discussion of gas lasers begins with a discussion of the plasma state. Plasma is called the fourth state of matter. When the temperature is below the melting point of a material, it is a solid. As temperature increases, and the temperature increases above the melting point of the material, it is a liquid. If temperature continues to increase beyond the boiling point of the material, it becomes a vapor. When temperature increases to the point where electrons are stripped from the atom, the formation of ion pairs indicates that the material has achieved the plasma state [1–7].

A2.1 Plasma Species

Plasma implies that electrons have been separated from neutral gas atoms and molecules. Plasma can be weak where the fraction of ions to neutral species can be very low 1×10^{-6} to 1×10^{-4} (true for gas discharges, lasers, etc.). Plasmas can be fully ionized where the fraction of ions to neutral species approaches ∞ (e.g., a plasma generated in a Tokamak fusion experiment). For plasma processing, the plasma is usually a weak plasma (an exception would be Electron Resonance Heated Plasmas). The components that are typically found include:

A2.1.1 Neutral Species

The gas atoms can be simple monoatomic – X (e.g., He, Ne, Ar, Kr, or Xe), diatomic – *XY* (e.g., H_2, N_2, O_2, etc.), tiatomic *XYZ* (e.g., CO_2, H_2O, etc.), or complex (e.g., CH_4, CH_3F, or other organic species). Neutral gases can have pressures ranging from as low as a millitorr (3.22×10^{13} atoms or molecules cm^{-3} at room temperature) to hundreds of atmospheres (on the order of 3.22×10^{21} atoms or molecules cm^{-3} at room temperature). In plasma processing the pressure is usually on the order of several torr (where 1 torr $= 3.22 \times 10^{16}$ atoms or molecules cm^{-3} at room temperature).

A2.1.2 Charged Particles

The charged particles in a plasma are electrons and ions. For a typical gas discharge the electron density falls between 1×10^{10} and 1×10^{14} electrons cm^{-3}. This electron density is very much dependent upon the power density (Wcm^{-3}) which creates the plasma. For electrostatic discharges the electron density is a linear function of the input power density. If a magnetic field is superimposed on a discharge, than the electron density is no longer a linear function of power density. In a magnetic plasma it is possible to use electron cyclotron heating to create fully ionized plasmas with virtually any gas species. This latter property has made electron cyclotron heated plasmas useful to the semiconductor industry.

Ions are more complex due to the nature of the neutral species. It is possible to form atomic and molecular ions (e.g., C^+, He^+ and $N_2{}^+$), or to have multiple ionized species (e.g., C^{++}, He^{++}, $N_2{}^{++}$, etc.).

A2.1.3 Excited Species and Photons

The neutrals and ions dominate the analysis of the electrical properties of plasma. However, the plasma is constantly trying to cool itself to reach equilibrium with the outside world. Two major mechanisms for cooling is recombination of ions and electrons and spontaneous emission. In these processes, large amounts of energy are carried away from the plasma by photons. These photons originate from excited states of the gas species. An excited state will be denoted by "*" in this text. It is possible to have excited neutral species such as X*, or XY* as well as excited states in an ion species such as X^{+*}. These types of excited states are important to the field of emission spectroscopy. Additionally, they are the key to the field of gaseous electronics and laser electronics. The focus of these disciplines is in some aspect emission spectroscopy.

A2.1.4 Interaction Between Species

With the types of species that are present in plasma, it is necessary to introduce how these species interact with one another. The important categories of reactions are summarized below. The reaction is always followed by its inverse reaction (e.g., electron collisional excitation followed by superelastic collisions). Electrons have an energy (e.g., e(E) which means that the energy of the electron is E), and a photon has a frequency ν.

Electron collisional excitation of an atom or molecule (this is one method of forming excited states in a discharge) can be written as:

$$e(E) + X \rightarrow X^* + e(E - \Delta E_1) \tag{A2.1}$$

$$e(E) + X^* \rightarrow X^{**} + e(E - \Delta E_2) \tag{A2.2}$$

where ΔE_1 is the energy lost by the electron in the first excitation reaction and ΔE_2 is the energy lost by the electron in the second excitation reaction.

Superelastic collisions of an excited atom and an electron resulting in a de-excited atom and fast electron according to Eq. A2.3 is:

$$e(E) + X^* \rightarrow X + e(E + \Delta E) \tag{A2.3}$$

The electron collisional ionization reaction can be written as:

$$e(E) + X \rightarrow X^+ + e(E - \Delta E) + e(E_2) \quad \text{(A2.4)}$$

$$e(E) + X^* \rightarrow X^+ + e(E - \Delta E) + e(E_2) \quad \text{(A2.5)}$$

The collisional recombination reaction is given by:

$$X^+ + e(E_1) + e(E_2) \rightarrow e(E_2 + \Delta E) + X^* \quad \text{(A2.6)}$$

Dissociation and ionization of molecules can be represented as:

$$\begin{aligned} &e + YZ \rightarrow e + Y^+ + Z^- \quad (Negative\ ion) \\ &e + YZ \rightarrow e + Y + Z \end{aligned} \quad \text{(A2.7)}$$

$$e + YZ \rightarrow Y + Z^- \quad \text{(A2.8)}$$

Radiative attachment to form a negative ion occurs according to the following reaction:

$$e + X \rightarrow X^- + h\upsilon \quad \text{(A2.9)}$$

Photo detachment can be written as:

$$X^- + h\upsilon \rightarrow e + X \quad \text{(A2.10)}$$

Dissociative recombination is given by:

$$e + YZ^+ \rightarrow Y + Z^* \quad \text{(A2.11)}$$

Charge transfer takes place according to the following equation:

$$X + Y^+ \rightarrow X^+ + Y \quad \text{(A2.12)}$$

Ion-ion recombination occurs by the following reaction:

$$X^- + Y^+ \rightarrow XY \quad \text{(A2.13)}$$

Excitation exchange and collisional de-excitation are shown in the following reaction:

$$X^* + Y \rightarrow X + Y^* \quad \text{(A2.14)}$$

$$X^* + Y \rightarrow X + Y + Kinetic\ energy \quad \text{(A2.15)}$$

Penning ionization is given by:

$$X^* + Z \rightarrow X + Z^+ + e \tag{A2.16}$$

Elastic collisions are shown the following reactions:

$$e(E_1) + X(E_2) \rightarrow e(E_1 - \Delta E) + X(E_2 - \Delta E) \tag{A2.17}$$

$$e(E_1) + e(E_2) \rightarrow e(E_1 - \Delta E) + e(E_2 + \Delta E) \tag{A2.18}$$

$$X(E_1) + Y(E_2) \rightarrow X(E_1 - \Delta E) + Y(E_2 + \Delta E) \tag{A2.19}$$

A2.1.5 Electrons

Electrons are the most important species since they usually control the electrical and optical properties of the plasma plus they typically dominate all other processes. The important collisional processes can be graphically illustrated as shown in Fig. A2.1.

One of the most significant aspects of plasma is the electron number density, and the electron energy. In general different classes of ionized gases have different electron properties as shown in Fig. A2.2. The electron density tends to be linear with power density in electrostatic discharges. In a magnetized plasma such as that used in fusion research, the electron density is much higher than that of an electrostatic discharge with the same power densities.

A2.1.5.1 Debye Shielding

If plasma is formed in a charge neutral media, then the bulk plasma will remain charge neutral. Locally, on an atomic level, electrons and ions are free to flow. The question then is can a local charge imbalance occur. To address this question we can examine a plasma in which a test charge q is placed in a plasma and in which the electron density equal the ion density ($n_e = n_+$). The Coulomb potential due to the test charge is given by:

$$V(r) = \frac{q}{4\pi\varepsilon_o r} \tag{A2.20}$$

Electrons in the plasma will tend to move towards the test charge and ions will move away from the test charge. This will tend to "shield" the test charge from the rest of the plasma. To see the shielding effect, first assume that the electrons are in thermodynamic equilibrium at a temperature T_e and that the ions are in thermodynamic equilibrium at a temperature T_+. Given that the potential near the charge obeys Eq. A2.20, then equations for n_e and n_+ become:

$$n_e(r) = n\exp\left(-\frac{eV}{kT_e}\right) \tag{A2.21}$$

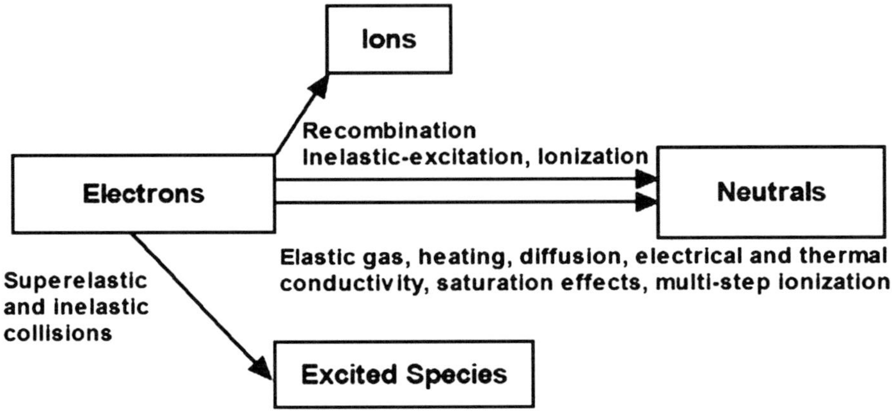

Fig. A2.1 Important electron interactions [7]

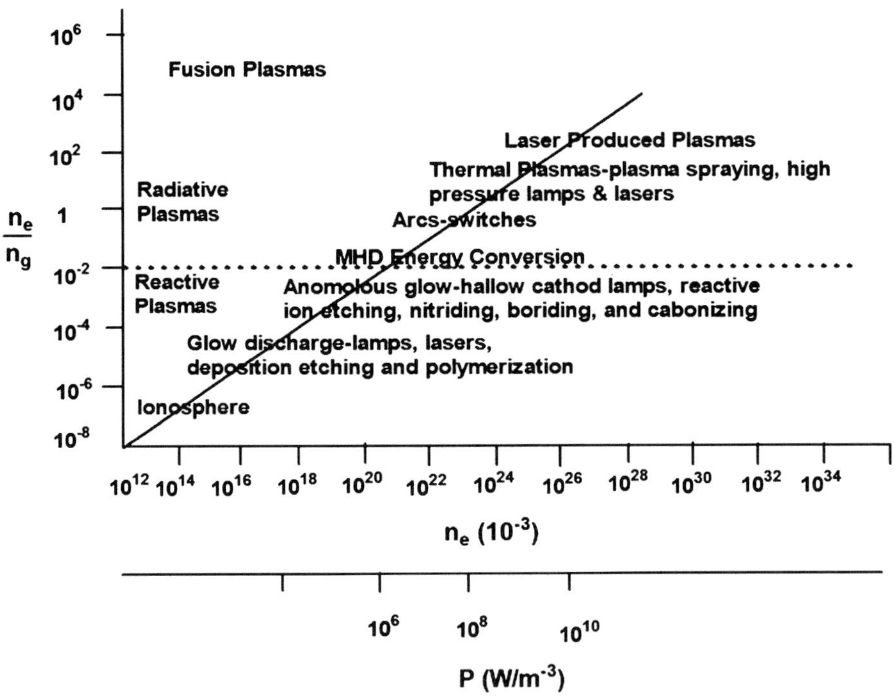

Fig. A2.2 Different classes of ionized gases

and,
$$n_+(r) = n\exp\left(-\frac{eV}{kT_+}\right) \tag{A2.22}$$

where e is the electronic charge.

The screened potential is then determined by Poisson's equation, which can be written as:

$$\nabla^2 V = n\left[e^{\left(\frac{eV}{kT_e}\right)} - e^{-\left(\frac{eV}{kT_+}\right)}\right]\frac{\varepsilon}{\varepsilon_o} \tag{A2.23}$$

If the local potential energy is less than the thermal energy of the particles $eV/kT_e \ll 1$, and $eV/kT_+ \ll 1$, then Poisson's equation becomes:

$$\frac{1}{r^2}\frac{d}{dr}\left\{r^2\frac{dV}{dr}\right\} = \frac{ne^2}{k\varepsilon_o}V\left[\frac{1}{T_e} + \frac{1}{T_+}\right] = \frac{V}{\lambda_D^2} \tag{A2.24}$$

λ_D^2 is called the Debye length. The solution of Poisson's equation is given by the following expressions:

$$V(r) = \frac{q}{4\pi\varepsilon_o r}e^{-\frac{r}{\lambda_D}} \tag{A2.25}$$

$$\lambda_D = \left[\frac{kT_e\varepsilon_o}{ne^2}\right]^{1/2} = 69\left[\frac{T_e}{n}\right]^{1/2} \tag{A2.26}$$

where, T_e is in Kelvin and n is in m^{-3}.

Within the Debye length, the space charge effects are significant, outside of the Debye length; the moving charges shield the local region from space charge effects.

A2.1.5.2 Plasma Frequency

As was seen in the previous discussion of Debye length, the plasma is made up of ions and electrons which are able to move freely. Since electrons are much lighter than ions, they will have a higher velocity than ions. In normal thermal motion, it is expected that electrons will separate from ions. As electrons begin to separate, a charge potential develops between the fast moving electrons and the slow moving ions. Eventually, the potential becomes so large that the electrons slow down and reverse direction. The electrons then move towards the ions accelerated by the charge potential. When the electrons reach the ions, the potential drops, thus acceleration ceases. However, the electrons have developed a large velocity and then begin to overshoot the ions in the opposite direction. As the electrons move further out the charge potential begins to build up and slow the electrons down. Eventually, the electrons will stop and reverse their direction once again. Thus

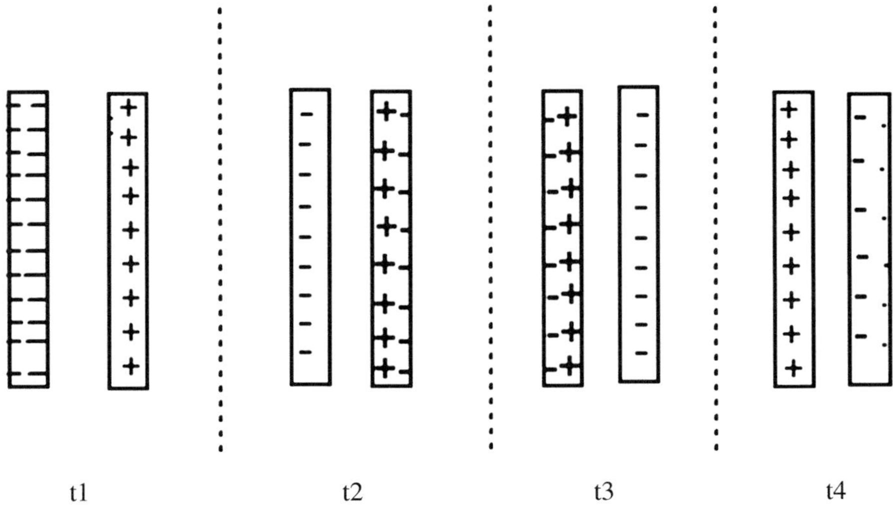

Fig. A2.3 Electron motion relative to ion motion at times t1, t2, t3, and t4

electrons oscillate about the ions at a frequency called the plasma frequency. Electron motion relative to ion motion is depicted in Fig. A2.3.

This oscillatory behavior can be described by the electron equation of motion given in Eq. A2.27.

$$F = -eE = \frac{-ne^2\delta}{\varepsilon_0} \tag{A2.27}$$

where δ is the distance between ions and electrons.

Therefore,

$$m\frac{d^2\delta}{dt^2} = \frac{-ne^{2\delta}}{\varepsilon_0} \tag{A2.28}$$

and the solution to the electron equation of motion is given by:

$$\delta = \delta_0 e^{j\omega_p t} \tag{A2.29}$$

where ω_P is the plasma frequency and is given by:

$$\omega_p{}^2 = \frac{ne^2}{m\varepsilon_o} = 3.18 \times 10^3\ n$$

If an electric field is introduced, which has a driving frequency of ω (here a bold letter indicates a vector quantity), then the expression becomes (note that a bold symbol is used to indicate a vector):

$$\mathbf{E} = \mathbf{E}_o \, e^{j\omega t} \tag{A2.30}$$

and, the motion of an electron changes. The equation of force for plasma, using the Lorentz force for charged particles, with a driving electric field is,

$$m\frac{dv}{dt} = m\frac{d^2\mathbf{x}}{dt^2} = -e\,\mathbf{E} - e\,\mathbf{v} \times \mathbf{B} \tag{A2.31}$$

For now eliminate the magnetic field using following expressions:

$$m\frac{d^2x}{dt^2} = \mathbf{E}_o e^{j\omega t} \tag{A2.32}$$

$$\mathbf{x} = \mathbf{x}_o e^{j\omega t} \tag{A2.33}$$

$$\text{or,} \quad \mathbf{x} = \frac{-e\mathbf{E}}{m\omega^2}$$

The dipole moment of the plasma is expressed by:

$$\mathbf{p} = -e\mathbf{x} = \frac{-e^2\mathbf{E}}{m\omega^2} \tag{A2.34}$$

Making the assumption that the electron density is uniform at a value n, and that all electrons act alike, then the total polarization is given by:

$$\mathbf{P} = \frac{-ne^2\mathbf{E}}{m\omega^2} \tag{A2.35}$$

It may be noted that the total polarization is equal to the summation over all moments. This can now be related to the dielectric constant from Maxwell's equations through following expressions:

$$\mathbf{D} = \varepsilon\mathbf{E} = \varepsilon_0\varepsilon_R\mathbf{E} = \varepsilon_0\mathbf{E} + \mathbf{P} \tag{A2.36}$$

$$\mathbf{D} = \varepsilon_0\left[1 - \frac{-ne^2}{m\omega^2}\right]\mathbf{E} \tag{A2.37}$$

$$\varepsilon_R = 1 - \frac{-ne^2}{m\varepsilon_0\omega^2} = 1 - \frac{{\omega_p}^2}{\omega^2} \tag{A2.38}$$

The physical interpretation of plasma is that it is a dielectric. If $\omega > \omega_p$, then ε_R is positive and the electromagnetic wave of frequency ω can propagate through the plasma. If $\omega < \omega_p$, the ε_R is negative which says an electromagnetic wave of frequency ω cannot propagate in the plasma.

A2.2 Elementary Theory of Discharges

In the derivation of the plasma frequency, it was assumed that the electron velocity is directional, and that the electrons had no interactions. Electrons will clearly have a random direction in real plasma. However, it should be noted that if you examine only the average directed velocity **u** (or *drift velocity*) of electrons under the influence of an external field using a full kinetic theory approach, you will get the same result.

The effect of particle interactions is much more complex and will be considered in this section.

A2.2.1 The Langevin Equation

The simplest approach to describing complex electron motion is the Lorentz gas model. In this model it is assumed that the electrons are free to move in a stationary uniform background of ions and neutrals which in turn react with the electrons. This reaction is described by a viscous damping force. This is a good model for weakly ionized gases, but is not good for fully ionized gases. The Langevin equation results from these assumptions and can be expressed by:

$$\frac{d(m\mathbf{u})}{dt} = -e[E + \mathbf{u} \times \mathbf{B}] - m\mathbf{u}v_m \qquad \text{(A2.39)}$$

The term ν_m is the collision frequency for momentum transfer. Thus every ν_m times per second the electron loses all of its directed momentum.

A2.2.2 Basic Definitions

A2.2.2.1 Mobility

The electron mobility is a constant by which the electron drift velocity is related to the applied electric field which is responsible for that velocity. In a extremely viscous gas, this constant is low, in a mildly viscous gas this constant is large. The drift velocity (**u**) is,

$$\mathbf{u} = -\mu_e \mathbf{E} \qquad \text{(A2.40)}$$

Assuming that the plasma is not subject to magnetic fields, and **E** obeys Eq. A2.30, than substituting $-\mu_e \mathbf{E}$ for **u** in the Langevin equation,

$$j\omega m\mathbf{u} = -e\mathbf{E} - m\mathbf{u}v_m \qquad \text{(A2.41)}$$

Therefore the drift velocity can be found,

$$\mathbf{u} = \frac{-e\mathbf{E}}{m(v_m + j\omega)} \tag{A2.42}$$

Thus, the electron mobility is,

$$\mu_e = \frac{-e}{m(v_m + j\omega)} \tag{A2.43}$$

In the case of neutral gases in a DC discharge, the $j\omega$ term is 0 and ν_m is independent of electron velocity. For example the collision frequency for H_2 and He are given below,

$$\begin{aligned} \nu_m &= 4.8 \times 10^9 \times \mathrm{P}\,(\mathrm{H_2}) \\ \nu_m &= 2.3 \times 10^9 \times \mathrm{P}\,(\mathrm{He}) \end{aligned} \tag{A2.44}$$

where P is the gas pressure at 300 K in torr.

Using Eq. A2.44, electron mobility and drift velocity for electron neutral collisions can be calculated for H_2 and He gases from following expressions,

$$u\,(m/s) = 37\frac{E}{P} \quad (\text{volts/m-torr } \mathrm{H_2})$$

and,

$$u\,(m/s) = 76\frac{E}{P} \quad (\text{volts/m-torr He})$$

A2.2.2.2 Conductivity

The mobility of an electron is directly related to the conductivity of the electron (σ). The conduction current density due to an electric field is given by:

$$\mathbf{J} = n_e e\mathbf{u} = n_e e\mu_e\mathbf{E} \tag{A2.45}$$

$$\mathbf{J} = \sigma\mathbf{E} \tag{A2.46}$$

$$\mathbf{J} = \frac{-n_e e^2\mathbf{E}}{m(v_m + j\omega)} \tag{A2.47}$$

Therefore,

$$\sigma = \frac{-n_e e^2}{m(v_m + j\omega)} \tag{A2.48}$$

A2.2.3 Energy Balance, Electron Temperature and Energy Relaxation

In an ionized medium, we are interested in determining the electron temperature and the relaxation time of the electron. We approach these issues by implementing an energy balance. The energy gained in a plasma due to the presence of an electric field is given by:

$$J \bullet E = \sigma \mathbf{E}^2 = \frac{ne^2}{m v_m} \mathbf{E}^2 \tag{A2.49}$$

The energy lost due to electron neutral collisions can be calculated using the statistical mechanics relationship for average energy in a Maxwell Boltzmann Distribution (3nkT/2). Here assume that the average energy of an electron is,

$$\overline{E_e} = \frac{3}{2} nkT_e$$

and the average energy of a neutral gas atom is,

$$\overline{E_n} = \frac{3}{2} nkT_n.$$

The energy loss rate from electron to neutral is related to collision rate times the difference between the average electron energy minus the average neutral energy,

$$\frac{dE_{e-n}}{dt} = \delta \nu_m \left(\overline{E_e} - \overline{E_n} \right),$$

where $\delta = 2\,m/M$ is the fraction of energy lost per collision.

So, the collisional energy loss from electron to neutral is,

$$\frac{dE_{e-n}}{dt} = n \delta v_m 3k \frac{(T_e - T_n)}{2} \tag{A2.50}$$

where $3\,k(T_e - T_n)/2$ is the excess energy of the electrons above the background gas energy $3kT_n/2$.

The rate equation for the electron energy is expressed as:

$$\frac{d}{dt}\left(\frac{3}{2} nkT_e\right) = \sigma E^2 - n \delta v_m 3k \frac{(T_e - T)}{2} \tag{A2.51}$$

In a DC electric field, Eq. 1.63 becomes,

$$\sigma E^2 - n \delta v_m 3k \frac{(T_e - T)}{2} = 0 \tag{A2.52}$$

where T is the temperature of background gas,

and,

$$T_e = T + \frac{2e^2 E^2}{3km\delta {v_m}^2} \tag{A2.53}$$

A result which will turn out very general is that T_e is a function of $\frac{E^2}{\nu_m{}^2}$. Since ν_m is a function of gas pressure, P, it is common to use the parameter E/P. If an electric field is turned off, then Eq. A2.51 becomes,

$$\frac{d}{dt}\left(\frac{3}{2}nkT_e\right) = -n\delta\nu_m 3k\frac{(T_e - T)}{2} \tag{A2.54}$$

and,

$$T_e = T + (T_e - T)_0 e^{-\delta \nu_m t} \tag{A2.55}$$

Therefore, the electron energy relaxation time is given by,

$$\tau_{el} = \left(\frac{1}{\delta \nu_m}\right) \tag{A2.56}$$

A2.2.3.1 Collisions

As previously discussed, energy is transferred in collisions. We will examine the microscopic behavior of colliding particles by first relating a collision frequency to the collisional cross section and then to the differential scattering cross section.

A2.3 Definitions

A2.3.1 Cross Section

The collisional cross section is given the symbol σ. This is a microscopic quantity which indicates a probability for an interaction. For example, if a billiard ball on a billiard table is put into motion in the +*x* direction but the location *y* is random and a group of billiard balls are lined up on side of the table (see Fig. A2.4), how can a probability for collision be calculated? (This problem is a simple one dimensional case.)

The way to approach this problem is to look at the impact parameter for the billiard balls. As illustrated in Fig. A2.4, a collision will occur if the moving billiard ball approaches another ball within 4 times the ball's radius: Hence the impact parameter, b, is equal to 2*r*. The interaction length for this one dimensional problem is then given by $\sigma_2 = 4r$. The probability of interaction can be calculated simply. You would first add up the number of stationary billiard balls and call this number *N*. If the moving billiard ball were to approach any of the stationary balls within the impact parameter, then this would result in a collision. The total length on the side of the table where the billiard balls are lined up in which the moving billiard ball would collide with a stationary billiard ball is given by the product of the number of stationary billiard balls times the impact parameter (4*Nr*). The probability for interaction is then 4*Nr* divided by the length of the side (*L*) of the table. The probability in given by Eq. A2.57:

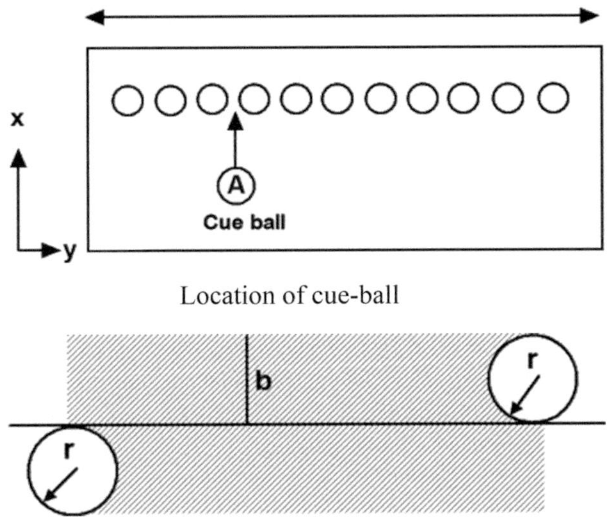

Fig. A2.4 Billiard table with a single ball put into motion along a random direction

$$P = \frac{4Nr}{L} \tag{A2.57}$$

For example, if $r = 3$ cm, $N = 1$, and $L = 300$ cm, then the probability for interaction is 0.04. Clearly, billiards must be a difficult game since it depends on more than random chance.

In the case of plasma, the problem is far more complex because all of the species have motion in 3-dimensions. The cross section however, provides us with the basic tool to attack this difficult problem. If an electron interacts with an atom, the simplest approach is to treat the atom like a billiard ball. The cross sectional area of a billiard ball is calculated from Eq. A2.58 .

$$\sigma = \pi a^2 \tag{A2.58}$$

This simple model can be useful. For example, consider a monoenergetic beam of electrons with a cross sectional area A and a flux $\Gamma = nv$ that enters a gas mixture with a density N atoms m^{-3} (see Fig. A2.6). Without going into details on columbic collisions between the electron beam and the electrons orbiting the atoms, a useful description of a cross section can be developed.

Each gas atom has a cross section σ. The question then is what fraction of the flux is scattered out of the beam in a distance of dx? The fraction of the flux which is scattered out of the beam is equal to the fraction of the area of the beam which is blocked by the gas atoms within the distance of dx. Flux is the directional current of the beam as expressed by Eq. A2.59.

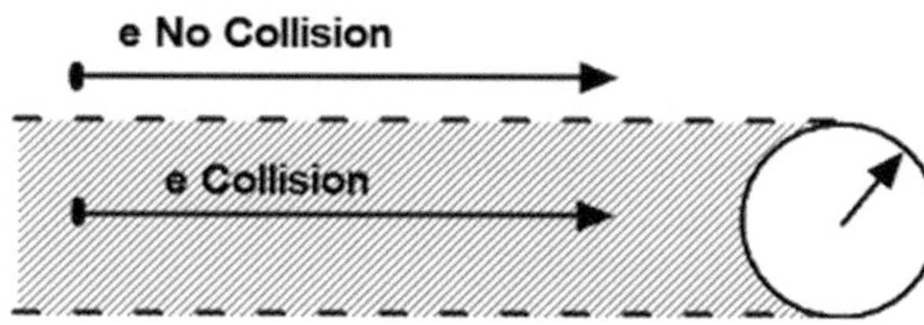

Fig. A2.5 The billiard ball model for electron atom collisions

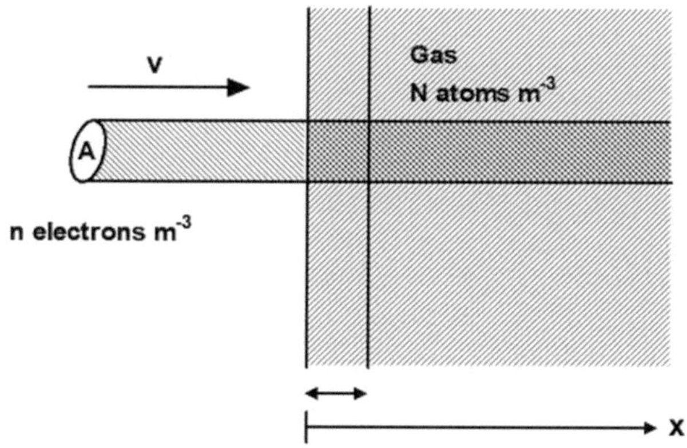

Fig. A2.6 An electron beam interacting with a gas

$$\frac{d\Gamma}{\Gamma} = \frac{number\ of\ gas\ atoms\ in\ dx\ times\ \sigma}{area\ of\ beam} \tag{A2.59}$$

Thus,

$$\frac{d\Gamma}{\Gamma} = \frac{-NA\sigma dx}{A} = -N\sigma dx \tag{A2.60}$$

Solving the above equation, Eq. A2.61 is obtained.

$$\Gamma(x) = \Gamma_0 e^{-N\sigma x} \tag{A2.61}$$

The term $e^{N\sigma x}$ is the probability that any given electron will penetrate a distance x into the gas.

A2.3.2 Mean Free Path

The mean free path is the average distance that an electron, in the above example, travels before colliding with a gas atom. From Eq. A2.61, the mean free path can be found, (cross section tells how probable it is that an electron can get through gas without a collision – or how electron is scattered) and is given below:

$$\lambda = \frac{\int_0^\infty x e^{-N\sigma x} dx}{\int_0^\infty e^{-N\sigma x} dx} = \frac{1}{N\sigma} \quad \text{(A2.62)}$$

A2.3.3 Collision Frequency

The collision frequency represents the average number of collisions per second for an electron of velocity v traveling through a gas of density N. The number of collisions an electron undergoes per second is ν. Therefore, the relationship between electron velocity (v), collision frequency (ν) and mean free path (λ) is,

$$\nu = \frac{\mathrm{v}}{\lambda} \quad \text{(A2.63)}$$

The relationship in Eq. A2.63 can be related to the gas density (N) and collision cross section (σ) realizing that $\lambda = 1/N\sigma$,

or,
$$\nu = N\sigma\mathrm{v} \quad \text{(A2.64)}$$

References

1. Ginzburg VL, Gurevich AV (1969) Nonlinear phenomena in a plasma located in an alternating electromagnetic field. Sov Phys Uspekhi 3:115–146
2. Denes LJ, Lowke JJ (1973) V-I characteristics of pulsed CO_2 laser discharges. Appl Phys Lett 23:130–132
3. Cobine JD (1958) Gaseous conductors theory and engineering applications. McGraw-Hill Books, New York
4. Chen F (2006) Introduction to plasma physics and controlled fusion, vol 1, 2 ed. Plenum Press, New York
5. Dolan T (1981) Fusion research volume I principles. Pergamon Press, Oxford
6. Huba JD (2013) NRL plasma formulary. In: Division BP (ed) Naval Research Laboratory. Washington, DC
7. Cherrington BE (1980) Gaseous electronics and gas lasers. Elsevier, Burlington

A3 Maxwell Boltzmann Distribution Function

A3.1 Maxwellian Distribution

A thermal distribution is given by the Maxwell-Boltzmann distribution function. This distribution function was determined by statistical mechanics. For a system that is in thermodynamic equilibrium, the energy dependent distribution function is expressed by:

$$f(E) = 2\sqrt{\frac{1}{\pi}}(kT)^{-3/2}E^{1/2}e^{-E/kT} \tag{A3.1}$$

Equation A3.1 is used in generating Fig. A3.1 which is the distribution function for a deuterium plasma heat to temperatures of 10, 15, 20 and 30 keV.

The velocity dependent form of the distribution function is shown in Eq. A3.2.

$$f\left(v_x, v_y, v_z\right) = \left(\frac{m}{2\pi kT}\right)^{3/2} e^{-m\left(v_x{}^2+v_y{}^2+v_z{}^2\right)/2kT} \tag{A3.2}$$

Distribution functions are used to calculate statistically averaged quantities of scalar functions or vector functions. In order to calculate average quantities, it is necessary to first simplify the differential form of velocity space, $d^3\boldsymbol{v}$. For a spherical geometry, $d^3\boldsymbol{v}$ can be expressed in terms of the magnitude of the velocity vector or speed (v).

$$d^3\mathbf{v} = \mathrm{v}^2 \sin\theta d\theta d\phi d\mathrm{v} \tag{A3.3}$$

So, to find the speed independent particle density, the distribution function is integrated over speeds (from 0 to ∞) as shown in Eqs. A3.4 and A3.5.

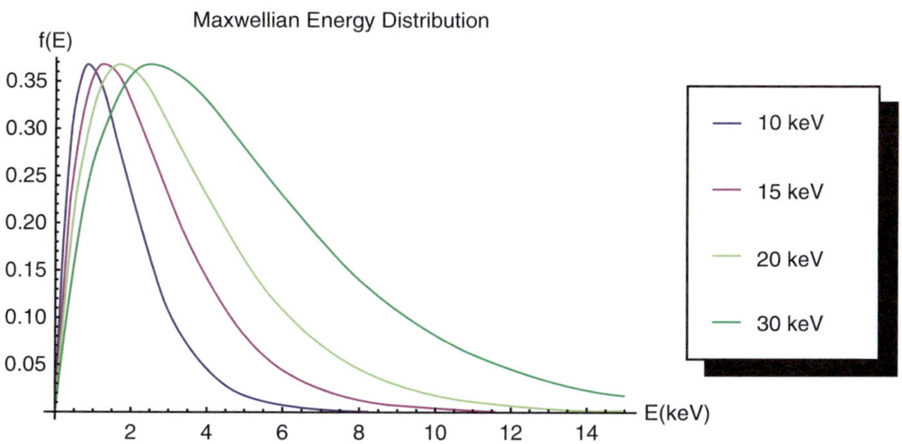

Fig. A3.1 Maxwellian energy distribution for a plasma with temperatures of 10, 15, 20 and 30 keV

$$n\left(\vec{r}, t\right) = \int_0^{\infty} \int_0^{2\pi} \int_0^{\pi} f\left(\vec{v}, \vec{r}, t\right) v^2 \sin\theta \, d\theta \, d\phi \, dv \tag{A3.4}$$

and,

$$n\left(\vec{r}, t\right) = \int_0^{\infty} f\left(\vec{r}, \vec{v}, t\right) v^2 4\pi \, dv = \int_0^{\infty} f\left(\vec{r}, v, t\right) dv \tag{A3.5}$$

$$\text{where,} \quad f\left(\vec{r}, v, t\right) = f\left(\vec{r}, \vec{v}, t\right) v^2 4\pi.$$

The latter definition is used to find average values of scalar quantities. For example the average of a scalar function P(v, t) which is dependent upon the speed v, such as $\sigma(v)v$, is found using Eq. A3.6.

$$\langle P(v, t) \rangle = \frac{1}{n} \int_0^{\infty} P(v, t) f(v, t) \, dv \tag{A3.6}$$

The distribution function can also be used to average a vector function $(Q(\tilde{v}, t))$. The average of the vector function is found using Eq. A3.7.

$$\left\langle Q\left(\vec{v}, t\right) \right\rangle = \frac{1}{n} \int_0^{\infty} \int_0^{\infty} \int_0^{\infty} Q\left(\vec{v}, t\right) f\left(\vec{v}, t\right) dv_x dv_y dv_z \tag{A3.7}$$

A4 Thermal Neutron Cross Sections for Some Elements and Compounds (from Reactor Physics Constants ANL 5800 1963 http://www.osti.gov/scitech/servlets/purl/4620873) for 2200 m/s Neutron Speed

Table A4.1 Microscopic cross sections for some elements and compounds for a neutron speed of 220 m/s

Atomic No.	Element or compound	Atomic Wt.	Density (g cm^{-3})	Nuceli cm^{-3} × 10^{-24}	σ_a (barns)	σ_s (barns)	σ_t (barns)	σ_f (barns)
1	H	1.008	0.000089	0.000053	0.33	38	38	
	H_2O	18.016	1	0.0335	0.66	103	103	
	D_2O	20.03	1.1	0.0331	0.001	13.6	13.6	
2	He	4.003	0.000178	0.000026	0.007	0.8	0.807	
3	Li	6.94	0.534	0.0463	71	1.4	72.4	
4	Be	9.013	1.85	0.1236	0.01	7	7.01	
	BeO	25.02	3.025	0.0728	0.01	6.8	76.8	
5	B	10.82	2.45	0.1364	755	4	759	
6	C	12.011	1.6	0.0803	0.004	4.8	4.8	
7	N	14.008	0.0013	0.000053	1.88	10	11.9	
8	O	16	0.0014	0.000053	0.0002	4.2	4.2	
9	F	19	0.0017	0.000053	0.001	3.9	3.9	
10	Ne	20.183	0.009	0.000026	2.8	2.4	5.2	
11	Na	22.991	0.971	0.0254	0.525	4	4.53	
12	Mg	24.32	1.74	0.0431	0.069	3.6	3.67	
13	Al	26.98	2.699	0.0602	0.241	1.4	1.64	
14	Si	28.09	2.42	0.0522	0.16	1.7	1.86	
15	P	30.975	1.82	0.0354	0.2	5	5.2	
16	S	32.066	2.07	0.0389	0.52	1.1	1.62	
17	Cl	35.457	0.0032	0.000053	33.8	16	49.8	
18	Ar	39.944	0.0018	0.000026	0.66	1.5	2.16	
19	K	39.1	0.87	0.0134	2.07	1.5	3.57	
20	Ca	40.08	1.55	0.0233	0.44	3	3.44	
21	Sc	44.96	2.5	0.0335	24	24	48	
22	Ti	47.9	4.5	0.0566	5.8	4	9.8	
23	V	50.95	5.96	0.0704	5	5	10	

24	Cr	52.01	7.1	0.0822	3.1	3	6.1	
25	Mn	54.94	7.2	0.0789	13.2	2.3	15.5	
26	Fe	55.85	7.86	0.0848	2.62	11	13.6	
27	Co	58.94	8.9	0.091	38	7	45	
28	Ni	58.71	8.9	0.0913	4.6	17.5	22.1	
29	Cu	63.54	8.94	0.0848	3.85	7.2	11.05	
30	Zn	65.38	7.14	0.0658	1.1	3.6	4.7	
31	Ga	69.72	5.91	0.0511	2.8	4	6.8	
32	Ge	72.6	5.36	0.0445	2.45	3	5.45	
33	As	74.91	5.73	0.0461	4.3	6	10.3	
34	Se	78.96	4.8	0.0366	12.3	11	23.3	
35	Br	79.916	3.12	0.0235	6.7	6	12.7	
36	Kr	83.8	0.0037	0.000026	31	7.2	38.2	
37	Rb	85.48	1.53	0.0108	0.73	12	12.7	
38	Sr	87.63	2.54	0.0175	1.21	10	11.2	
39	Yt	88.92	5.51	0.0373	1.313	4.3	5.613	
40	Zr	91.22	6.4	0.0423	0.185	8	8.2	
41	Nb	92.91	8.4	0.0545	1.16	5	6.16	
42	Mo	95.95	10.2	0.064	2.7	7	9.7	
43	Tc	98			22			
44	Ru	101.1	12.2	0.0727	2.56	6	8.56	
45	Rh	102.91	12.5	0.0732	149	5	154	
46	Pd	106.4	12.16	0.0689	8	3.6	11.6	
47	Ag	107.88	10.5	0.0586	63	6	69	
48	Cd	112.41	8.65	0.0464	2450	7	2457	
49	In	114.82	7.28	0.0382	191	2.2	193	

(continued)

Table A4.1 (continued)

Atomic No.	Element or compound	Atomic Wt.	Density (g cm^{-3})	Nuceli cm^{-3} × 10^{-24}	σ_a (barns)	σ_s (barns)	σ_t (barns)	σ_f (barns)
50	Sn	118.7	6.5	0.033	0.625	4	4.6	
51	Sb	121.76	6.69	0.0331	5.7	4.3	10	
52	Te	127.61	6.24	0.0295	4.7	5	9.7	
53	I	126.91	4.93	0.0234	7	3.6	10.6	
54	Xe	131.3	0.0059	0.000027	35	4.3	39.3	
55	Cs	132.91	1.873	0.0085	28	20	48	
56	Ba	137.36	3.5	0.0154	1.2	8	9.2	
57	La	138.92	6.19	0.0268	8.9	15	24	
58	Ce	140.13	6.78	0.0292	0.73	9	9.7	
59	Pr	140.92	6.78	0.29	11.3	4	15.3	
60	Nd	144.27	6.95	0.29	46	16	62	
61	Pm	145			60			
62	Sm	150.35	7.7	0.0309	5600	5	5605	
	Sm_2O_3	348.7	7.43	0.000000128	16,500	22.6	16522.6	
63	Eu	152	5.22	0.0207	4300	8	4308	
	Eu_2O_3	352	7.42	0.000000127	8740	30.2	8770	
64	Gd	167.26	7.95	0.0305	46,000			
65	Tb	158.93	8.33	0.0316	46			
66	Dy	162.51	8.56	0.0317	950	100	1050	
	Dy_2O_3	372.92	7.81	0.000000126	2200	214	2414	
67	Ho	164.94	8.76	0.032	65			
68	Er	167.27	9.16	0.033	173	15	188	
69	Tm	168.94	9.35	0.0333	127	7	134	
70	Yb	173.04	7.01	0.0244	37	12	49	
71	Lu	174.99	9.74	0.0335	112			
72	Hf	178.5	13.3	0.0449	105	8	113	

73	Ta	180.95	16.6	0.0553	21	5	26	
74	W	183.86	19.3	0.0632	19.2	5	24.2	
75	Re	186.22	20.53	0.0644	86	14	100	
76	Os	190.2	22.48	0.0712	15.3	11	26.3	
77	Ir	192.2	22.42	0.0703	440			
78	Pt	195.09	21.37	0.066	8.8	10	18.8	
79	Au	197	19.32	0.0591	98.8	9.3	107.3	
80	Hg	200.61	13.55	0.0407	380	20	400	
81	Ti	204.21	11.35	0.033	0.17	11	11.2	
82	Pb	207.21	11.35	0.033	0.17	11	11.2	
83	Bi	209	9.747	0.0281	0.034	9	9	
84	Po	210	9.24	0.0265				
85	At	211						
86	Rn	222	0.0097	0.000026	0.7			
87	Fr	223						
88	Ra	226.05	5	0.0133	20			
89	Ac	227			510			
90	Th	232.05	11.3	0.0293	7.56	12.6	20.2	
91	Pa	231	15.4	0.0402	200			
92	U	238.07	18.9	0.04783	7.68	8.3	16	
	UO_2	270.07	10	0.000000223	7.6	16.7	24.3	
	^{233}U		18.9	0.04783				524

(continued)

Table A4.1 (continued)

Atomic No.	Element or compound	Atomic Wt.	Density (g cm^{-3})	Nuceli $cm^{-3} \times 10^{-24}$	σ_a (barns)	σ_s (barns)	σ_t (barns)	σ_f (barns)
	^{235}U		18.9	0.04783	669.25	16.95	686.2	577
	^{238}U		18.9	0.04783				
93	Np	237			170			
94	Pu	239	19.74	0.0498	1026	9.6	1036	741
95	Am	242			8			
96	Cm	245						
97	Bk	249			500			
98	Cf	249			900			
99	E	253			160			
100	Fm	256						
101	Mv	260						

Note: 1 barn = 1×10^{-24} cm^{-2}

Table A4.2 Macroscopic cross sections and properties for some elements and compounds for a neutron speed of 220 m/s

Atomic No.	Element or compound	Atomic Wt.	Density (g cm^{-3})	Nuceli $cm^{-3} \times 10^{-24}$	$1-\bar{\mu}_0$	Log energy decrement	Σ_a cm^{-1}	Σ_s cm^{-1}	Σ_t cm^{-1}
1	H	1.008	0.000089	0.000053	0.3386	1	0.000017	0.002	0.002
	H_2O	18.016	1	0.0335	6.76	0.948	0.022	3.45	3.45
	D_2O	20.03	1.1	0.0331	0.884	0.57	0.000033	3.45	3.45
2	He	4.003	0.000178	0.000026	0.8334	0.425	0.0000002	0.000221	0.000021
3	Li	6.94	0.534	0.0463	0.9047	0.268	3.29	0.065	3.35
4	Be	9.013	1.85	0.1236	0.9259	0.209	0.00124	0.865	0.865
	BeO	25.02	3.025	0.0728	0.939	0.173	0.00073	0.501	0.501
5	B	10.82	2.45	0.1364	0.9394	0.171	103	0.346	104
6	C	12.011	1.6	0.0803	0.9444	0.158	30.0608	0.385	0.385
7	N	14.008	0.0013	0.000053	0.9524	0.136	0.000099	0.0005	0.0006
8	O	16	0.0014	0.000053	0.9583	0.12	0	0.00021	0.00021
9	F	19	0.0017	0.000053	0.9649	0.102	0.0000001	0.0002	0.0002
10	Ne	20.183	0.009	0.000026	0.9667	0.0968	0.000073	0.000062	0.000135
11	Na	22.991	0.971	0.0254	0.971	0.0845	0.013	0.102	0.115
12	Mg	24.32	1.74	0.0431	0.9722	0.0811	0.003	0.155	0.158
13	Al	26.98	2.699	0.0602	0.9754	0.0723	0.015	0.084	0.099
14	Si	28.09	2.42	0.0522	0.9762	0.0698	0.008	0.089	0.097
15	P	30.975	1.82	0.0354	0.9785	0.0632	0.007	0.177	0.184
16	S	32.066	2.07	0.0389	0.9792	0.0612	0.02	0.043	0.063
17	Cl	35.457	0.0032	0.000053	0.981	0.0561	0.002	0.0008	0.003
18	Ar	39.944	0.0018	0.000026	0.9833	0.0492	0.000017	0.000039	0.000056
19	K	39.1	0.87	0.0134	0.9829	0.0504	0.028	0.02	0.048
20	Ca	40.08	1.55	0.0233	0.9833	0.0492	0.01	0.07	0.08
21	Sc	44.96	2.5	0.0335	0.9852	0.0438	0.804	0.804	1.61

(continued)

Table A4.2 (continued)

Atomic No.	Element or compound	Atomic Wt.	Density (g cm^{-3})	Nuceli $cm^{-3} \times 10^{-24}$	$1-\bar{\mu}_0$	Log energy decrement	Σ_a cm^{-1}	Σ_s cm^{-1}	Σ_t cm^{-1}
22	Ti	47.9	4.5	0.0566	0.9861	0.0411	0.328	0.226	0.555
23	V	50.95	5.96	0.0704	0.9869	0.0387	0.352	0.352	0.704
24	Cr	52.01	7.1	0.0822	0.9872	0.0385	0.255	0.247	0.501
25	Mn	54.94	7.2	0.0789	0.9878	0.0359	1.04	0.181	1.22
26	Fe	55.85	7.86	0.0848	0.9881	0.0353	0.222	0.933	1.15
27	Co	58.94	8.9	0.091	0.9887	0.0335	3.46	0.637	4.1
28	Ni	58.71	8.9	0.0913	0.9887	0.0335	0.42	1.6	2.02
29	Cu	63.54	8.94	0.0848	0.9896	0.0309	0.0326	0.611	0.973
30	Zn	65.38	7.14	0.0658	0.9897	0.0304	0.072	0.237	0.309
31	Ga	69.72	5.91	0.0511	0.9925	0.0283	0.143	0.204	0.347
32	Ge	72.6	5.36	0.0445	0.9909	0.0271	0.109	0.134	0.243
33	As	74.91	5.73	0.0461	0.9911	0.0264	0.198	0.77	0.475
34	Se	78.96	4.8	0.0366	0.9916	0.0251	0.45	0.403	0.853
35	Br	79.916	3.12	0.0235	0.9917	0.0247	0.157	0.141	0.298
36	Kr	83.8	0.0037	0.000026	0.9921	0.0236	0.00081	0.00019	0.00099
37	Rb	85.48	1.53	0.0108	0.9922	0.0233	0.008	0.13	0.138
38	Sr	87.63	2.54	0.0175	0.9925	0.0226	0.021	0.175	0.195
39	Yt	88.92	5.51	0.0373	0.9925	0.0223	0.049	0.112	0.16
40	Zr	91.22	6.4	0.0423	0.9927	0.0218	0.008	0.338	0.347
41	Nb	92.91	8.4	0.0545	0.9928	0.0214	0.063	0.273	0.0336
42	Mo	95.95	10.2	0.064	0.9931	0.0207	0.173	0.448	0.621
43	Tc	98			0.9932	0.0203			
44	Ru	101.1	12.2	0.0727	0.9934	0.0197	0.186	0.436	0.622
45	Rh	102.91	12.5	0.0732	0.9935	0.0193	10.9	0.366	11.3
46	Pd	106.4	12.16	0.0689	0.9937	0.0187	0.551	0.248	0.799

47	Ag	107.88	10.5	0.0586	0.9938	0.0184	3.69	0.352	4.04
48	Cd	112.41	8.65	0.0464	0.994	0.0178	114	0.325	114
49	In	114.82	7.28	0.0382	0.9942	0.0173	7.3	0.084	7.37
50	Sn	118.7	6.5	0.033	0.9944	0.0167	0.021	0.132	0.152
51	Sb	121.76	6.69	0.0331	0.9945	0.0163	0.189	0.142	0.331
52	Te	127.61	6.24	0.0295	0.9948	0.0155	0.139	0.148	0.286
53	I	126.91	4.93	0.0234	0.9948	0.0157	0.164	0.084	0.248
54	Xe	131.3	0.0059	0.000027	0.9949	0.0152	0.00095	0.00012	0.001
55	Cs	132.91	1.873	0.0085	0.995	0.015	0.238	0.17	0.408
56	Ba	137.36	3.5	0.0154	0.9951	0.0145	0.018	0.123	0.142
57	La	138.92	6.19	0.0268	0.9952	0.0143	0.239	0.403	0.642
58	Ce	140.13	6.78	0.0292	0.9952	0.0142	0.021	0.263	0.283
59	Pr	140.92	6.78	0.29	0.9953	0.0141	0.328	0.116	0.444
60	Nd	144.27	6.95	0.29	0.9954	0.0138	1.33	0.464	1.79
61	Pm	145			0.9954	0.0133			
62	Sm	150.35	7.7	0.0309	0.9956	0.0133	173	0.155	173
	Sm_2O_3	348.7	7.43	0.000000128	0.974	0.076	211	0.289	211
63	Eu	152	5.22	0.0207	0.9956	0.0131	89	0.166	89.2
	Eu_2O_3	352	7.42	0.000000127	0.978	0.063	111	0.383	111
64	Gd	167.26	7.95	0.0305	0.9958	0.0127	1403		
65	Tb	158.93	8.33	0.0316	0.9958	0.0125	1.45		
66	Dy	162.51	8.56	0.0317	0.9959	0.0122	30.1	3.17	33.3
	Dy_2O_3	372.92	7.81	0.000000126	0.993	0.019	27.2	2.7	30.4
67	Ho	164.94	8.76	0.032	0.996	0.0121	2.08		
68	Er	167.27	9.16	0.033	0.996	0.0119	5.71	0.495	6.2

(continued)

Table A4.2 (continued)

Atomic No.	Element or compound	Atomic Wt.	Density (g cm^{-3})	Nuceli $cm^{-3} \times 10^{-24}$	$1 - \overline{\mu}_0$	Log energy decrement	Σ_a cm^{-1}	Σ_s cm^{-1}	Σ_t cm^{-1}
69	Tm	168.94	9.35	0.0333	0.9961	0.0118	4.23	0.233	4.46
70	Yb	173.04	7.01	0.0244	0.9961	0.0115	0.903	0.293	1.2
71	Lu	174.99	9.74	0.0335	0.9962	0.0114	3.75		
72	Hf	178.5	13.3	0.0449	0.9963	0.0112	4.71	0.0359	5.07
73	Ta	180.95	16.6	0.0553	0.9963	0.011	1.16	0.277	1.44
74	W	183.86	19.3	0.0632	0.9964	0.0108	1.21	0.316	1.53
75	Re	186.22	20.53	0.0644	0.9964	0.0107	5.71	0.93	6.64
76	Os	190.2	22.48	0.0712	0.9965	0.0105	1.09	0.783	1.87
77	Ir	192.2	22.42	0.0703	0.9965	0.0104	30.9		
78	Pt	195.09	21.37	0.066	0.9966	0.0102	0.581	0.66	1.24
79	Au	197	19.32	0.0591	0.9966	0.0101	5.79	0.55	6.34
80	Hg	200.61	13.55	0.0407	0.9967	0.0099	15.5	0.814	16.3
81	Ti	204.21	11.35	0.033	0.9968	0.0096	0.006	0.363	0.369
82	Pb	207.21	11.35	0.033	0.9968	0.0096	0.001	0.253	0.256
83	Bi	209	9.747	0.0281	0.9968	0.0095	0.001	0.253	0.256
84	Po	210	9.24	0.0265	0.9968	0.0095			
85	At	211			0.9968	0.0094			
86	Rn	222	0.0097	0.000026	0.997	0.009			
87	Fr	223			0.998	0.0089			
88	Ra	226.05	5	0.0133	0.997	0.088	0.266		
89	Ac	227			0.9971	0.0088			
90	Th	232.05	11.3	0.0293	0.9971	0.0086	0.222	0.369	0.592
91	Pa	231	15.4	0.0402	0.9971	0.0086	8.04		

92	U	238.07	18.9	0.04783	0.9972	0.0084	0.367	0.397	0.765
	UO_2	270.07	10	0.000000223	0.9887	0.036	0.169	0.372	0.542
	^{233}U		18.9	0.04783	0.9972	0.0084			
	^{235}U		18.9	0.04783	0.9972	0.0084			
	^{238}U		18.9	0.04783	0.9972	0.0084			
93	Np	237			0.9972	0.0084			
94	Pu	239	19.74	0.0498	0.9972	0.0083	51.1	0.478	51.6
95	Am	242			0.9973	0.0082			
96	Cm	245			0.9973	0.0081			
97	Bk	249			0.9973	0.0081			
98	Cf	249			0.9973	0.0079			
99	E	253			0.9974	0.0079			
100	Fm	256			0.9974	0.0078			
101	Mv	260			0.9974	0.0077			

$\overline{\mu}_0$ is the average value of the cosine μ_0 of the scattering angle θ measured in the laboratory system [1]. The term $1 - \overline{\mu}_0$ is known as the transport scattering factor [2]

References

1. Henry AF (1975) Nuclear-reactor analysis. MIT Press, Cambridge
2. ANL (1963) In: Argonne National Laboratory (ed) Reactor physics constants, 2nd ed. U. S. Government Printing Office, Washington, DC, United States Atomic Energy Commission

Index

M. Prelas, *Nuclear-Pumped Lasers*, DOI 10.1007/978-3-319-19845-3

L

M

N

MIX
Papier aus verantwortungsvollen Quellen
Paper from responsible sources
FSC® C105338

If you have any concerns about our products,
you can contact us on
ProductSafety@springernature.com

In case Publisher is established outside the EU,
the EU authorized representative is:
Springer Nature Customer Service Center GmbH
Europaplatz 3, 69115 Heidelberg, Germany

Printed by Libri Plureos GmbH
in Hamburg, Germany